Liquid Crystals
Experimental Study of Physical Properties and Phase Transitions

This book describes in detail various experimental techniques used in the study of liquid crystals. It will be indispensable for students embarking on liquid crystal research as well as established workers in the field. Each chapter in the book is dedicated to an important experimental technique used in the study and characterization of liquid crystalline systems. The techniques include x-ray diffraction, DSC, ac heat capacity, miscibility, NMR, quasielastic light scattering, polarization microscopy, x-ray reflectivity, freely suspended film experiments, optical measurements of birefringence, spontaneous polarization and other physical properties. Use of these techniques on liquid crystals requires modifications of traditional experiments, special sample handling and data analysis not described anywhere else. In addition, general routes used to synthesize liquid crystals and tools to characterize liquid crystal phases are described. Attempts have been made to show structure–property relationships for well-known systems.

This book will be of particular interest to graduate students in physics and chemistry as well as established researchers in the fields of liquid crystals and soft condensed matter.

S ATYENDRA K UMAR received his M.Sc. Honors School degree in physics with specializations in solid state physics and electronics in 1974 from Panjab University in India. He then joined the graduate school of the University of Nebraska and obtained an M.S. degree in physics in 1975. His Ph.D. dissertation work on high-resolution x-ray scattering studies of liquid crystals was done at the University of Illinois at Urbana-Champaign and completed in 1981. After working as a research associate at the Massachusetts Institute of Technology for three years, he joined Tektronix Incorporated in Beaverton, Oregon, developing a high-resolution liquid crystal based projection system. In 1987 he was appointed to the faculty of Kent State University where he now holds the rank of Professor in the Department of Physics and Chemical Physics Interdisciplinary Program of the Liquid Crystal Institute. In the past 12 years at Kent, he has directed research projects with grants from the National Science Foundation, US Department of Energy, American Chemical Society, and several companies.

LIQUID CRYSTALS

Experimental Study of Physical Properties
and Phase Transitions

SATYENDRA KUMAR

Author and Editor

with additional contributions by
J. Brock
D. Finotello
M. Fisch
C. Garland
J. Ho
M. Neubert
B. Padulka
P. Photinos
S. Sinha
P. Ukleja

CAMBRIDGE
UNIVERSITY PRESS

CAMBRIDGE UNIVERSITY PRESS
Cambridge, New York, Melbourne, Madrid, Cape Town,
Singapore, São Paulo, Delhi, Tokyo, Mexico City

Cambridge University Press
The Edinburgh Building, Cambridge CB2 8RU, UK

Published in the United States of America by Cambridge University Press, New York

www.cambridge.org
Information on this title: www.cambridge.org/9780521187947

First published 2001
First paperback edition 2011

A catalogue record for this publication is available from the British Library

Library of Congress Cataloguing in Publication data

Liquid crystals: experimental study of physical properties and phase transitions / edited
by Satyendra Kumar.
p. cm.
Includes bibliographical references and index.
ISBN 0 521 46132 4 (hb)
1. Liquid crystals–Experiments. 1. Kumar, Satyen.
QD923.L537 2001
530.4′29–dc21   99-087679

ISBN 978-0-521-46132-0 Hardback
ISBN 978-0-521-18794-7 Paperback

Additional resources for this publication at www.cambridge.org/9780521187947

Contents

There is a plate section located between pages 52 and 53.
These plates are available for download in colour from
www.cambridge.org/9780521187947

Preface

There are several excellent textbooks (e.g., de Gennes and Prost; Vertogen and de Jeu; and Chandrasekhar) and reference books (such as Collings and Patel; Bahadur; Kumar; and Koo and Wu) on liquid crystals. But none of them fulfills a long-felt need for a source of general, non-superfluous, and practical information regarding the various experimental research techniques and how they are applied to liquid crystals. The present text largely fulfills that void. For established researchers, this book is expected to provide insight into techniques beyond the realm of their expertise. The goal has been to give in-depth technical details as well as illustrative examples where the techniques have been successfully applied. This book should serve as a starting point for graduate students, postdoctoral research associates, and scientists from other sub-fields of physics, chemistry, and polymer science.

This book presents details on the experimental techniques most widely applied to the study of liquid crystals. The range of topics covered includes: synthesis; characterization by differential scanning calorimetry; optical textures under polarization microscopy; bulk, freely suspended film, and surface x-ray scattering; static and quasielastic light scattering; nuclear magnetic resonance; adiabatic and ac heat capacity; and methods to measure optical birefringence, spontaneous polarization, and other physical properties. In addition, attempts have been made to show structure–property relationships for selected systems. The chapters are authored by scientists who are internationally recognized as experts and among the best in the profession.

This text does not attempt to review the latest research or the most important results, neither does it give exhaustive details of theory, chemistry, or applications of liquid crystals, for fear of losing sight of the main objective. For brevity's sake, no attempt was made to cover all the aspects that can be easily and more appropriately found elsewhere in the literature. The bibliographies at the end of each chapter include only the most directly related work.

It should be specifically noted that the nomenclature of liquid crystalline phases that initially evolved was based on the chronology of their discovery. However, as the various phases revealed the details of their symmetry and structure, the classification scheme was revised twice. This has caused some confusion especially regarding the more ordered smectic phases. We have used a streamlined classification scheme and notation discussed in Chapter 1 and Section 2.1. A special note should be taken of the fact that deviations from this scheme had to be made in Chapters 2 and 10. This became necessary because the hexatic-B phase was not distinguished from the crystal-B phases in early studies. They were both designated as the smectic-B phase. In a large number of materials, the precise structure of smectic-B phases is still not known. For this reason, the original designation of smectic-B has been retained, where necessary, in these two chapters. I wish to extend my sincere thanks to the authors for being very generous with their time in developing these unique chapters. I am very grateful to Mike Fisch for his invaluable and timely help with several chapters and the editorial work. I am in debt to Sadhna, my wife, for being supportive and relieving me from many household duties to work on this project.

Satyendra Kumar
Kent State University

1

Introduction to liquid crystals

MICHAEL R. FISCH AND SATYENDRA KUMAR

Department of Physics and Liquid Crystal Institute, Kent State University, Kent, OH 44242, USA

1.1 What is a liquid crystal?

We all remember learning that there are three states of matter: solid, liquid and gas. This, however, is not the whole story. There are situations in which more than just these three phases exist. For now, consider the large class of organic molecules which do not show a single transition from solid to liquid, but rather a series (more than one) of transitions between the solid and the normal (isotropic) liquid as their temperature is raised. These new phases have mechanical, optical, and structural properties between those of the crystalline solid and the corresponding isotropic liquid. For this reason, these phases are referred to as *liquid crystalline* phases, and the materials which form them upon a change in phase are often referred to as thermotropic liquid crystals. A more proper name is mesomorphic (or intermediate) phases.

Liquid crystalline properties are exhibited by several different types of systems. In addition to certain classes of organic molecules, micellar solutions of surfactants, main and side chain polymers, and a large number of biological systems are known to be liquid crystalline. Several textbooks [1–4] on liquid crystals have discussed these topics to varying degree. The purpose of the present chapter is not to paraphrase what is in these texts but rather to lay the foundation for understanding the topics covered in different chapters of this book. A number of publications which emphasize and review specific topics in significant detail are available and are strongly recommended as supplemental reading [5–10]. Following de Gennes and Prost [1], we will define a liquid crystal as an intermediate phase which (a) has liquid-like order in at least one direction, and (b) possesses a degree of anisotropy, which is characteristic of some sort of order. The latter requirement is typically met if the molecules (and, in our later discussion, other objects such as micelles) which form liquid crystals are anisotropic, either rod-like (prolate) or disk-like (oblate). However, one must remember that while all liquid crystalline phases are formed by anisotropic objects not all anisotropic molecules form liquid crystalline phases [11, 12].

1.2 Ordering of anisotropic objects

Objects which have shapes other than spherical may possess three types of order giving rise to different liquid crystalline phases. Let us first define the types of order that are necessary to discuss the phases involved. The simplest of these is *orientational order* (OO). This is possible when the symmetry axes of the ordering objects are on average parallel to a well-defined spatial direction, **n**, known as the *director*. The degree of order, denoted by the *orientational order parameter*, S, is the thermal average $\langle 3/2 \cos^2\theta - 1/2 \rangle$, where θ is the angle between a molecule's symmetry axis and the director. The orientational order parameter is typically a function of temperature.

The second type of order is *positional* or *translational order* (PO). When PO is present, the system remains invariant under translation by an integer number of lattice translation vectors. The arrangement of the basic units (e.g., molecules) and the mass density (and, consequently, the electron density) exhibit periodicity. In the simplest case, this periodicity is one-dimensional, and may be conveniently represented by a density function of the form:

$$\rho(\mathbf{r}) = \langle \rho \rangle + \text{Re}[\Psi \exp(i\mathbf{k}\cdot\mathbf{r})],$$

where $\langle \rho \rangle$ is the average electron density, Re means real part of, Ψ is the complex amplitude, and $\mathbf{k}$ is the wavevector. It is important to note that a liquid crystal system does not have to possess translational order in all three directions. In particular, for the 'fluid' smectic (-A and-C) phases, which possess positional order only in one dimension, the *smectic order parameter* representing the density variation is written as

$$|\Psi(\mathbf{r})| = \text{Re}[\rho_1 \exp(i 2\pi z/d)],$$

where d is the thickness of smectic layers which are assumed to be perpendicular to the z direction [1].

A third type of order is *bond orientational order* (BOO) [13]. A bond, in the present context, is not a chemical bond but a line in space joining two adjacent molecules. If the orientation of these bonds is preserved over a long range, then a system possesses BOO. It is encountered in a category of smectic phases called the *hexatic smectics*. In these phases, the molecules within a smectic plane possess BOO and, since the molecules can best pack

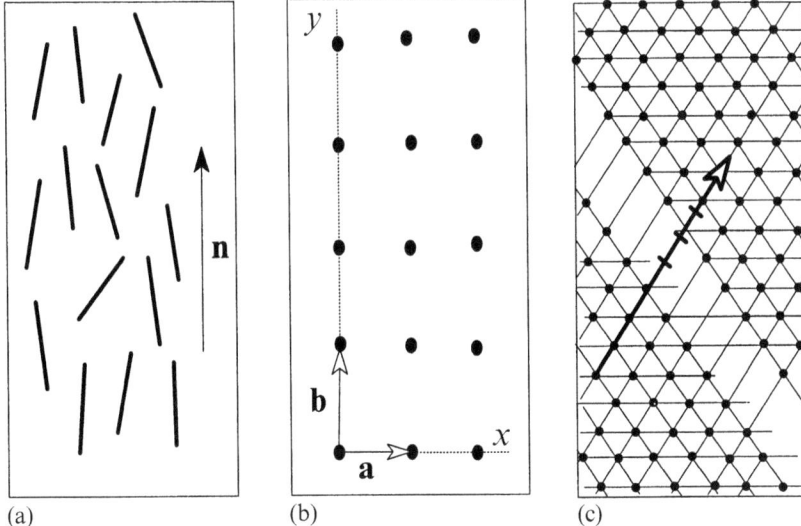

Figure 1.1. Schematic representation of the three basic types of order: (a) orientational order parallel to the director **n**, (b) positional order in two dimensions with lattice vectors **a** and **b**, and (c) bond orientational order. In the last case, translation from one lattice site along the arrow by an integer multiple of lattice spacing shows a lack of translation symmetry.

in a hexagonal fashion, the orientation of *bonds* possess six-fold symmetry. A complex order parameter, ψ_6, similar to the smectics density wave has been used to account for BOO;

$$\psi_6 = \mathrm{Re}[I_6\exp(6i\phi)],$$

where I_6 is a complex amplitude, and ϕ is the azimuthal angle with respect to the layer normal. Systems can possess long range BOO without long range positional order PO, but the reverse is not true. It is possible for PO to become short range due to the presence of dislocations and disclinations which can leave the BOO unperturbed. The three types of order discussed above are shown in two dimensions in Fig. 1.1.

1.3 The concept and use of symmetry in liquid crystals

One of the strongest guiding principles in liquid crystal science has been symmetry, i.e., physical properties of a liquid crystalline phase depend on the symmetry of the liquid crystal phase. Thus, the least ordered phase is the most symmetric *isotropic* (I) phase which exhibits isotropic behavior similar to regular liquids such as water. There is, then, a plethora of mesophases of lower symmetry than the I phase before the least symmetric crystalline phases belonging to the 230 space groups [14] which characterize crystals are encountered. This progression may be described in terms of the three types of order described above.

Consider imposing the simplest, i.e. orientational, order on a collection of molecules. The resulting structure is known as the *nematic* (N) phase. Such a phase is described by the director and the orientational order parameter defined above. If the molecules are on average parallel to **n**, the value of S is positive and ranges between 0 and 1. On the other hand, if the OO is such that molecules, on average, are perpendicular to **n**, then S is negative and its values range from 0 to $-\frac{1}{2}$. Evidently, a Landau–Ginzburg type expression for its free energy must not be invariant with respect to a change in the sign of S as its positive and negative values represent physically different systems. This is usually ensured by including a cubic term in the free energy. A consequence of this symmetry requirement is that the transition from I to an orientationally ordered state, or the N phase, must be first order. Indeed, the I–N transition is always found to be first order with a large enthalpy of transition. The physical properties of a nematic are described by a symmetric second rank tensor.

The presence of PO in one direction leads to the smectic-A (SmA) phase. However, there can be no long range order in one dimension [15], so this is really a quasi long range order (QLR). Transitions between the isotropic liquid and the SmA phase are first order while transitions between the N and the SmA phase may be either first or second order depending upon the coupling between the OO and PO, which in turn depends on the width of the N phase. A liquid crystal with a wide N phase is more likely to exhibit a second order N–SmA transition. Further discussion of this topic can be found in Chapters 3 and 7.

A smectic phase has at least two unique directions, the director and the layer normal. In the SmA phase, these two directions are collinear. But there are phases in which the director makes an angle with the layer normal as in the smectic-C (SmC) phase. In the chiral smectic-C (SmC*) phase, the director maintains a constant angle with respect to the layer normal while describing a helical path as the sample is traversed along the direction normal to the smectic planes. The chiral phases invariably are ferroelectric in nature. The details of the structure of different phases will be discussed later in this chapter.

Various smectic phases have BOO and PO to varying degree within a smectic plane and it is convenient to consider the extent of order that separates a smectic phase from the rest. The BOO and PO can be either short

Untilted thermotropic liquid crystal phases

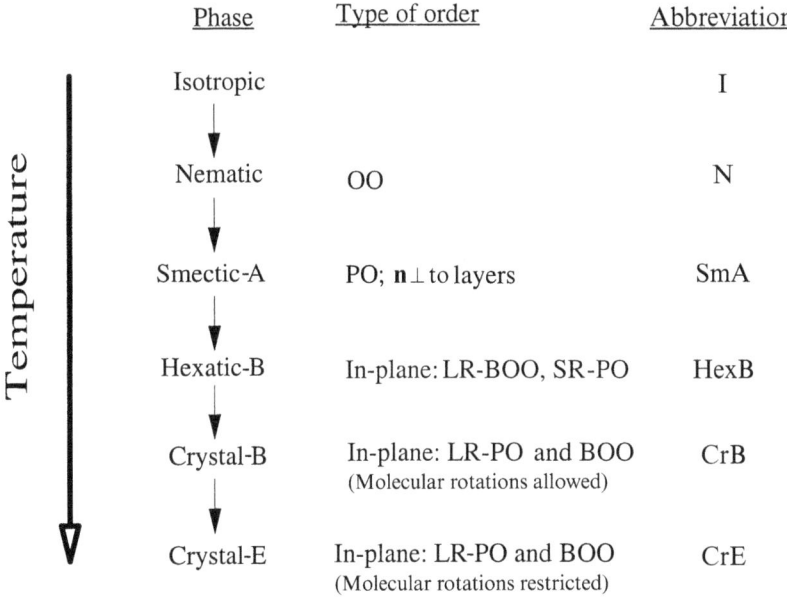

Phase	Type of order	Abbreviation
Isotropic		I
Nematic	OO	N
Smectic-A	PO; $\mathbf{n}\perp$ to layers	SmA
Hexatic-B	In-plane: LR-BOO, SR-PO	HexB
Crystal-B	In-plane: LR-PO and BOO (Molecular rotations allowed)	CrB
Crystal-E	In-plane: LR-PO and BOO (Molecular rotations restricted)	CrE

Figure 1.2. An illustration of a possible phase progression in a liquid crystal. The most symmetric and least ordered phase is at the top, and the most ordered and least symmetric phase is at the bottom.

range (SR) as is observed in liquids or long range (LR) as observed in crystals. Three distinct possibilities exist:

1. Both the BOO and PO are short range – such smectics can be considered as fluid smectics, examples being SmA, SmC, and SmC*, etc.
2. Long range BOO but short range PO – these are referred to as the hexatic smectic phases. Smectic-I, -F, and -HexB are some examples.
3. Long range BOO and PO – such phases are very close to being crystalline phases except that the molecules undergo rotational diffusion. They used to be known as smectic-B, -E, etc. but nowadays they are referred to as the crystal-B, -E, -G, etc. phases. (See Section 2.1 for recent terminology.)

Various (untilted) smectic phases, with the director parallel to the layer normal, are given different names as summarized in Fig. 1.2 [8].

The use of symmetry goes beyond noting which phase is more symmetric than another and helping one discern the symmetry of physical properties of the corresponding phase. It also allows one to make analogies to phase transitions in other non-liquid crystalline systems. For example, as de Gennes [1, 16] has observed, the introduction of the smectic order parameter allows one to draw a strong analogy between the SmA phase and the superconducting phase. The tilted SmC phase has been compared [17] with superfluid helium, and the SmA to SmC transition predicted to exhibit critical behavior similar to that of helium or the XY model. Similarly, the transition between SmA and hexatic-B is described by a two-dimensional order parameter and may similarly be described by the superfluid analogy. The use of symmetry in liquid crystals will become evident in different chapters of this book. Suffice it to state that symmetry has been a crucial and indispensable tool in our understanding of the physics of liquid crystalline phases. At the same time, the field of liquid crystals has been used as a testing ground of theoretical ideas that are not manifest in other systems with such elegance and simplicity.

1.4 Liquid crystal phases formed by rod-like molecules

In this section we will briefly discuss the liquid crystalline phases formed by simple entities: organic molecules which can be viewed as short rigid rods of a length to diameter ratio of roughly 3–8. Such molecules exhibit various mesophases at different temperatures and hence are generically referred to as thermotropic liquid crystals. The chemistry and structure–property relationships of this technologically and scientifically interesting group of materials are discussed in greater detail in Chapter 10 by Neubert. A common feature of all molecules of this type is that they all comprise a central rigid core connected to a flexible alkyl chain at one or both ends.

Without proof or reference we will state some of the physical properties of these phases. A detailed discussion of these properties and how they can be measured will be found in the various chapters of this book and the referenced literature. However, one should note that the size of a typical molecule which forms a thermotropic liquid crystal is such that x-rays work as an ideal and direct probe of the structure of these materials. Thus, a particularly good starting point is Chapter 3 on x-ray scattering by Kumar and the book by Pershan [8].

1.4.1 Non-tilted phases

This is a sub-class of thermotropic liquid crystalline phases which are free from chirality and molecular tilt. However, these phases differ from each other in the type and extent of order and the symmetry they possess.

The nematic phase

As discussed above, the simplest liquid crystalline phase is the nematic phase. There are several types of nematic phases but all of them can, to a first approximation, simply be thought of as liquids which have long range orientational order, OO, but lack PO and BOO. An example of a nematic made of rod-like molecules is shown in Fig. 1.1(a). The various types of nematics have slightly different properties based on the details of their molecular structure and chemical behavior. The two primary types of nematic are uniaxial and biaxial.

The uniaxial nematic is characterized by the following features [2]: (a) no PO, so, no BOO; (b) OO parallel to the director **n**; (c) the direction of **n** in space is arbitrary, and typically imposed by outside forces such as electric/magnetic fields; (d) **n** and −**n** are equivalent; and (e) molecules which form nematics are either archiral (identical to their mirror images) or racemic (contain equal numbers of left and right handed molecules).

The biaxial nematic phase is also characterized by the above properties. However, this liquid crystalline phase does not possess cylindrical symmetry about **n**. This phase possesses two unique directions perpendicular to **n** rendering it biaxial. It should be pointed out that, despite some reports, there are no known thermotropic biaxial nematic phases [18]. The existence of biaxial phases has been confirmed in lyotropic liquid crystals (Section 1.8)

The smectic-A phase

The name 'smectic', was coined by G. Friedel to describe certain mesophases that feel slippery like soap when touched, and was originally associated with what is now known as the smectic-A phase. The important feature which distinguishes smectic phases is that they have a layered structure. In fact, as mentioned above, PO and BOO have all been observed among molecules lying in a smectic plane resulting in over 20 smectic liquid crystalline phases.

The simplest is the SmA phase made of non-chiral and non-polar molecules and characterized by a one-dimensional layered structure (or PO), in which each layer is essentially a two-dimensional liquid as shown in Fig. 1.3(a). However, arguments [15] show that one-dimensional liquids can not exhibit long range PO. Nevertheless, PO almost does exist, and this system is said to exhibit quasi long range order or have Landau–Peierls' instability. This phase is uniaxial, the layers are essentially incompressible, and the long axes of molecules within the layer, on the average, are perpendicular to the layers. The SmA phase is characterized by short range PO and short

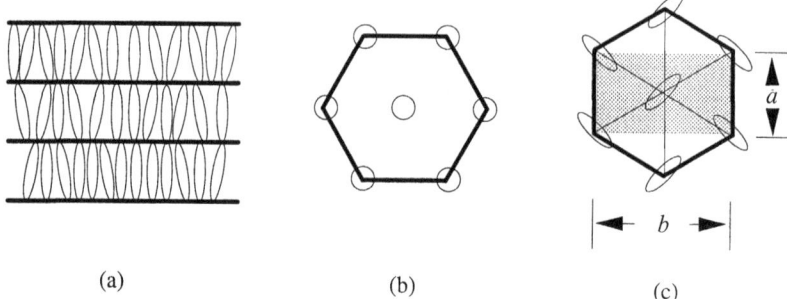

| (a) | (b) | (c) |

Figure 1.3.   Some liquid crystal phases formed by untilted rod-like molecules. (a) The smectic-A phase; note that there is quasi-long range PO normal to the layers, OO with the mean orientation of the molecules perpendicular to the layers and short range PO and BOO within the layers. (b) A view of the short range hexagonal PO of the molecules within a layer of a HexB phase. The molecules are perpendicular to the smectic plane as indicated by the circles. (c) The local herringbone arrangement of molecules in a layer of the crystal-E phase. The ellipses represent a view of the rigid benzene rings in the molecules. The lattice has a rectangular unit cell.

range BOO in the layers, and quasi long range PO perpendicular to the planes. It may be thought of, somewhat crudely, as a stack of two-dimensional fluid layers.

The hexatic-B or hexatic smectic-B phase

These two names for the same phase, abbreviated as HexB, describe a phase which is characterized by a layered structure, just like the SmA phase, *and* long range BOO within the layers. BOO is the essential property of a hexatic phase. This is a very special class of smectics characterized by quasi long range PO in the direction perpendicular to the layers, short range PO within the smectic layers (although typically longer than in the SmA), and most importantly BOO. In the HexB phase, the molecules are locally hexagonally packed, and the resulting six-fold BOO is maintained for macroscopic distances. The HexB is a uniaxial phase. In three dimensions, the transition from SmA to HexB phase is expected to be in the same universality class as super-fluid helium. However, this is not the case. The explanation is that there is a strong coupling between the mass density and the hexatic order parameter, and fluctuations in the order parameter influence its behavior near the transition [2]. A schematic of the BOO of a hexatic phase is shown in Fig. 1.1(c). A top view of the molecules within a smectic plane, which indicates that the molecules are perpendicular to the planes, is shown in Fig. 1.3(b).

The hexatic phases have been studied quite extensively in two-dimensional freely suspended films. In these studies, liquid crystal films were drawn over a hole in a metal or glass plate. These films have smectic layers aligned almost perfectly parallel to their physical surface rendering the data interpretation easier than in bulk samples. X-ray diffraction studies of the growth of hexatic order in such films will be discussed in greater detail in Chapter 8 by Brock. The transition between SmA and HexB (in two dimensions) can be second order. The Halperin–Nelson–Young theory [19] has been applied to describe this transition. This theory predicts that in two dimensions dislocations destroy PO but not BOO. This is indeed the case.

The crystal-B phase

There are several crystalline smectic phases; that is, smectics which are very close to three-dimensional crystals. They differ from true crystal phases in one important aspect. The molecules in them have freedom of rotation about their long axis, i.e., their thermal motion is not completely frozen out. Such phases, in which the position of the molecules is fixed but their motion is not arrested, should more correctly be classified as *plastic crystalline* phases. The reader who is interested in greater detail in this area should read Pershan [8] and references therein. The common feature of these phases is that the average molecular orientation is normal to smectic layers and within each layer the molecules are ordered in a triangular lattice and possess long range PO and, of course, BOO. The stacking of these triangular arrays varies from material to material and restacking transitions have been observed.

The crystal-E phase

As in the crystal-B phase, the molecules within a smectic-E plane are arranged on a triangular (or hexagonal) lattice and are perpendicular to the smectic layers. This phase differs from the crystal-B phase in the rotational motion of molecules. In the crystal-E phase, the thermal motion of molecules is reduced to the extent that they arrange themselves in a herringbone pattern within a smectic layer. The intralayer molecular packing is shown in Fig. 1.3(c). With this type of packing their continuous rotational freedom is hindered. NMR [20] studies have concluded that molecules can undergo correlated six-fold jump rotational diffusional motion.

1.4.2 Tilted phases

There is a complete set of smectic phases in which the long axis of molecules (or the director) is not perpendicular to the layer normal, but makes

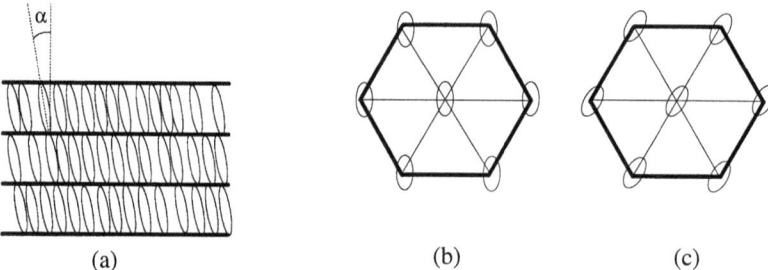

Figure 1.4. Schematic representation of some tilted phases formed by rod-like molecules. (a) The SmC phase; the layers are indicated by the lines and the molecules by the ellipses. Short range order in hexatic smectic-F and hexatic smectic-I phases are shown in (b) and (c) respectively. Here the major axis of ellipse indicates the molecular tilt direction with respect to the hexagon axes.

a relatively large angle. Just like the untilted phases discussed above, the various tilted phases possess different structures and order.

The smectic-C phase

The smectic-C, SmC, phase is similar to the SmA phase in that it is a layered structure and each layer may be thought of as a two-dimensional liquid film with no BOO or PO. However, in this case the molecules are on the average tilted with respect to the normal to the layers, i.e., **n** and smectic layer normal are not collinear, Fig. 1.4(a). Furthermore, the tilt angle, α, that the molecular long axis makes with the layer normal, is a strong function of temperature. In the case of a transition from SmA to SmC phase at temperature T_{AC}, smectic layer spacing at a temperature T below the transition $d(T) = d(T_{AC}) \cos \alpha$. The angle α which is an order parameter for this phase can range from zero to as high as 45–50° deep in the SmC phase. The temperature dependence of molecular tilt [21] in the SmC phase of terephthal-bis-(4n)-butylaniline (TBBA) is shown in Figure 1.5. Most of the SmA to SmC transitions are second order but first order transitions in fluorinated compounds have recently been observed [22]. A consequence of the tilting of the molecules is that this phase exhibits biaxial optical and physical properties.

The smectic-F and smectic-I phases

Tilted phases with in-plane hexatic order have also been observed. These may be thought of as tilted analogs of the HexB phase. However, with a hexagonal arrangement within smectic planes, the molecules can tilt along two distinct directions with respect to the hexagonal lattice. In the HexF phase, the tilt is in a direction perpendicular to the sides of the hexagon. If molecules

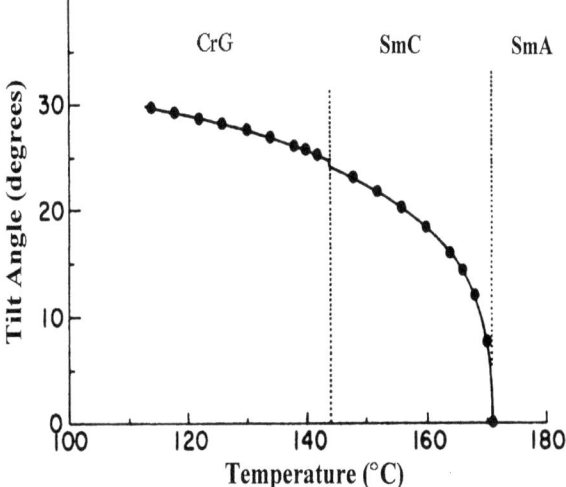

Figure 1.5.   Temperature dependence of molecular tilt in the SmC phase of TBBA determined by high-resolution x-ray diffraction.

tilt towards a corner, it is known as the HexI phase. The descriptions of these phases often leave one confused. When both phases are formed by the same material, the HexI is always the higher temperature phase. The short range in-plane order of these three phases is shown in Fig. 1.4(b) and (c). These phases are also the topic of discussion for Chapter 8 on freely suspended film measurements and further details can be found there.

The crystal-G and crystal-J phases

The crystal-G and crystal-J phases are the phases normally obtained at temperatures lower than the HexF and HexI phases. The molecular arrangement within a smectic plane possesses long range PO and BOO. In both these phases and their higher temperature hexatic versions, the molecules are tilted with respect to the layer normals by approximately 25 to 30°. The local order is hexagonal which is distorted due to molecular tilt, with molecules having rotational freedom comparable to the untilted crystal-B phase.

The crystal-H and crystal-K phases

The crystalline H and K phases are tilted versions of the crystal-E phase. The molecules in the crystalline H (K) phase are tilted in a manner similar to the crystal-G (J) and HexF (HexI) phases, i.e., along the direction perpendicular to a side (towards a corner) of the underlying hexagonal structure. The general sequence of tilted phases with decreasing temperature is

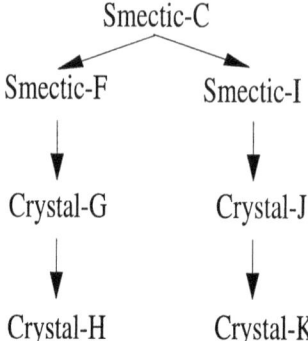

Figure 1.6. A typical phase sequence for tilted phases of rod-like molecules.

shown in Fig. 1.6. However, the actual order of phases varies from material to material.

1.4.3 Chiral tilted phases

Molecules which are not identical to their mirror image are said to be chiral. Molecular chirality has profound influence on liquid crystal properties. There exists a complete *designer line* of liquid crystal phases with chirality, as discussed in the following pages. The extension of the schemes discussed above on the basis of symmetry properties to the phases formed by chiral molecules is rather straightforward.

The cholesteric or chiral nematic phase

The phase which, in most regards, is similar to the nematic phase but with chiral structure is known as the cholesteric phase. The name is based on the fact that derivatives of the infamous cholesterol were initially found to exhibit this phase. This phase forms if the molecules which form the liquid crystal are either intrinsically chiral or if chiral dopants are added to a non-chiral (regular) nematic. The chirality of the system leads to a helical distortion, in which the director **n** rotates continuously in space along, say, the z-direction, with a spatial period of π/p. Here p is the pitch of the cholesteric which can be positive or negative, corresponding to right or left handedness, and has a magnitude which can be a strong function of temperature. The temperature dependence of their pitch has been exploited in making thermometers. The pitch of a cholesteric can also be increased (unwound) with the help of an applied electric field. It is important to note that for systems which have a small twist (long pitch), locally a cholesteric

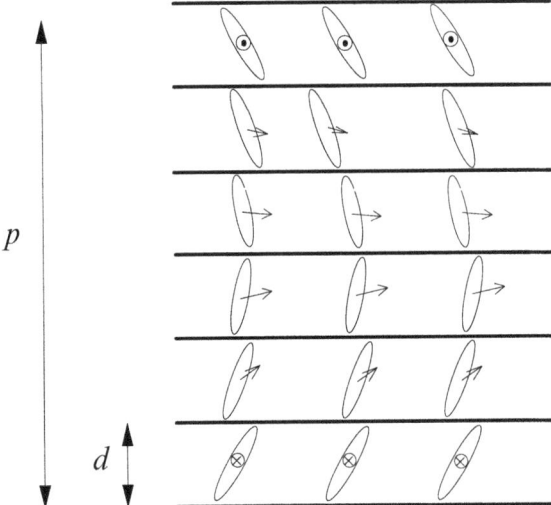

Figure 1.7. A schematic diagram of the SmC* phase showing the precession of the tilt angle as one moves along the layer normal. The arrows attached to the molecules point to represent the direction of the molecular electric dipole moment which renders these phases ferroelectric. d is the layer spacing and p is the pitch.

liquid crystal 'looks' like a nematic. Of course, this picture breaks down for short pitched cholesterics. We will discuss this situation later in the section on defect phases.

Chiral smectic phases

The smectic phases formed by chiral molecules are designated with a * over the letter specifying the analogous non-chiral phase. For example, the chiral smectic-A is abbreviated as SmA*. Many of the physical properties of a chiral phase, e.g., the linear electro-optical effect [14] of the SmA* phase, are not present in the SmA phase. Tilted phases occur for chiral molecules. The simplest is called the fluid smectic-C*, SmC*, phase. This extra degree of freedom obtained from chirality causes a distortion in the SmC structure, and the direction of tilt precesses around the layer normal as one moves from layer to layer. A schematic picture of the SmC* is shown in Fig. 1.7. This phase is ferroelectric in nature as the molecules have a transverse permanent dipole moment. This property has been the basis of many new technologies and the subject of intense study in its own right [23–25]. Ferroelectric phases will be discussed in Section 1.4.5 in more detail.

The precession of tilt angle from layer to layer is also seen in other chiral smectic and crystalline phases. They generally show the same phase sequence as shown in Fig. 1.6. Because the molecules are chiral in analogy with the SmC* all of these phases are also labeled with a star. Thus we obtain HexF*, HexI*, CrG*, CrJ*, CrH* and CrK* phases.

1.4.4 Frustrated or polar smectics

Liquid crystalline compounds with a large longitudinal permanent electric dipole moment exhibit a very rich mesomorphism including several smectic-A phases and reentrant nematic phases. The longitudinal dipole moment arises because of the presence of chemical groups such as NO_2 or CN. Chemical formulae and mesomorphism of some of the often studied materials are given in Fig. 1.8. This behavior was first observed unexpectedly in a mixture by Sigaud and coworkers [26]. The term frustrated stems from a microscopic frustrated spin model of Berker *et al.* [27]. At present, more than ten smectic phases have been observed in these systems.

The molecular arrangement in the four simplest frustrated phases, namely the N, smectic-A_1, -A_2, and -A_d (abbreviated as N, SmA_1, SmA_2, and SmA_d) phases, is shown in Fig. 1.9 along with their characteristic x-ray diffraction patterns. Here, the molecules are drawn as straight lines with an arrowhead at one end to represent the dipole. Two diffuse spots observed in the x-ray diffraction pattern of these nematic phases set them apart from the nematic phase of non-polar materials. The dipoles have no orientational preference in the SmA_1 phase. In the SmA_1 phase, quasi-Bragg reflection and a diffuse peak are observed. The diffuse peak condenses into a sharp reflection in the SmA_2 phase as the antiferroelectric order develops accompanied by a dipole density modulation of periodicity $2l$, l being the molecular length. The dipole density modulation has periodicity of $2l' < 2l$ in the SmA_d phase, in which molecules partially overlap. The two quasi-Bragg reflections in the SmA_d phase appear at slightly higher values of the scattering vectors than in the SmA_2 phase. In this phase, a diffuse peak corresponding to liquid-like short range order at a length scale of l persists.

Prost, Lubensky, and coworkers [28] developed a mean-field theory for these systems using two two-component smectic order parameters Ψ_1 and Ψ_2 [$\Psi_i(\mathbf{r}) \sim \Psi_{i0} \exp(i\mathbf{q}_i \cdot \mathbf{r})$, $i = 1, 2$] that characterize long range head-to-head antiferroelectric order of molecules and mass density modulations, respectively. The wavevectors $\mathbf{q}_1$ and $\mathbf{q}_2$, at which Ψ_1 and Ψ_2 condense, are in general not required to be collinear. The two order parameters would prefer to condense at their natural wavevectors $\mathbf{k}_1$ ($= 2\pi/2l'$) and $\mathbf{k}_2$ ($= 2\pi/l$) in the absence of coupling between them. Here, $2l'$ ($2l' \leq 2l$) is the effective length

R – O –⟨O⟩– OOC –⟨O⟩– OOC –⟨O⟩– CN

DB₇OCN :   R = C₇H₁₅

CrK ——132.2——▶ SmA ——182.0——▶ N ——247.0——▶ I

DB₈OCN :   R = C₈H₁₇

CrK ——128.5——▶ SmA ——199.4——▶ N ——242.8——▶ I

8OCB

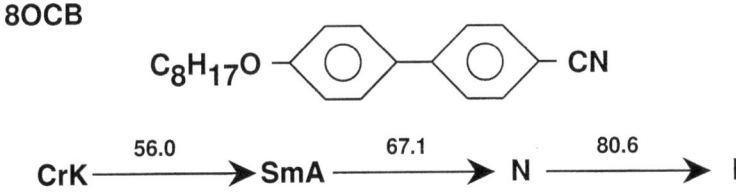

C₈H₁₇O –⟨O⟩–⟨O⟩– CN

CrK ——56.0——▶ SmA ——67.1——▶ N ——80.6——▶ I

8OBCAB

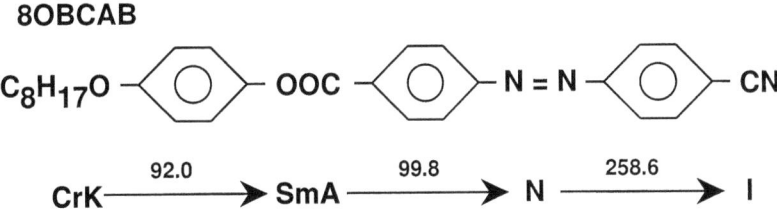

C₈H₁₇O –⟨O⟩– OOC –⟨O⟩– N = N –⟨O⟩– CN

CrK ——92.0——▶ SmA ——99.8——▶ N ——258.6——▶ I

Figure 1.8.   Molecular formulae and phase sequence of some frustrated smectic liquid crystals.

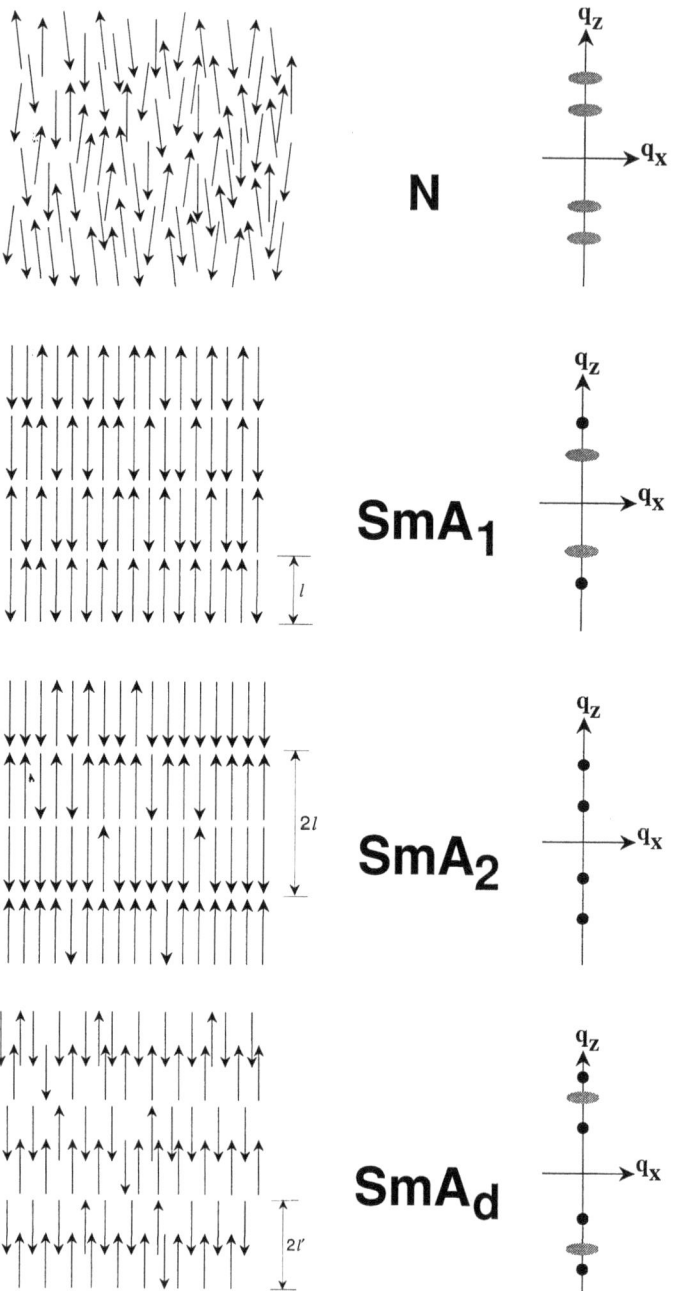

Figure 1.9. Schematic representation of frustrated liquid crystal phases and corresponding x-ray diffraction patterns.

of an antiferroelectrically ordered pair of molecules which can be thought of as a dimer. The excess free energy density can be written as

$$f - f_0 = a_1(T - T_{c1})|\Psi_1|^2 + D_1|(\Delta + k_1^2)\Psi_1|^2 + C_1|\nabla_\perp \Psi_1|^2 + u_1|\Psi_1|^4$$
$$+ a_2(T - T_{c2})|\Psi_2|^2 + D_2|(\Delta + k_2^2)\Psi_2|^2 + C_2|\nabla_\perp \Psi_2|^2 + u_2|\Psi_2|^4$$
$$+ 2u_{12}|\Psi_1|^2|\Psi_2|^2 - w\mathrm{Re}\Psi_1^2\Psi_2^* - v\mathrm{Re}\Psi_1\Psi_2^*.$$

Here T_{ci} are the two mean-field transition temperatures. The elastic terms with coefficients D_i describe the spatial modulation and favor $q_1^2 = k_1^2$ and $q_2^2 = k_2^2$. The terms with $\nabla_\perp$, a gradient in the plane perpendicular to $\mathbf{n}$, favor that $\mathbf{q}_1$, $\mathbf{q}_2$, and $\mathbf{n}$ be parallel to each other. The terms $\Psi_1\Psi_2^*$ and $\Psi_1^2\Psi_2^*$ favor locking of two wavevectors at $\mathbf{q}_1 = \mathbf{q}_2$ and $\mathbf{q}_1 = \mathbf{q}_2/2$, respectively. These competing tendencies give rise to a variety of phases in mixtures of compounds with very different molecular lengths. In binary mixtures, the quantities Ψ_1 and Ψ_2 depend on the concentration, which influences the strength of coupling terms and the natural wavevectors. The cubic term dominates $l' \sim l$ but the harmonic term becomes most important when $l' \sim l/2$.

The formation of various phases follows quite naturally from this theory. When $\Psi_1 = \Psi_2 = 0$, a nematic phase is obtained which possesses neither a mass density nor a polarization density modulation. It has only short range PO along with the director or, in other words, has smectic order fluctuations corresponding to both wavevectors which appear as two sets of diffuse peaks in x-ray diffraction experiments. In the monolayer SmA_1 phase, only the mass density wave develops, so $\Psi_1 = 0$ and $\Psi_2 \neq 0$. The quasi-Bragg reflection from this phase appears at $2q_0 = 2\pi/l$ and a diffuse nematic-like peak at $2\pi/2l'$. When $\Psi_1 \neq 0 \neq \Psi_2$, a number of interesting cases arise. If both density waves are collinear and $l' \cong l$, then the cubic coupling term enforces commensurability, requiring $q_2 = 2q_1 = 2\pi/2l (= q_0$, say) and the bilayer SmA_2 phase results. On the other hand, if $l' < l$, as would be the case if molecules were overlapping, no locking of the wavevectors is possible and the partial bilayer SmA_d phase is obtained. Obviously, this theory provides a very satisfactory explanation for the formation of these and several other (SmC_1, SmC_2, etc. when molecules are tilted) phases which we shall not discuss in any detail here.

Numerous other possibilities exist arising from (i) letting the magnitude of the two collinear wavevectors take different relative values giving rise to the possibility of incommensurate SmA phases [29] which have not yet been discovered. Only the existence of incommensurate crystal smectic phase has been confirmed [30], and (ii) letting them become non-collinear permitting the formation of phases in which one of the density waves is in a

direction parallel to the smectic planes. Depending upon whether the molecules are normal or tilted with respect to the smectic layers, the SmÃ or the
SmC̃ antiphase forms [31]. The existence of these phases has experimentally
been confirmed.

1.4.5 Ferroelectric, antiferrolectric, and ferrielectric smectic phases

One of the more remarkable aspects of the SmC* phase is that it is ferroelectric. In the present context this means that a non-zero spontaneous
polarization develops in this phase. Typically the polarization is a very
strong function of temperature near the transition (from higher temperature phase) to the SmC* phase and somewhat slower farther from the transition. Recall that the SmC* phase consists of tilted molecules and typically
the polarization is coupled to the tilt angle which also is zero at the transition to the (untilted) SmA phase. In fact, one often writes:

$$\theta = \theta_0 (T_C - T)^{\alpha},$$
$$P = P_0 (T_C - T)^{\beta}.$$

where θ is the tilt angle, P is the spontaneous polarization, θ_0 and P_0 are constants, T_c is the transition temperature, T the temperature, and α and β are
constants, theoretically predicted to be equal to 0.5. The above is somewhat
deceiving. Recall that in our discussion of the SmC* phase we stated that
the chirality leads to a twisting of the molecules and hence there is a helical
arrangement of the molecules as shown in Fig. 1.7. In a macroscopic
sample, without surfaces or external electric fields, the molecular polarization will follow this helix and result in zero total spontaneous polarization.
Hence the more correct name is helielectric phases [32]. However, if the helix
is unwound, which happens if a strong electric field is applied, there is a
spontaneous non-zero polarization which can be measured. Measurement
of P is discussed in Chapter 4 on physical properties.

Surface stabilization of the SmC* phase is the basis of surface stabilized
ferroelectric liquid crystal (SSFLC) display devices first proposed by Clark
and Lagerwall [33]. These cells, which are very thin, use surface anchoring
to provide alignment of the liquid crystal. The cell is switched between two
equivalent energy states, with very different optical properties when viewed
through crossed polarizers, by reversing the direction of the applied electric field. This type of cell has achieved switching speed as fast as 0.1 μs
which is three to four orders of magnitude faster than the nematic displays
[33]. Other details and a discussion of the chemistry of these materials is
available elsewhere [34].

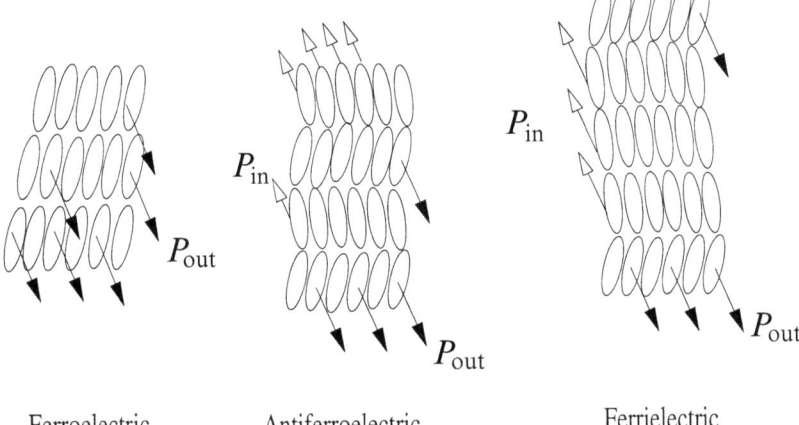

Ferroelectric   Antiferroelectric   Ferrielectric

Figure 1.10. Schematic pictures of the ferroelectric SmC* phase, an antiferroelectric SmC* phase and a ferrielectric SmC* phase. The open arrows indicate that
the spontaneous polarization of the molecular layer is into the page, the dark
arrows indicate that the spontaneous polarization is out of the page.

Recently, two more phases closely related to the SmC* have been discovered. These are somewhat like the frustrated smectic phases in which local
dipoles become antiparallel. In these phases, the layer spacing and the polarization direction are not related in the same manner as in the SmC* phase.
These phases, by analogy to the situation in magnets, are called antiferroelectric liquid crystals and ferrielectric liquid crystals. Figure 1.10 schematically depicts the SmC* phase, the proposed structure of the
antiferroelectric phase, and the simplest ferrielectric phase. All phases
possess chirality which has not been shown in Fig. 1.10 to clarify local
molecular and dipolar arrangements. The SmC* phase is characterized by
the spontaneous polarization of each layer pointing in the same direction.
The antiferroelectric phase is characterized by alternating directions of the
spontaneous polarization and the molecular tilt in adjacent layers. Thus,
the average tilt angle and the average spontaneous polarization in this
phase are both zero. A sufficiently strong electric field will switch the antiferroelectric phase to the ferroelectric phase. It is significant to note that
antiferroelectrics repeat their helical structure every 180° of rotation about
the layer normal compared to 360° for the ferrolectric phase. The ferrielectric phase is characterized by an uneven number of layers stacked with
tilt to the left and to the right. Thus there is a net but low spontaneous polarization, since there are an uneven number of molecules with polarization
in the two opposite directions. Recent work has shown that many shades of

ferrielectric phases exist differing in the proportion of layers with the two different tilt directions. Readers are referred to the literature for more details on these phases [35].

1.4.6 Blue and twist grain boundary phases

One of the advances in recent years in liquid crystal research has been the discovery of defect phases. You might recall that in crystals defects such as vacancies, dislocations, grain boundaries and the like exist, and can sometimes be removed by special preparation of the material. In liquid crystals defects also exist [36]. These defects have several specific geometries, and are energetically unfavored. In most experiments these defects are accidents of sample preparation and one tries to remove them by annealing, careful and clean sample preparation, or other techniques. However, in sufficiently high chirality materials defects form spontaneously and are an essential feature of the liquid crystalline phase. There are two common examples, both of which we will briefly discuss: the blue phases (BP) and the twist grain boundary phases (TGB). The BPs may be thought of as double twist order permeated by a lattice of line defects. The TGB phase occurs when twist is frustrated by layer structures. The resulting structure has periodic grain boundaries.

Recall that the chiral nematic phase is locally a nematic with a slow (on a molecular scale) rotation in the director. This phase occurs when the pitch is large or the chirality is low. At the other extreme, when the chirality is high, this picture is not correct. The presence of the high chirality makes it possible for the system to have a lower energy state by having a double twist structure rather than the single twist. However, this is not the whole story; the free energy can be further reduced by forming many spatially ordered double twist structures. The resulting defect phases possess simple cubic, body centered cubic, and amorphous structures [37, 38]. Usually these phases occur only over a small temperature range between the isotropic and the chiral nematic phases, and are called blue phases because the first discovered phase appears blue in reflected light.

The twist grain boundary phases formed by high chirality molecules have a structure similar to the SmA phase and are more accurately denoted as the TGB_A phases. The layered structure of the SmA excludes twist. However, just as in the blue phases, high chirality subtly changes the energetics of the system leading to a different type of structure. In this case the free energy is minimized by introducing grain boundaries at periodic intervals. This phase was theoretically predicted by Renn and Lubensky [39] and almost simulta-

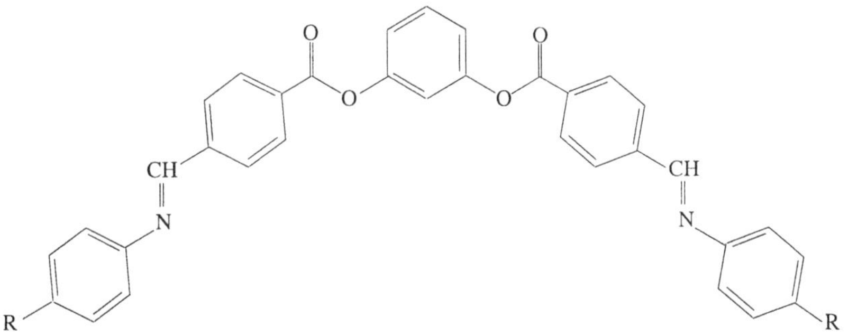

Figure 1.11. The chemical structure of a bownana molecule synthesized by Matsunaga. The R groups are hydrocarbon chains; R is C_8H_{17} and C_9C_{19} for the two molecules synthesized.

neously experimentally observed by Goodby *et al*. [40]. A tilted analog made of SmC-like layers has also been observed. It is known as the TGB_C phase. These phases extend the analogy between superconductors and liquid crystal phases (SmAs). The TGB_A phase is analogous to a type II superconductor, while the SmA phase is analogous to a type I superconductor.

1.5 Liquid crystals formed from bownana molecules

A recent development that has sparked great interest has been the phase behavior of bow, banana, boomerang or bent-core molecules. A more appropriate word to include all bent shapes, perhaps, is *bownana* shape. This was stimulated by the synthesis work of several groups including Matsunaga, Samulski, and Chien [41]. An example of the type of molecules involved is shown in Fig. 1.11.

The excitement concerning the smectic phases of these materials is that this molecule is achiral but macroscopically chiral liquid crystal phases have been identified and studied. The chiral domains spontaneously form upon cooling from the isotropic phase, and domains of opposite handedness have been observed in the same material [42]. In these materials three distinct symmetry-breaking events are possible upon cooling from the isotropic phase. These are: formation of layers, i.e., QLR-PO as in a smectic phase; orientation order of the molecules with non-zero tilt with respect to the layers order; and a third 'polar' ordering of the bownanas that form the cores of these molecules. The latter is believed to be due to the bent-core geometry. Link *et al*. [42] have named this the $SmCP_A$ phase.

The $SmCP_A$ phase is characterized by: (i) layer-by-layer antiferroelectric structure; (ii) spontaneous breaking of the achiral symmetry; (ii) both racemic and homogeneously chiral stacking of the layers, the racemic having the lower free energy; (iv) the coexistence of chiral domains of both handedness which maintain their handedness during switching in response to an applied field; and (v) 'sergeants-and-soldiers' biasing of the global chirality upon addition of chiral dopants. This phase has been observed to form thin films of two or three layers. Currently there are several research groups working on these materials. The structure and nomenclature of these materials have not yet been finalized.

1.6 Liquid crystals formed from disk-shaped molecules

The liquid crystalline phases formed by disk-shaped molecules are referred to as the discotic phases. They were first synthesized and identified in 1977 [2, 43]. Structurally they generally form either nematic or columnar phases. The simplest columnar phase consists of stacked disks, forming a one-dimensional liquid-like structure; the columns themselves form a two-dimensional lattice. Hexagonal, rectangular, and other lattice types have been identified. Tilted phases have also been reported. The discotic nematic phase is denoted by N_D where the subscript D is to avoid confusion with the normal nematic phase. This phase is characterized by the short axis of the molecule aligning parallel to the director. The N_D phase is diamagnetically and optically negative. This is in contrast with the rod-like nematics which are mostly optically and diamagnetically positive. Figure 1.12 illustrates the N_D phase and the columnar phases.

The structure of the columnar phases merits some discussion. There is no PO along the columns. The columns themselves are arranged with PO and form a two-dimensional array which is either a rectangular or a hexagonal lattice. There is orientational order (at least for the cores of these molecules) within the columns. The rectangular or D_r phase has columns which occupy a rectangular lattice. The molecules are oriented away from the column axis in this phase; the resulting tilt forms a herringbone pattern such as shown in Figure 1.12(c). The columns of the hexagonal, D_h, phase are arranged in a hexagonal pattern. The molecules are tilted within the columns, but there is no coordinated azimuthal orientation of the molecules in the columns.

1.7 Polymeric liquid crystals

Another class of thermotropic mesogens is polymer liquid crystals. These structures consist of mesogenic subunits (either rod-like or disk-like) which are connected together with flexible links forming what are known as the

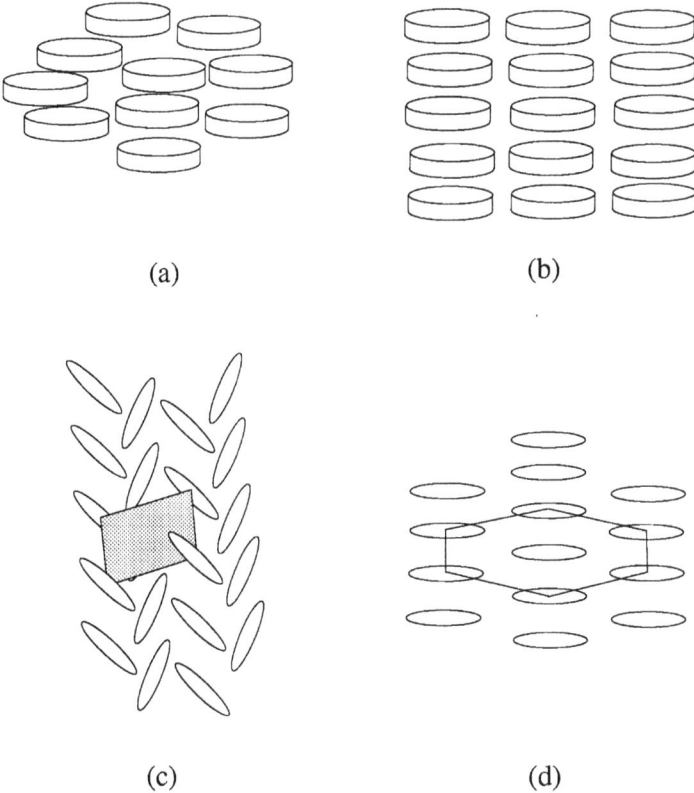

(a)   (b)

(c)   (d)

Figure 1.12.   The structure of the N_D (a) and columnar (b) phases. In columnar phases there is PO in the columns. The rectangular columnar phase, D_r, is shown in (c) and the hexagonal columnar phase, D_h, is shown in (d). In these two parts of the figure the molecules are tilted. This is illustrated by drawing the disks as ellipses.

main-chain polymer liquid crystals (PLC). Alternatively, the mesogenic subunits can be attached to the polymer chains as side groups or pendates. These are known as the side-chain PLCs. The nature of the liquid crystal-line phases which form depends on the backbone, spacers, flexible links, etc. These are reviewed in several references [4, 44]. A schematic presentation of these PLCs is shown in Fig. 1.13.

1.8 Lyotropic liquid crystals

1.8.1 Colloidal lyotropic phases

Solutions of biomolecules such as proteins and DNA and sufficiently con-centrated solutions of surfactants can form another interesting class of liquid crystals. Since the phase behavior is most easily induced by changes

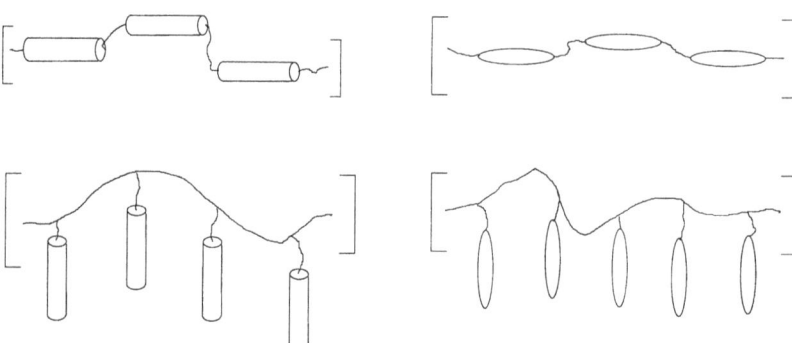

Figure 1.13. Schematic representation of polymeric liquid crystals. The rods and ellipses represent rod-like and disk-like mesogens, respectively. The wavy lines represent the polymer chains or segments. The brackets, [ ], indicate that the unit is repeated many times. The top two figures represent main-chain PLCs, the bottom two side-chain PLCs.

in concentration (although temperature is still an important variable) these are referred to as lyotropic liquid crystals. These mesophases [1, 2] were first considered by Onsager [45] and Flory [46]. Representative examples are synthetic polypeptides, precipitated metal oxides, and rigid polymers in appropriate solvents. Typically these molecules achieve length to diameter ratios of $\sim$10–20. Liquid crystal properties of solutions of DNA molecules, some proteins, and viruses have also been extensively studied. Of particular interest is tobacco mosaic virus (TMV) which has a length of approximately 3000 Å and a diameter of 200 Å and is known to form phases analogous to the nematic and smectic phases of thermotropic liquid crystals.

1.8.2 Self-assembled structures

In addition to small molecules, polymer liquid crystals, and large chemically bonded objects such as viruses, another simple class of materials exhibits liquid crystalline phases. This class of materials consists of amphiphilic molecules in a solvent in which they spontaneously form lyotropic micellar systems. An example of such a system is a solution of soap in water. A typical soap molecule consists of a polar head and one or more hydrocarbon tails. When a sufficient number of such molecules is dissolved in a solvent, the lowest free energy state is a state in which the hydrocarbon tails segregate to shield themselves from the polar water environment. This leads to the formation of aggregates of molecules. Molecules in such aggregates, called micelles, are not covalently bonded and can assume several

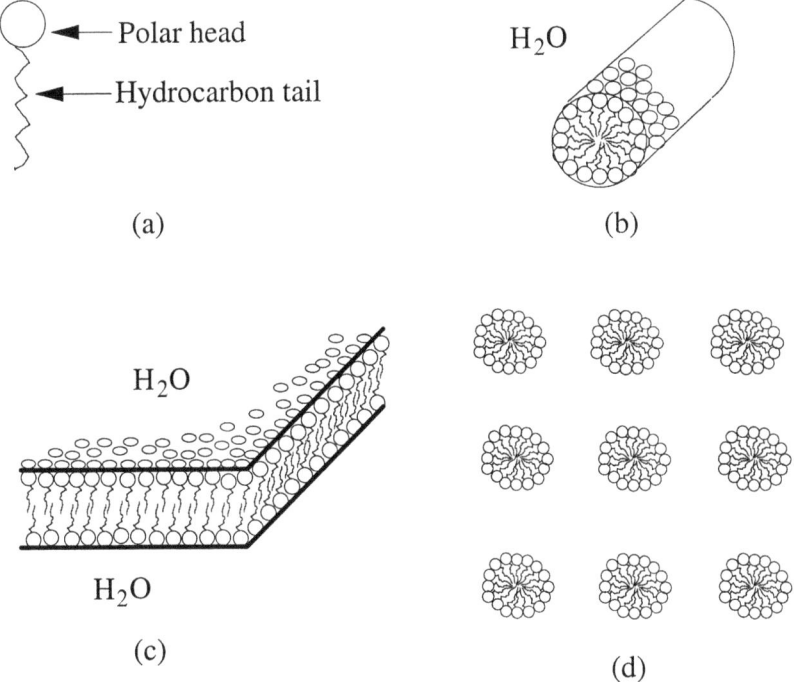

Figure 1.14.   (a) A 'cartoon' of a typical amphiphilic molecule. (b) An example of a rod-like micelle. (c) A slice of a bilayer structure. This is an example of a lamellar phase. (d) A cubic phase formed by spherical aggregates.

different geometries depending on the thermodynamic conditions and chemical nature of the molecules. These aggregates can be rod-like or disk-like which can be orientationally and/or positionally ordered to exhibit a wide range of liquid crystalline phases including isotropic, nematic, lamellar, hexagonal, and cubic phases [1, 5]. A typical surfactant and a few simple phases of amphiphilic materials are shown in Fig. 1.14.

This area of liquid crystal research, in fact, precedes the age of thermotropic liquid crystals. Early work [47] demonstrated the existence of mesophases with PO in one (lamellar or smectic phases), two (hexagonal and rectangular phases), and three (cubic) dimensions. The observation of the nematic phase came more recently [48]. It typically consists of rod-like or disk-like micelles with their symmetry axis orientationally ordered. This phase is observed between the isotropic phase and a smectic-like phase called the lamellar phase of disk-like micelles. For cylindrical micellar systems, it appears between the isotropic and the hexagonal phase. These phases can be observed over a rather large concentration range typically from 0.1 to 0.5 weight percent as well as over a wide temperature range. The Onsager theory explains the isotropic to nematic phase transition in these materials rather well. Finally, the nematic composed of rod-like micelles is called the canonic [5] nematic, N_C. A review of research in this field has been prepared by Boden [49].

References

1. P. G. de Gennes and J. Prost, *The Physics of Liquid Crystals*, 2nd Edn., Oxford University Press, New York (1993).
2. S. Chandrasekhar, *Liquid Crystals*, 2nd Edn., Cambridge University Press, New York (1992).
3. G. Vertogen and W. H. de Jeu, *Thermotropic Liquid Crystals, Fundamentals*, Springer-Verlag, New York (1988).
4. A. Blumstein (Ed.), *Polymeric Liquid Crystals*, Plenum Press, New York (1985).
5. P. J. Collings and J. S. Patel (Eds.), *Handbook of Liquid Crystal Research*, Oxford University Press, New York (1997).
6. R. Blinc and R. Zeks, *Ferroelectric and Antiferroelectric Liquid Crystals and Electro-optic Applications*, World Scientific, New Jersey (1999).
7. S. Kumar, *Liquid Crystals in the Nineties and Beyond*, World Scientific, New Jersey (1995).
8. P. S. Pershan, *Structure of Liquid Crystal Phases*, World Scientific, New Jersey (1988).
9. I.-C. Khoo and S.-T. Wu, *Optics and Nonlinear Optics of Liquid Crystals*, World Scientific, New Jersey (1993).
10. P. Tolédano and A. M. Figueiredo Neto, *Phase Transitions in Complex Fluids*, World Scientific, New Jersey (1998).
11. K. C. Chu and W. L. McMillan, *Phys. Rev. A* **15**, 1181 (1977).
12. E. B. Priestley, P. J. Wojtowicz, and P. Sheng, *Introduction to Liquid Crystals*, Plenum Press, New York (1975).
13. R. J. Birgeneau and J. D. Litster, *J. Phys. (Paris)* **39**, L-399 (1978); A. Aharony, R. J. Birgeneu, J. D. Brock, and J. D. Litster, *Phys. Rev. Lett.* **57**, 1012 (1986).
14. *International Tables for X-ray Crystallography*, Vol. A–C, 2nd Edn., Kluwer Academic Publishers, Netherlands (1999).
15. R. E. Peierls, *Helv. Phys. Acta, Suppl.* **7**, 81 (1934); M. A. Caille, *C. R. Acad. Sci. Ser. B* **274**, 891 (1972).
16. P. G. de Gennes, *Solid State Commun.* **10**, 753 (1972).
17. P. G. de Gennes, *C. R. Acad. Sci.* **274**, 758 (1972) and *Mol. Cryst. Liq. Cryst.* **21**, 49 (1973).
18. M. J. Freiser, *Phys. Rev. Lett.* **24**, 1041 (1970); R. Alben, *Phys. Rev. Lett.* **30**, 778 (1973); D. W. Allender, M. A. Lee, N. Halfiz, *Mol. Cryst. Liq. Cryst.* **124**, 45 (1985); J. Melthete, L. Liebert, A. M. Levelut, and Y. Galerne, *C. R. Acad. Sci.* **303**, 1073 (1986); G. Shinuoda, Y. Shi, and M. E. Neubert, *Mol. Cryst. Liq. Cryst.* **257**, 209 (1994).
19. B. I. Halperin and D. R. Nelson, *Phys. Rev. Lett.* **41**, 121 (1978); D. R. Nelson and B. I. Halperin, *Phys. Rev. B* **19**, 2457 (1979); A. P. Young, *Phys. Rev. B* **19**, 1855 (1979).

20. See for instance: B. Deloche, J. Charvolin, L. Liébert, and L. Strzelecki, *J. Phys. (Paris)* **36**, C1-21 (1975), G. J. Krüger, H. Spiesecke, and R. Van Steenwinkel, *J. Phys. (Paris)* **37**, C3-123 (1976), and R. Blinc, M. Luzar, M. Vilfan, and M. Burgar, *J. Chem. Phys.* **63**, 3445 (1975).

21. S. Kumar, *Phys. Rev. A* **23**, 3207 (1981).

22. T. P. Rieker and E. P. Janulis, *Phys. Rev. E* **52**, 2688 (1995); M. Hird and K. J. Toyne, *Mol. Cryst. Liq. Cryst.* **323**, 1 (1998).

23. N. A. Clark and S. T. Lagerwall, *Appl. Phys. Lett.* **36**, 899 (1980); J. W. Goodby, *Ferroelectric Liquid Crystals: Principles, Properties, and Applications, Ferroelectricity and Related Phenomena* vol. 7, Gordon and Breach Science Publishers, New York (1991).

24. P. Patel, D. Chu, J. W. West, and S. Kumar, *SID Digest* **XXV**, 845 (1994); V. Vorflusev and S. Kumar, *Ferroelectrics* **213**, 117 (1998).

25. X.-Y. Wang, J.-F. Li, E. Gurarie, S. Fan, T. Kyu, M. E. Neubert, S. S. Keast, and C. Rosenblatt, *Phys. Rev. Lett.* **80**, 4478 (1998).

26. G. Sigaud, M. F. Achard, F. Hardouin, and H. Gaspraux, *J. Phys. (Paris) Colloq.* **40**, C3-356 (1979).

27. J. O. Indekeu and A. N. Berker, *Phys. Rev. A* **33**, 1158 (1986); J. O. Indekeu, A. N. Berker, C. Chiang, and C. W. Garland, *ibid.* **35**, 1371 (1987).

28. P. Barois, J. Prost, and T. C. Lubensky, *J. Phys. (Paris)* **46**, 391 (1985); J. Prost, *Adv. Phys.* **33**, 1 (1984), and references therein.

29. P. Patel, S. Kumar, and P. Ukleja, *Liq. Cryst.* **16**, 351 (1994).

30. J. T. Mang, B. Cull, Y. Shi, P. Patel and S. Kumar, *Phys. Rev. Lett.* **74**, 4241 (1995).

31. Y. Shi, G. Nounesis, and S. Kumar, *Phys. Rev. E* **54**, 1570 (1996); C. R. Safinya, W. A. Varady, L. Y. Chiang, and P. Dimon, *Phys. Rev. Lett* **57**, 432 (1986).

32. H. R. Brand, P. E. Cladis, and P. L. Flinn, *Phys. Rev. A.* **31**, 361 (1985).

33. N. A. Clark and S. T. Lagerwall, *Appl. Phys. Lett.* **36**, 899 (1980); N. A. Clark and S. T. Lagerwall, *Ferroelectrics* **59**, 25 (1984).

34. B. Bahadur (Ed.), *Liquid Crystals Applications and Uses*, 3 vols. World Scientific, New Jersey (1990–1992); J. S. Patel and J. W. Goodby, *Mol. Cryst. Liq. Cryst.* **144**, 117 (1987).

35. See for instance: A. D. Chandani, Y. Ouchi, H. Takezoe, A. Fukuda, K. Terashima, K. Furukawa, and A. Kishi, *Jpn. J. Appl. Phys. Lett.* **28**, 1261 (1989); E. Gorecka, A. D. L. Chandani, Y. Ouchi, H. Takezoe, and A. Fukuda, *Jpn. J. Appl. Phys.* **29**, 131 (1990); S. Inui, S. Kawano, M. Saito, H. Iwane, Y. Takanishi, K. Hiraoka, Y. Ouchi, H. Takezoe, and A. Fukuda, *Jpn. J. Appl. Phys. Lett.* **29**, 987 (1990); A. Fukuda, Y. Takanishi, T. Isozaki, K. Ishikawa, and H. Takezoe, *J. Mater. Chem.* **4**, 997 (1994).

36. M. Kleman, *Points, Lines and Walls: In Liquid Crystals, Magnetic Systems and Various Ordered Media*, Wiley Interscience, New York (1983).

37. There are many reviews of this area. See for instance: P. P. Crooker, *Liq. Cryst.* **5**, 751 (1989).

38. P. H. Keyes, *Mater. Res. Bull.* **16**, 32 (1991).

39. S. R. Renn and T. C. Lubensky, *Phys. Rev. A.* **38**, 2132 (1988).

40. J. W. Goodby, M. A. Waugh, S. M. Stein, E. Chin, R. Pindak, J. S. Patel, *J. Am. Chem. Soc.* **111**, 8119 (1989).

41. Y. Matsunaga and S. Miyamoto, *Mol. Cryst. Liq. Cryst.* **237**, 311 (1993); K. J. Semmler, T. J. Dingemans, and E. T. Samulski, *Liq. Cryst.* **24**, 799 (1998); L. C. Chien, private communication.

42. D. R. Link, G. Natale, R. Shao, J. E. Maclennan, N. A. Clark, E. Körblova, and D. M. Walba, *Science* **278**, 1924 (1997).
43. S. Chandrasekhar, B. K. Sadashiva, and K. A. Suresh, *Pramana* **9**, 471 (1977); S. Chandrasekhar, B. K. Sadashiva, K. A. Suresh, N. V. Madhusudana, S. Kumar, R. Shashidhar, and G. Venkatesh, *J. Phys. (Paris) Colloq.* **40**, C3-120 (1979).
44. T. Tsuruta (Ed.), *Biopolymers, Liquid Crystalline Polymers and Phase Emulsions*, Advances in Polymer Science Series 126 (1996).
45. L. Onsager, *Ann. NY Acad. Sci.* **51**, 627 (1949).
46. P. J. Flory, *Proc. R. Soc. London Ser. A* **234**, 73 (1956).
47. S. Safran, *Statistical Thermodynamics of Surfaces, Interfaces, and Membranes*, Addison-Wesley, Reading (1994); G. J. T. Tiddy, *Phys. Rep.* **57**, 1 (1980).
48. There are numerous references, for example: J. Charvolin, *J. Chim. Phys.* **80**, 16 (1983); A. Saupe, S. Y. Xu, S. Plumley, Y. K. Zhu, and P. J. Photinos, *Physica A* **174**, 195 (1991).
49. N. Boden in *Micelles, Membranes, Microemulsions and Monolayers*, W. Gelbart, A. Ben-Shaul, and D. Roux (Eds.) Springer-Verlag, New York, (1994).

2

Characterization of mesophase types and transitions

MARY E. NEUBERT

Glenn H. Brown Liquid Crystal Institute, Kent State University, Kent, OH 44242, USA

The colour plates referred to in this chapter can be downloaded from www.cambridge.org/9780521187947

The initial determination of mesomorphic properties for a new compound is usually made by the chemist who synthesized the materials. Two widely used methods that require relatively inexpensive equipment are hot-stage polarizing microscopy and thermal analysis via differential thermal analysis (DTA) or differential scanning calorimetry (DSC). A third method, normally requiring interaction with another scientist and expensive equipment, is x-ray crystallography [1, 2a]. This method usually defines the structure of the phase and is used to confirm the identifications provided by the first two methods or to characterize a new mesophase. Phase transitions can be further studied by numerous additional methods, such as adiabatic calorimetry. Only the first two methods will be discussed as they are the only ones normally employed by the synthesis chemist.

To be able to determine mesmorphic properties accurately, an understanding of the relationship between liquid crystals and other states of matter is needed. A summary of the relationships between solids and liquids can be found in the literature [3]. Those pertinent to the discussion here are the following:

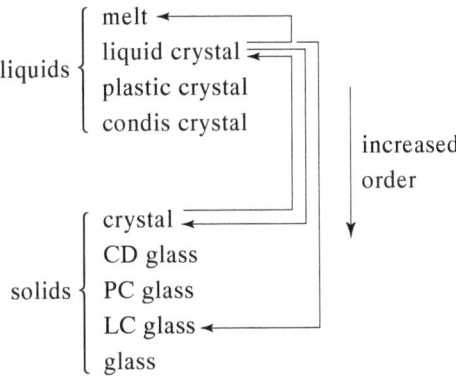

It is quite common to observe crystal-to-crystal changes in materials designed to have mesomorphic properties. This is not surprising since designing such molecules requires achieving a delicate balance between flexibility and rigidity to allow for stepwise melting and the observance of various mesophases. Such molecules may then be able to adopt different types of packing in the crystalline phase as well as in mesophases.

A variety of crystal-to-crystal changes are possible. They can occur on heating fresh (virgin) crystals before melting takes place or only on cooling. These crystals can convert to the virgin crystals or they might not within the time of a transition study. Of course this depends on how long a sample sets. Sometimes it will take a while and reheating too soon may give a much broader melting transition than if the same is allowed to set overnight. Different crystal forms can also show different melting temperatures. Up to three crystal forms are sometimes observed; more can rarely be detected easily. The following schematic diagrams show some of the various possibilities with the highest temperature crystals always being labeled CrK_1 and M = mesophase.

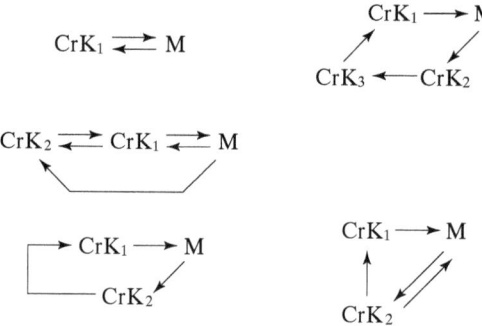

An understanding of the phenomenon of supercooling is also essential for accurately determining mesomorphic properties. Isotropic liquids usually *supercool* before they crystallize under non-equilibrium conditions. This is also true of mesophases. This means that *the crystallization temperature will usually be lower than the melting temperature*. With some compounds, this supercooling will be small ($\sim 1\ °C$) and therefore appear to be non-existent, but in most compounds it is large enough to be easily detected. In some compounds, it can be quite large. It tends to be larger in branched-chain mesogens than in straight-chain ones. Since supercooling represents an unstable condition, it also means that the crystallization temperature is dependent on such things as cooling rate, vibration and previous history of the sample. Transitions between mesophases usually do not involve supercooling. Thus, supercooling can serve as a means for differentiating between crystals and mesophases. As is true in structure–property relationships, there can be exceptions, but these are rare. The first goal of determining mesomorphic properties should be to determine if the phase transitions that are observed involve mesophases rather than crystal phases. An absolutely essential part of identifying mesophases is to eliminate the possibility that a three-dimensional crystalline phase is present.

The supercooling of fluids makes it possible to observe additional mesophases called monotropic phases. Such mesophases are not observed on heating from crystals, but can be seen on cooling a sample below the melting temperature. *This transition is reversible as long as it occurs before crystallization.* Thus, the best definition for monotropic phases is that they are *mesophases that occur below the melting temperature*; not the ones that occur only on cooling:

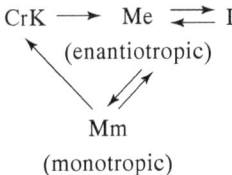

An enantiotropic mesophase is one that occurs above the melting temperature and is therefore seen on melting the sample from crystal (or a lower temperature mesophase) to the istropic liquid or the lowest temperature mesophase and is always seen on cooling. Thus, if a mesophase is seen on heating, it must also be seen on cooling. With some compounds, a monotropic mesophase occurs only slightly below the melting temperature, within the experimental error of the temperature sensor of the measuring device. This makes it difficult to determine whether these are enantiotropic or monotropic, but carefully repeated results often makes this possible. Sometimes the mesophase ranges are so small that the best that can be done is to call the melting transition a $CrK \rightarrow M_{1,2}$ transition, for example, CrK–SmH. This is also true of some mesophase transitions.

It is also possible to have two crystal forms: one of which melts to a mesophase on heating giving an enantiotropic mesophase and another that does not melt to this phase, but instead forms it on cooling giving a monotropic phase. This can occur when the two crystal forms have two different melting temperatures.

2.1 Microscopic textures

The pioneering work of Arnold, Sackmann and Demus characterized many mesogenic textures and developed a classification system from contact mixture studies and phase diagrams based on the rule that identical mesophases will be miscible in all proportions [4, 5a, 6–8]. The simplest method is the contact mixture study. A small sample of the new material is placed on one side of the viewing area of a slide. A known sample is placed nearby leaving a small gap between the two in the center. A cover slip is added, the slide inserted into a heating stage and heated until both samples melt to the mesophase being characterized. If the two mesophases are different, a boundary will occur in the center. This can be easily seen in Fig. 2.1 (color) between the nematic and smectic-A phases. However, it is smectic phases which usually need to be identified. Although some boundaries between smectics are obvious, such as smectics-A and -C, others are more difficult, such as smectics-A and -B. It can also be difficult to determine if a boundary exists between two smectic phases when both give mosaic textures. Another problem is that both mesophases must occur in approximately the same temperature range. Additionally, the two mesophases may seem to be miscible in all proportions in contact mixture studies, but actually show a small region of immiscibility in a more accurate phase diagram. A detailed phase diagram was needed to show that the phase in TBBA that was initially thought to be a biaxial (tilted), B phase was not a B but a CrG phase [4a, 5a, 7–9]. Still, these two methods have been used extensively in conjunction with x-ray crystallography to characterize smectic phases by type, their order in a sequence and to develop a classification system [5b, 10]. Mesogenic series showing numerous smectic phases were particularly useful. These include the anils [5c, 11], dianils [11], pyrimidines [11] and biphenyl esters [12]. These studies were reviewed and the classification updated at successive International Liquid Crystal Conferences [13, 14].

During these characterization studies, mesophases were given various letter designations according to the order in which they were discovered. As long as only a few groups used this system, no assignment problems developed. However, with more researchers automatically assigning the next letter to what they felt was a new mesophase, problems began to occur. Identifications had to be changed to avoid confusion. For some phases, such as crystals, a wide variety of designations have been used (C, Cr, Crys, K, Kr). Additionally, there are different ways of indicating these letters for smectics: smectic-A, SmA, S_A, SA or A. A group of scientists is now trying to develop a universal set of standard designations. For the discussion here, the following terminology will be used:

I	Isotropic
N	Nematic
N*	Chiral nematic (formerly cholesteric 'Ch' phase)
SmA	
SmC	Fluid smectic-A, -C, and -C* phases
SmC*	
HexB	
HexF	Hexatic phases (See Section 1.4 for discussion)
HexI	
HexL	

Crystalline phases with molecular rotational-diffusional motion

CrB	Previously known as crystalline smectic-B (SmB)
CrG	Previously known as crystalline smectic-H
CrJ	Previously known as crystalline smectic-J

Crystalline phases with molecular jump-diffusional motion

CrE	Previously known as smectic-E
CrH	Previously known as smectic-G
CrK	Previously known as smectic-K

Crystalline solid phases

CrK_1	Low temperature crystal
CrK_2	Higher temperature crystal

In this chapter, we will use SmB to denote phases HexB and CrB, as in early literature. This is done to avoid unnecessary confusion and also because many of them have not yet been identified as a hexatic or crystalline phase.

The current classification system gives the following sequence of non-chiral mesophases with decreasing order [15]:

CrH–CrK–CrE–CrG–CrJ–CrB–HexL–HexF–HexI–HexB–SmC–SmA–N–I

decreasing order

$\longrightarrow$

Known transitions between phases have been summarized [7, 15] as well as the structures of these phases [15, 16].

The order of these phases is very useful in identifying mesophases when polymorphism (more than one mesophase) occurs. For example, a smectic B phase can occur below any of the phases SmC, SmA and N, but not below HexF. The only exception to this rule occurs in reentrants where a second nematic phase can occur below a SmA phase or a second SmA below a reentrant nematic phase. This can occur only in certain types of structures, all of which have a strong dipole along the molecular axis.

Despite the availability of three books [17–19] and numerous articles containing excellent photographs of mesophase textures [6, 20, 21b, 22, 23a], specific instructions for the novice first trying to observe and identify these textures are not available. Most new researchers will simply place a sample on a slide, add a cover slip, place this into a microscope heating stage and heat the sample. Sometimes this yields beautiful textures, but often it does not. This is because the alignment of the molecules in relationship to the slide determines what texture will be observed. *It is important to remember that if you cannot obtain a texture good enough to identify the characteristic features of a particular mesophase, you cannot use textures to identify a phase.* A totally mixed, aligned texture such as that shown in Fig. 2.2 (color) is useless for identifying mesophases. Texture depends on alignment and there are three types that can occur relative to the slide: parallel to it (homogeneous), perpendicular (homeotropic) or a mixture of these two. Two major factors affect this alignment: the structure of the molecule itself, and the surfaces of the slide and cover slip. Many mesogens will readily align in either direction, sometimes on the same slide. Usually the molecules will align parallel to the slide if both the slide and cover slip are first rinsed with acetone and then wiped with a tissue in a direction parallel to the long direction of the slide and cover slip with both being wiped in the same direction. This often yields homogeneous alignment when the slide is placed on the stage at a temperature close to the clearing temperature and cooled. Starting at the isotropic liquid end of the phase sequence generally will produce better textures than beginning at the crystal end, since mesophases often like to retain some of the texture characteristics of the preceding phase (paramorphism) which in the latter case is the solid phase. This is the way to obtain all those beautiful focal conic fan textures shown in numerous books and papers. Color, seen only under polarized light in colorless mesogens, is dependent on thickness and will be more intense where the sample is thinner, such as along the sample edges or near the clearing temperature. A thinner sample that covers the entire viewing area is easier to obtain with samples having a low viscosity. The viscosity is determined by the molecular structure. Thinner samples also tend to produce better focal conic fan textures. Good textures are often difficult to achieve in polymer materials because of their high viscosity. Thicker, more viscous samples also tend to trap air bubbles which are recognized by the appearance of perfectly circular black holes seen in a texture.

A major problem with most of the texture photos found in the literature is that they rarely present a sequence that shows exactly the same sample area. The advantages of such a series to a novice can be seen in the sequence shown in Fig. 2.3 (color). This makes it possible to compare the subtle differences often found in liquid crystalline textures. It is also easier to detect these differences in a sample area which consists of large domains (such as fans, Fig. 2.3(b)) than smaller ones. Often a sample will show various textures in different areas. The viewer needs only to gently move (jarring a focal conic texture can cause it to form smaller fans) the slide until a good viewing area free of air bubbles can be found. Most of the time, this is possible; but sometimes a new sample needs to be prepared. With some samples, obtaining good fans is more difficult. A higher magnification is sometimes useful, but there are times the viewer simply has to accept a less than perfect texture. Usually, a little time spent searching for a good area will be successful and rewarded with an easier identification. Good fan textures obtained on cooling may convert to poor textures on reheating the crystals formed on cooling due to paramorphism and loss of alignment (Fig. 2.3(h, j)). Once again, such textures are useless for identification.

The homeotropic texture can be extremely useful in providing confirmation of phase identifications. A homeotropic alignment can often be obtained by moving the cover slip gently with a small spatula, while the sample is in a fluid phase (I, N, SmA preferably). Sometimes this needs to be repeated to achieve a good homeotropic texture. A perfect homeotropic texture would be totally black, but different from that for the isotropic liquid. It is rare to obtain such an alignment using this method, but many effective aligning agents, such as lecithins and polyvinyl alcohol (PVA), are useful for achieving good alignment. More information on aligning agents can be found in the literature [24]. Figure 2.4 (color) shows a typical homeotropic texture having imperfect alignment. This plain, black texture can be observed only in uniaxial mesophases. There are only three non-chiral, uniaxial mesophases: nematic and SmA and SmB. Only smectic phases which do not have their collective molecules tilted in the smectic layers are uniaxial. Further confirmation of this uniaxial nature can be obtained by observing a uniaxial conoscopic cross using a Bertrand lens (Fig. 2.5 (color)). The polarizing microscope used for studying mesophase textures must be fitted with this special lens. Observation of the conoscopic cross requires a good homeotropic alignment of the mesophase (better than that shown in Fig. 2.4) and sufficient light to see it. This means that the objectives, a diaphragm and condenser must all be fully open. Crystals too can be uniaxial, but this requires that a single crystal be aligned with its axis perpendicular to the slide. More information on conoscopic studies and how the Bertrand lens works can be found in Refs. 19 and 25. The reason these conoscopic crosses can be seen in mesophases is because the mesophase acts like a single crystal that is aligned with its optical axis perpendicular to the slide.

If the mesophase is not a uniaxial one, it will show either a schlieren texture which is usually gray or a mosaic/platelet texture. Smectic-C phases will show a gray schlieren texture similar to that shown in Fig. 2.6. (color). Using the Bertrand lens, this texture will show a biaxial cross which in a thin sample appears primarily as a circle of light containing a black arc off to the side, sometimes seen as colored concentric arcs as shown in Fig. 2.7 (color). Variations of biaxial crosses can be seen in thicker samples. Since seeing a uniaxial cross depends on the sample alignment, *the absence of a uniaxial cross does not mean the mesophase is biaxial.* It could simply mean that the molecules are not aligned properly. Once alignment is achieved in one mesophase, such as the nematic phase, it often is retained in the other mesophases on cooling. For example, in a N–SmA–SmC–SmB sequence a homeotropic texture with an uniaxial cross in the nematic phase would be retained in both the SmA and B phases, but convert to a schlieren texture with disappearance of the uniaxial cross in the SmC phase. However, *alignment is not always retained.* It is not unusual to see a homeotropic nematic texture give a SmA fan texture and vice versa or for the fan texture to become more homeotropic on repeated heating and cooling. Thus, *it cannot be assumed that alignment is always retained* on heating or cooling a mesophase.

Focal conic fan textures are particularly useful in identifying mesophases, since they have specific characteristics for each type of mesophase. These are not observed for nematic phases, but they can occur in chiral nematic (cholesteric) phases. They occur in homogenously aligned samples. Sharp, focal conic fans occur primarily in SmA and SmB phases (Figs. 2.3(b,f)), but can also occur in some more highly ordered smectic phases due to paramorphism. In the SmC phase, these fans convert to broken fans which often are less sharp (Figs. 2.3(c,d)). Several discussions of the focal conic texture can be found in the literature [24, 26].

Equally important to the specific texture for identification are the changes that occur in the phase transitions that take place between these textures. The formation of droplets that grow and then coalesce when the isotropic liquid is cooled indicates a nematic phase (Fig. 2.8 (color)); coalescing batonnets indicate either a SmA or a SmC phase (Figs. 2.3(a,b)); and formation of a mosaic texture indicates either a SmB (Fig. 2.9 (color)) or a more highly ordered smectic phase.

Heating a nematic phase near the clearing temperature will often give a more highly colored texture as the phase becomes more fluid, and a 'scintillation' type effect occurs due to the fluctuation of the nematic phase director. This scintillation is rarely observed in any other mesophase. The appearance of arcs across the backs of the fans in either a SmA or a SmC phase (Fig. 2.3(e)) indicates the appearance of a CrB phase. These transition bars disappear once the CrB phase is totally formed (Fig. 2.3(f)), differentiating this transitional texture from that of the CrE phase (Fig. 2.10 (color)). On cooling, the lines in the CrE texture do not disappear, but the texture often becomes more mosaic. Sometimes crystals will grow in a SmB phase before the transition bars can disappear, decreasing the certainty of the identification.

A smectic-C phase that occurs below a smectic-A phase is obvious because the transition is not sharp as many mesophase transitions. This is due to the presence of a temperature dependent tilt angle. The fan texture (Fig. 2.3(c)) will continue to change as the tilt angle changes with the result that the fan texture becomes more 'muddied' and less fan looking (Fig. 2.3(d)). This change is more obvious in the schlieren texture, which shows wave after wave of alternating light and dark gray textures (Fig. 2.11 (color)). A sharp SmC–SmA transition temperature is more easily obtained by observing the uniaxial cross disappearing. The conversion of the uniaxial to the biaxial cross is well illustrated in Ref. 27.

A smectic-C phase that occurs below a nematic phase, on the other hand, does not have a large temperature dependence of the tilt angle (generally larger) and, therefore, little texture change is observed. Usually, no obvious change occurs in the broken fan texture as the temperature changes, but some change does occur in the schlieren texture. This is considerably less dramatic than that observed in a SmC below a SmA phase. Actually, an N–SmC transition has its own characteristic textural change. On cooling the nematic phase, the marbled texture (Fig. 2.12(a)) begins to appear like a mixed aligned smectic phase (Fig. 2.12(b)), but then converts to either a nematic-like texture (Fig. 2.12(c)) or to a broken fan texture (Fig. 2.13 (color)). The nematic-like texture (it may also appear as a grainy texture) is the only one seen in the 4-alkoxybenzoic acids. It also occurs in the chlorophenylendiamines, where it was initially thought to be a second nematic phase [26], but this series also shows a broken fan texture (Fig. 2.13) as well. X-ray crystallography later showed that these second nematic phases were actually smectic-C phases. The transitional texture is quite obvious in the benzoic acids, but is less obvious in other compounds, such as the phenylbenzoates. Another characteristic of this N–SmC transition is that it is often difficult or impossible to obtain a homeotropic texture in the nematic phase or a schlierien texture in the smectic-C phase.

Two additional textures need to be discussed. The marbled texture (Fig.

2.14 (color)) occurs only in the nematic phase. It shows threads (lines) in some form or another. The nematic schlieren texture already mentioned is often highly colored (Fig. 2.15 (color)) and consists of disclination points from which two or four brushes radiate. It is this texture which is most commonly associated with the nematic phase, but in reality the marbled texture is seen much more frequently in pure, single mesogens. The nematic schlieren texture is quite different from that observed for the smectic-C phase (Fig. 2.6).

Another common texture is the mosaic one. This occurs in many smectic phases with orders of a SmB or higher, and can also occur in crystals in both homogenously and homeotropically aligned samples. Thus, it is less characteristic of a specified phase. A SmB phase will form a mosaic rather than a fan texture (Fig. 2.9) [28] when it occurs below a nematic or isotropic phase. In this phase, a fan texture occurs only when it comes below a SmA or SmC phase, both of which never show a mosaic texture. Thus, when a mosaic texture occurs on cooling either the isotropic liquid or the nematic phase it must be either a smectic phase with an order higher than that of a SmC phase or a crystalline phase. An example of the mosaic texture for a CrE phase is shown in Fig. 2.16 (color).

As the order of smectic phases increases, the differences between the textures of these phases and often those for crystals become more subtle. This can make it difficult to determine if crystallization has occurred, or if the phase formed is a more highly ordered smectic phase. This often occurs for smectic-B phases below a smectic-C, making it difficult to determine when crystallization has ocurred. Another problem is that crystal to crystal changes can occur. These can be obvious, occurring rapidly, or subtle and very slow. This can make it difficult to differentiate such changes from mesophase transitions. Thus, it is essential that the liquid crystal microscopist be familiar with the textures of crystals as well as those of mesophases. A variety of crystal textures are shown in Fig. 2.17 (color) from the fresh (never melted, virgin) crystals (Fig. 2.17(a)) to mosaics (Fig. 2.3(h)) and the beautiful fans (Fig. 2.17(b)) and finally, the commonly observed fine texture (Fig. 2.17(c)).

Several characteristics of crystals and their formation are helpful. Crystals can show paramorphism, while retaining some of the texture of the previous mesophase (Fig. 2.18 (color)). Usually, crystal growth is either very rapid with a swoop across the viewing area, or slow – plodding along, taking its good old time. Mesophases usually form quickly over the entire viewing area. Needles or broad spread fans are definitely crystals. Also useful is the previously mentioned supercooling phenomenon. Immediately reheating a phase that has formed on cooling can often determine whether the phase is a mesophase or a crystalline phase. *Note again that a mesophase seen on heating from crystals must also be seen on cooling.* There are no exceptions to this rule. Additionally, monotropic *mesophases can be reheated as long as crystallization does not occur.* Sometimes crystallization will occur on reheating. This is due to a time factor and is usually not caused by a temperature change. Mesophases will not do this; crystals will since crystallization depends on conditions such as time. A mesophase will convert back to the preceding phase on reheating within no more than 0.8 °C depending on the experimental error of the heating stage; whereas crystals often take at least a degree at a rate of 2 °C/min. A heating temperature is usually a more accurate value than a cooling one, since the heating rate is programmed and cooling usually is not. Most crystals will show a noticeable difference, i.e. larger than 1 °C. Some will be too close to call. A DSC scan is then more useful for determining crystallization.

As the order of mesophases increases, they become more crystalline-like. Many smectic phases, including the most common B phase crystal-B, are now considered to be disordered crystalline phases. At some point, the order could conceivably be enough to cause these phases to act like crystals. Still, these mesophases do not have the third degree of order assigned to most crystals. The discussion here uses the term crystals to mean only those with three degrees of order and mesophase to mean any phase with 1–2 degrees of order.

Once it is determined that a smectic phase is a mesophase rather than a crystalline phase, identification can still be difficult using texture alone. Thus, x-ray crystallography has been used to determine the structures of all known smectic phases. A comparison of an x-ray pattern of an unknown smectic phase with that for a known phase can be used for identification purposes.

Texture can vary considerably within a single phase either in a single sample or in different ones. Even a typical smectic-A fan texture has a number of variations (Fig. 2.3(b) and 2.19 (color)). Alignment can vary in a single sample giving a mixture of textures (Fig. 2.20 (color)). Unlike the totally mixed aligned texture shown in Fig. 2.2, this texture can be useful in identifying mesophases. Textures for chiral mesophases can be different from those seen in racemic compounds. For example, the chiral nematic (cholesteric) phase can show oily streaks (Fig. 2.21(a) (color)), planar (Fig. 2.21(b)) and focal conic fan (Fig. 2.21(c)) textures. Other types of mesogens, such as discotics and lyotropics, have their own characteristic textures.

All of these variations provide an excellent means for identifying mesophases, but for the novice, the many subtle differences can be difficult to see. For example, the fan textures for the smectic-A and smectic-B phases appear the same (Figs. 2.3(b,f)), but the skilled observer will see more structural details in the smectic-A fans than in the smectic-B ones. The actual viewing of textures makes it easier to differentiate textures because still photographs do not show the movement that occurs when a transition takes place. Such movement can either signal that a transition is about to occur or simply indicate a change in texture. The change in a smectic-C schlieren texture has already been discussed. It is not unusual to see the threads moving in a nematic texture as well. Two attempts have been made to capture the movement occurring both in textures and phase transitions; either on film [28] or videotape [29].

Books discussing textures, films and videos can all be very useful, but there is only one way to develop the skill of accurate identification of meso-phases. This is to spend a lot of time looking at many different samples having various phases. The human brain is still the most superior device for recording, storing and recalling information; and the human eye is still the most sensitive instrument for observing the fine nuances found in meso-phase textures. Combining these two incredible instruments makes it possible to create a highly skilled liquid crystal microscopist. But this does not come without a great deal of patience, an immense expenditure of time, and *practice, practice, practice.* Access to a number of mesogens is also required. Table 2.1 lists suggested mesogens that can be used. Many of these are commercially available, and when this is the case the chemical name used is given. More extensive lists can be found in Refs. 18 and 30. The novice is advised to obtain as many of these mesogens as possible and study their textures, starting with those having one or two phases and working up to the more difficult textures and transitions. Spend time looking at these samples by moving the slide, moving the cover slip and trying a new sample. Look at these before trying to identify the phases you see in a new compound. *Remember to eliminate the known phases before declaring you have a new one!* Nematic, smectics-A and -C occur far more frequently than the more highly ordered smectics. This development of a useful skill provides an additional benefit. It is fun! You will experience the pure pleasure of exploring the beautiful world of liquid crystal textures. Just remember to take your eyes off the microscope occasionally (at least once an hour) to give them a rest or you will find yourself totally exhausted. With enough practice you will experience the pleasure of having the confidence that you now know how to identify the textures. Be careful; there are always exceptions to the rules so that even the expert can be fooled. The real expert is the one who is reluctant to assign a definite identification to a phase until enough information is available to support a convincing assignment. Such a person is willing to say, 'I cannot be sure until x-ray data are obtained.' A wrong identification can mislead numerous researchers who do additional studies leading to erroneous conclusions. It is better to leave a phase unidentified than to misidentify it!

For accurate characterization of mesophases, a good microscope fitted with a heating stage is essential. Skimping on optics, unless you plan to study only the nematic phase, is unwise. A Bertrand lens is essential for conoscopic studies. This is best added into a trinocular microscope. The third tube can also be used for photography or video. Video is very useful as a teaching tool. The microscope must have enough working space for the heating stage, and to be able to focus using a large objective. A rotating objective changer having three objectives ($\times 10$, $\times 20$ and $\times 32$ make a good selection) is preferred. The lenses in these must be able to withstand temperatures up to 300 °C without lens distortion occurring. On the Mettler stage, a glass heat shield also helps to protect the objective. A rotating stage is not essential, but a polarizer–analyzer is. A condenser with iris is essential along with a good light source. Top of the line Leitz, Nikon and Olympus can all be used.

Several programmed heating stages are now available. These use either a microprocessor or a computer for programming heating rates. Although this provides advantages in recording data and it makes it possible to do simultaneous microscopy and DSC, they frankly are not as easy to use for texture identification. Here heating rates are changed rapidly and inconsistently in ways that are time consuming to continually reprogram. The old push-button Mettler FP-2 is much easier to use for making many unplanned rate changes. Unfortunately, the currently available models no longer have this great feature. In some heating stages, the sample is totally enclosed and cannot be moved from outside the stage. These are pretty useless since the microscopist needs to be able to move the slide a lot and at times the cover slip. The narrow glass slides designed for the Mettler are preferred over the wider standard slides, since these allow for more movement in the heating stage. Some stages have a mechanical means for moving the slide, including the Mettler. Although useful, they do not offer the dexterity provided by using human fingers to move the slide. Most stages can be used to study one or two mesophases, but characterization studies really require a more versatile stage.

A heating stage not only needs a good heating system, but also a good cooling system. Both mesophases and crystallization can occur below 40 °C. Cooling becomes quite slow below 40 °C if some coolant is not used and impossible below room temperature. We modified the old Mettler heating stage to allow for the introduction of cold anhydrous nitrogen right near the sample area [31]. The sample area may show some temperature differentials, but this is not a serious problem for characterization. Care

Table 2.1 *Mesogens useful for microscope texture identification.*

Single mesophase
Nematic

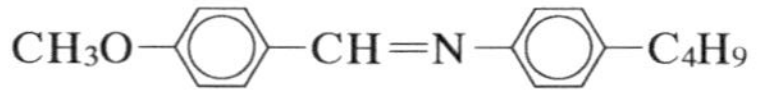

CH_3O—⬡—$CH=N$—⬡—C_4H_9 CrK 22 N 47 I

p-methoxybenzylidene-p-butylaniline (**MBBA**) – Frinton
N-(4-methoxybenzylidene)-4-butylaniline – Aldrich

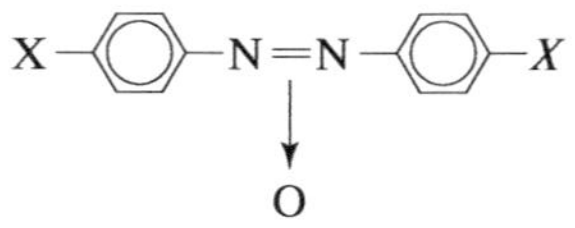

X—⬡—$N=N$—⬡—X
              ↓
              O

$X = C_5H_{11}$ 4,4'-dipentylazoxybenzene – Frinton   CrK 22 N 65 I
$X = MeO$ azoxyanisole – Aldrich   CrK 118 N 136 I

C_5H_{11}—⬡—⬡—CN   CrK 24 N 35.3

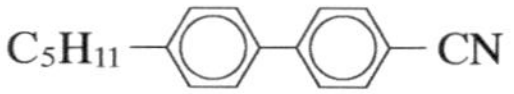

4-n-pentylcyanobiphenyl (**5-CB**) – E. M. Merck
4'-pentyl-4-biphenylcarbonitrile – Aldrich

R—⬡—CO_2—⬡—CN $R = C_4$ CrK 66.1 I 41.1 N
                     $R = C_7$ CrK 44.0 N 56.5 I

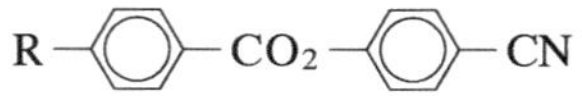

C_8H_{17}—⬡—COS—⬡—C_8H_{17}

(**8S5**) CrK(virgin) $\xrightarrow[23.5]{39.4}$ N $\xrightarrow{22.6}$ CrK$_1$ $\xrightarrow{19.8}$ CrK$_2$
                                              33.7

Smectic-A

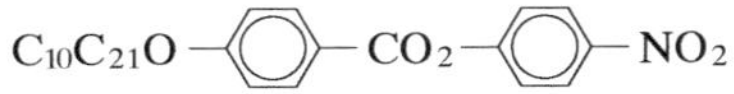

$C_{10}C_{21}O$ —⟨benzene⟩— CO_2 —⟨benzene⟩— NO_2

p-nitrophenyl-*p*-decyloxybenzoate – Frinton

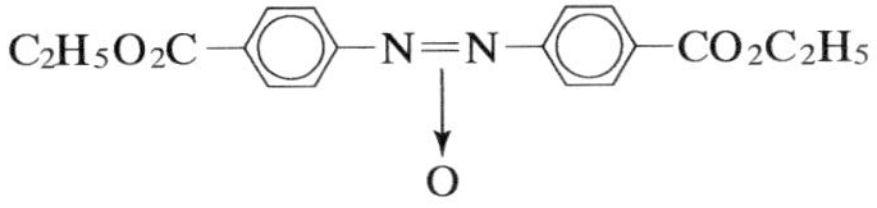

CrK 55 SmA 77.5 I

$C_2H_5O_2C$ —⟨benzene⟩— $N{=}N$ —⟨benzene⟩— $CO_2C_2H_5$
$\downarrow$
O

CrK 114.9 SmA 123.3 I 105 CrK

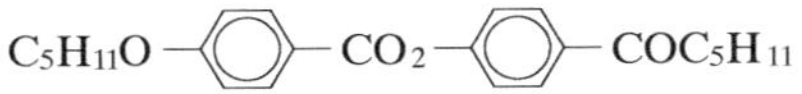

$C_5H_{11}O$ —⟨benzene⟩— CO_2 —⟨benzene⟩— COC_5H_{11}

CrK$_2$ 88.2 CrK$_1$ 90.5 SmA 111.5 I, SmA 77.3 CrK$_2$

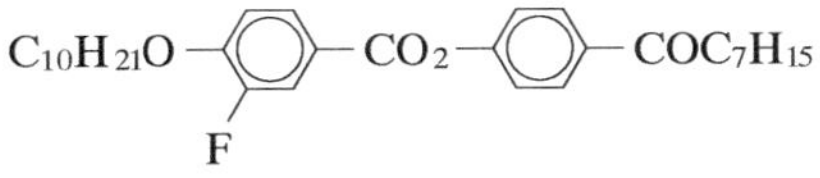

$C_{10}H_{21}O$ —⟨benzene(F)⟩— CO_2 —⟨benzene⟩— COC_7H_{15}
F

CrK$_3$ 97.2 CrK$_2$ 100.2 CrK$_1$ 103.2 SmA 117.3 I, CrK$_2$ 100.1 SmA

Smectic-B

$C_6H_{13}O$ —⟨benzene⟩— $CH{=}N$ —⟨benzene⟩— I

CrK 88 SmB 112 I, SmB 52 CrK

$C_{10}H_{21}$ —⟨benzene⟩— COS —⟨benzene⟩— $C_{10}H_{21}$

CrK 52.3 SmB 59.6 I, SmB 35.9 CrK

Smectic-C

$C_{11}H_{23}O$ —⟨benzene⟩— CO_2 —⟨benzene⟩— $OC_{14}H_{29}$

CrK 82 SmC 89.2 I, SmC 70.6 CrK

Table 2.1 (*cont.*)

CrG

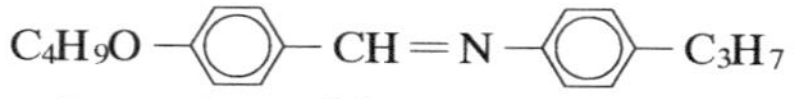

C$_4$H$_9$O—⬡—CH=N—⬡—C$_3$H$_7$ CrK 54 CrG 82.5 I

p-butoxybenzylidene-*p*-propylaniline (40.3) – Frinton

Two mesophases

Nematic and smectic-A

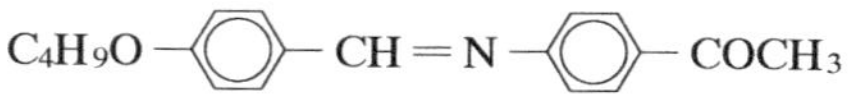

C$_4$H$_9$O—⬡—CH=N—⬡—COCH$_3$ CrK 86.1 SmA 102.0 N 113.9 I

C$_7$H$_{15}$—⬡—CO$_2$—⬡—C$_5$H$_{11}$ CrK 41.7 SmA 44.1 N 60.9 I, SmA 19.8 CrK

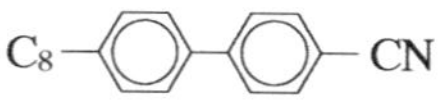

C$_8$—⬡—⬡—CN   CrK 24.5 SmA 34.7 N 4.3 I, SmA −10 CrK

Nematic and smectic-C

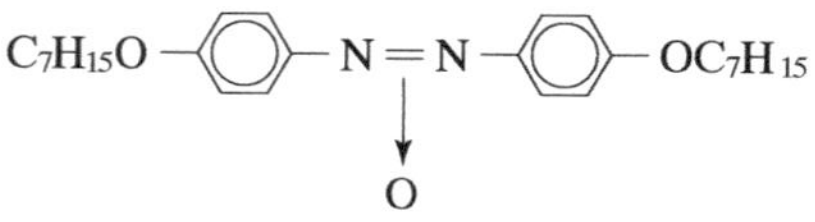

C$_7$H$_{15}$O—⬡—N=N—⬡—OC$_7$H$_{15}$ CrK 69 SmC 91.0 N 121.5 I

     ↓
     O

4,4'-bis(heptyloxy)azoxybenzene (HOAB) – EKC

C$_8$H$_{17}$O—⬡—CH=N—⬡(Cl)—OC$_8$H$_{17}$ CrK 61.7 N 178 I, N 60.0, SmC 56.4 CrK

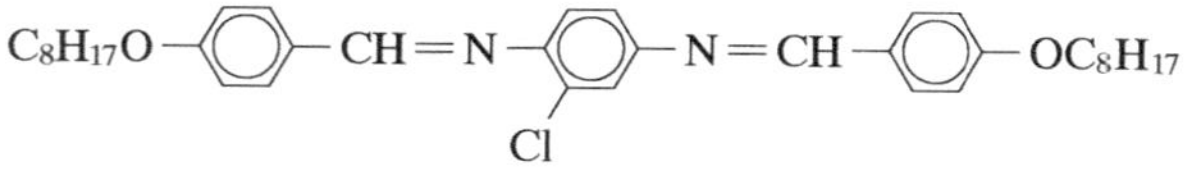

bis-(4'-*n*-octyloxybenzylidene)2-chloro-1,4-phenylenediamine – Frinton

C₈H₁₇O—⬡—CO₂H

4-octyloxybenzoic acid (OOBA) – Frinton

CrK₂ 72.5 CrK₁ 100 SmC 106 N₁ 131 N 146 I

C₇H₁₅O—⬡—CO₂—⬡—OC₈H₁₇

Nematic and smectic-B

CrK 60.7 SmC 61.1 N 88.1 I, SmC 46.7 CrK

C₅H₁₁—⬡—CH=N—⬡—C₄H₉

p-pentylbenzylidene-*p*-butylaniline (5.4) – Frinton

CrK 0 SmB 24 N 30.1 I

C₈H₁₇O—⬡—⬡—CO₂CH(Me)C₂H₅

Smectics-A and -C

CrK 75 SmC 41 SmA 69 I

C₄H₉O—⬡—CH=N—⬡—Cl

Smectics-A and -B

CrK 67.2 SmC 69.3 I, SmC 68.2 HexB 62.5 CrK

C₁₀H₂₁—⬡—CO₂—⬡—OC₁₆H₃₃

Smectics-C and -B

CrK 67.2 SmC 69.3 I, SmC 68.2 SmB 62.5 CrK

C₅H₁₁—⬡—⬡—C₅H₁₁

Smectic-B and CrE

CrK 12 CrE 47 SmB 52 I

Table 2.1 (*cont.*)

More than two mesophases

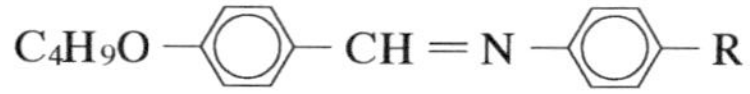

C_4H_9O—⟨benzene⟩—$CH=N$—⟨benzene⟩—R

p-butoxybenzylidene-*p*-alkylaniline – Frinton

R = C$_4$ (40,4)-butyl
CrK 9 CrG 38.5 SmB 44.2 SmA 45.1 N 74.3 I
R = C$_8$ (40,8)-octyl
CrK 48 CrG 53 HexB 69 SmC 70.3 SmA 83 I

R—⟨benzene⟩—$CH=N$—⟨benzene⟩—$N=CH$—⟨benzene⟩—R

terephthalylidene bis(*p*-alkylaniline) – Frinton

R = C$_4$ (-butyl) TBBA
CrK 113 CrG 144.5 SmC 172 SmA 199 N 235 I, CrG 89.2 CrH 74.0 Sm
R = C$_9$ (nonyl)–TBNA
CrK 57.3 CrG 132.5 HexF 155.5 HexI 157.5 SmC 192.7 SmA 199 I

$C_7H_{15}O$—⟨benzene⟩—$CH=N$—⟨benzene⟩—$N=CH$—⟨benzene⟩—OC_7H_{15}

bis(*p*-heptyloxybenzylidene)-*p*-phenylenediamine

CrK 127 SmK 130 CrG 146 SmA 154 CrJ 157 HexF 164 SmC
197 N 241 I

$C_8H_{17}O$—⟨benzene⟩—O_2C—⟨cyclohexane⟩—CO_2—⟨benzene⟩—OC_8H_{17}

CrK$_2$ 72.1 CrK$_1$ 91 Sm 92.4 SmB 110.4 SmC 119 SmA 178 I Sm 88.6
CrK$_1$ 57 CrK$_2$

Chiral

For N* – cholesteryl esters – EKC, Aldrich, Frinton

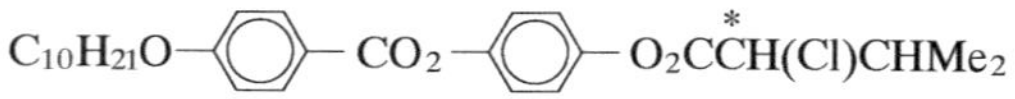

$C_{10}H_{21}O$—⟨benzene⟩—CO_2—⟨benzene⟩—$O_2CCH(Cl)CHMe_2$
$\overset{*}{}$

4-(R)-(−) or (S)-(+)-2-chloro-3-methylbutyryloxy-phenyl-4-(decyloxy) benzoate – Aldrich
CrK 72 I 67 SmA 45.8 SmC*

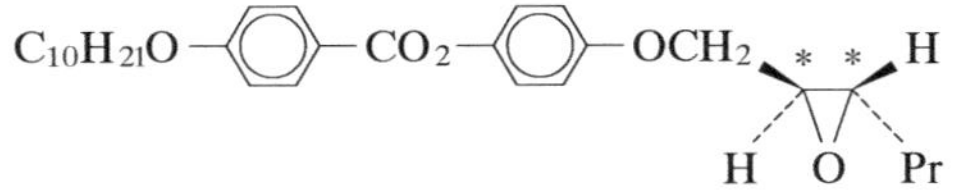

$C_{10}H_{21}O$—⬡—CO_2—⬡—OCH_2—[epoxide structure with H, Pr, H, O and * marks]

4-((S,S)-2,3-epoxyoxyloxy)phenyl-4-decyloxy)benzoate – Aldrich

CrK 75 SmC 80 SmA 81 N 97 I

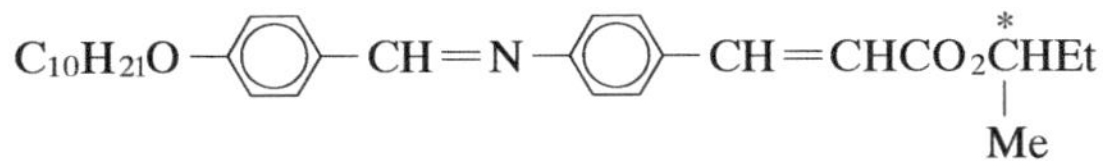

$C_{10}H_{21}O$—⬡—$CH=N$—⬡—$CH=CHCO_2\overset{*}{C}HEt$
$\phantom{C_{10}H_{21}O—⬡—CH=N—⬡—CH=CHCO_2CH}|$
$\phantom{C_{10}H_{21}O—⬡—CH=N—⬡—CH=CHCO_2CH}Me$

(S)(+)-2-methylbutyl-4-(4-decyloxybenzylidineamino)-cinnamate – Aldrich

p-decyloxybenzylidene-p-aminocinnamic acid 1-2-methylbutyl ester (DOBAMBC) – Frinton

CrK 76 SmC* 92 SmA 117 I, SmC 63 SmB

Notes:

Frinton = Frinton Laboratories Inc., Vineland, NJ.
Aldrich = Aldrich Chemical Co., Milwaukee, WI.
EKC = Eastman Kodak Company, Rochester, NY.
E. M. Merck = E. M. Merck Chemicals, Hawthorne, NY.

CrK = crystal; I = isotropic liquid; N = nematic; Sm = unidentified smectic. Identified smectics are indicated by letters. Transition temperatures given in °C.

must be taken to avoid condensation of water on the lens. This is easily avoided by always making sure that at least a small stream of dry nitrogen is running through the stage at temperatures below room temperature. Sublimation can be another problem at temperatures above 200 °C. Unlike condensation, sublimation of the sample depends on the compound being studied and is unavoidable. *It is not eliminated by the presence of a small fan.*

No matter what heating stage is used, like any temperature-measuring device, it must be calibrated. Calibration involves comparing the melting temperatures of pure known samples from the United States Bureau of Standards, such as melting point standards, with observed values for the heating stage [32]. This will give the temperature corrections, if any, that need to be made. There is a problem using these crystalline samples for standards. Some have crystal-to-crystal changes making melting temperatures more broad compared to clearing temperatures. Actually, a set of stable mesogens with clearing temperatures ranging from *c.* 40 to 300 °C would make a better series of standards to use for this purpose. The combination of a good temperature controller on the heating stage and a high magnification on the microscope is essential for detecting narrow temperature range phases.

For a new material where the mesomorphic properties are not known, the following microscope procedure is recommended for a complete, accurate characterization of these properties:

1. Prepare a slide of the sample as described earlier.
2. Place this slide on the heating stage at room temperature.
3. Heat rapidly (*c.* 10 °C/min) to gain a rough idea where the transition temperatures are.
4. Continue to heat the sample to the isotropic liquid, stop to allow the stage to equilibrate and cool 2 °C/min to either a mesophase or crystals.
5. If a good homogeneous texture is not formed, try preparing a new slide and adding it to the heated stage and cool.
6. Record the temperature of the phase formed and reheat at 2 °C/min. Record the transition temperature for conversion to the isotropic liquid. This will be either a clearing temperature, if a mesophase was formed, or a melting temperature if crystals were formed.
7. Cool again at 2 °C/min. Continue to cool the first mesophase until another one forms. Reheat again at 2 °C/min to obtain the heating transition temperature.
8. Continue this heating–cooling process until crystals are formed.
9. Compare textures with those for known materials to identify mesophases.
10. Crystal formation indicates the end of the total mesophase range.
11. Reheat the crystals at 2 °C/min to obtain the melting temperature.

12. Continue to heat to a fluid phase (SmC, SmA, N are best). Move the cover slip gently while the sample is hot.
13. Repeat the heating–cooling sequence studying the homeotropic/schlieren textures and conoscopic crosses to obtain confirmation of identifications from homogeneous textures. Additional runs through these heating and cooling cycles may be necessary to obtain a complete characterization.
14. If there is uncertainty about identification obtain DSC and/or x-ray crystallographic data.

2.2 Thermal analysis

Another method for determining mesomorphic properties involves the use of thermal analysis equipment. Since heat is required to induce a phase transition between two phases having different levels of order, determining this heat can be a useful means for detecting phase transitions. Three types of calorimetric thermal analyses have been used in the liquid crystal area: classic adiabatic calorimetry (CAC), differential thermal analysis (DTA) and differential scanning calorimetry (DSC). Classic adiabatic calorimetry is the most precise and versatile method. It was the sole method used in early determinations of enthalpy values for liquid crystals [33]. It is, however, a very time consuming method requiring expensive equipment and, initially, requiring large quantities of material. Later modifications made it possible to use smaller samples [34]. It is still used today to determine heat capacities or for more detailed studies of phase transitions. However, the development of DTA and DSC instruments made it possible to determine values faster, more conveniently and at a lower cost albeit less accurately. This method is now widely used to evaluate large numbers of new mesogens [35].

The difference between the DTA and DSC methods is simply in the way the heat of transition is measured and recorded. In DTA, this heat is measured directly so that in a heating scan, the peaks are negative:

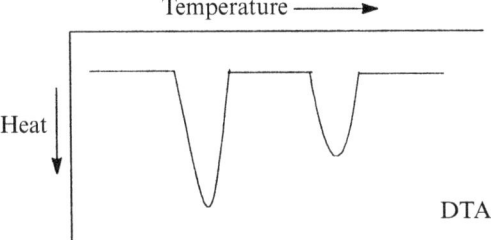

whereas in DSC, heat is added to compensate for the heat absorbed so that the peaks are positive:

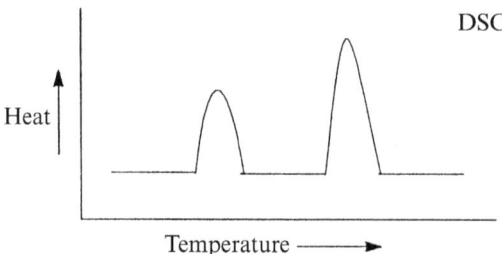

General thermal analysis reviews discuss equipment and techniques that are available [36, 37]. A couple of reviews discuss the early work done with liquid crystals [2b, 23b]. DSC instruments seem to now be more popular than DTA ones. Thus, this discussion will concentrate on the DSC method as another means for organic chemists to determine mesomorphic properties.

Since the enthalpy of phase transitions can be determined by DSC, this method provides information which cannot be obtained by microscopy. A comparison of phase transition data determined by DSC and microscopy has been made [38]. Generally, DSC is poor at detecting low enthalpy transition such as second order ones, but better for observing some crystal-to-crystal changes. DSC cannot be used to identify mesophases as can be done with microscopy. Thus, the two methods complement each other. It is therefore not surprising that some instruments combine both of these techniques [2b, 39]. DSC has been combined with x-ray diffraction equipment as well [40].

Other data can also be obtained from DSC scans. The entropy of transition can be obtained from the relationship of $\Delta H = T\Delta S$. Because of this relationship, discussions of enthalpy value trends also apply to entropy values. Other information obtained from DSC includes purity [4b, 41], phase transition order [42, 43], confirmational disorder [3] and molecular shapes, [44], although not all of these employ simple DSC scans.

Although small enthalpy transitions are difficult to observe, they can sometimes be seen by using larger samples, increasing the sensitivity, or changing the heating rate. Small enthalpy transitions include the following: SmC–SmA, CrG–HexF, SmA–SmÃ and Ch–blue phase. There is no latent heat for the second order SmC–SmA transition, but there is a small enthalpy change due to the change in the tilt angle [43]. Sometimes this can be detected by using a sensitivity of 0.02 mcal/sec per in and a heating rate of 5 °/min [38].

Today, obtaining a DSC scan is relatively easy. The instrument must first be calibrated. This is usually done with indium since metals can often be obtained with a higher purity than organic materials. However, highly purified liquid crystals have also been used [45]. An accurately weighed sample is placed in an aluminium pan, crimped tightly closed with a sealing press and placed in the sample chamber of the DSC instrument. Instructions are provided with each type of instrument for obtaining a good DSC scan. Since the peaks are often not symmetrical, the center of a peak cannot be used as the transition temperature. The determination of transition temperatures is discussed in the literature [36, 46], but today's instrument is operated by a computer data station which determines both the transition temperature and the enthalpy of the transition. Determining ΔH values involves drawing the best base line [47]. Since mesophase transitions may involve pre- or post-transitional effects and/or second order transitions of small ΔH values, determining accurate ΔH values can be difficult [42, 48]. Thus, accuracy tends to decrease with decreasing ΔH values. Before the use of computers for ΔH calculations, the error in ΔH values from author to author could be as much as 10% [2b, 23b]. However, a computerized TADS system on a DSC can give ΔH values in good agreement with CAC studies [2b, 48].

Once a good curve is obtained, then it must be accurately interpreted as a number of factors affect the peak shape and area. The first step should be to eliminate the possibility that a peak is due to a crystal-to-crystal change. These changes, as well as cystallization, depend on both the cooling rate and the history of the sample as discussed earlier in the microscope section. Of course, the appearance of the DSC curve will be affected by these transitions. It is important to remember that due to supercooling, most melts will crystallize at a lower temperature than the solid melts. As with microscopy, this can be useful in differentiating crystal changes and crystallization from mesophases. The first peak that occurs in a cooling DSC scan below the melting temperature and is not immediately reversible is a crystallization peak. A reversible peak can be due either to a mesophase transition or a crystal-to-crystal change. Usually, the largest peaks in both the heating and cooling curves are due to melting the solid and crystallization respectively. Smaller peaks below these are usually due to crystal-to-crystal changes. However, there are exceptions to this. If the crystalline phase that melts to a mesophase has nearly the same order, then the ΔH of melting may be small. This sometimes occurs when melting takes place through a series of several crystal forms. The same thing is true of crystallization. Crystal changes may occur at the same temperature in both the heating and cooling curves, or they might not. Mesophase transitions will generally occur at the same temperature within experimental error, if the mesophase is reheated immediately.

Sometimes a mesogen can produce two crystal forms that will melt at two different temperatures. This also will result in a different DSC scan upon immediately reheating a cooled sample. Some mesophases will take a long time to crystallize; such as a smectic B phase that occurs below a smectic C phase. Such samples may have to set overnight before a second heating scan is done in order to obtain a melting curve. Some mesogens form glasses, another transition that occurs below the crystallization one.

In order to obtain as accurate and complete a thermal history as possible by DSC, a good approach is to first heat the virgin crystals (usually obtained by recrystallization) at a rate of 10 °C/min until the sample converts to the isotropic liquid, cool until crystals form, reheat to the isotropic liquid, cool to room temperature (or below if necessary), allow the sample to remain at room temperature for a long time and then reheat. If a sample is reheated before crystals are obtained, this scan will look different than the original one since the melting peak will not be observed. Sometimes crystals are formed, but these slowly change to anther form. Reheating this sample will give a different melting peak (often a broader one). Sometimes a crystal-to-crystal change can occur before the melting transition. This may be a true temperature change, but often it is simply a time effect. It is common to find that the ΔH from the melting of virgin crystals is larger than that of the crystallized melt. This often is because sufficient time has not been allowed to obtain the ΔH of all the crystal changes. Thus, the most accurate value for the ΔH of melting is that obtained from a virgin sample. This is why Merck has chosen this melting ΔH as the one for evaluating its compounds [35]. Examples of DSC curves of several types of crystal changes can be found in the literature [49].

Typical DSC scans which illustrate many of these points are given in Figs. 2.22–24. Transition temperatures for these mesogens can be found in Table 2.1; other examples can be found in the literature [2c]. In Fig. 2.22, the SmA–N transition is barely detectable; it is likely this transition would be missed using DSC scans only. A good example of the complexity of solid transitions found in melting and crystallization can be seen in Fig. 2.23. Three crystal forms were observed both in the microscopy and DSC scans, but these vary according to sample history. Figure 2.23(b) shows the supercooling effect which confirms that all of these phases are crystals and not mesophases. The low ΔH transitions for the more highly ordered smectic phases in TBBA are shown in Fig. 2.24(a). Note that in this normal scan,

Fig. 2.1 A contact mixture showing the boundary between the nematic phase on the left and the smectic-A phase on the right.

Fig. 2.2 Undefined (mixed) texture for a poorly aligned smectic-A phase.

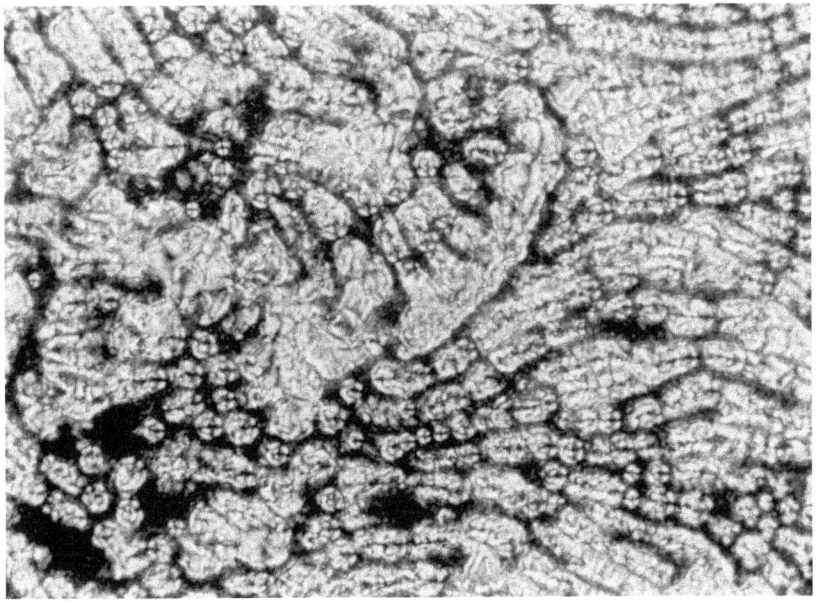

Fig. 2.3(a–j) Sequence of photographs showing the textures for a series of smectic phases with exactly the same viewing area.

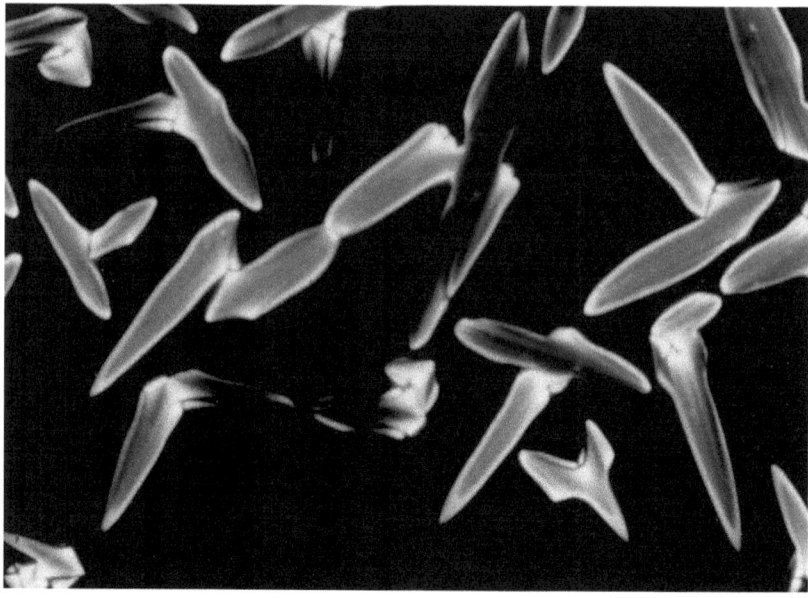

(a) Growth of smectic-A batonnets.

(b) Focal conic smectic-A fan texture fully grown.

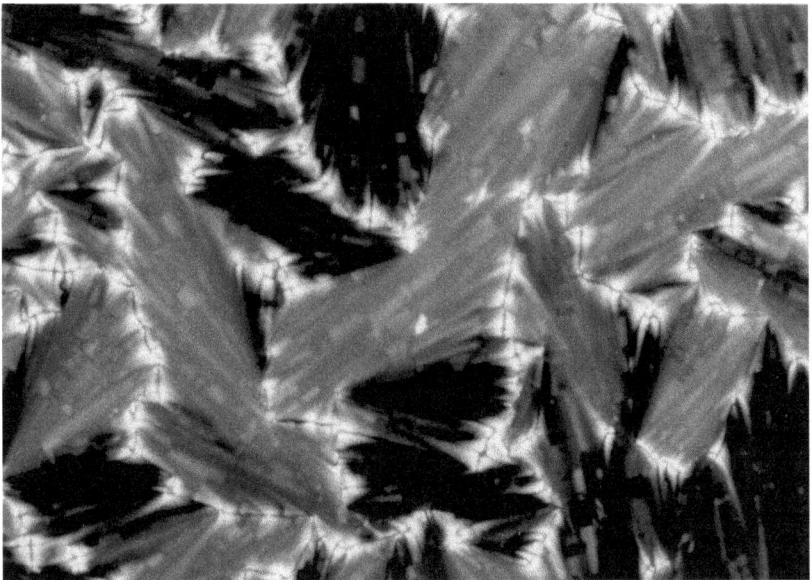

(c) Smectic-C broken fan texture near the SmA–SmC transition.

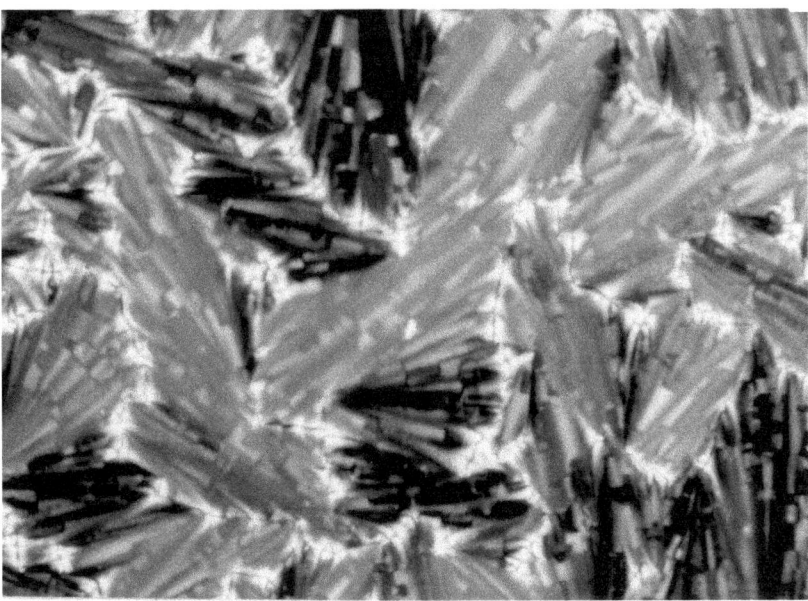

(d) Smectic-C fan texture near the lower end of its temperature range.

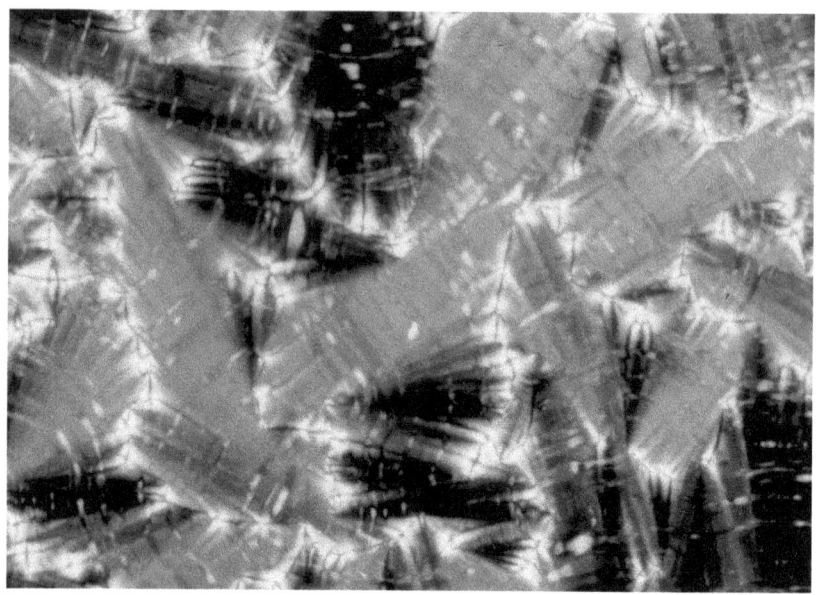

(e) Transition bars for the SmC–CrB transition.

(f) CrB focal conic fans.

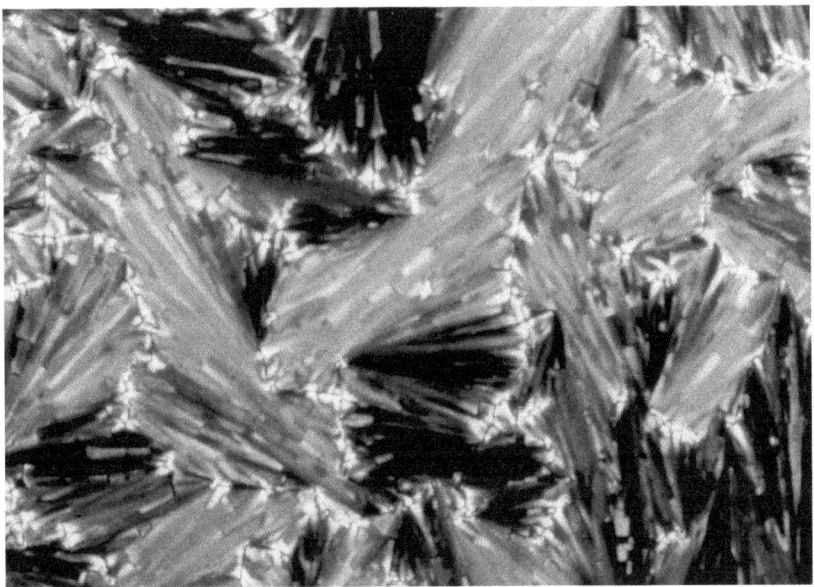

(g) Fan texture for a more highly ordered smectic phase.

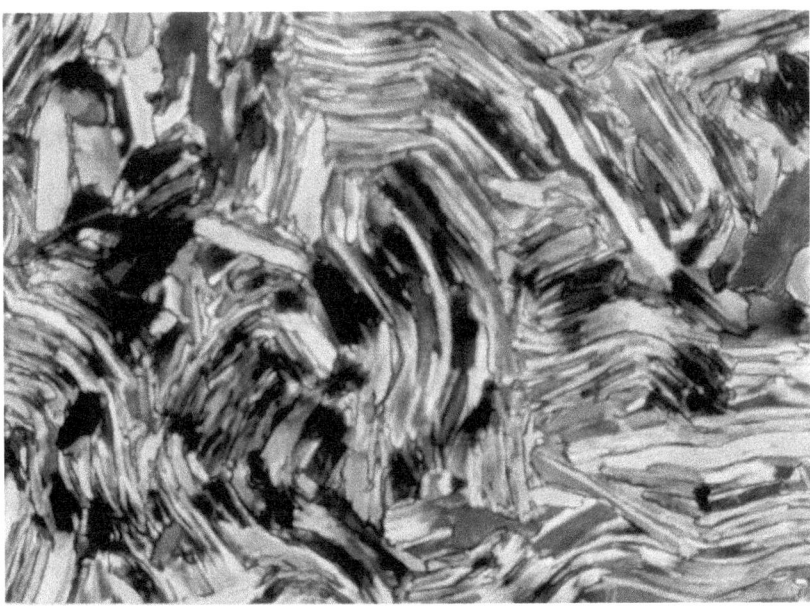

(h) Mosaic texture for first crystals (K_1) formed on cooling.

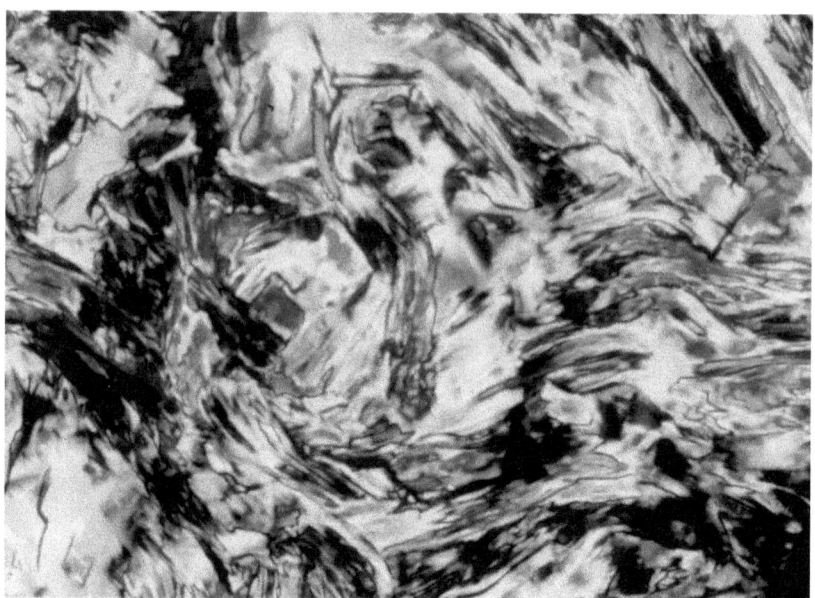

(i) Texture for the second crystal form (CrK$_2$).

(j) Unidentified smectic phase texture formed on heating CrK$_1$.

Fig. 2.4 Partial homeotropic alignment for a smectic-B phase.

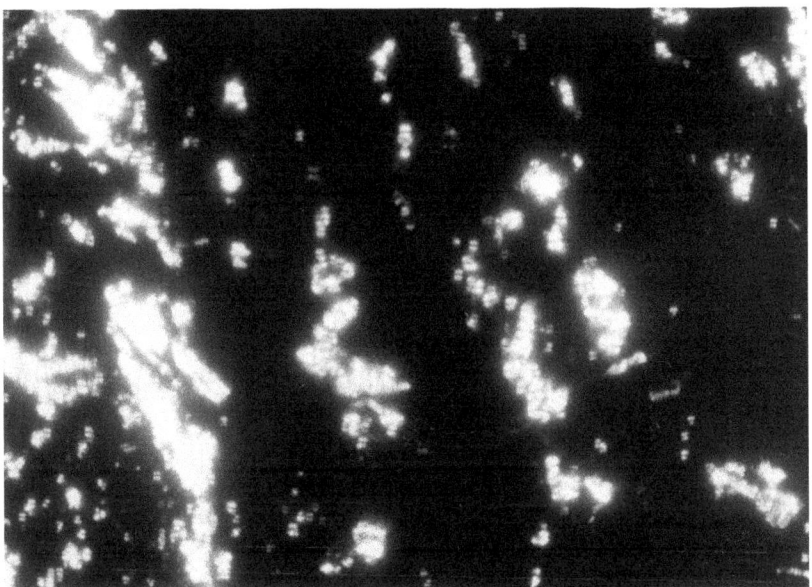

Fig. 2.5 Uniaxial conoscopic cross.

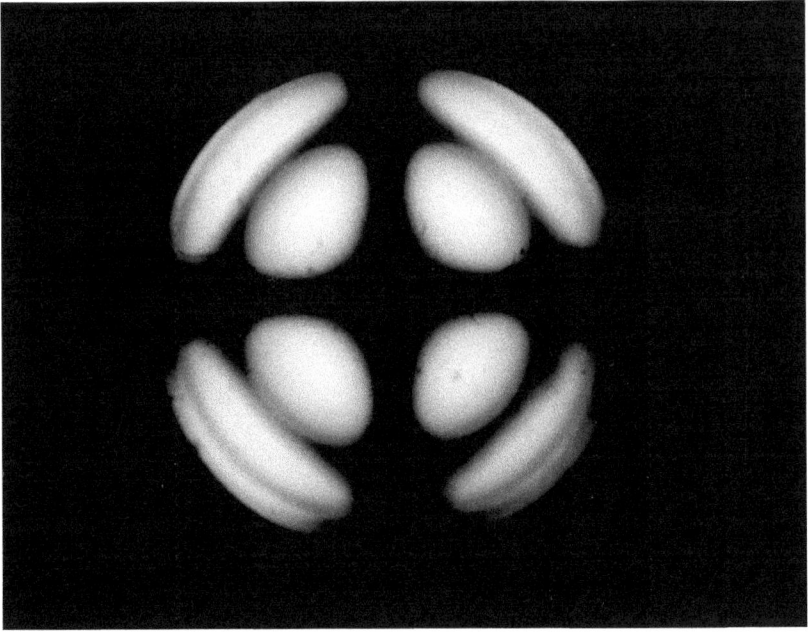

Fig. 2.6 Smectic-C schlieren texture.

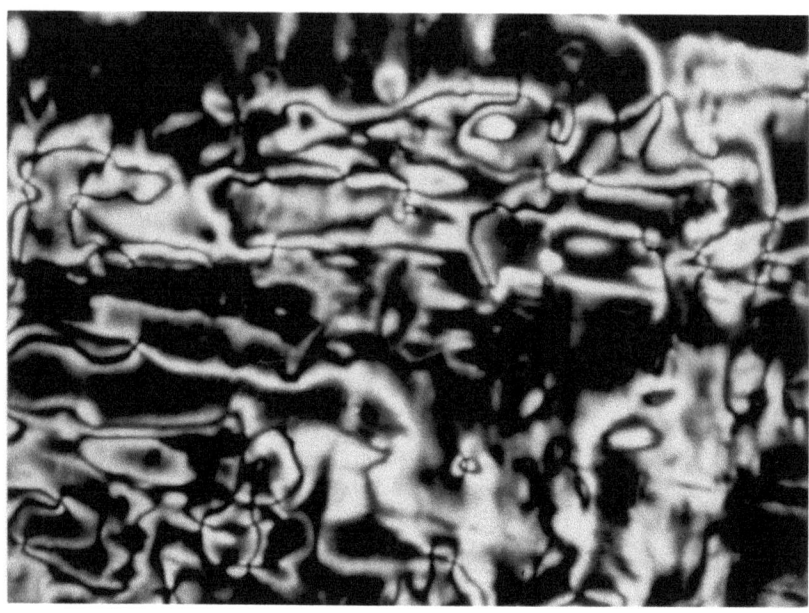

Fig. 2.7 Smectic-C biaxial cross.

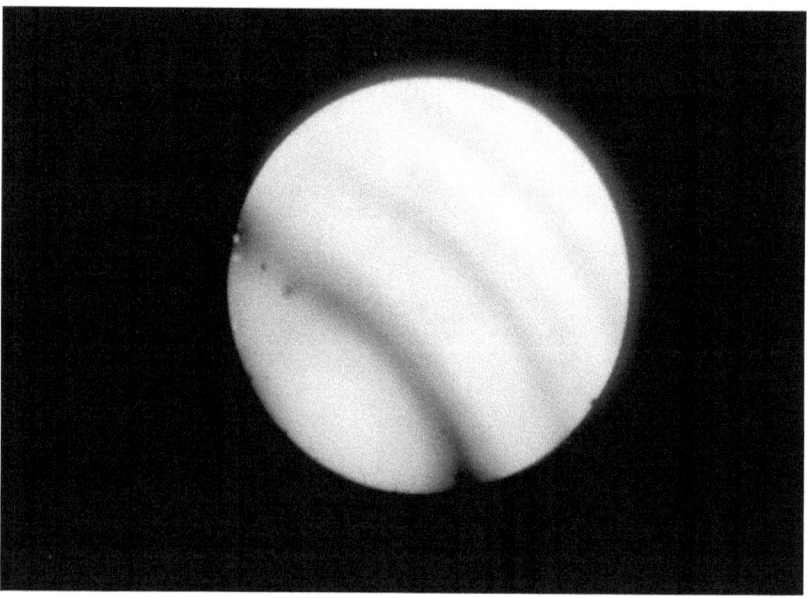

Fig. 2.8 Droplet formation on cooling an isotropic liquid to a nematic phase.

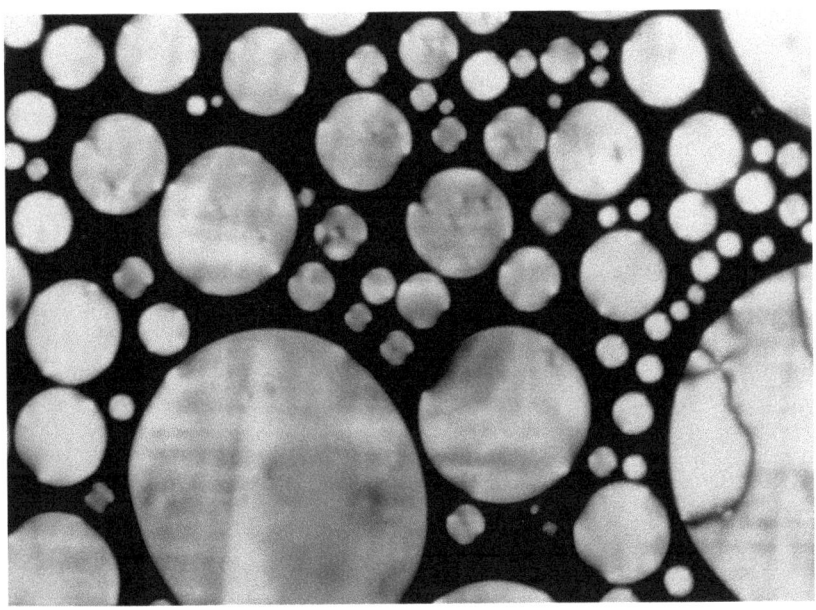

Fig. 2.9(a,b) Two examples of smectic-B mosaic texture.

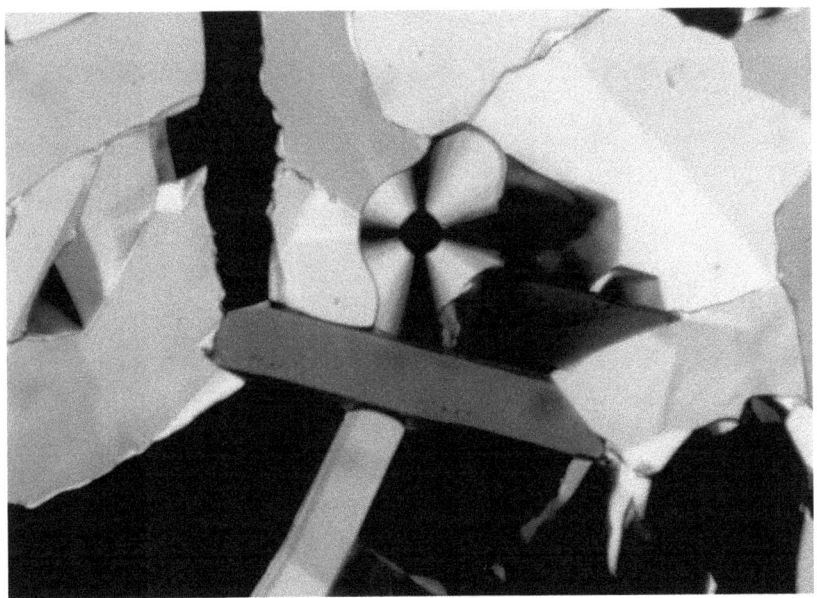

(a)

(b) Another example of smectic-B mosaic texture.

Fig. 2.10 CrE focal conic fan texture.

Fig. 2.11 Same area of a smectic-C schlieren texture at different temperatures.

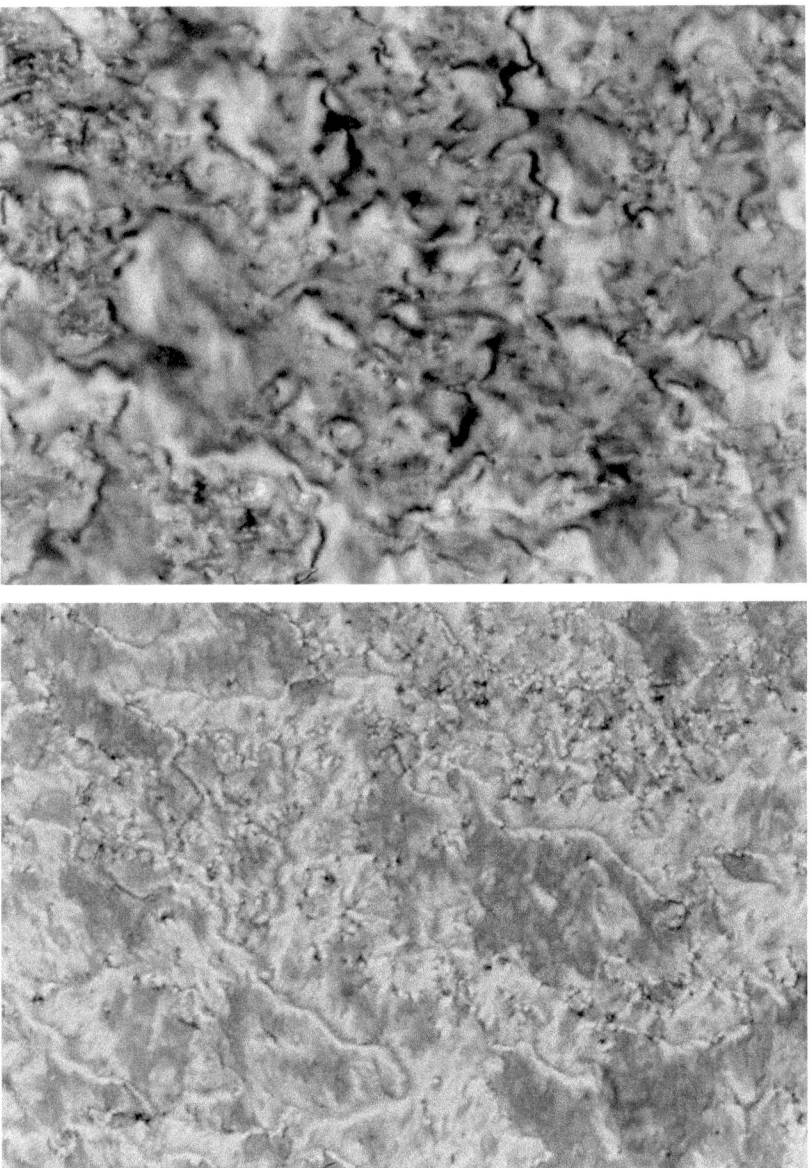

Fig. 2.12(a–c) Textures for a N–SmC transition.

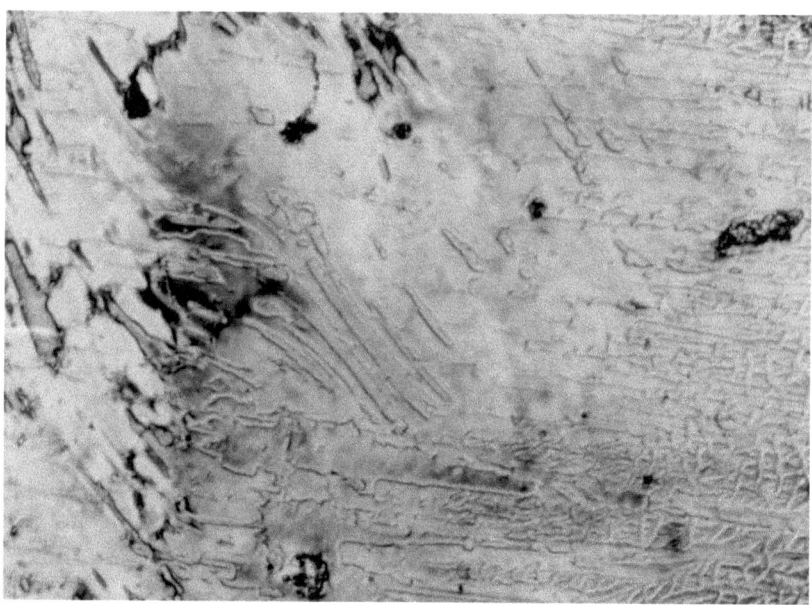

(a) Nematic marbled texture (all the same area).

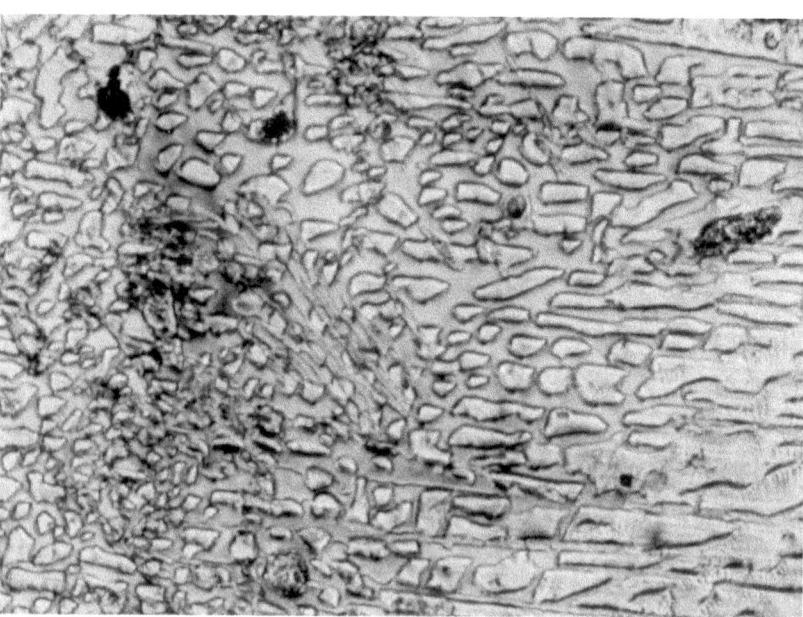

(b) Intermediate texture at the N–SmC transition.

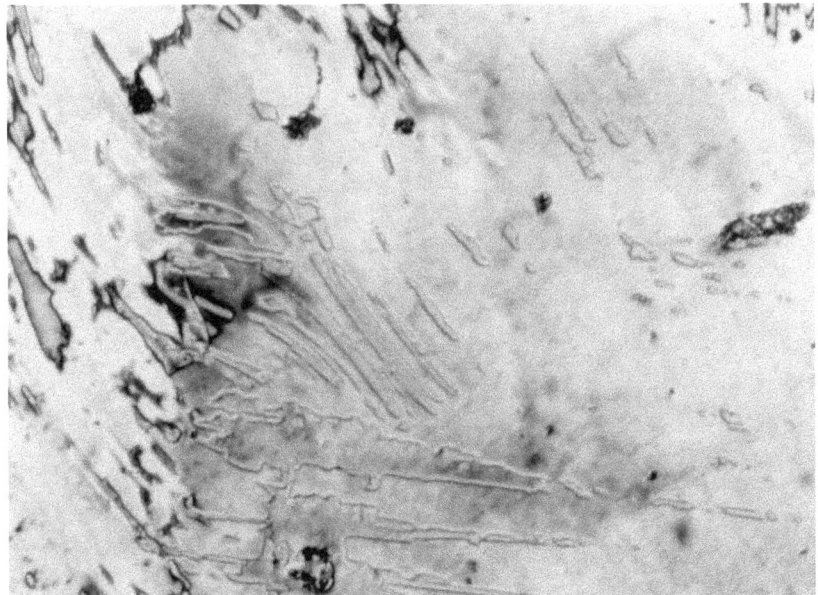

(c) Smectic-C texture.

Fig. 2.13 Broken fan texture for smectic-C phases that occur below a nematic phase.

Fig. 2.14 Typical nematic marbled texture.

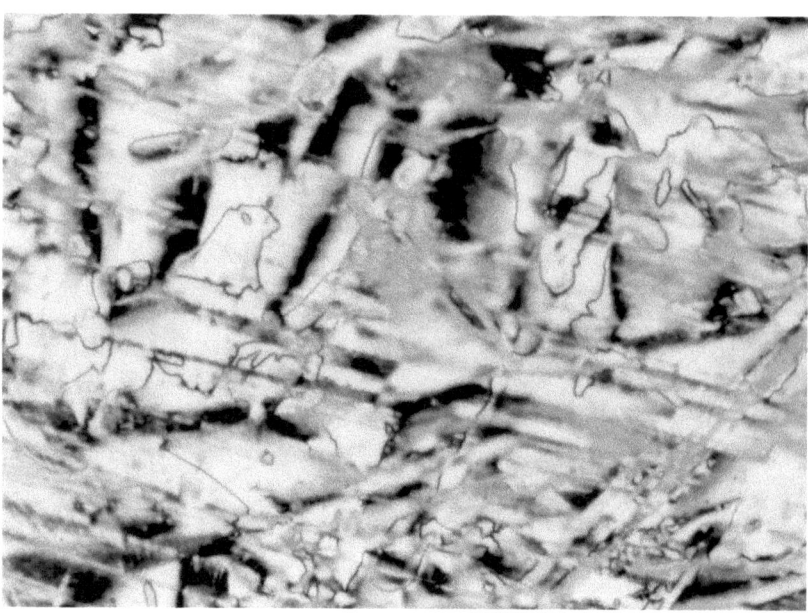

Fig. 2.15 Nematic schlieren texture.

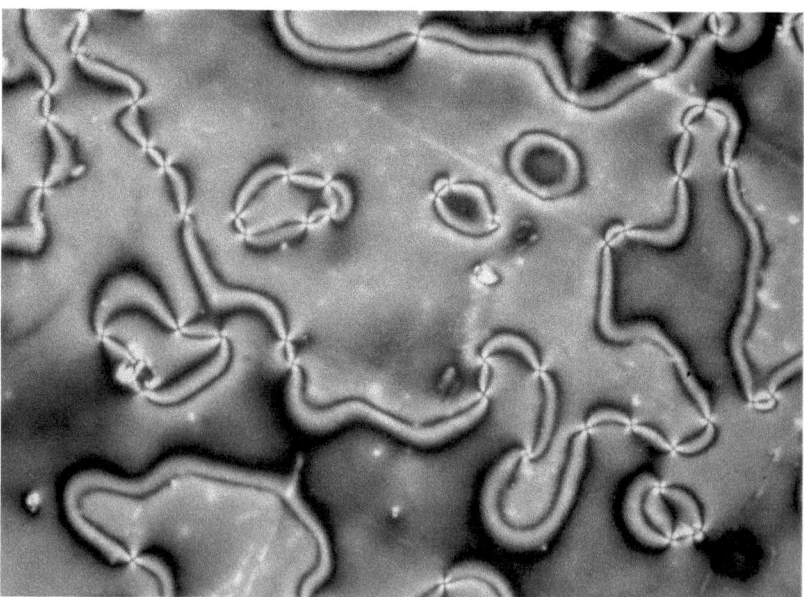

Fig. 2.16 CrE mosaic texture.

Fig. 2.17(a–c) Textures for various crystals.

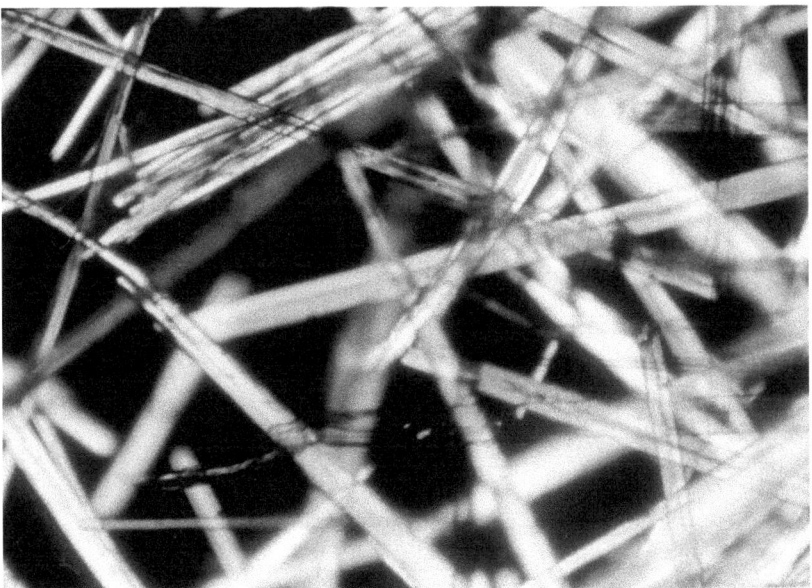

(a) Virgin (fresh) crystals.

(b) Fans formed on cooling a melt.

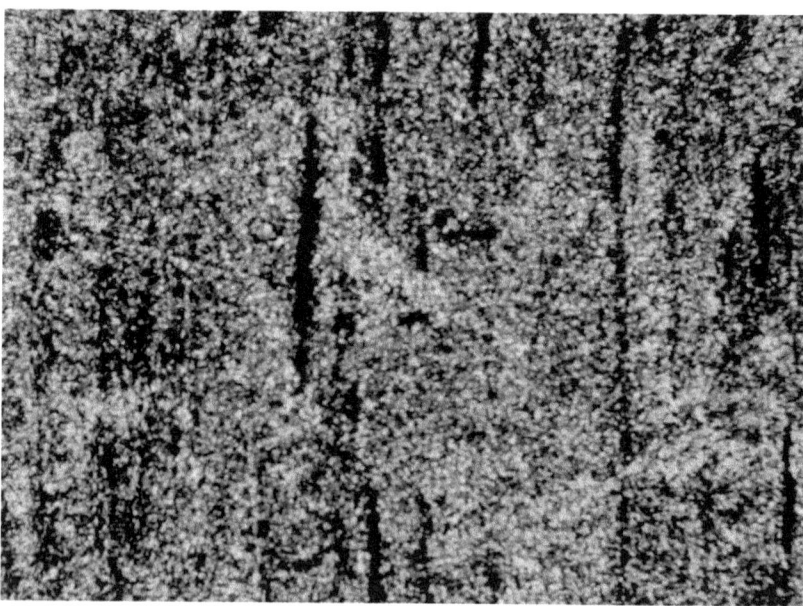

(c) Fine texture formed on cooling a melt.

Fig. 2.18 Crystals formed on cooling a smectic fan texture that show paramorphism.

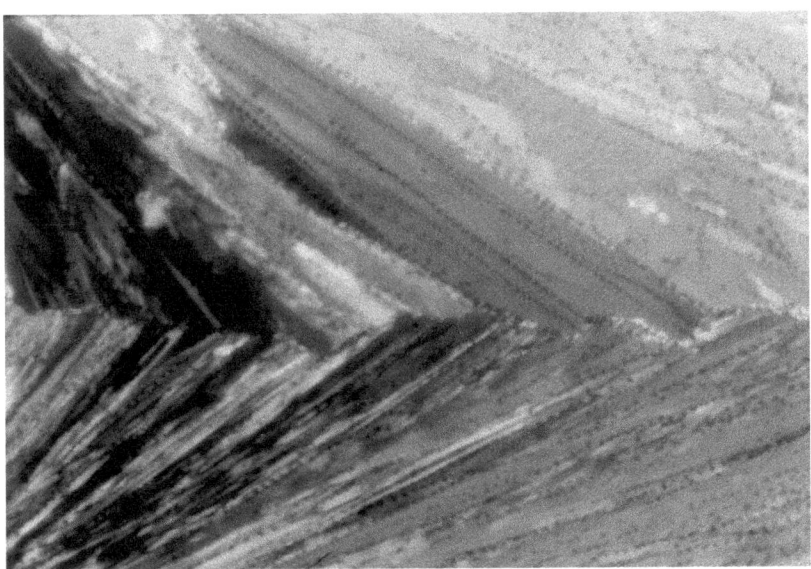

Fig. 2.19 Smectic-A focal conic fan texture.

Fig. 2.20 Texture showing both the smectic-C broken fans and the schlieren texture.

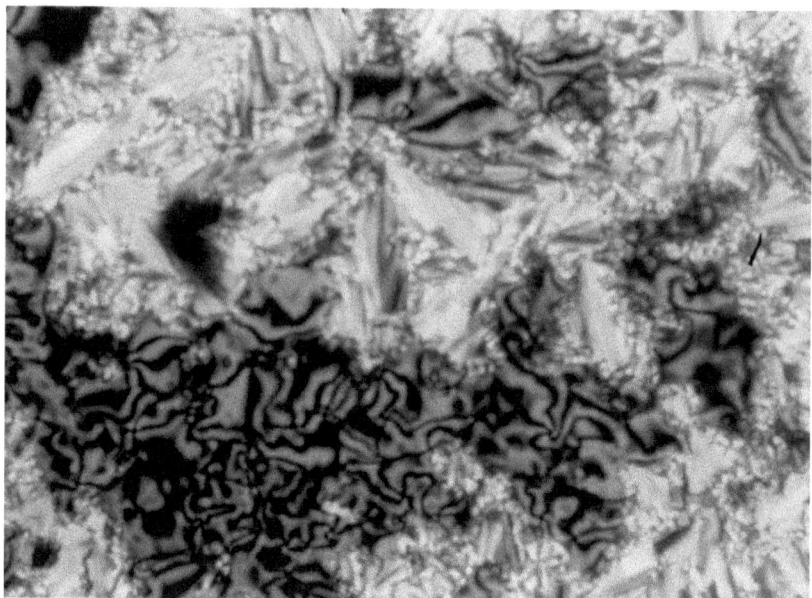

Fig. 2.21(a–c) Cholesteric textures.

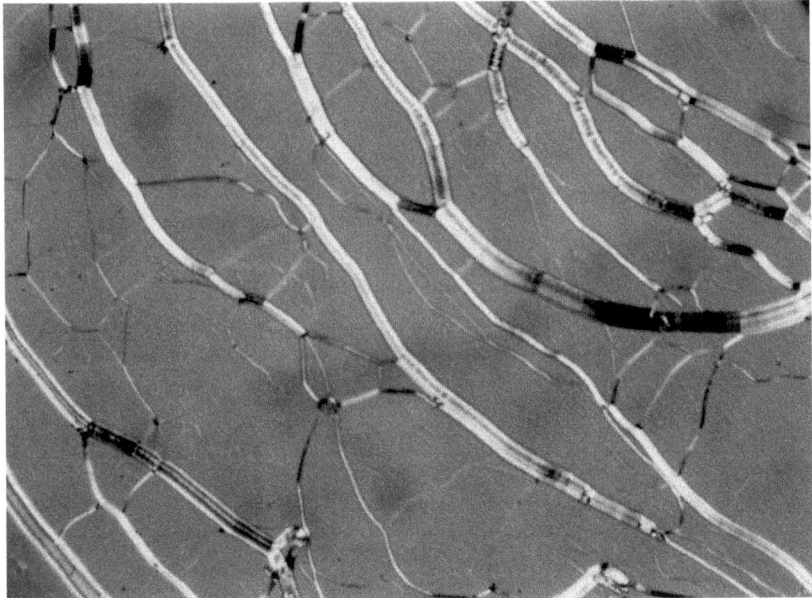

(a) Oily streaks.

(b) Planar.

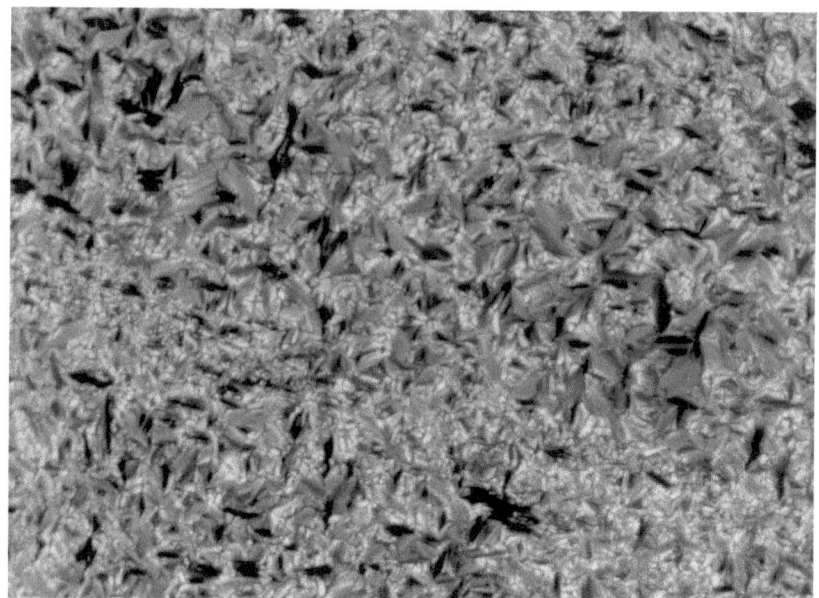

(c) Focal conic fans.

Fig. 3.6 Typical x-ray diffraction pattern of the isotropic phase.

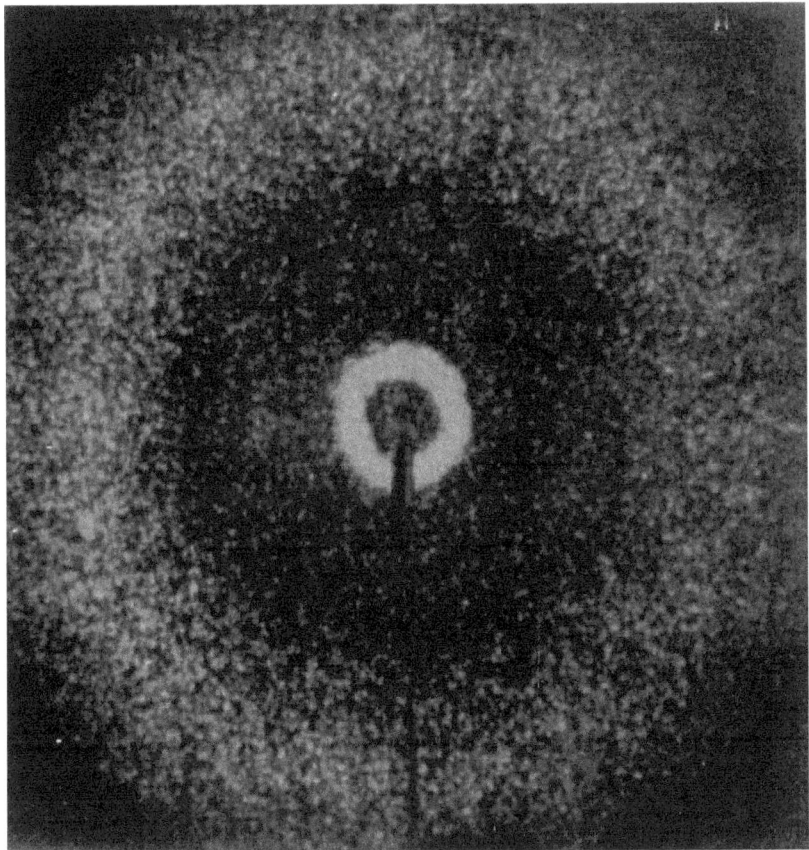

Fig. 3.7 Diffraction patterns of an aligned nematic phase. B marks the direction of the magnetic field.

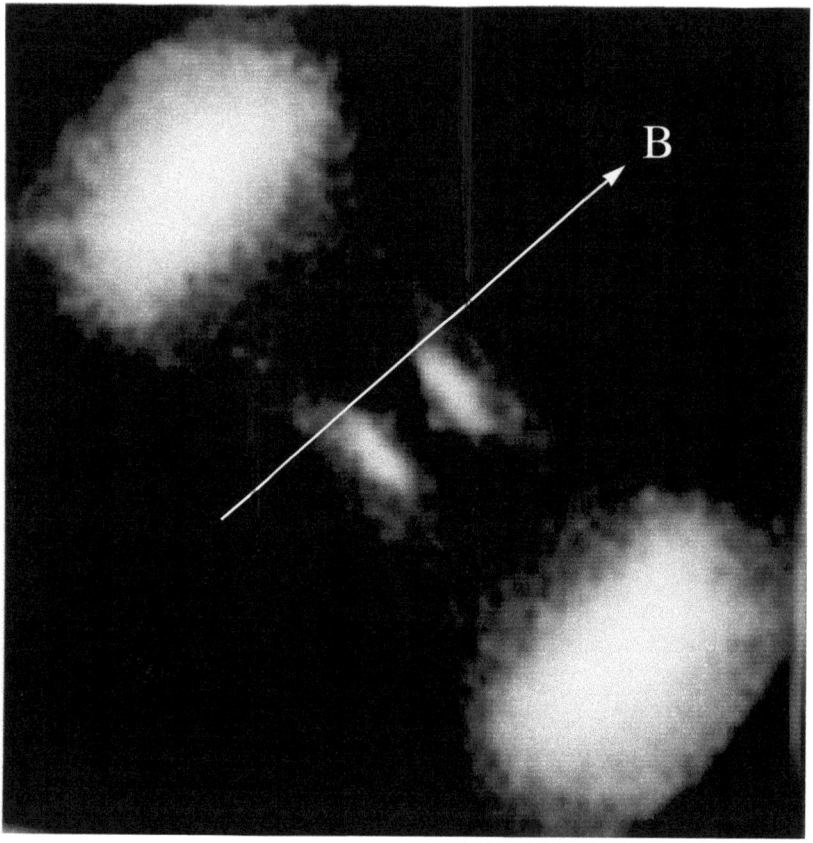

Fig. 3.8 Diffraction pattern of the SmA phase; the central diffuse arcs of the N phase have condensed into sharp Bragg peaks while the outer peaks, representing intermolecular distance in the direction perpendicular to their length, remain diffuse.

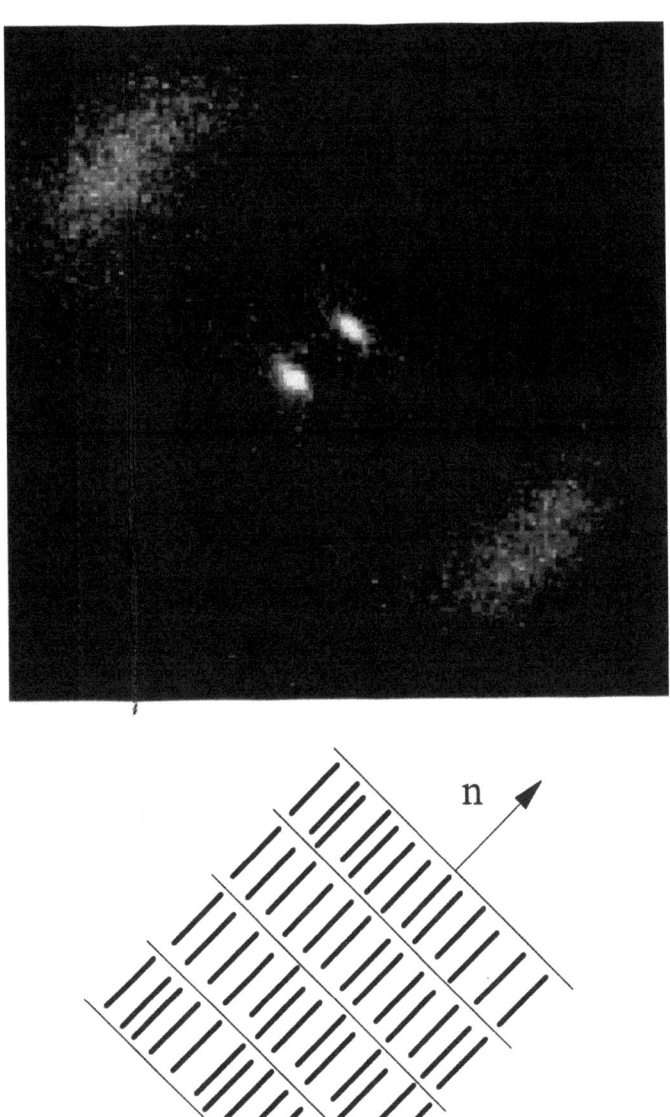

Fig. 8.13 Reflected color vs. film thickness. Crosses mark increments of five layers for $20 \leq n \leq 100$.

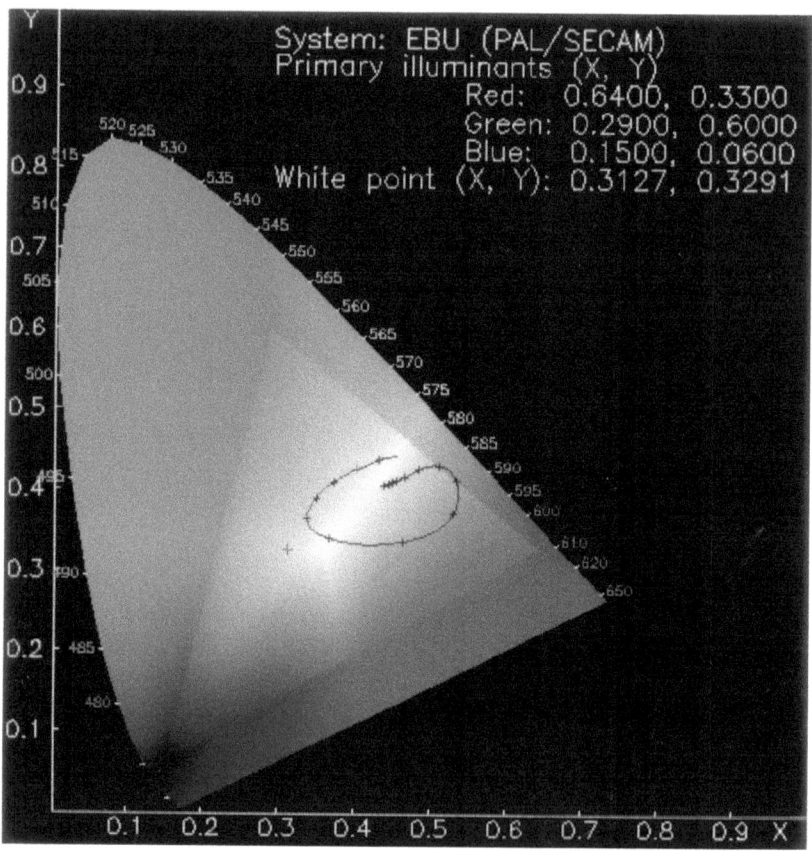

the SmC–SmA transition is not detected. In the cooling curve, (Fig. 2.24(b)) the monotropic smectic H phase occurs well below the melting temperature (112.9 °C) as does the crystallization (55.6 °C). No crystal changes were observed. Peaks for crystallization and melting have the largest ΔH values with ΔH crystallization being less than ΔH melting.

Relationships between molecular structures and enthalpy values have been studied just as the relationships between structure and transition temperatures have. These data are now available in the Vill database discussed in Chapter 10. Three factors need to be considered in comparing enthalpy values either within a homologous series or for different structures. Data coming from a variety of sources can show a large error variation. Secondly, different types of melting or clearing transitions will have different values. For example, the enthalpy for crystal to smectic-B should be smaller than crystal to isotropic liquid. Thirdly, the enthalpies of all crystal changes must be included in the enthalpy of melting. Even so, enthalpies of melting can change throughout a homologous series simply because melting occurs from a different crystal structure. It is common to observe changes in the crystal structures that melt to either a mesophase or an isotropic liquid throughout a homologous series. Of course, this is related to how the chain length can affect the packing of molecules in the solid phase. The only way to eliminate the effect of transition type is to compare the total enthalpies of melting, i.e., crystal–crystal–mesophases–isotropic liquid. This practice was done in the early work with homologous series of cholesteryl esters [23b].

Many of these early DSC studies done using cholesteryl esters showed rising melting curves with increasing chain lengths [2b, 23b]. However, most of these curves show considerable random variation in the shorter chain segments of the curves. It is now obvious that there is even more randomness in plots of enthalpy values than in transition temperature curves for homologous series. It is also clear that a longer chain length will ultimately lead to larger enthalpy values for the melting transitions in such series, sometimes rising even before the melting curve begins to rise [50]. A comparison of enthalpy with temperature plots for a number of homologous series shows the large variations that can occur [48].

Despite the large effect that chain length has on enthalpy values, attempts have been made in the literature to give a range of values for each type of mesophase transition without considering chain length [48]. Even in series showing only nematic phases, melting enthalpies increase with increasing chain lengths [51]. Still, using data for 391 mesogens, the following trends

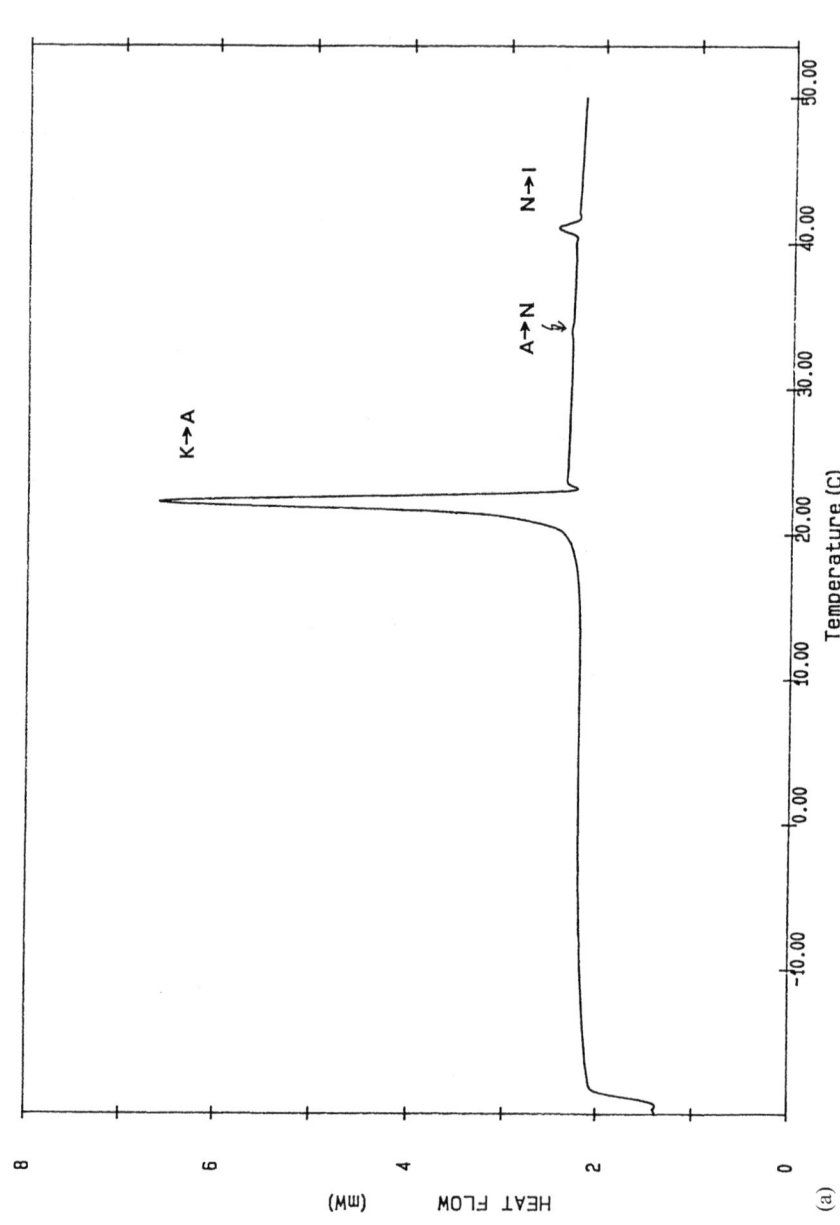

(a)

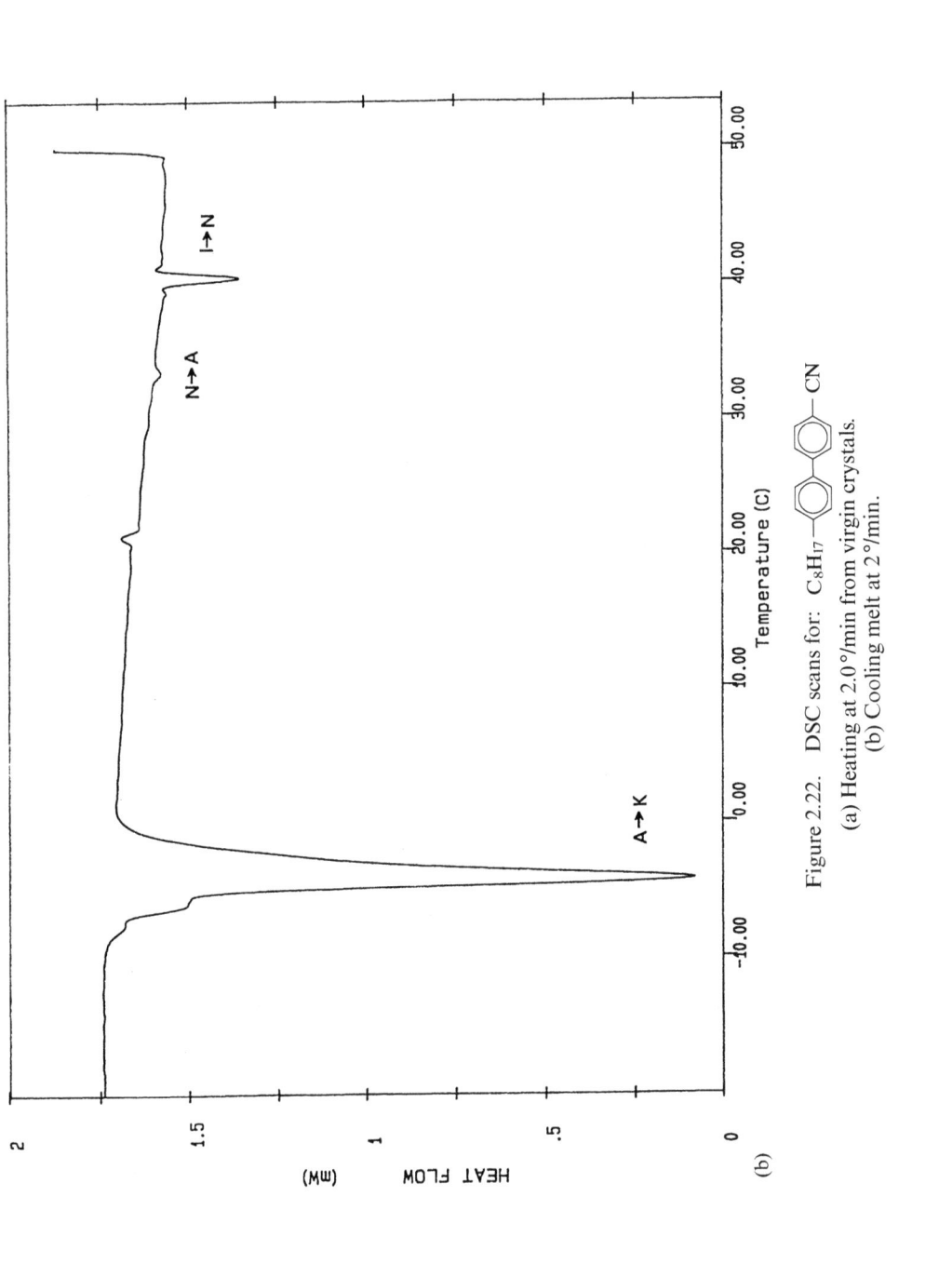

Figure 2.22.   DSC scans for: C$_8$H$_{17}$—⟨benzene⟩—⟨benzene⟩—CN

(a) Heating at 2.0°/min from virgin crystals.
(b) Cooling melt at 2°/min.

(b)

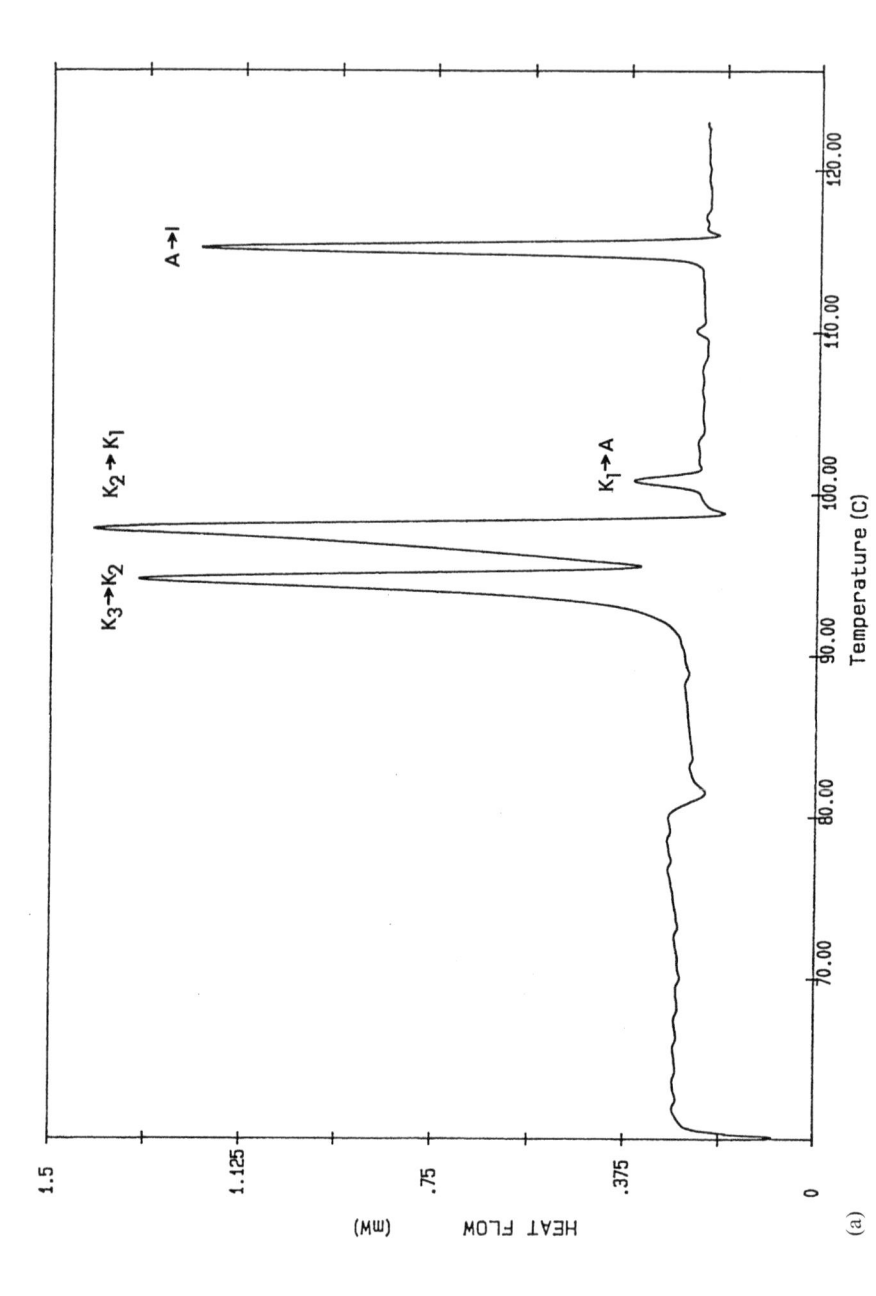

(a)

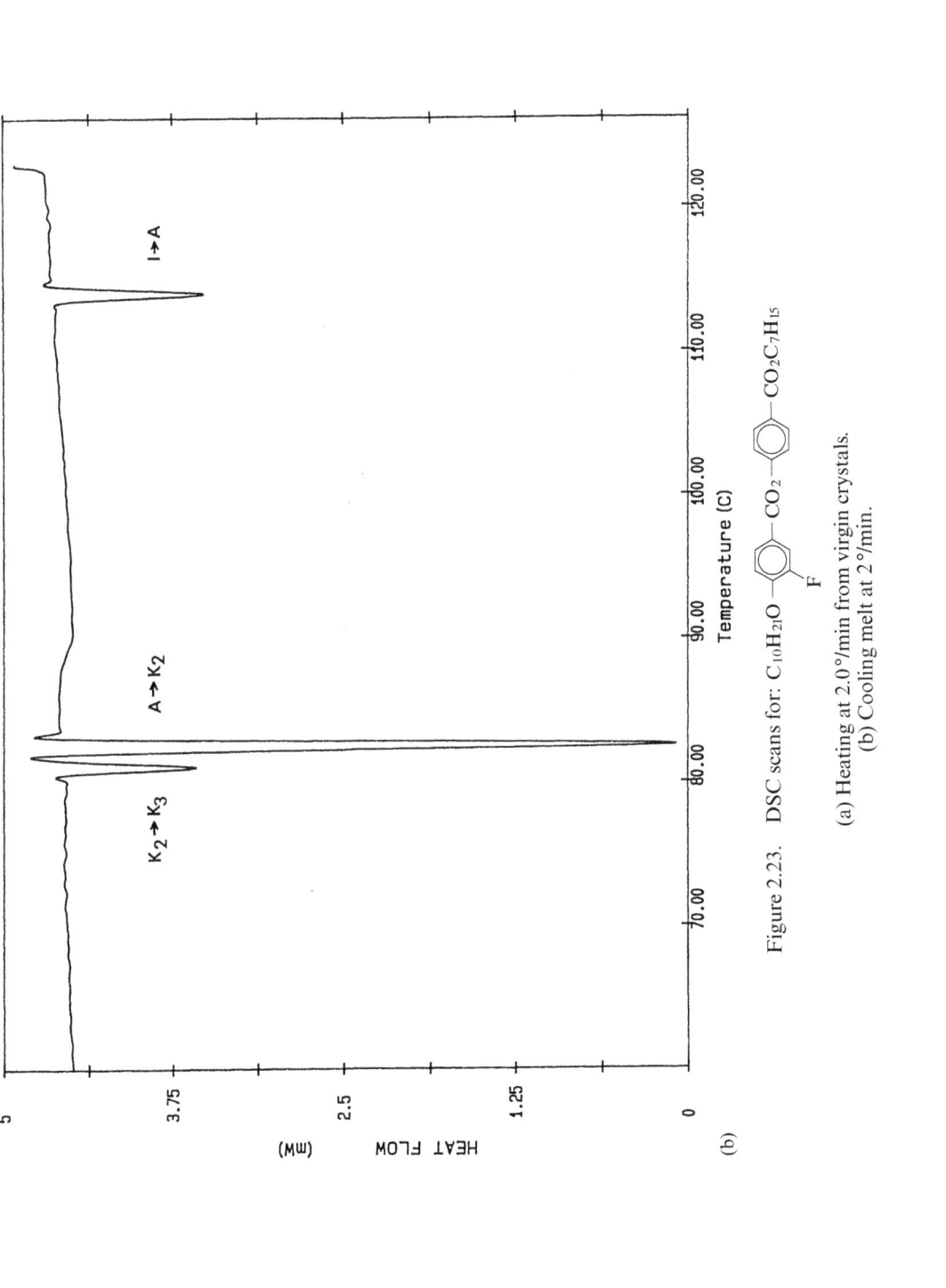

Figure 2.23.   DSC scans for: $C_{10}H_{21}O$ —⟨⟩— CO_2 —⟨⟩— $CO_2C_7H_{15}$

(a) Heating at 2.0°/min from virgin crystals.
(b) Cooling melt at 2°/min.

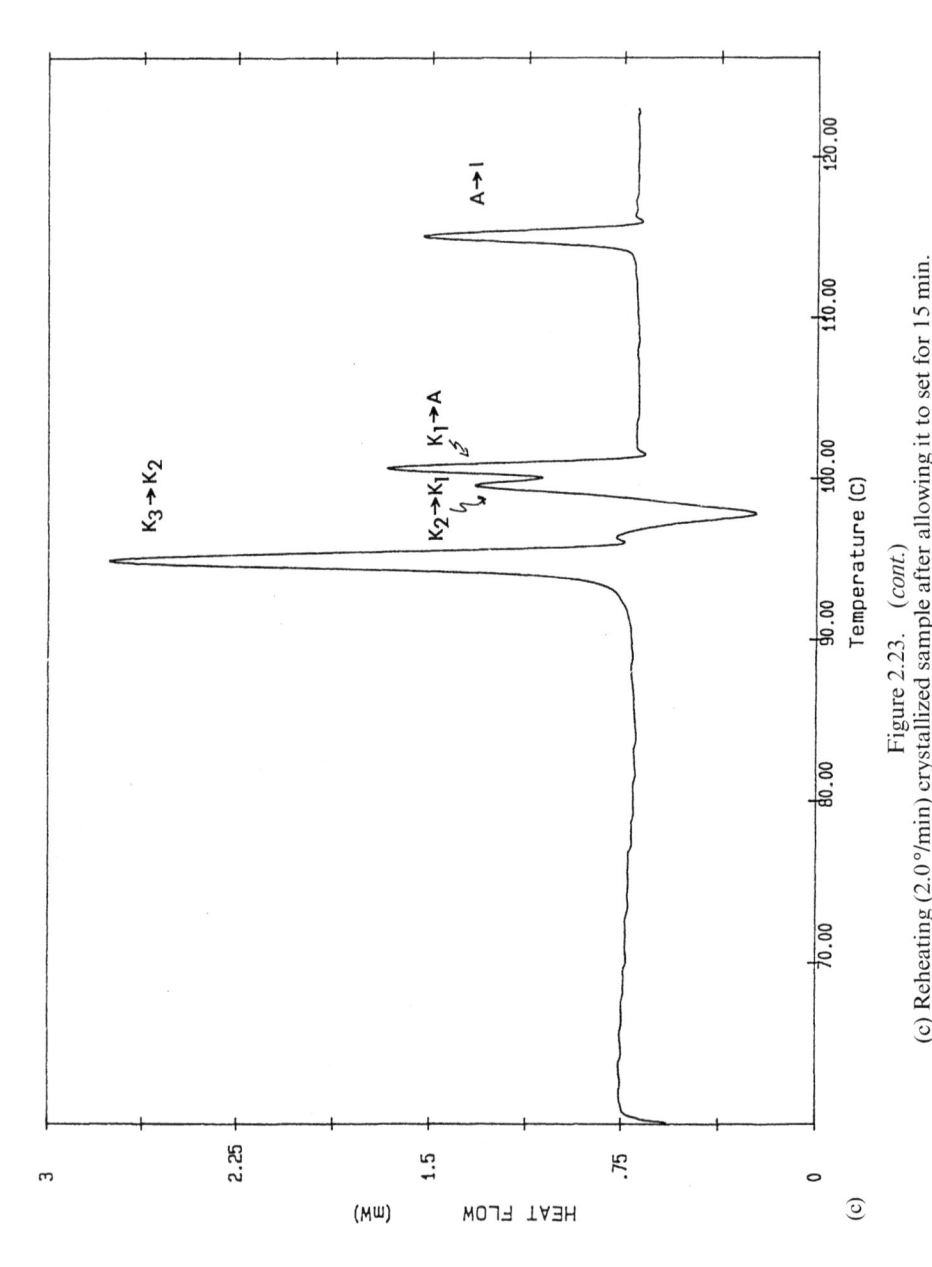

Figure 2.23. (*cont.*)

(c) Reheating (2.0 °/min) crystallized sample after allowing it to set for 15 min.

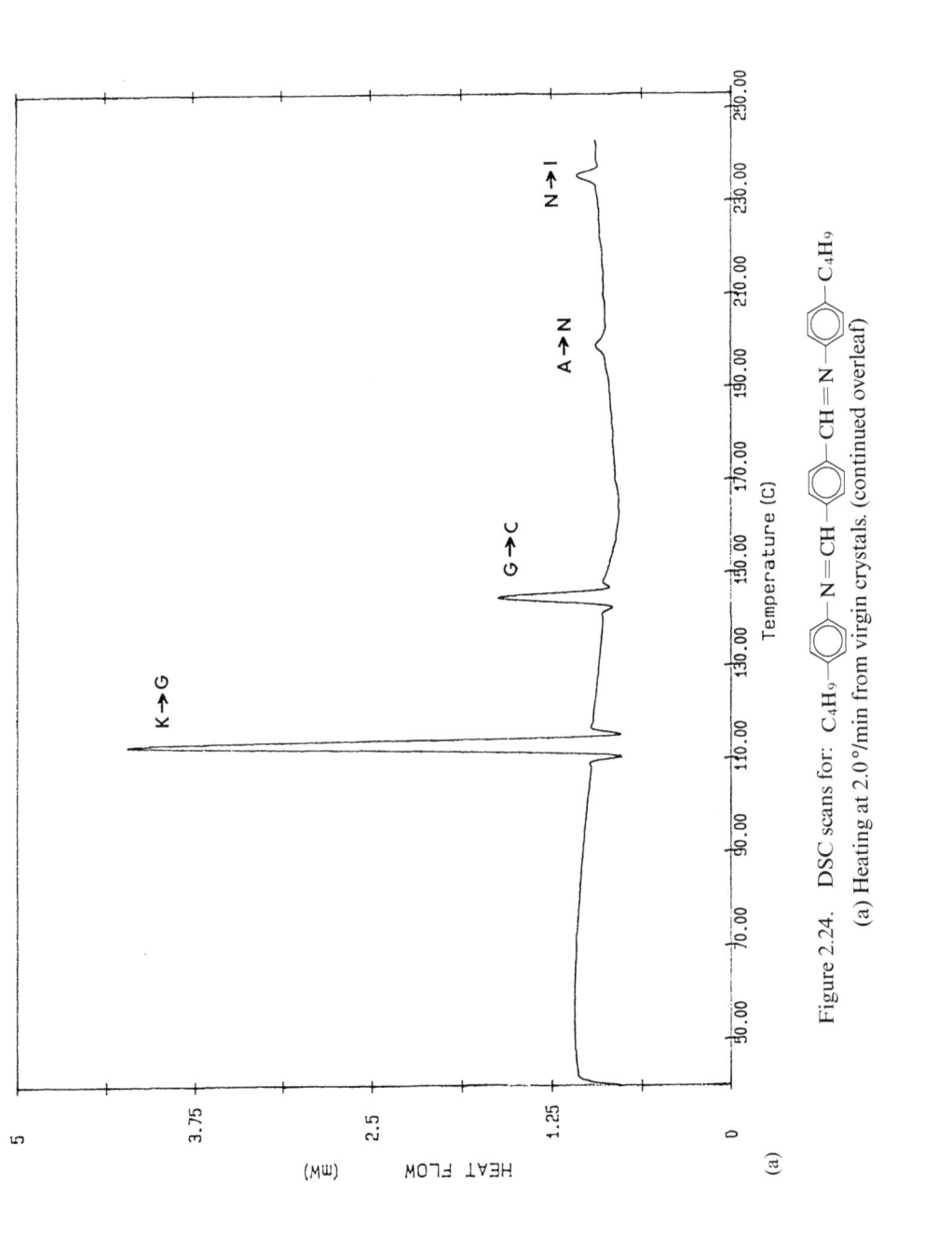

Figure 2.24. DSC scans for: C₄H₉—⟨⟩—N=CH—⟨⟩—CH=N—⟨⟩—C₄H₉

(a) Heating at 2.0 °/min from virgin crystals. (continued overleaf)

(a)

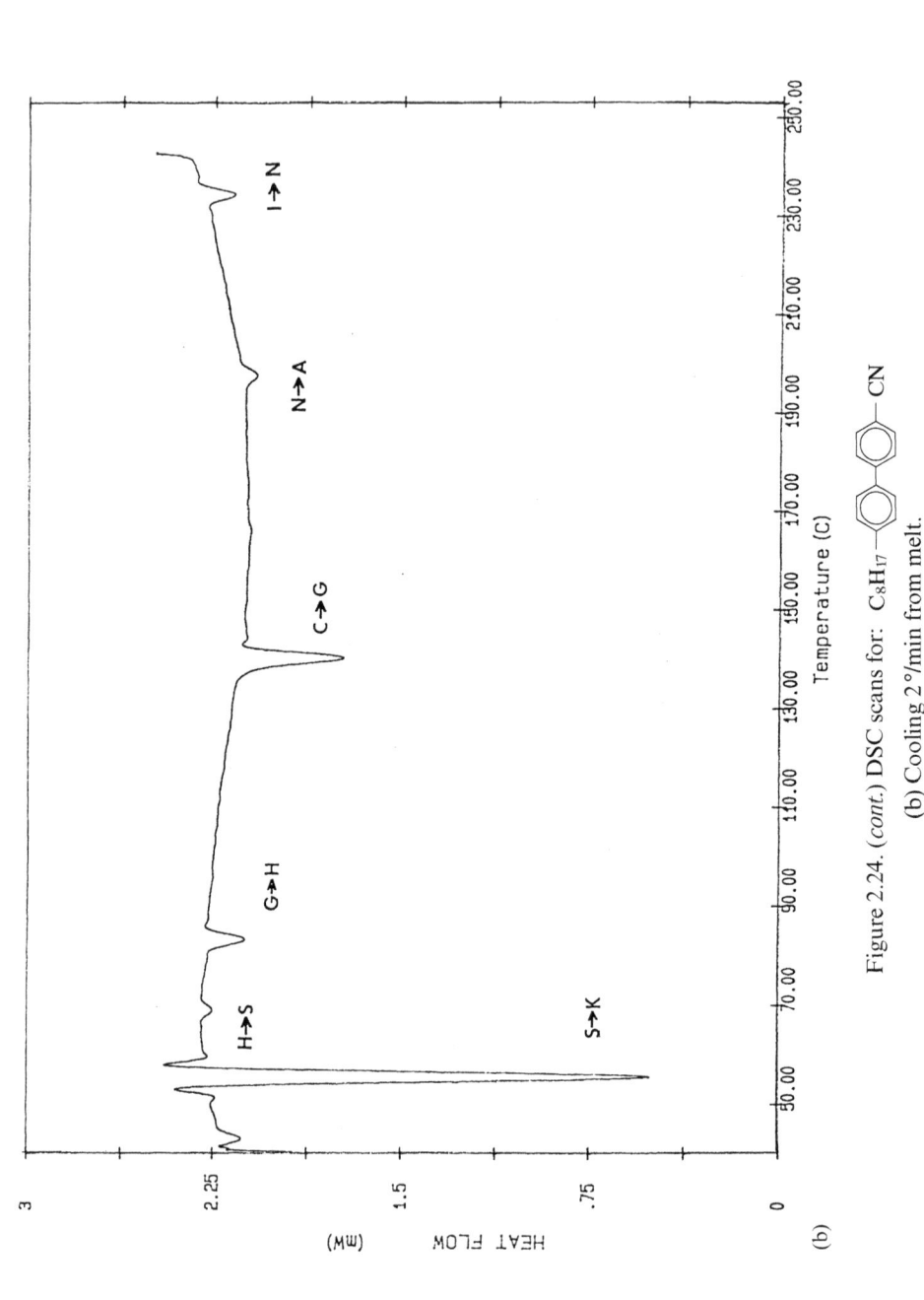

Figure 2.24. (*cont.*) DSC scans for: C₈H₁₇—⬡—⬡—CN

(b) Cooling 2 °/min from melt.

(b)

in enthalpy values were tabulated with the maximum values (kJ/mole) indicated:

C–M or I ≫ SmC–I > SmA–I > SmB–SmC > N–I ≈ SmC–N > SmA–N ≈ SmB–SmA
  117.3   42.7   12.6   10.5   9.64   4.61

Generally, the trend observed is that expected, i.e., melting values are larger than clearing and these are larger than mesophase transitions. One exception stands out – the SmB–SmC transition; perhaps because such transitions occur only in mid to long chain length mesogens. It would be interesting to see if this trend would hold with the considerably larger collection of enthalpy values now available in the Vill database. Still, the range of values for a particular transition can vary over a considerable range. For example, of the 391 compounds studied, melting enthalpies varied from 1.7 to 28 kcal/mole, N–I from 0.02 to 2.30 and S_c–I from 2.4 to 10.2. These ranges would likely be greater in our considerably larger, current collection of data. Obviously, overlapping can occur so that using enthalpy values to assign types of phase transitions is rarely useful except in a general sense (i.e. melting is usually the largest value).

Since enthalpy is directly related to the temperature, $\Delta H = T\Delta S$, it is not surprising that the enthalpy values will increase as temperature increases: this often happens as chain lengths increase. However, as previously discussed, melting enthalpies often increase in a homologous series when melting temperatures do not. This leads to consideration of the effect of entropy changes. As the chain length increases, the molecule becomes more flexible and disordered on going from the presumed all trans configuration in the highly ordered crystalline state to the more disordered mesophase or liquid phase. Obviously, some mesogenic structures would be more flexible and disordered at shorter chain lengths contributing to increasing the range of reported enthalpy values.

Other chain modifications can also affect enthalpy values [52]. Strongly polar liquid crystals tend to have larger melting enthalpy values, due to the strong lattice forces between the molecules in the solid state [21b, 35]. This is not surprising since highly polar mesogens often have higher melting temperatures. A comparison of the melting enthalpy values for a variety of polar mesogens indicates the following order for these values [53]:
OMe > CN > CH_3 > H > Cl > F > NO_2 > Br.

References

1. A. J. Leadbetter, in *The Molecular Physics of Liquid Crystals*, G. R. Luckhurst and G. W. Gray (Eds.), Academic Press, London (1979), Chapter 13; L. V. Azaroff, *Mol. Cryst. Liq. Cryst.* **60**, 73 (1980); A. de Vries, in *Physics of Liquid Crystalline Materials*, P. Mariani, F. Rustichelli and G. Torquati (Eds.), Gordon and Breach, Philadelphia (1988), Chapter 1.
2. *The Fourth State of Matter*, F. D. Saeva (Ed.), Marcel Dekker, New York (1979), a. A. de Vries, Chapter 1; b. E. M. Barroll II, Chapter 9; c. S. E. Petrie, Chapter 4.
3. B. Wunderlich, *Thermochim. Acta* **162**, 59 (1990).
4. *Advances in Liquid Crystal Research and Applications*, L. Bata (Ed.), Pergamon Press, Oxford (1980), a. H. Sackmann, p. 27; b. R. Dabrowski and K. C. Zuprynski, p. 125.
5. *Liquid Crystals of One- and Two-Dimensional Order and Their Applications*, Springer Series on Chemical Physics, Vol. II, W. Helfrich and G. Heppke (Eds.), Springer-Verlag, New York (1980), a. H. Sackmann, p. 19; b. D. Demus, J. W. Goodby, G. W. Gray and H. Sackmann, p. 268; c. J. W. Goodby, G. W. Gray, A. J. Leadbetter and M. A. Mazid, p. 3.
6. H. Sackmann and D. Demus, *Mol. Cryst. Liq. Cryst.* **21**, 239 (1973) and H. Sackmann, *Pure and Appl. Chem.*, **38**, 505 (1974).
7. D. Demus, S. Diele, S. Grande and H. Sackmann, *Adv. Liq. Cryst.* **6**, 1 (1983).
8. H. Sackmann, *J. Phys (Paris) Colloq.* **C3–40**, 5 (1979).
9. L. Richter, D. Demus and H. Sackmann, *Mol. Cryst. Liq. Cryst.* **71**, 269 (1981).
10. D. Demus, J. W. Goodby, G. W. Gray and H. Sackmann, *Mol. Cryst. Liq. Cryst.* **56**, 311 (1980).
11. A. Wiegeleben, L. Richter, J. Deresch and D. Demus, *Mol. Cryst. Liq. Cryst.* **59**, 329 (1980).
12. J. Budai, R. Pindak, S. C. Davey and J. W. Goodby, *J. Phys. (Paris) Lett.* **45**, L1053 (1984).
13. S. Haddawi, S. Diele, H. Kresse, G. Pelzl and W. Weissflog, *Cryst. Res. Technol.* **29**, 745 (1994).
14. H. Sackmann, *Liq. Cryst.* **5**, 43 (1989).
15. A. Adamczyk, *Mol. Cryst. Liq. Cryst. Sci. Technol.* A **249**, 75 (1994).
16. J. W. Goodby, *Mol. Cryst. Liq. Cryst.* **92**, 171 (1983).
17. D. Demus and L. Richter, *Textures of Liquid Crystals*, Verlag Chemie, Weinheim (1978).
18. G. W. Gray and J. W. Goodby, *Smectic Liquid Crytals, Textures and Structures*, Leonard Hill, Philadelphia (1984).
19. N. H. Hartshorne, *The Microscopy of Liquid Crystals*, Microscope Publications Ltd, London (1974).
20. A. Saupe, in *Liquid Crystals and Plastic Crystals*, Vol. I, G. W. Gray and P. A. Winsor, (Eds.), Ellis Horwood Ltd, Chichester; John Wiley & Sons, Inc., New York (1974), Chapter 2.
21. *Liquid Crystals Applications and Uses*, Vol. I, B. Bahadur (Ed.), World Scientific, Singapore (1990); a. D. Demus, Chapter 1; b. L. Pohl and U. Finkenzeller, Chapter 4.
22. D. Coates and G. W. Gray, *The Microscope* **24**, 117 (1976).
23. *Liquid Crystals and Plastic Crystals*, G. Gray and P. A. Winsor, (Eds., Vol. 2,

Ellis Horwood Publishers Ltd, Chichester (1974); a. N. H. Hartshorne, Chapter 2; b. E. M. Barrall II and J. F. Johnson, Chapter 10.

24. J. Cognard, *Mol. Cryst. Liq. Cryst. Supplement Series, Supplement 1, Alignment of Nematic Liquid Crystals and Their Mixtures*, Gordon and Breach Science Publisher, New York (1982); J. S. Patel, T. M. Lesle and J. W. Goodby, *Ferroelectrics* **59**, 137 (1984).
25. N. H. Hartshorne and A. Stuart, *Crystals and the Polarizing Microscope*, 4th Edn., Edward Arnold Ltd, London (1970).
26. G. H. Brown and W. G. Shaw, *Chem. Rev.* **57**, 1049 (1957).
27. S. L. Arora, J. L. Fergason and A. Saupe, *Mol. Cryst. Liq. Cryst.* **10**, 243 (1970).
28. Y. Bouligand, *J. Phys. (Paris) Colloq.* **C136**, 173 (1975).
29. M. E. Neubert, B. A. Williams, M. Willbourne and G. Mote, Preparation of a Colored Video Tape Textbook on the Identification of Liquid Crystalline Microscopic Textures, presented at the 12th ILCC, Friburg, Aug. 1988, Abstract No. AP40, p. 386.
30. H. Kelker and R. Hotz, *Handbook of Liquid Crystals*, Verlag Chemie, Weinheim (1980), Chapter 1.
31. M. E. Neubert, *Mol. Cryst. Liq. Cryst.* **31**, 253 (1975).
32. G. D. Woodard, *Microscope* **18**, 105 (1970).
33. H. Arnold, *Z. Phys. Chem. (Leipzig)* **225**, 45 (1964); ibid. **226**, 146 (1964); H. Arnold and P. Roediger, ibid. **231**, 407 (1966); H. Arnold, J. Jacobs and O. Sonntag, ibid. **40**, 177 (1969); H. Arnold, D. Demus, H-J. Koch, A. Nelles and H. Sackmann, ibid. **240**, 185 (1969); H. Arnold, *Mol. Cryst. Liq. Cryst.* **2**, 63 (1966).
34. K. Arvidsson, B. Folk and S. Sunner, *Chem. Scripta* **10**, 193 (1976); R. Müller, G. Hasl and H. Pauly, *J. Chem. Thermodyn.* **10**, 591 (1978).
35. U. Finkenzeller, R. E. Jubb and H. Schubert, E. Merck literature, *Physical Properties of Liquid Crystals V. Transitions Observed by Differential Scanning Calorimetry.*
36. T. Hatakeyama and F. X. Quinn, *Thermal Analysis–Fundamentals and Applications to Polymer Science*, John Wiley and Sons, New York (1994).
37. W. M. Wendlandt, *Thermal Methods of Analysis*, 2nd Edn., John Wiley & Sons, New York (1974).
38. J. R. Flick, A. S. Marshall and S. E. B. Petrie, in *Liquid Crystals and Ordered Fluids*, Vol. 2, J. F. Johnson and R. S. Porter (Eds.), Plenum Publishing Corp., New York (1974), p. 97.
39. K. S. Kunihisa, *Bull. Chem. Soc. Jpn* **46**, 2862 (1973); K. S. Kunihisa and T. Shinoda, ibid. **48**, 3506 (1975); K. S. Kunihisa and S. Hagiwara, ibid. **49**, 1204, 2658 (1976); K. S. Kunihisa, *Thermochim. Acta* **31**, 1 (1979); H. G. Wiedemann and G. Bayer, *Thermochim. Acta* **83**, 153 (1985) and *J. Therm. Anal.* **30**, 1273 (1985).
40. G. Ungar and J. L. Feijoo, *Mol. Cryst. Liq. Cryst.* **180B**, 281 (1990).
41. H. Staub and W. Perron, *Anal. Chem.* **46**, 128 (1974); G. Widmann and H. Sommerauer, *Am. Lab.* 106 (May, 1988) and *M. E. Brown, J. Chem. Ed.* **56**, 310 (1979).
42. P. Navard and J. M. Houdin, *J. Therm. Anal.* **30**, 61 (1985).
43. P. Narvard and R. Cox, *Mol. Cryst. Liq. Cryst.* **102**, 265 (1984).
44. C. Rein and D. Demus, *Liq. Cryst.* **16**, 323 (1994).
45. J. D. Menczel and T. M. Leslie, *J. Therm. Anal.* **40**, 957 (1993).
46. B. R. Ratna and S. Chandrasekhar, *Mol. Cryst. Liq. Cryst.* **162B**, 157 (1988).

47. C. M. Guttman and J. H. Flynn, *Anal. Chem.* **45**, 408 (1973).
48. D. Marzotko and D. Demus, *Pãrama*, Suppl. No. 1, 189 (1975); M. J. Richardson, *Thermochim. Acta* **229**, 1 (1993).
49. B. Perrenot and G. Widmann, *Thermochim. Acta* **234**, 31 (1994).
50. M. E. Neubert, M. E. Stahl and R. E. Cline, *Mol. Cryst. Liq. Cryst.* **89**, 93 (1982).
51. A. W. Levine and K. D. Tomeczek, *Mol. Cryst. Liq. Cryst.* **43**, 183 (1977).
52. G. L. Hillemann and G. R. van Hecke, *J. Phys. Chem.* **80**, 944 (1976).
53. A. G. Griffin, *J. Therm. Anal.* **12**, 335 (1977).

3

Structure: x-ray diffraction studies of liquid crystals

SATYENDRA KUMAR

Department of Physics and Liquid Crystal Institute, Kent State University, Kent, OH 44242, USA

The colour plates referred to in this chapter can be downloaded from www.cambridge.org/9780521187947

3.1 Introduction

X-ray diffraction provides one of the most definitive ways to determine the structure of liquid crystalline phases. This technique has played a key role in the identification of liquid crystalline phases and in the quantitative study of several liquid crystal phase transitions. Conventional analysis and data interpretation of x-ray crystallography can not be applied to liquid crystals. Although there are numerous books on theory and experimental methods of x-ray diffraction [1] and on liquid crystals, there is no single source that one can use to learn about the application of x-ray diffraction to liquid crystals and the interpretation of the data. This chapter is written to provide this basic information and to help a beginner develop an intuitive understanding of this technique.

The basic *mantra* of x-ray diffraction is the Bragg law which every science student encounters in a general or modern physics course. As shown in Fig. 3.1, this law states that x-rays reflected from adjacent atomic planes separated by a distance d of a crystal interfere constructively when the path difference between them is an integer multiple of the wavelength λ, i.e.,

$$2d \sin \theta = n\lambda.$$

Here n, an integer, is the order of the reflection. The reflected rays make an angle of 2θ with the direction of the incident beam. It should be borne in mind that, although this situation appears to be similar to optical reflection from a mirror, it is quite different. A beam of light will always be reflected from a mirror at an angle equal to the angle of incidence. X-ray 'reflection', on the other hand, is present only when the planes are oriented at the correct angle to satisfy the Bragg condition. There is no reflection at other angles. For this reason, we refer to this as x-ray diffraction.

One of the goals in an x-ray experiment is to determine the Bragg

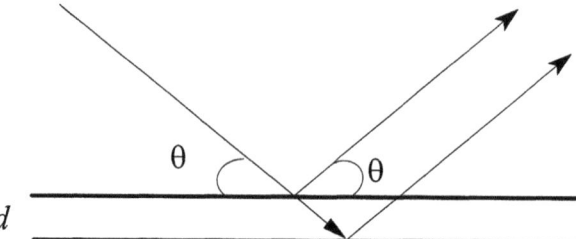

Figure 3.1.   X-ray diffraction from two planes in a crystal. The reflected rays from the upper and lower planes interfere constructively when the Bragg condition is satisfied.

angle(s) which can be related to the inter-planar distance or lattice constants. Conceptually, it appears simple but in actual experimental situations one needs to carefully consider the details especially in liquid crystal experiments. It is this requirement which has kept x-ray diffraction techniques from being commercially packaged for routine measurements of liquid crystals. As will become clear soon, an x-ray diffraction experiment provides information not only about the inter-planar distance, but also about the relative orientation and spatial orientational (better known as the *mosaic*) distribution of different sets of planes. It is perhaps the only technique to measure the changes in (or growth of) electron density (and thereby mass density) correlation lengths across phase transitions.

As is clear from Fig. 3.1 and becomes more evident in Fig. 3.2, one must be able to measure the angle 2θ between the incident and the diffracted beam of x-rays as well as orient a set of desired planes in the sample with respect to the x-ray beam to satisfy the Bragg condition. Experiments are conducted by performing different types of scans, each yielding specific structural information about the sample.

These scans can be grouped into either *real-space* or *reciprocal-space* [2] scans. In real-space scans, the sample orientation and/or the position of the detector are changed in a predetermined manner. In the reciprocal-space scans, the momentum transfer vector, **q**, is changed so that it takes specific trajectories through the reciprocal lattice corresponding to the structure of the sample at hand. This, as you might imagine, is accomplished by changing the scattering angle and the orientation of the sample in the laboratory frame of reference, in a coordinated and often complex manner. This task is made relatively painless with the use of computers which can carry out all the necessary calculations on-line and change various angles accordingly. In the following, we introduce the most commonly conducted scans in the study of liquid crystals. You will find them used here and in Chapter

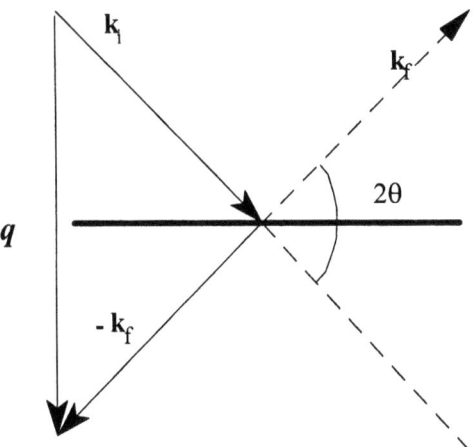

Figure 3.2.   Wavevectors for the incident ($\mathbf{k}_i$) and diffracted ($\mathbf{k}_f$) x-ray beams make an angle of 2θ. The momentum transfer or scattering vector $\mathbf{q} \equiv \mathbf{k}_i - \mathbf{k}_f$.

8 'Freely suspended film experiments' and Chapter 9 'X-ray surface scattering and reflectivity'.

3.2 Types of x-ray scans

3.2.1 Real-space scans

Real-space scans are those which are performed while changing the detector angle 2θ and sample orientation angles θ, φ, and χ defined in Fig. 3.3. One of the most common real-space scans is the θ–2θ-scan. In such a scan, the detector is at an angular position 2θ and the orientation of the diffracting set of crystal planes is at an angle θ in the scattering plane. When the Bragg condition is satisfied, the sample will Bragg scatter x-rays into the detector. This position is used to define the sample orientation θ to be half of the diffraction angle 2θ and thus to set the zero of θ. In this type of scan, the two angles are simultaneously varied over the desired range such that the sample's rotation angle is always half of the detector angle. A plot of diffracted intensity, versus the angle 2θ will exhibit peaks corresponding to the different lattice spacings belonging to the sets of crystallographic planes that lie parallel to each other and parallel to the bisector of the angle 2θ between the incident and diffracted beams which are perpendicular to the direction of the scattering vector $\mathbf{q}$ (as shown Fig. 3.3).

Several other scans are conducted in real space to obtain information about the orientation of the crystallographic planes. For example, if the

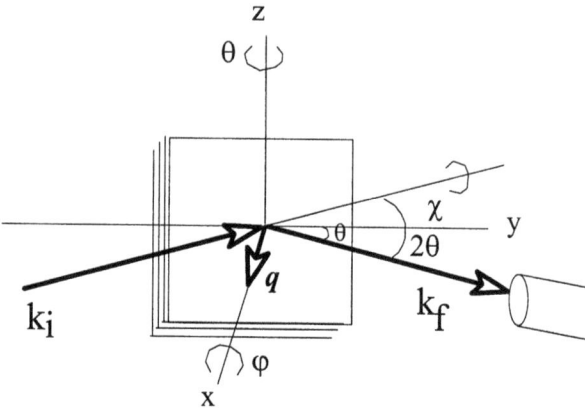

Figure 3.3.   The various angles of crystal rotation in an x-ray experiment. The directions x and y lie in the diffracting plane and z is perpendicular to the planes. Wavevectors for the incident and reflected x-rays make an angle of 2θ. The scattering vector $\mathbf{q} \equiv \mathbf{k}_i - \mathbf{k}_f$.

angular position of the detector is set to receive a diffracted beam from planes of a specific d-spacing and scans are conducted to measure the intensity as a function of sample rotation θ, it is referred to as the θ-scan or the ω-scan ($\omega = \theta - \theta_0$, where θ_0 is the position for maximum intensity). This scan provides an orientational distribution map of these crystal planes about the axis of θ-rotation, which is perpendicular to the scattering plane. Since such a scan provides information about the sample mosaic, it is also referred to as the *mosaic* scan. The θ-axis of rotation is chosen to be the z-axis in Fig. 3.3. There are two more (y and x) axes which lie in the scattering plane and one could conduct similar scans by rotating the sample about these axes. These are referred to as the φ- and χ-scan, respectively. Evidently, they provide additional information about the orientational distribution of the same planes.

3.2.2 Reciprocal-space scans

Reciprocal-space scans are the family of scans that are discussed in terms of the geometry of the reciprocal lattice. A discussion of the reciprocal lattice can be found in any standard solid state text book [2]. In the θ–2θ-scan discussed above, the magnitude of the scattering vector $\mathbf{q}$ is varied while leaving its direction unchanged. Since only increments that are parallel to the vector $\mathbf{q}$ are made, it is also known as the $\mathbf{q}_{\parallel}$-scan. As depicted in Fig. 3.4, one could also traverse the 'reciprocal lattice point' along the

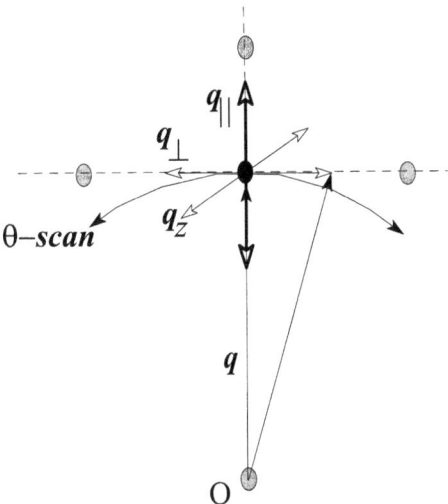

Figure 3.4. Various scans in the reciprocal space. Double arrowhead lines indicate the trajectories of **q**. An θ-scan is also shown to distinguish it from the $q_\perp$-scan.

two directions perpendicular to the vector **q**. Such scans, for example the $q_\perp$-scan, require that the sample be rotated about the axis perpendicular to the scattering plane and the magnitude of **q** be slightly changed to keep its trajectory coincident with the perpendicular direction shown as the dashed horizontal line. Such scans require on-line computations and coordinated change of different angles. One could also conduct scans in which the component of **q** perpendicular to the scattering plane is changed. These are known as the **q**$_z$-scans. In Fig. 3.4, the arc-like trajectory, in the reciprocal space, taken by the θ-scan is also shown.

The position of the peak in a **q**$_\parallel$-scan depends on the separation between the crystal planes. The mathematical function needed to describe the shape of the peak (commonly known as the *line shape*) depends on the type and spatial range of the structural order. As the extent or range of the spatial order grows, say, with temperature, the width of the peak in three directions determined by the **q**$_\parallel$-, **q**$_\perp$-, and **q**$_z$-scans becomes narrower. The three reciprocal-space scans are often used to determine the extent of *positional order correlations* known as *correlation lengths*. Of course, this requires a very well aligned single domain sample. The peak widths for a perfect crystal depend on its size in the three directions. For an imperfect or powder crystal, they are determined by the average mono-domain (crystallite) size.

3.3 Experimental details

X-ray scattering experiments on liquid crystals are considerably more difficult to perform than conventional experiments using crystalline samples. They have fewer reflections and their interpretation requires a deeper understanding of the system. Furthermore, liquid crystals require special handling, temperature control, and alignment for successful experiments.

A liquid crystal sample is prepared in a manner that depends on its properties, and experimental details. First, the sample should be free of solvents and absorbed oxygen because their presence dramatically accelerates chemical decomposition at elevated temperatures. A sample is *degassed* by repeated melting and freezing (thaw cycle) in vacuum to remove solvents and absorbed gases. For low resolution experiments, the liquid crystal can be filled in a thin-wall glass capillary specifically designed for x-ray diffraction. If the sample is volatile, the ends of the capillary need to be sealed with vapor-free epoxy or fused with flame without adversely heating the sample. Capillaries make very good sample cells for simple powder experiments. Often, the filling process causes the liquid crystal to align in the flow direction. Alternative methods, needed for high-resolution work, include sealing the sample between two thin sheets of polymer (e.g. Mylar or Kapton for temperatures below 250 °C) or beryllium using an o-ring made of chemically inert material such as Teflon. Care should be taken to ensure that the liquid crystal does not come in contact with epoxy or metal surfaces. They are known to accelerate sample decomposition.

The size of the sample should be large enough to permit unobstructed passage of the incident x-ray beam at all possible sample orientations for planned scans. Thickness of the sample should be equal to one *absorption length* of x-rays in the liquid crystal to permit 63% of the incident flux to pass through to yield optimum scattering intensity. The absorption length can be easily calculated from the density and mass absorption coefficient of the constituting elements [3]. As a rule of thumb, for most hydrocarbon liquid crystals the absorption length is approximately 1.5 mm for CuK_α ($\lambda = 1.5418$ Å). It can be one order of magnitude smaller if the sample contains heavy elements. In such cases, one needs to invariably use more energetic x-rays, e.g., molybdenum, MoK_α ($\lambda = 0.71069$ Å) radiation.

It is imperative to align the liquid crystal to obtain useful information. Unaligned samples are a poor choice for phase identification. An *ideal* powder sample will give the same diffraction pattern in the isotropic, nematic, and smectic-A phases! Samples must be at least partially aligned. This is normally accomplished with the help of an external field as discussed below. The external field invariably competes with surface anchoring potential imposed by cell boundaries. The effectiveness of the external field is significantly enhanced if the boundary conditions also favor the same orientation as the field.

Most liquid crystal samples can be aligned by heating them to their isotropic phase and then slowly cooling into a liquid crystalline phase in the presence of an external magnetic field. The molecules in the nematic phase are free to reorient under the torque produced by the magnetic field due to the diamagnetic anisotropy, $\Delta\chi$, of the liquid crystal molecules. It should be pointed out that even materials without a nematic phase can be aligned by this method. The effectiveness of the field in aligning a specific sample depends on the magnitude and sign of $\Delta\chi$. Samples with aromatic cores have a positive $\Delta\chi$ and they align relatively easily with the directors parallel to the field. Because of their two to three orders of magnitude smaller $\Delta\chi$, lyotropic liquid crystals are hard to align with the magnetic field. Moreover, many of them tend to have negative $\Delta\chi$, and the director aligns in a plane perpendicular to the field. In such cases, it is necessary to spin the sample about an axis perpendicular to the magnetic field to break the degeneracy in that plane. As the sample is spun, the local directors which lie perpendicular to the spin axis will become parallel to the field at some time during spinning. The field then reorients these directors parallel to the spin axis. This aligns the director along the spin axis throughout the sample. Fields smaller than ~ 5 kG, typically produced by cobalt– samarium permanent magnets [4] or electromagnets normally available in a laboratory, are effective only in aligning a nematic phase. Once the sample enters a smectic phase, the magnetic field's effect can be safely disregarded. If necessary, the field can be turned off or the sample removed from the field without losing the alignment.

In some cases, it may be preferable to use an electric field for alignment. One of the problems with electric fields is that one needs to apply *ac* fields [5] because a *dc* field often gets annulled by the opposing field produced by ionic impurities which migrate to oppositely charged electrodes. The second problem is that the field has to be relatively high, $\sim 10^5$–10^6 V/m. For bulk samples, this requires voltages in the range 500–1000 V. A third complication with the use of electric field is that there is always a small but finite current flow through the sample. This can result in internal heating which could be counterproductive in experiments where precision temperature control is required.

After crossing a phase transition, the degree of alignment is almost always diminished. Within a smectic phase, the alignment deteriorates

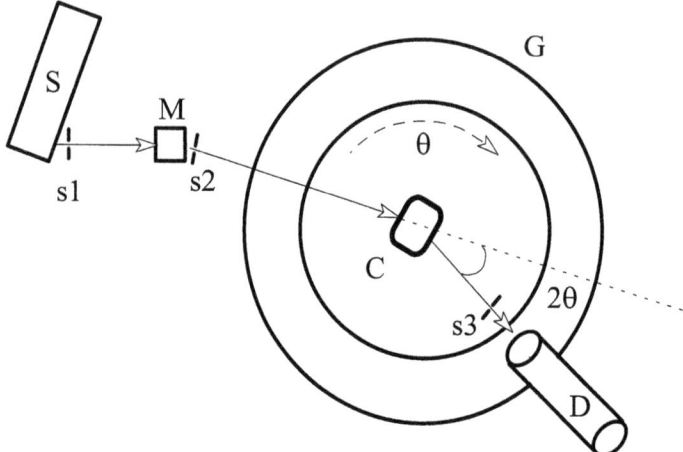

Figure 3.5.   Schematic of x-ray diffraction experimental set-up. Here: S, source;
s 1–s 3, collimating slits; C, sample chamber; M, monochromator; G, goniometer;
D, detector; θ and 2θ mark the inner and outer rotation stages for rotations of the
sample and the detector, respectively.

rapidly if the layer spacing decreases when temperature is lowered as in the
SmC phase. On the other hand, the alignment recovers dramatically if the
layer spacing increases with decreasing temperature, for example, in the
SmA$_d$ phase. Bulk liquid crystal samples are known to exhibit a surprising
'memory effect', i.e., an aligned sample, if allowed to crystallize and then
heated back to a liquid crystalline phase, is found to be partially aligned.

X-ray diffraction has so far been used for measurements of (a) static
properties: structure determination and characterization of liquid crystal
phases using low resolution experiments, and (b) dynamics: study of critical phenomena using high resolution. We first discuss low-resolution
experiments and characteristic features of various liquid crystalline phases
in x-ray diffraction patterns. In a later section, we will discuss measurements of smectic correlation lengths near the N to SmA phase transition
as an illustration of x-ray measurements of dynamical quantities.

A typical x-ray diffraction experimental setup, as shown in Fig. 3.5, consists of a source of x-rays, monochromator, sample chamber with capabilities of aligning and controlling sample temperature mounted on a
goniometer, followed by a detector. The most common source of x-rays is
a generator using a sealed electron impact tube. A rotating anode x-ray generator can be used when high x-ray flux is needed such as in high-resolution
experiments. In both cases, x-rays are emitted by a metal target, usually
copper, whose K$_\alpha$ doublet is emitted at approximately 8 keV. Rotating

anode sources give nearly one order of magnitude higher x-ray flux than sealed tubes. It is necessary to use more penetrating radiation, such as the 17.4 keV K_α line of molybdenum, for strongly absorbing samples. High-precision experiments are increasingly being performed at synchrotron x-ray facilities sources [6] with an x-ray flux $\sim 10^{12}$ times higher than a rotating anode source and freedom to select any wavelength from their wide continuous spectrum.

The beam emerging from a source is always polychromatic. A single wavelength is selected using the monochromator, M. The monochromator may be as simple as a thin foil of metal; for example, Ni which has an absorption edge at wavelengths just below the desired CuK_α line. Ni absorbs wavelengths shorter than 1.54 Å much more strongly, thereby increasing the relative intensity of the K_α line. These are known as the 'filter' monochromators [1, 7]. The purity of the wavelength selected with the help of a filter is not very high. Another common and better method utilizes a single crystal of graphite which is set to Bragg reflect the desired wavelength. Graphite crystals are also available *singly-bent* or *doubly-bent* to focus an otherwise divergent beam in one or two dimensions, respectively, at a distance dependent on their radius of curvature in a manner similar to focusing of light by spherical mirrors. These bent monochromators not only provide a single wavelength, but also enhance the intensity. High-resolution experiments use perfect single crystals of germanium or silicon. In applications where fine collimation or low thermal diffuse background is essential, asymmetrically cut and channel-cut crystals are, respectively, employed.

The sample is placed in a chamber C with capabilities of controlling the temperature and aligning with the help of a magnetic or electric field. Rare-earth (cobalt–samarium) magnets can produce fields of several kilogauss with the help of properly designed yoke and tapered pole pieces. One must seriously think about the direction of the field to be applied. If smectic layer spacing is to be measured, make sure that the field lies in the scattering plane parallel to the scattering vector or roughly perpendicular to the incident beam. On the other hand, if the in-plane structure is to be probed, then the field must be along one of the two directions perpendicular to the scattering vector, i.e., either along the incident beam or perpendicular to the scattering plane.

The sample chamber is mounted at the center of the goniometer on the θ-circle which permits rotation about a vertical axis. The detector D is attached to the 2θ-circle and can be rotated about the same vertical axis independently of the sample. Commercial goniometers come in different configurations ranging from two to six axes (or circles) of rotation about different axes. The optimum configuration for general liquid crystal structure determination is the two-circle goniometer. The x-ray optics is arranged to steer the x-ray beam through the point of interaction of the axes of rotation (i.e., common center) of all the circles. The sample is rigidly mounted to lie at this point.

The choice of the detector depends on the information sought. To obtain an overview of the diffraction pattern, two-dimensional detectors or their older cousin, the photographic film are ideally suited. For quantitative work, a single element detector, such as scintillation detector, is essential. The position of a single element detector is swept over a range of diffraction angles (2θ) while the sample is rotated by θ to generate, as defined above, the 2θ–θ-scan to measure different lattice spacings in the sample. Once the layer spacings have been determined, one can park the detector at the diffraction angle to receive the diffracted beam from a specific set of planes and the sample rotated (or rocked) about the θ-axis to obtain an orientational distribution map of planes in the sample. A combination of these two types of scans can be very conclusive in characterizing a particular phase.

3.4 Diffraction patterns of liquid crystal phases

The highest temperature phase that all liquid crystalline materials have is the isotropic phase. This phase lacks long range positional or orientational order. However, even in classical liquids such as water, nearest neighbor molecules exhibit a short range positional order that gives rise to short range mass density correlations. The diffraction peaks from liquids are very weak and broad. The inverse of the peak width gives the spatial range of the order, in this case an intermolecular distance. Clearly, an isotropic phase can not be aligned and its diffraction pattern consists of diffuse concentric rings as shown in Fig. 3.6 (color).

The ring near the center (i.e., at small scattering angles) corresponds to the effective molecular length which, at a finite temperature, is always smaller than the length calculated using the space-filling model. This, evidently, is due to the thermal motion of different parts of the organic molecules. The large angle ring corrresponds to the average width of the constituting molecules. It almost always appears at approximately 20.5° and gives the characteristic hydrocarbon chain diameter of ~4.5 Å. Actual measured dimensions for different compounds will, of course, be slightly different.

3.4.1 The nematic phase

The diffraction pattern of an *unaligned* nematic phase is not much different from the isotropic phase pattern. Fortunately, the nematic phase reorients under the influence of an external field in a manner similar to the ordering of magnetic domains in a ferromagnetic material when an external field is applied. The director of a nematic phase with $\Delta\chi>0$ orients parallel to the magnetic field. Consequently, a liquid-like peak corresponding to the length of the molecule appears along the direction of the field. The diffuse peak arising from intermolecular spacing appears in the perpendicular direction as shown in Fig. 3.7 (color). When a nematic phase is cooled in an external field, the two sets of diffuse peaks become better defined because of the increasing value of the order parameter S. Early experiments on liquid crystals used these patterns to calculate the value of S.

In the case of the nematic phase with $\Delta\chi<0$, the molecules, and hence the director, lie in a plane perpendicular to the field. The diffraction patterns appear to be rotated by $\pi/2$ and the inner pair of reflections is relatively weak as the director is confined to a plane and only the regions where **n** is perpendicular to the field and the scattering plane contribute to this peak.

It should be remembered that the above description is valid only for nematics formed by rod-like objects such as thermotropic liquid crystals and lyotropic liquid crystals with cylindrical micelles. In lyotropic liquid crystals with discotic micelles, the large angle peak corresponds to the thickness of the micelle and the small angle peak to the diameter of the micelles. Again, the diffraction pattern's orientation relative to the field will depend on the sign of $\Delta\chi$.

The diffraction pattern of the nematic phase at lower temperatures, close to the transition to an underlying smectic phase, depends upon the symmetry of that underlying phase. If the underlying phase is a SmA phase, the two sets of reflections will be orthogonal to each other, but the inner pair will become sharp as the temperature is lowered. On the other hand, if the underlying phase is a SmC phase, in which the molecules are tilted with respect to the layer normal, then the small and large angle reflections appear at an oblique angle with respect to each other. This happens because of the development of pretransitional short range order corresponding to the lower temperature phase. In the early days, these changes were misunderstood and thought to be arising from a different (*cybotactic*) nematic phase. Caution must be taken while interpreting such diffraction results. Several times in the past, they have been mistaken for the much sought after biaxial thermotropic nematic phase [8].

An especially confusing situation can arise if the molecules have different dimensions in the three directions. Each dimension will appear as a distinct ring in the isotropic and nematic phases. When such materials are subjected to an external field, one dimension is usually aligned along the field and the other two are confined to the plane perpendicular to the field. They appear as two sets of diffuse crescents along the direction perpendicular to the field, easily misleading researchers unfamiliar with x-ray diffraction to conclude the existence of a biaxial nematic phase. At the time of this writing, the existence of the biaxial nematic phase has been confirmed only in lyotropic and polymeric liquid crystals. No monomer thermotropic materials have so far been found to exhibit the biaxial nematic phase.

3.4.2 The smectic-A and smectic-C phases

As discussed in Chapter 1, the smectic phases are characterized by a one-dimensional density modulation. In the SmA phase, the density modulation is along the director or the molecular long axis. Consequently, diffuse peaks observed at small angles in the magnetically aligned nematic phase condense into sharp quasi-Bragg peaks. There is no change in reflections in the orthogonal direction as the molecules are still randomly arranged in the smectic planes. In the SmA phase, the two sets of reflections are in directions orthogonal to each other as shown in Fig. 3.8 (color). As the temperature is lowered in the SmA phase, thermal motion of molecules is reduced and the alkyl chains become stiffer. The net result is that the effective molecular length and hence the smectic layer spacing, d, increases, as can be seen from the results for terephthal-bis-(n)-propylaniline (TB3A) [9] in Fig. 3.9. Only high-resolution techniques are capable of detecting changes so small. The measured rate of increase depends on the material and on whether there is a nematic or an isotropic phase at higher temperature. The rate of increase in d is typically in the range of 0.000 25 to 0.005 Å/K.

The SmC phase has a layered structure in which the smectic layer normal and the director are no longer collinear. At the SmA to SmC transition, molecules start to tilt with respect to the layer normal and d decreases correspondingly. The molecular tilt is the order parameter for the SmC phase. It is used to describe the transition from the SmA to the SmC phase. The tilt angle, calculable from the smectic layer spacing as $\cos^{-1}(d/d_A)$, where d_A is the smectic layer spacing at the transition, follows a material dependent power law. Molecular and phenomenological models [10] of this transition provide a satisfactory understanding of the molecular tilt and the phase transition.

A single domain of the SmC phase and its diffraction pattern are

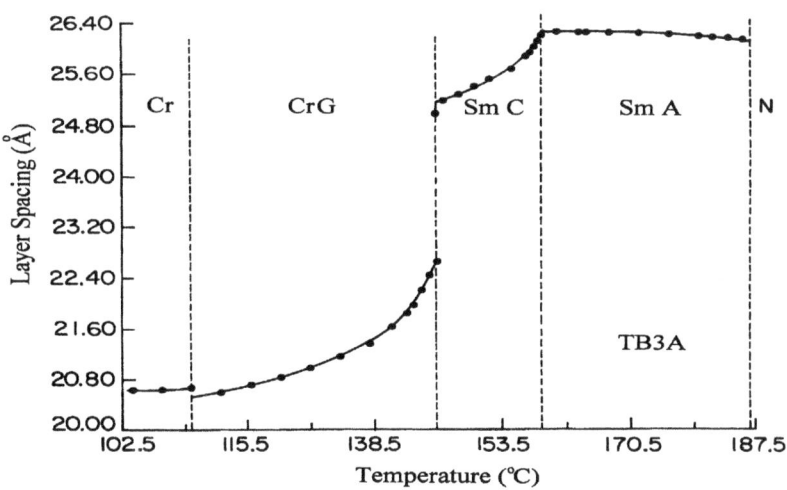

Figure 3.9.   Temperature dependence of smectic layer spacing in the SmA, SmC, and CrG phases of TB3A.

schematically shown in Fig. 3.10. The tilt angle α increases as the distance from the transition is increased. In a real experiment, where there is normally no control on the tilt direction, the situation is far from this idealized and simple picture. One or all of the following can happen:

1. The smectic layers can remain unchanged while the molecules tilt (Fig. 3.11a with respect to the layer normal. Since there is no azimuthal preference for the tilt, the tilt occurs in different azimuthal directions in different parts of the sample. Relative to the x-ray beam, molecules tilt either to the left or to the right. The diffraction pattern consists of two sets of arc-shaped reflections at large angle. Since these are diffuse peaks, it is usually difficult to see this splitting clearly. Most of the time, two reflections with uniformly distributed intensity appear along the large angle ring.

2. It is also equally likely for molecules to retain the same orientation as in the SmA phase. Then layers tilt either to the right or left or both (Fig. 11b). Such effects give rise to two pairs of smectic reflections at small angle. The angle between these pairs is twice the molecular tilt α.

3. Both of the above scenarios can occur simultaneously in different parts of the sample. Experimentally, this is the most often encountered situation and one observes two pairs of reflections.

4. In addition to the possibilities above, one can expect the layers and the molecules to partially share the burden of tilt. One then can expect both the small and large angle peaks to move from their initial positions.

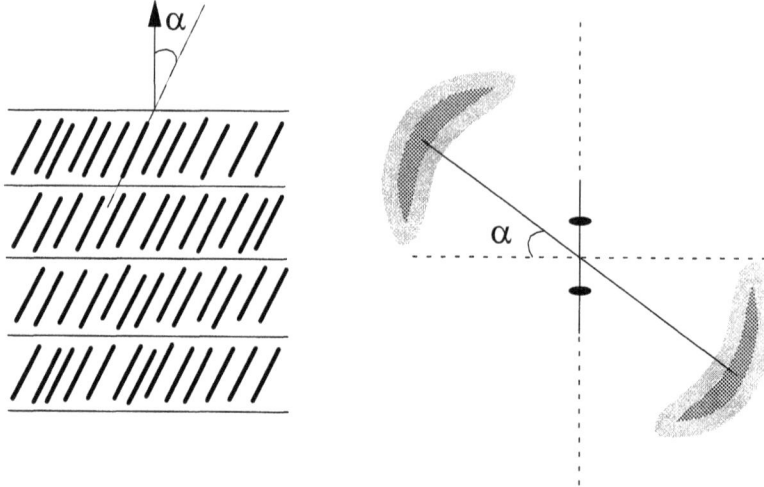

Figure 3.10.   Diffraction pattern of the SmC phase.

Because of the temperature dependent layer spacing, the layers in the SmC phase buckle to fill the same volume with thinner layers. This is referred to as the undulation instability. Obviously, if the smectic layers do not remain flat, the mosaicity of the sample increases. As a consequence, the sample becomes increasingly misaligned as the temperature is lowered in the SmC phase. The peak intensity drops precipitously becoming one to two orders of magnitude smaller than in the SmA phase.

One can measure the molecular tilt in two ways. The angle between the pairs of small angle reflections, mentioned in (2) above, can be measured and used to calculate the tilt angle α. Alternatively, the change in layer spacing can be used to determine α. The effective d becomes smaller by a factor of $\cos \alpha$. Knowing the layer spacing in the SmA phase, one can precisely calculate tilt as small as 0.5° in a high-resolution experiment.

3.5 The power of mosaic scans

Mosaic scans are a very powerful tool in the identification of liquid crystalline phases. As mentioned above, one must have at least a partially aligned liquid crystalline phase before it can be identified or the structure determined. When a phase is aligned, the width of its mosaic scan becomes narrower providing a measure of the degree of alignment. This is a very important piece of information not only for the identification of phases but also to determine if a phase is a single phase or a coexistence of two or more phases.

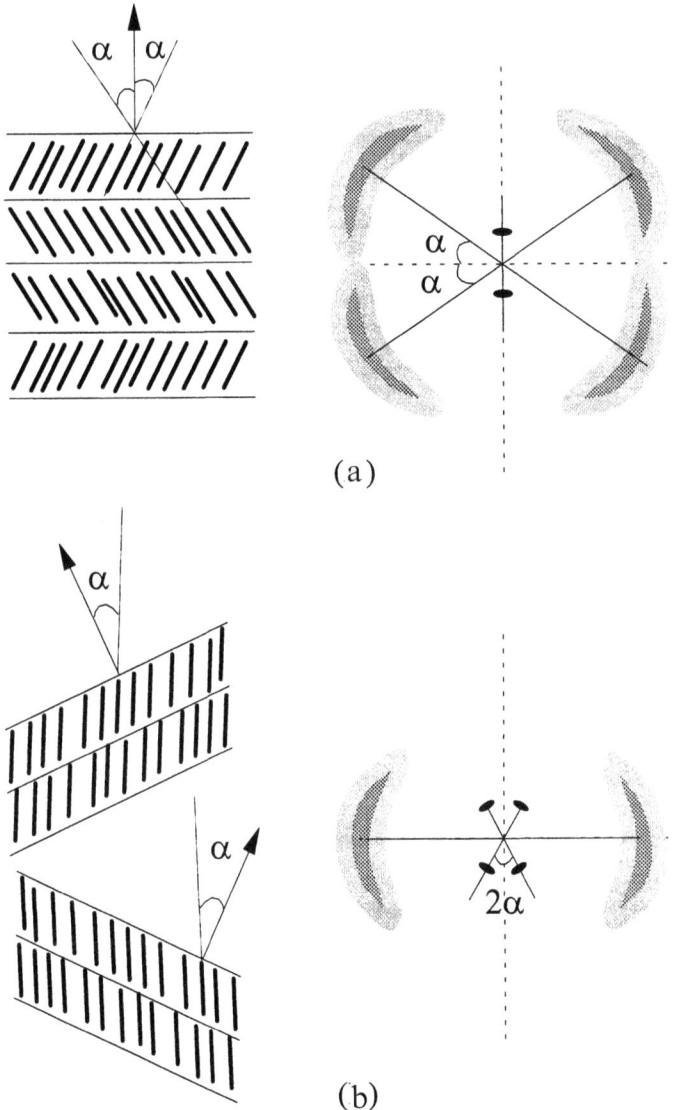

(a)

(b)

Figure 3.11. (a) The SmC phase with layers maintaining the same orientation as in the SmA phase and the resulting x-ray diffraction pattern. (b) The SmC phase with the molecules keeping the same orientation as in the SmA phase but the smectic layers tilting, and the corresponding diffraction pattern. Here, the x-ray beam is incident perpendicularly into the plane of the paper.

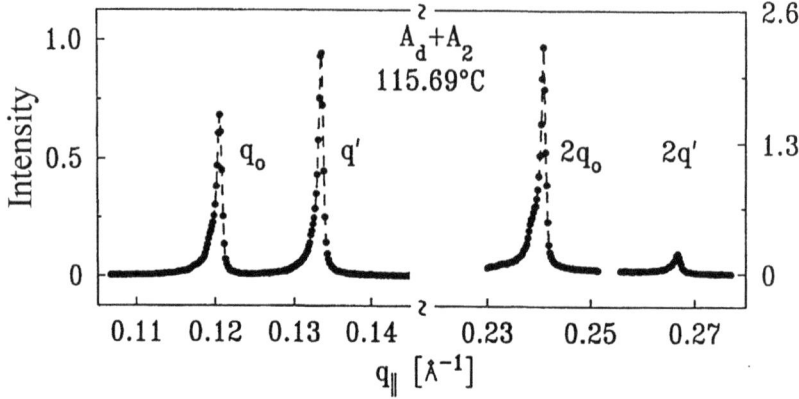

Figure 3.12.   Longitudinal scans for a mixture of DB7OCN + 8OCB at 115.69 °C. Peaks at $2q_o$ and $2q'$ are second harmonics of the peaks at $q_o = 0.1206$ Å^{-1} and $q' = 0.1335$ Å^{-1}, respectively.

As an illustrative example, consider the DB7OCN + 8OCB mixture which exhibits frustrated smectic phases [11]. In the three (SmA$_1$, SmA$_2$, and SmA$_d$) phases, condensed (liquid-like) peaks emanate from the two density waves depending upon whether they are (not) condensed. When the two density waves condense in the SmA$_2$ and SmA$_d$ phases, they are supposed to be collinear, i.e., their wavevectors are parallel to each other and parallel to the direction of alignment. The best test/proof of whether they are really collinear can be obtained through mosaic scans [12].

Figure 3.12 shows the result of 2θ–θ-scan at 115.69 °C. The first and the second multiples of the reflections at wavevectors q_0 and q' are clearly visible. The intensities of the corresponding peaks reflect the difference in the extent to which the corresponding density waves have developed. The question 'Are these characteristic peaks of a new phase?' naturally arises. It turns out that the mosaic scans provide the most unequivocal answer to this question. The mosaic scans taken at q_0 and $2q_0$ peaks, shown in the upper panel of Fig. 3.13, are identical in their shape and width suggesting that they are indeed originating from parallel sets of planes in the same (physical) volume of the sample. Also, mosaic scans conducted on the peaks at q' and $2q'$ are identical as is clear from their nearly identical shape and finer features, shown in the lower panel in Fig. 3.13. The minor differences between the two peaks arise from the fact that they were taken at different times and the system was slowly evolving with time [12]. What the mosaic scans provide is the evidence that the smectic layers giving rise to these two sets of reflections are quite different in their spatial distribution and must

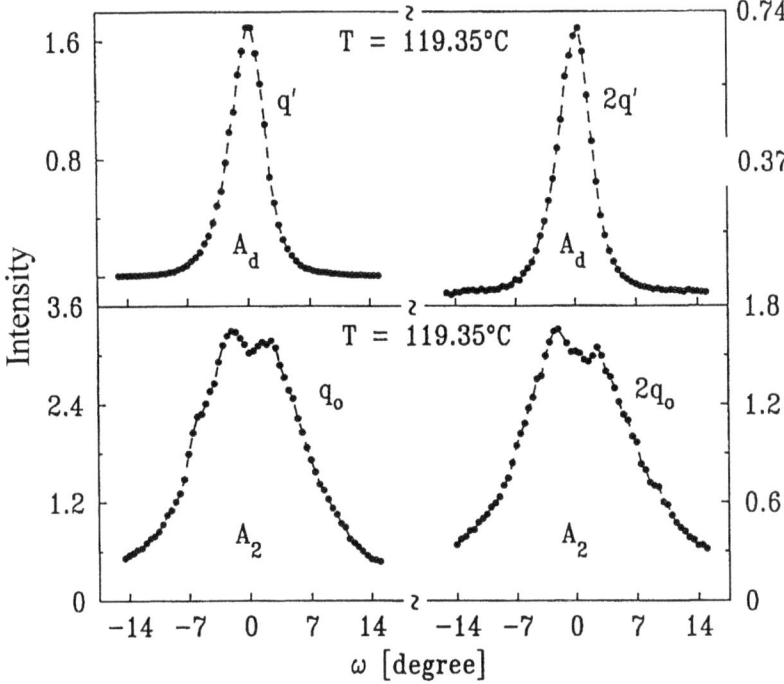

Figure 3.13. Mosaic scans taken at the two sets of peaks corresponding to the SmA$_d$ and SmA$_2$ phases in a mixture of DB7OCN and 8OCB.

originate from different volume elements of the sample. The fact that the two sets of scans are centered at $\omega = 0$ indicates that the corresponding density waves are parallel to the direction initially determined by the magnetic field used to align the nematic phase at higher temperatures. In other words, there are two distinct phases in the sample at this temperature. Indeed the coexistence is observed for a wide range of temperature in this two component system. When the temperature is lowered below the coexistence range, only the SmA$_2$ peaks at q_0 and $2q_0$ with identical mosaicity survive. It is strongly recommended that to gain a more complete appreciation of the power of mosaic scans one read Refs. 12 13, and references therein.

The mosaic scans, in this manner, can be used to establish purity or coexistence of a thermodynamic phase not just in liquid crystals but in other systems as well. It should be pointed out that one does not necessarily need high-resolution to obtain this information. A trained eye should be able to decipher similar qualitative information from diffraction patterns recorded with photographic techniques. In a situation similar to the above case, the diffraction spots recorded on a film will have different shapes specifically in the direction(s) perpendicular to the radial direction.

3.6 Hexatic and crystal smectic phases

The hexatic phases discussed in the first chapter are intermediate between the *liquid* smectic (SmA and SmC) and crystalline smectic phases. To unequivocally determine their structure, one must have a well-oriented sample. Here, well-oriented means not only that the smectic planes are oriented parallel to each other, but also that the lattice axes of the in-plane structure be well defined. Evidently, it is a challenging task to prepare well-oriented samples in more than one direction. One method that has been successfully used with the tilted hexatic phases involves the use of the freely suspended film technique discussed in detail in Chapter 8. An alternative technique, adopted in early days of liquid crystal research, relies on preparation of single crystals of the underlying crystalline solid phase and then slowly melting them by heating into the hexatic phase(s) [14]. However, due to low viscosity of these phases, the sample rapidly misaligns and the results are only semi-quantitative.

Identification of a hexatic phase by simple diffraction methods is less complex than the two methods mentioned above but not completely reliable. It is possible to start with a bulk sample magnetically aligned in a high temperature phase (or even a powder sample). As a hexatic phase is approached, the in-plane structure slowly develops resulting in increased scattering at large angles in the proximity of the 20.5° *in-plane* ring [15]. This ring gradually becomes sharp. In the case of a tilted hexatic phase, since the reflections from (1 1 0) and (2 0 0) are no longer degenerate, internal structure of the large angle ring is observed. This is indicative of a short range positional order within the smectic planes. This pattern is distinguishable from the crystalline smectic phases (B, E, G, H, K, etc.) [16] in which many other peaks appear due to long range in-plane order. In these crystalline phases, there is a tendency for all ring patterns to develop a granular appearance due to the formation of microscopic crystal domains.

3.7 High-resolution x-ray diffraction

High resolution in x-ray experiments is achieved with the use of a pair of single crystals, one as a monochromator and the other as an analyzer. Thanks to the semiconductor industry, excellent quality single crystals of silicon and germanium are available at relatively low costs. Crystals are cut with their physical surface parallel to the selected set of crystal planes, most often the (1 1 1) planes. They are mechanically polished using an aqueous suspension of fine abrasive aluminum oxide particles or diamond paste and then chemically etched to improve the efficiency. These crystals can easily give 80% reflectance. Sometimes asymmetrically cut or channel cut (also known as a multiple bounce monochromator) crystals [1] are used.

When a well-collimated polychromatic beam of x-rays is made incident on a single crystal, say of silicon, the Bragg diffracted beam of a specific wavelength emerges at a well-defined angle, determined by the Bragg law. For incidence angle close to the Bragg angle, the beam penetrates the sample and only a finite number of crystal planes near the surface are probed in the process. As a result, the reflected beam has a finite angular width known as the Darwin width [17] which depends on the nature of the crystal as well as the selected set of planes.

The angular divergence of the reflected beam depends on the Darwin width of the crystal which is much narrower than possible with the use of slits. Furthermore, while the physical cross-section of the beam can be relatively large, high resolution is achieved without sacrificing the intensity as would be the case with the use of narrow slits. For example, the (1 1 1) reflection from germanium has an efficiency approaching 80% and a width $\Delta(2\theta) \sim 0.007°$ or $\Delta q = 4 \times 10^{-4} \text{Å}^{-1}$. Silicon crystals give approximately 2.5 times better resolution.

The x-ray diffractometer is placed such that the beam reflected by the monochromator crystal passes through the point of intersection of θ-, ϕ-, and χ-circles. The sample is placed at this point so that the scattering volume remains the same under various rotations during scans.

The second crystal (referred to as the analyzer) which matches the monochromator is mounted on the 2θ (or the detector) arm. The pair of monochromator and analyzer crystals is most often used in the *non-dispersive* geometry in which the two crystals reflect the x-ray beam in opposite senses. Reflection from the analyzer crystal recombines the $K_{\alpha 1}$ and $K_{\alpha 2}$ lines of a conventional source separated by the monochromator. In this geometry, all the measured diffraction and dispersion is caused by the sample, rendering the interpretation and analysis of the data simpler. The dispersive geometry is useful in certain situations that have not so far arisen in the field of liquid crystals. Another advantage of this geometry is that a stray photon arriving at the analyzer which did not originate at the sample will be incident at an angle different from the Bragg angle and therefore will not be detected. This naturally reduces the background and improves the signal to

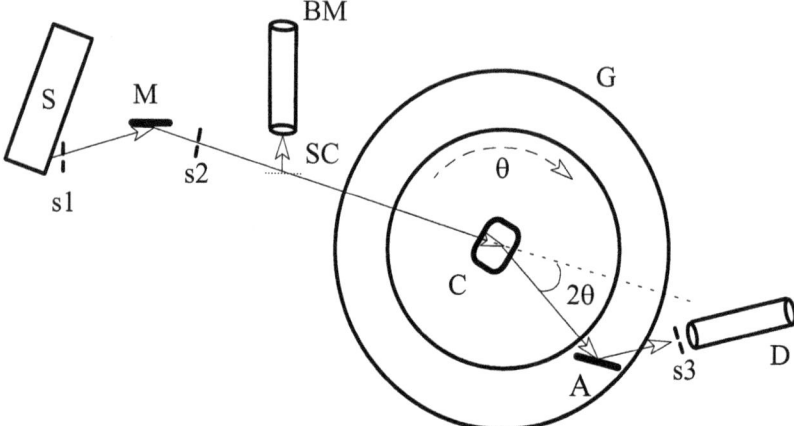

Figure 3.14.   High-resolution experimental set-up. Here: S, x-ray source; S1, S2, and S3, collimating slits; M and A are monochromator and analyzer crystals; SC, a film target which scatters a small fraction of the beam into the beam monitor BM; D, detector; G, goniometer; and C, sample chamber.

noise ratio. Evacuated flight paths are used in high-resolution experiments to reduce the absorption and scattering by air. This minimizes the background noise. Typical background count rates are less than 1 photon/s arising, in part, from the photocell's dark current and cosmic radiation.

A set-up for a high-resolution experiment using a scintillation detector is shown in Fig. 3.14. The beam diffracted by the sample Bragg scatters from the analyzer into the detector. The analyzer A and the detector D are mounted on the same stage (not shown) so that they move together during a scan. Often it is advantageous to place a thin film SC of, say Mylar or Kapton, which scatters a small fraction of the incident intensity into the beam monitor BM to measure the incident x-ray flux. This detector is referred to as the *monitor*. With its help, the intensity of scattered x-rays can be corrected for any changes in the operating power of the source and all scans scaled to the same incident flux. Additionally, if scans are performed in the *monitor mode*, where the counting of photons is done until the monitor accumulates a fixed number of counts instead of a time interval, it is possible to temporarily interrupt a scan or change source power in the middle of a scan, if need be, without affecting a scan.

A scan of the direct beam, also known as a *zero-arm* scan, with Ge monochromater and analyzer is shown in Fig. 3.15. The width of a zero-arm scan gives the instrumental resolution in the scattering plane. Since the actual line shape measured in an experiment is dependent on the resolution func-

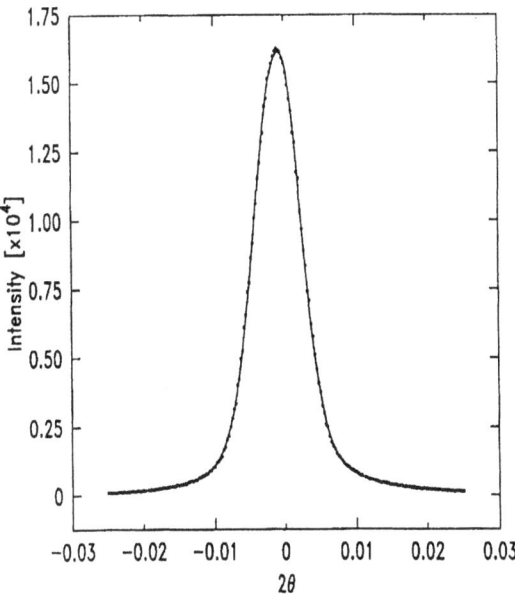

Figure 3.15.   Zero-arm scan with Ge-Ge resolution. The solid line represents fit to
a sum of three Lorentzians.

tion of the instrument, it should be fully determined prior to conducting the experiment. The instrument resolution function consists of three functions which characterize the resolution in the three directions. Depending on the value of the Bragg angle θ for a specific peak, the resolution parallel and perpendicular to the scattering vector has widths of $\Delta q_\perp = \Delta q \sin\theta$ and $\Delta q_\parallel = \Delta q \cos\theta$, in a manner similar to components of a vector $\Delta\mathbf{q}$. The resolution in the direction perpendicular to the scattering plane depends on the convolution of angles subtended at the sample by the (effective) heights of the slits before and after the sample. It can be either calculated from the experimental geometry or measured by putting a perfect crystal in place of the sample and conducting a χ-scan. Such a scan conducted with a silicon crystal is shown in Fig. 3.16. The in-plane resolution function is usually modelled by a sum of three Lorentzians [18, 19] (fit in Fig. 3.15) while Gaussian shape fits the best to the out-of-plane resolution. The shape of the resolution functions used in data analysis will depend on the actual experimental set-up and is taken to be the one that gives the best fit, with the smallest number of adjustable parameters, to the (in- and out-of-plane) scans performed on known (perfect) samples.

In liquid crystal studies, the most often probed reflection arises from the

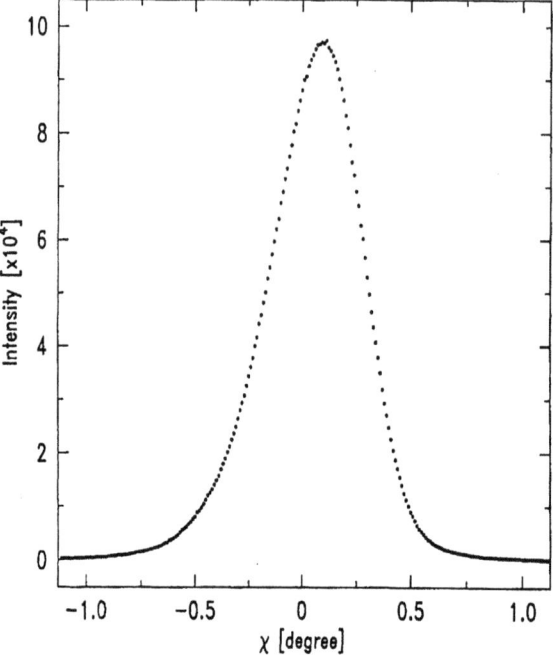

Figure 3.16. χ-scan performed with silicon crystal in sample position.

smectic layers. It appears in the range of $2\theta = 1°$ to $4°$. At these angles, $\Delta q_{\perp} \cong 10^{-1} \Delta q_{\parallel} = \Delta q$. Unfortunately, the 10 times narrower resolution along the in-plane perpendicular direction does not necessarily translate into more precise information along that direction because, in most cases, the smectic phase is not perfectly aligned, i.e. it has *mosaicity* or mis-oriented layers. The $q_{\perp}$-scans are taken along the direction tangential to the ω-scans (see Fig. 3.4). Consequently, the width of perpendicular scans is a projection of the mosaic width onto that direction. Thus the *effective* resolution function is determined by the sample mosaic which is the narrowest $q_{\perp}$-scan, and invariably the scan conducted closest to the transition temperature when the correlations become infinite. The effective $q_{\perp}$-resolution function is most conveniently described by a Gaussian function.

To summarize, the resolution function $R(\mathbf{q})$ is a product of three functions: (1) R_x along the longitudinal direction, normally a sum of three Lorentzians; (ii) R_y along the perpendicular direction describing the sample mosaicity with a Gaussian; and (iii) R_z in the vertical direction and modeled by a Gaussian function. The experimentally measured spectra and line shape also depend on the form factor. The measured intensity $I(\mathbf{q})$ can be written as

$$I(\mathbf{q}) = F(\mathbf{q})S(\mathbf{q}) \otimes R(\mathbf{q}),$$

where $F(\mathbf{q})$ and $S(\mathbf{q})$ are the form and structure factors, respectively, and the symbol $\otimes$ means convolution. The structure factor is a function of positional order correlation lengths. The form factor depends on the shape of the constituting units which are rod-shaped molecules in thermotropic liquid crystals. However, while the often used assumption that the effect of the form factor can be disregarded for thermotropic liquid crystals as it is a slowly changing function remains valid, it does not hold true in general. It has been shown [20] that it should be taken into account for more complex lyotropic liquid crystals. Another factor that influences the measured lineshape is energy broadening. Energy broadening arises from the fact that various lines in the x-ray spectra of conventional sources have finite widths. For example, the $K_{\alpha1}$ and $K_{\alpha2}$ lines of copper have widths of 3.2 and 3.8 eV respectively.

3.8 Positional order correlations at the nematic to smectic-A phase transition

The nematic to SmA phase transition is one of the most interesting and carefully studied liquid crystal transitions because of its similarity [21] with superfluid (i.e., helium and superconductor) transitions. To paraphrase, it belongs to the same *universality class* [22] as the superfluid systems. It must be pointed out that in spite of the similarities, it has several unique features not found in the other systems. For example, contrary to popular belief, the smectic order is not truly long range. A Landau–Peierls instability exists in the SmA phase. Furthermore, the smectic order grows with temperature at different rates in the three spatial directions. The smectic correlations grow faster along the director than in the perpendicular direction. In simpler language, as the nematic phase is cooled towards the phase transition temperature T_c, the number of layers in a spontaneously formed smectic-like correlation volume grows faster than the area of the layers. What is of interest and the subject of numerous investigations is the rate of growth of the sizes $\xi_\parallel$ and $\xi_\perp$, or correlation lengths in the directions parallel and perpendicular to the director (or layer normal) as the transition is approached from above. It is found that the correlation lengths diverge with a single power law, i.e., $\xi_{\parallel,\perp} \sim t^{-\nu_{\parallel,\perp}}$; here $t = [(T - T_c)/T_c]$ is the reduced temperature. The goal of experimental studies has been to measure the correlation lengths and the associated exponents, recognize the trends and universal features, and interpret the results in the light of predictions of various theoretical calculations [23]. Since x-rays couple to the electron density and thus to the mass density, x-ray scattering is the technique of choice to directly probe the growth of smectic order.

There has been a wealth of high-precision light, x-ray scattering and heat capacity data accumulated over the past 25 years for this transition in a large number of compounds. But the transition still remains poorly understood. Our goal in this section is not to review the results or provide an understanding of the transition itself but to give a glimpse of how the high-resolution x-ray experiments have been performed and what the results obtained from them have been able to contribute to its understanding.

For the purposes of illustration, we will use the results of a previous study [24] of a mixture of the sixth (90 wt%) and seventh (10 wt%) members of the dialkylazoxybenzene (DAAOB) [25] for which the nematic phase exists over a wide temperature range of ~40 K and the transition is second order. The tricritical point, where this transition becomes first order, is at much higher concentrations of the seventh homolog. The experimental results for this system can be considered to be free from the influence of the tricritical point. The samples were degassed to remove solvents and absorbed gases by repeated melting and freezing in vacuum to enhance thermal stability. They were sealed between two 8 μm thick sheets of Mylar. The stretch direction of Mylar sheets, which is a consequence of their manufacturing process, was kept parallel to each other and parallel to the applied magnetic field used to align the director in a direction roughly perpendicular to the x-ray beam and in the scattering plane, as shown in Fig. 3.17. Temperature of the sample was raised to the isotropic phase and then lowered slowly in the presence of a magnetic field.

As a rule of thumb, the measurements must be made over three decades of reduced temperature to obtain reliable values of the critical exponents. This requires that the sample temperature be controlled to a precision of at least ± 1 mK. The data should be taken starting from approximately 3 K above the transition to less than 3 mK to cover an approximate range of reduced temperature from 10^{-2} to 10^{-5}. Furthermore, it is highly desirable to make measurements at a large number of points evenly spaced in ln (t). At every temperature, one needs to conduct a $q_{\parallel}$- and a $q_{\perp}$-scan. Since the value of T_c is not normally known with mK precision and it almost always drifts with time due to increasing decomposition, the first couple of scans

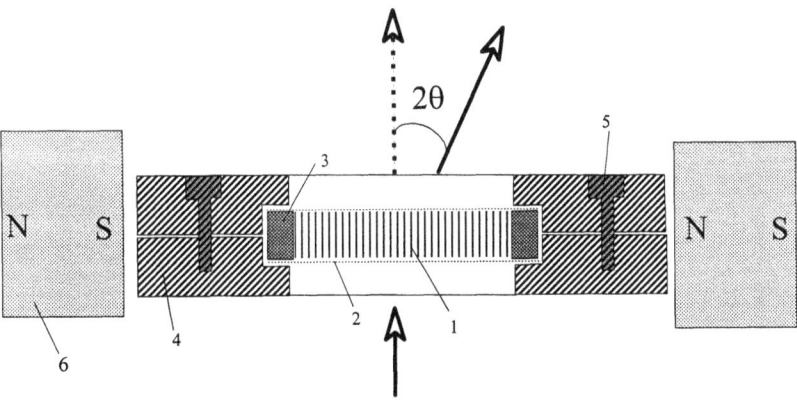

Figure 3.17. Sample assembly: 1, aligned smectic layers; 2, Mylar sheets; 3, Teflon o-ring; 4, aluminium cell; 5, screws for sealing the cell; 6, permanent magnets; arrows represent the incident and transmitted beams, and the diffracted beam at an angle of 2θ.

are used to take stock of the situation by determining T_c, its rate of drift, and the degree of alignment. The latter is measured from the minimum width of the $q_\perp$-scans. The width $\Delta q_\perp$ of the peaks obtained in these scans continues to get smaller until the transition is crossed as shown in Fig. 3.18. On the SmA side, $\Delta q_\perp$ increases and in general changes irregularly with temperature because of the misorienting effects of the sample's thermal expansion and inability of the field to maintain the alignment. If these three parameters are in acceptable range, which depends on how ambitious an experimentalist is, a series of serious scans is then performed. It is important to keep track of the time at which each scan is done. This is essential if the transition temperature is to be corrected for the drift in T_c.

The data is analyzed by fitting to the convolution of three resolution functions, energy broadening, and the structure factor which has the form

$$S(\mathbf{q}) = \frac{\sigma_0}{1 + (q_\parallel \xi_\perp)^2 + (q_\perp \xi_\perp)^2 + C(q_\perp \xi_\perp)^4}.$$

Here, σ_0 is the smectic susceptibility; q and ξ with appropriate labels represent the scattering wave-vector components and the correlation lengths in the directions parallel and perpendicular to the smectic layer normal. The fourth power term in the denominator with coefficient C is included in this otherwise Lorentzian line shape because it has previously been empirically found [26] to be necessary to fit the data. Near the transition, the peaks obtained in the two-dimensional reciprocal space are very sharp and their

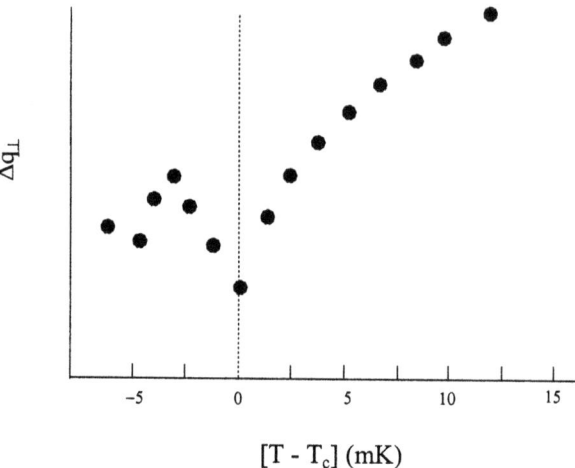

$[T - T_c]$ (mK)

Figure 3.18. Temperature dependence of $\Delta q_\perp$ scan near the transition temperature.

widths become resolution limited making it difficult to conduct the $q_\parallel$-scan at exactly $q_\perp = 0$ and the $q_\perp$-scan at the precise center of the $q_\parallel$-peak. This is because the smallest angle steps, typically $0.0025°$, that can be taken during an experiment are large! As a result, the width and thus the correlations will not be accurate. It becomes imperative [24] to analyze the two scans simultaneously to overcome this difficulty.

The scans and fits obtained at different reduced temperatures for the DAAOB system are shown in Fig. 3.19. From the fits to experimental data, the two correlation lengths, the smectic susceptibility, and values of coefficient C are extracted. The correlation lengths are multiplied with the scattering vector q_0, corresponding to the peak position, to make them dimensionless and plotted as a function of reduced temperature on a log–log scale along with σ_0 as shown in Fig. 3.20. The straight lines passing through the data represent a single power law behavior of σ_0, $\xi_\parallel$, and $\xi_\perp$ with powers γ, $\nu_\parallel$, and $\nu_\perp$, respectively.

The two correlation lengths are found to be different from each other. Their values far from the transition temperature are comparable to the size of molecules. These values correspond to the *bare* values, $\xi_\parallel^0$ and $\xi_\perp^0$, of the correlation lengths. Their values grow at a relatively rapid but different pace with different powers of reduced temperature. The fact that they possess different values and grow at different rates are hallmarks of this anisotropic critical behavior. The values obtained for the sample under discussion are listed in Table 3.1.

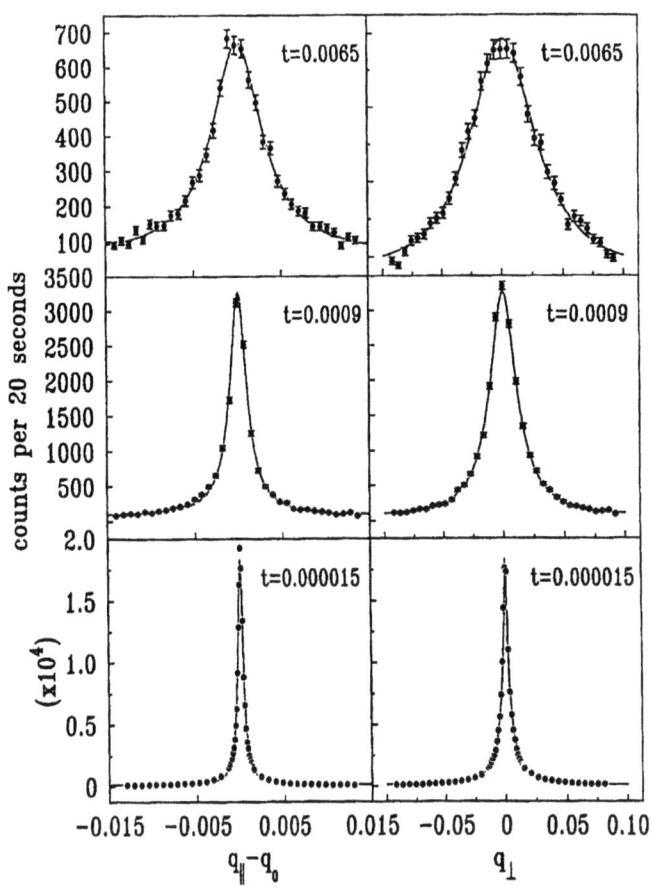

Figure 3.19. $q_{\parallel}$- and $q_{\perp}$-scans at three reduced temperatures. Peaks become sharper in both directions but the $q_{\perp}$-peaks remain nearly 10 times broader than $q_{\parallel}$-scans.

It should be pointed out that the values of the correlation lengths and the three exponents are found to depend on the material. Furthermore, the coefficient of the fourth order term in the structure factor, C, is a function of reduced temperature. Its value becomes smaller near the transition. No two materials appear to have the same values for all exponents. A number of elegant and cumbersome theories have been developed in attempts to explain the critical behavior at this transition. They either predict isotropic (i.e., $\nu_{\parallel} = \nu_{\perp}$) or anisotropic (i.e., $\nu_{\parallel} = 2\nu_{\perp}$) divergence. The experimental results are not in agreement with either prediction.

While it may appear that the situation is hopeless, there seems to be some

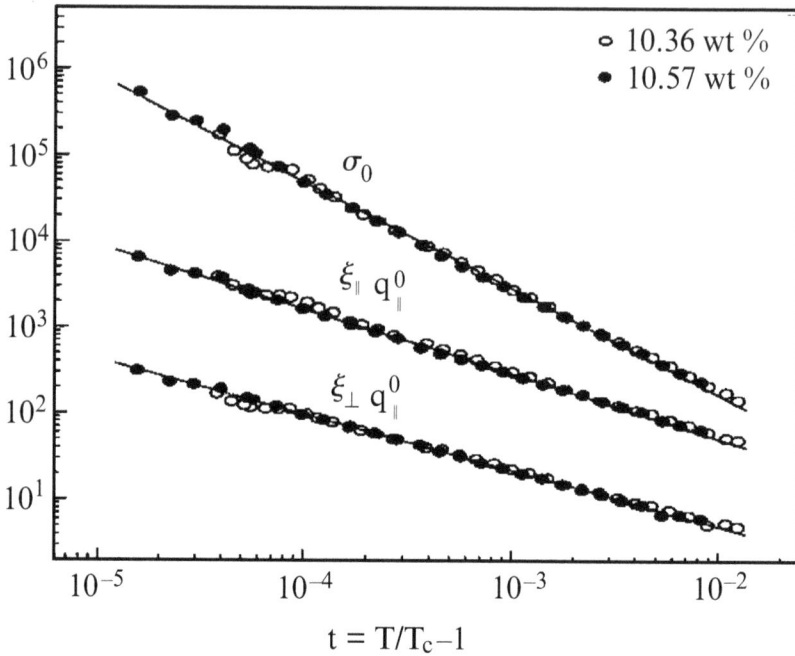

Figure 3.20.   Temperature dependence of parallel and perpendicular correlation
lengths and smectic susceptibility near the N-SmA transition.

Table 3.1

$\xi^o_\parallel$	$\xi^o\perp$	γ	$\nu\parallel$	$\nu\perp$
7.2Å	1.0Å	1.24±0.05	0.75±0.03	0.65±0.03

order that is beginning to emerge from the data. (i) The values of the exponents monotonically increase with decreasing value of the McMillan ratio
$r = T_{NA}/T_{NI}$ [23], i.e., with increasing width of the nematic phase. Here, T_{NA}
and T_{NI} are the transition temperatures from the N to SmA and the N to I
phase. (ii) As the value of r decreases, the net anisotropy, measured as the
ratio of the two exponents, decreases [24] but never reaches unity. Of course,
the reason is that as the nematic width increases the coupling between the
smectic and nematic order parameters decreases. Consequently, the transition moves farther from the tricritical point which is known to exist for
systems with a narrow nematic range. (iii) In spite of all the apparent chaos,
the hyperscaling relation $\nu_\parallel + 2\nu_\perp = 2 - \alpha$, is satisfied within experimental
errors by the two correlation lengths and the specific heat (α) exponents of

nearly all materials. For a complete discussion of this problem, the reader is referred to excellent references [27].

While x-ray diffraction has provided an enormous body of structural information on liquid crystals, new x-ray techniques are being applied to liquid crystalline systems. These advances are aided by the construction of new and more powerful facilities such as second and third generation synchrotron sources. Resonant x-ray scattering [28] and coherent x-ray scattering [29] are two of several new techniques being developed which should reveal important new details about this fascinating class of materials. In this rapidly advancing field, readers interested in x-ray studies should keep abreast with recently published papers in scientific journals, such as *Physical Review E, Physical Review Letters*, and *Liquid Crystals*.

References

1. B. D. Cullity, *Elements of X-ray Diffraction*, 2nd Edn., Addison Wesley Publishing Co., Inc., Reading, MA (1978).
2. Any book on solid state physics should have adequate description of the reciprocal space, e.g., C. Kittel, *Introduction to Solid State Physics*, 7th Edn., John Wiley & Sons, Inc., New York (1999).
3. *International Tables for X-ray Crystallography*, Vol. C, 2nd Ed., Kluwer Academic Publishers, Netherlands (1999).
4. Available from places such as *Edmund Scientific* in different sizes and shapes.
5. B. Bahadur, in *Liquid Crystals, Applications and Uses*, vol. 1, B. Bahadur (Ed.), World Scientific, New Jersey (1990) p. 199.
6. J. Baruchel, J. L. Hondeau, M. S. Lehmann, J. R. Regnard, and C. Schlenker (Eds.), *Neutron and Synchrotron Radiation for Condensed Matter Studies*, Springer-Verlag, Berlin (1993).
7. A. J. C. Wilson, *Elements of X-ray Crystallography*, Addison-Wesley Publishing Co., Reading, MA (1970).
8. M. J. Freiser, *Phys. Rev. Lett.* **24**, 1041 (1970); R. Alben, *Phys. Rev. Lett.* **30**, 778 (1973); D. W. Allender, M. A. Lee, N. Hafiz, *Mol. Cryst. Liq. Cryst.* **124**, 45 (1985); J. Melthete, L. Liebert, A. M. Levelut, and Y. Galerne, *C. R. Acad. Sci. Paris* **303**, 1073 (1986); G. Shinouda, Y. Shi, and M. E. Neubert, *Mol. Cryst. Liq. Cryst.* **257**, 209 (1994).
9. S. Kumar, *Phys. Rev. A* **23**, 3207 (1981).
10. K. C. Chu and W. L. McMillan, *Phys. Rev. A* **15**, 1181 (1977); B. W. van der Meer and G. Vortogen, *J. Phys. (Paris), Colloq.* **40**, C3-222 (1979); C. C. Huang and S. C. Lien, *Phys. Rev. Lett.* **47**, 1917 (1981).
11. B. R. Ratna, R. Shashidhar, and V. N. Raja, *Phys. Rev. Lett.* **55**, 1476 (1985).
12. S. Kumar, Li Chen, and V. Surendranath, *Phys. Rev. Lett.* **67**, 322 (1991).
13. S. Kumar and P. Patel, *Cond. Matt. News* **2**, 9 (1993); P. Patel, S. Kumar, and P. Ukleja, *Liq. Cryst.* **16**, 351 (1994).
14. J. J. Benattar, J. Doucet, M. Lambert, and A. M. Levelut, *Phys. Rev. A* **29**, 2505 (1979); J. Doucet and A. M. Levelut, *J. Phys. (Paris)* **38**, 1163 (1977).
15. E. Gorecka, L. Chen, W. Pyzuk, A. Krowczynski, and S. Kumar, *Phys. Rev. E* **50**, 2863 (1994).
16. A. J. Leadbetter, M. A. Mazid, B. A. Kelley, J. W. Goodby, and G. W. Gray,

Phys. Rev. Lett. **43**, 630 (1979); P. A. C. Gane, A. J. Leadbetter, and P. G. Wrighton, *Mol. Cryst. Liq. Cryst.* **66**, 247 (1981).

17. R. W. James, *The Optical Principles of Diffraction of X-rays*, Ox Bow Press, Woodridge, (1948) pp. 52–66.

18. L. Chen, *High-resolution X-ray Diffraction Studies of the Nematic to Smectic-A Phase Transition and the Frustrated Smectic-A Phases*, Ph.D. Dissertation, Kent State University (1991).

19. C. Safinya, Ph.D. Dissertation, MIT (1981).

20. S. T. Shin, J. D. Brock, M. Sutton, J. D. Litster, and S. Kumar, *Phys. Rev. E (Rapid Commun.)* **57**, 5644 (1998).

21. P. G. deGennes, *Solid State Commun.* **10**, 753 (1972).

22. M. E. Fisher, S. K. Ma, and B. G. Nickel, *Phys. Rev. Lett.* **29**, 917 (1972); S. K. Ma, *Phys. Rev. A* **7**, 2172 (1973); M. Suzuki, *Prog. Theor. Phys.* **49**, 1106 (1973); M. E. Fisher and A. Aharony, *Phys. Rev. Lett.* **30**, 559 (1973).

23. W. L. McMillan, *Phys. Rev. A.* **4**, 1238 (1971); K. K. Kobayashi, *Phys. Lett. A* **31**, 125 (1970); T. C. Lubensky and J. H. Chen, *Phys. Rev. A* **17**, 366 (1978); C. Dasgupta and B. I. Halperin, *Phys. Rev. Lett.* **47**, 1556 (1981); D. R. Nelson and J. Toner, *Phys. Rev. B* **24**, 363 (1981); B. Andereck and B. R. Patton, *Phys. Rev. Lett.* **69**, 1556 (1992).

24. L. Chen, J. D. Brock, J. Huang, and S. Kumar, *Phys. Rev. Lett.* **67**, 2037 (1991).

25. E. F. Gramsbergen and W. H. deJeu, *J. Chem. Soc. Faraday Trans.2* **84**, 1015 (1988).

26. J. Als-Nielsen, R. J. Birgeneau, M. Kaplan, J. D. Litster, and C. R. Safinya, *Phys. Rev. Lett.* **39**, 352 (1972).

27. C. W. Garland and G. Nounesis, *Phys. Rev. E* **49**, 2964 (1994); M. A. Anisomov, *Mol. Cryst. Liq. Cryst.* **162A** (1988); H. K. M. Vithana, G. Xu, and D. L. Johnson, *Phys. Rev. E* **47**, 3441 (1993).

28. P. Mach, R. Pindak, A. M. Levelut, P. Barois, H. T. Nguyen, C. C. Huang, and L. Fjurenlid, *Phys. Rev. Lett.* **81**, 1015 (1998).

29. S. B. Dierker, R. Pindak, P. M. Fleming, I. K. Robinson, and L. Berman, *Phys. Rev. Lett.* **75**, 449 (1995).

4

Physical properties

PANOS PHOTINOS

Southern Oregon University, Ashland, OR 97520, USA

Introduction

The physical properties of a liquid crystal depend on the phase it is in. Consider a room temperature nematic (or cholesteric) liquid crystal in a vial. The sample will have the appearance of an ordinary fluid that scatters light strongly. If observed between crossed polarizers, the sample will appear bright and colorful. If the vial is tilted and/or rotated between the polarizers, blue, yellow and other color patterns will be observed, especially near the surface of the liquid. These color patterns, which depend on the direction of flow, are also observed in some other materials. This may be verified by placing a sheet of crystalline mica between crossed polarizers. Similar patterns could also be observed by stretching a film of plastic wrap between crossed polarizers. The liquid crystal sample, the plastic wrap and the mica sheet are very different, at least in terms of their rigidity. The mica is rigid, since the distance between molecules is essentially fixed. The plastic wrap deforms easily indicating that the molecules can move. Lastly, the liquid crystal can be poured and forms droplets just like water. However, they exhibit similarities in their optical appearance between crossed polarizers. What is the common feature of these three materials which is responsible for their colorful appearance?

In solid crystals, the positions and (for larger molecules) the orientations of the molecules obey regular patterns, the molecules exhibit both translational and orientational order (see Chapter 1). The stretched plastic wrap and the liquid crystal are both composed of large molecules and may exhibit orientational order. For example, stretching the plastic film aligns the long polymer molecules, and naturally, the optical properties in the direction of the stretch will be different from the perpendicular direction. The nematic or cholesteric liquid crystal also has orientational order, while the isotropic phase of the same material does not. Thus it is the order that is responsible for the observed behavior. The majority of liquid crystalline phases can possess translational order in addition to orientational order.

The orientational order in liquid crystals derives from the anisotropic shape of the molecules. From the fundamental point of view, it is important to understand the macroscopic properties of these materials in terms of their microscopic or molecular characteristics. Theories based on microscopic properties of the molecules have been successful in accounting for some of the physical properties, such as the magnetic behavior, but have not reached a high enough level of sophistication to explain other properties such as phase transition temperatures, elastic behavior, etc.

In what follows we will discuss some of the physical properties of primarily the nematic phase, which is the simplest of all liquid crystal phases. The description given here should serve as a step ladder to better understanding of the standard textbooks on liquid crystals, such as Chandrasekhar [1], de Gennes and Prost [2], de Jeu [3] and Vertogen and de Jeu [4], and the text by Collings and Patel [5]. In this chapter we will describe the basic equations and the principles of the experimental methods used to measure and analyze the properties of liquid crystals. In order to emphasize the concepts, the descriptions are simplified. For the specifics of a given experiment or technique the reader should read the original work, or other more recent publications. We will also review orientational order, the notion of anisotropy, and the mathematical tools used to represent physical properties of liquid crystals. We will then consider the following properties and their measurement in more detail: dielectric, optical and diamagnetic properties; elastic and viscosity coefficients; electrical and thermal conductivities and density.

The equations are written in SI units. This fact should be kept in mind when comparing results from other works, especially when discussing electric and magnetic properties.

4.1 Definitions and physical coefficients for anisotropic materials

The director, n

In a perfectly aligned nematic liquid crystal, all of the molecules point in the same direction. What determines this direction? First, consider an unaligned nematic. If one could observe a volume of approximately one micrometer in radius around a point in the nematic, one would find that the molecules are all pointing *preferentially* in the same direction. This is in spite of the fact that at any given temperature T the molecules possess kinetic (thermal) energy of the order of $k_B T$ (k_B is the Boltzmann constant). Over an 'infinitely long time' (say, a few μs), on average the molecular axes tend to point in a certain direction which is called the *local* director. A few micrometers away, the molecular axes favor a different direction. Obviously, the molecules are orientationally ordered, but the preferred direction varies from point to point throughout the bulk sample. In fact, the preferred direction could assume all possible orientations. This situation is similar to that observed in an unmagnetized ferromagnet where each domain has a characteristic direction, but the overall magnetization is zero. In analogy with ferromagnets, by the application of appropriate external fields or alignment layers, the nematic sample can be macroscopically aligned so that the preferred direction becomes the same for the entire monodomain sample. The unit vector **n** will be used to specify the preferred direction in a nematic and we will refer to it as the director. The director indicates the axis of rotational symmetry of the nematic.

The nematic order parameter

Assume that the direction of the preferred orientation at one point in the sample is along the z-axis. Assume further, that the long axis of a molecule makes an angle θ with respect to the preferred direction, as shown in Fig. 4.1.

The orientations θ and $\theta + \pi$ are equivalent. This means physically that a molecule standing upside-up is identical to it being upside-down as it does not affect the energy of the system. This also means that our measure of order, i.e., the nematic order parameter, must be an even function of the angle θ. The most common and a simple choice of nematic order is S, defined as

$$S = \frac{1}{2} \langle 3\cos^2\theta - 1 \rangle, \qquad (4.1)$$

where $\langle \cdots \rangle$ indicates average value. $S = 0$ corresponds to the isotropic phase, where all orientations of the molecular axis are equally probable. The nematic phase is defined as one with a non-zero value of S. This requires the molecules to orient preferentially along a specific direction in space. Based on Eq. (4.1), S can take values between 0 and $-\frac{1}{2}$, corresponding to molecules which prefer to orient perpendicular to the director and 0 and 1. In nematics with positive values of S, the molecules prefer to orient parallel to the director. In both cases, the numerical value of S decreases upon approaching the isotropic phase, and vanishes discontinuously at the transition to the isotropic phase. This corresponds to a temperature T_{NI} also

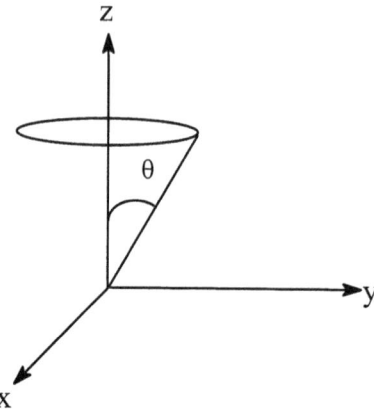

Figure 4.1. The coordinate system of an oriented uniaxial liquid crystal. The director is parallel to the z-axis and a molecule's long axis makes an angle θ with the director.

known as the clearing point. The typical behavior of the order parameter S with temperature is shown in Fig. 4.2.

The preferential alignment of molecules in the nematic phase makes the physical properties of nematic liquid crystals direction-dependent. For example, the value of the index of refraction for polarization directions parallel and perpendicular to the director are different. Anisotropy is present in the microscopic physical properties of the material. Suppose, for example, that we apply a magnetic field to a nematic sample. We will assume that on the molecular level the induced molecular magnetic moment is larger when the field is parallel to the molecular axis than when it is perpendicular. This will be the case if the numerical value of the diamagnetic susceptibility parallel to the molecular axis is larger than in the perpendicular direction. The difference between the two values of the susceptibilities defines the susceptibility anisotropy at the molecular level. The molecular anisotropy manifests itself at the macroscopic scale also. We will discuss the relationship between the susceptibility and the orientational order and the molecular parameters later. First, we will summarize the mathematical representation for the susceptibility and other direction-dependent properties. We will also review the procedure to transform these quantities between different coordinate systems.

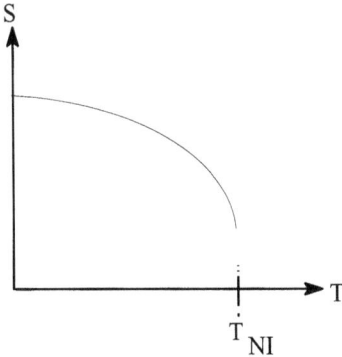

Figure 4.2. Typical nematic order parameter, S, as a function of temperature for a thermotropic liquid crystal. Note that S discontinuously vanishes at T_{NI} and is zero in the isotropic phase.

Representing direction-dependent quantities

In this section we will review how to represent the simpler direction-dependent properties using the electric susceptibility as an example. When an electric field, $\mathbf{E}$, is applied to a dielectric, the material develops an induced polarization density, $\mathbf{P}$, and to a first approximation (for weak electric fields)

$$\mathbf{P} = \varepsilon_0 \chi^e \mathbf{E}, \tag{4.2}$$

where χ^e is the electric susceptibility and ε_0 is the permittivity of free space. In the isotropic phase, i.e., above its clearing point, χ^e is a single scalar quantity, and $\mathbf{P}$ is parallel to $\mathbf{E}$. In the nematic phase $\mathbf{P}$ is not necessarily parallel to $\mathbf{E}$ and χ^e can no longer be specified by a single scalar. It is a tensor of rank 2, which physically reflects the fact that the induced polarization depends on the direction of the applied field relative to the symmetry axes of the nematic. A tensor of rank 2 in three dimensions has nine components. However, the symmetry of the nematic phase requires that some of these components be related to each other. For the ordinary (uniaxial) nematic liquid phase, the number of independent components is reduced to two.

The principal axes system

In the uniaxial nematic phase, we expect that $\chi^e_{\parallel}$, the susceptibility parallel to $\mathbf{n}$, is different from $\chi^e_{\perp}$, the susceptibility perpendicular to $\mathbf{n}$, and all directions perpendicular to $\mathbf{n}$ are equivalent. It is easy to construct the tensor χ^e for this case, with $\mathbf{n}$ along the z-axis:

$$\chi^e = \begin{bmatrix} \chi_\perp^e & 0 & 0 \\ 0 & \chi_\perp^e & 0 \\ 0 & 0 & \chi_\parallel^e \end{bmatrix} \tag{4.3}$$

The tensor is symmetric and diagonal, and has only two independent components. Note that for **E** along any of these principal directions **P** and **E** are parallel. The direction of the induced polarization **P** is not generally parallel to the applied field **E**. This is a general feature of anisotropic materials. If we could choose the coordinate system along the symmetry axes of the phase, the so-called principal axes, the susceptibility, and other second rank tensor quantities which we shall discuss later, will become diagonal, as χ^e did in Eq. (4.3). Diagonal forms are easier to manipulate and, understandably, the literature refers to the elements of the diagonal representation.

The lab coordinate system

In many cases the geometry of the experiment selects a lab coordinate system (or frame) that does not correspond to the principal axes of the nematic phase. Data interpretation is easier in a particular frame, and it is useful to know how to transform quantities from the principal axes to the given lab coordinate system, and vice versa. This is discussed in detail in most mathematical physics texts [6, 7]. Briefly, we will use **P** and **E** (without any subscripts) to denote the polarization and electric field in the principal axes system, and add the subscript L to indicate their values in the lab frame. It is well known that one can go from one coordinate system to another by performing a rotation described by an orthogonal matrix $\mathcal{R}$. The inverse of $\mathcal{R}$ is $\mathcal{R}^{-1}$ and we have $\mathcal{R} \cdot \mathcal{R}^{-1} = \mathcal{R}^{-1} \cdot \mathcal{R} = I$, where I is the 3×3 identity matrix, and $\mathcal{R}^{-1}$ is obtained by transposing the rows and columns of $\mathcal{R}$. The vector quantities in the principal axes and the lab system are related by the transformation equations:

$$\mathbf{P}_L = \mathcal{R} \cdot \mathbf{P} \quad \text{and} \quad \mathbf{E}_L = \mathcal{R} \cdot \mathbf{E}.$$

The susceptibility tensor in the lab coordinate system is given by

$$\chi_L^e = \mathcal{R} \cdot \chi^e \cdot \mathcal{R}^{-1}. \tag{4.4}$$

This equation shows how to transform a second rank tensor from the principal axes system to the lab frame, and it is known as a similarity transformation. A useful example is when **n** forms an angle θ wrt the z-axis of the lab system and lies in the yz-plane, as shown in Figure 4.3. In this case one can write

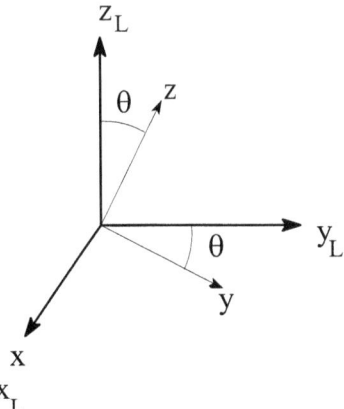

Figure 4.3.   The principal axes and the lab coordinate systems. The principal axes of the material have been rotated by an angle θ about the x-axis; thus z and **n** are parallel.

$$\mathcal{R} = \begin{pmatrix} 1 & 0 & 0 \\ 0 & \cos\theta & \sin\theta \\ 0 & -\sin\theta & \cos\theta \end{pmatrix} \tag{4.5}$$

and

$$\chi_L^e = \mathcal{R} \cdot \chi^e \cdot \mathcal{R}^{-1} = \begin{bmatrix} \chi_\perp^e & 0 & 0 \\ 0 & \chi_\perp^e\cos^2\theta + \chi_\parallel^e\sin^2\theta & (\chi_\parallel^e - \chi_\perp^e)\sin\theta\cos\theta \\ 0 & (\chi_\parallel^e - \chi_\perp^e)\sin\theta\cos\theta & \chi_\perp^e\sin^2\theta + \chi_\parallel^e\cos^2\theta \end{bmatrix}. \tag{4.6}$$

This equation is very useful, and will be applied repeatedly in this chapter, to relate second rank tensor quantities in the lab coordinate system to those in the sample. All that is required is to replace $\chi_\parallel^e$ and $\chi_\perp^e$ by the corresponding values of the second rank tensor representing the physical quantity under consideration.

Anisotropy

The difference $\chi_\parallel^e - \chi_\perp^e$ which appears in the off-diagonal terms is called the dielectric *susceptibility anisotropy*:

$$\Delta\chi^e \equiv \chi_\parallel^e - \chi_\perp^e \tag{4.7}$$

and plays an important role in the study of liquid crystals. To a good approximation, $\Delta\chi^e$ is proportional to S, and measurements of $\Delta\chi^e$ can be

used to determine the order parameter S. Analogous definitions of anisotropy are used for other physical properties, e.g., conductivity, diffusion,
etc. Anisotropies can also be defined for each molecule. However, it should
be noted that the *molecular anisotropies do not depend on S*.

Uniaxial and biaxial systems

The susceptibility and physical properties of *uniaxial* systems (i.e., systems
with one axis of cylindrical symmetry) can be represented by a second rank
tensor, having only two independent components. *Biaxial* systems, on the
other hand, have further reduced symmetry from the isotropic case and
must be represented in the principal axes frame by three parameters. To
help visualize the situation in a biaxial system, consider what happens when
the molecules in the liquid crystal have a long side and a broad side. In such
a case, the molecules may not be free to rotate about their long axes. In fact
it may be that the molecules in the liquid crystal phase have a preferred
direction for the long axes and a preferred direction for the broad sides (or
one of the short axes)! One then needs additional parameters to describe
the order in this phase. In fact, we will need to define S as a *tensor order
parameter*, which is described in the specialized texts [1, 2, 3, 4].

Relationship to sample preparation
Monodomain samples

For measurements of physical properties, monodomain samples are often
required. Aligned thin film monodomain samples can be prepared between
two substrates with proper treatment of their surfaces. These techniques
include coating and/or rubbing the surface, or oblique evaporation of thin
oxide films on to the glass. In some cases, the monodomain liquid crystal
sample is readily achieved by applying an external magnetic or electric field.
This is usually the case with nematic phases. For most thermotropic nematics, a field of 1 T is quite sufficient. If the nematic phase has positive
diamagnetic anisotropy, $\Delta\chi > 0$, then the director aligns parallel to the magnetic field. If $\Delta\chi < 0$, then the director aligns perpendicular to the applied
field. In the latter case, there are infinite directions perpendicular to the
applied field, and special techniques such as spinning the sample about
an axis perpendicular to an externally applied field are often necessary. The
alignment may deteriorate with time, especially in thin samples where
the alignment imposed by the boundaries may be competing with the field
alignment. In addition, during the course of an experiment, other externally applied fields (such as an electric field to measure conductivity), or

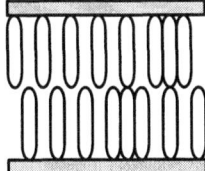 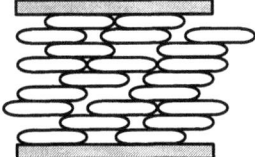

Figure 4.4. Geometry of the director field for homeotropic (left) and homogeneous (right) alignment of liquid crystals.

changes in the temperature, may disturb or even re-align the sample to a considerable extent. As a general rule, whenever visual inspection is available, the sample alignment should always be verified by direct observation under polarized light. The relative ease of aligning a nematic can be exploited for aligning other phases. The alignment of phases exhibited by a liquid crystal at lower temperatures can be achieved by aligning it in the nematic phase first, and then cooling to the desired phase. This technique works well if the phase transitions are continuous.

Homeotropic, homogeneous, and planar alignment

For simplicity consider samples prepared between two parallel glass plates. The following terminology is used to specify the liquid crystal configuration within the cell:

1. *homeotropic* alignment: the direction of **n** is perpendicular to the glass plates.
2. *homogeneous* alignment: the direction of **n** lies along a specific direction in the plane of the glass plates,
3. *planar* alignment: the direction of **n** lies in the plane of the glass plates with equal probability in all in-plane directions.

The geometry of the homeotropic and homogeneous alignments is shown in Figure 4.4.

The alignment imposed at the glass boundaries prevails throughout the volume of the sample, if the glass surfaces are properly treated and the sample is not too thick (~ 1 mm or less). Practical recipes for glass cleaning, and surface treatment for homogeneous and homeotropic alignments are given in the second chapter of Ref. 3.

4.2 Dielectric properties

When a dielectric material is placed between the plates of a capacitor, the capacitance increases by a factor ε, which is the dielectric permittivity of the material, measured with respect to the permittivity of air, ε_0 (which is essentially the same as that of vacuum). Under these conditions the dielectric material becomes polarized, and the net polarization is related to the strength of the applied field through the susceptibility of the medium by χ^e Eq. (4.2).

Measurement of dielectric constants
Dielectric anisotropy

If the dielectric material is a nematic liquid crystal, then the capacitance measurement will yield a value of $\varepsilon_\parallel$ when the director **n** is normal to the plates, and a value of $\varepsilon_\perp$ when it is parallel to the plates. The difference between the two values is the dielectric anisotropy:

$$\Delta\varepsilon = \varepsilon_\parallel - \varepsilon_\perp. \tag{4.8}$$

$\Delta\varepsilon$ will vanish in the isotropic phase, and will be maximum when the nematic is perfectly ordered, i.e., when $S = 1$. It is also a function of frequency. Like χ^e, the permittivity for the nematic is also a second rank tensor. The dielectric permittivity ε is related to χ^e by

$$\varepsilon = I + \chi^e, \tag{4.9}$$

where I is the unit 3×3 matrix. The expressions for χ^e in the principal axes and in the laboratory system can be readily obtained using the above equation and Eqs. (4.3) and (4.6), respectively. At this point it is natural to ask whether χ^e or the directly measurable ε can be related to the molecular properties and the order parameter. This is indeed the case, however, the derivation is rather complex. The difficulty is that the internal electric field must be related to the external electric field. An elementary discussion is given in Kittel [8]. The analysis of nematic liquid crystals in the Vuks approximation can be found in Ref. 3, Chapters 4 and 5. In this approximation, the two principal dielectric constants can be related to the order parameter, S, and the longitudinal and transverse polarizabilities of the molecule. A 'typical' trend for dielectric constant versus temperature is shown in Fig. 4.5.

The need for ac voltages

To measure the dielectric permittivity a nearly ideal parallel plate capacitor can be constructed using two glass plates coated with a thin conductive layer, e.g., In_2O_3, acting as electrodes. One could calibrate the capacitor, by first filling it with a material of known permittivity, or simply leave it in air.

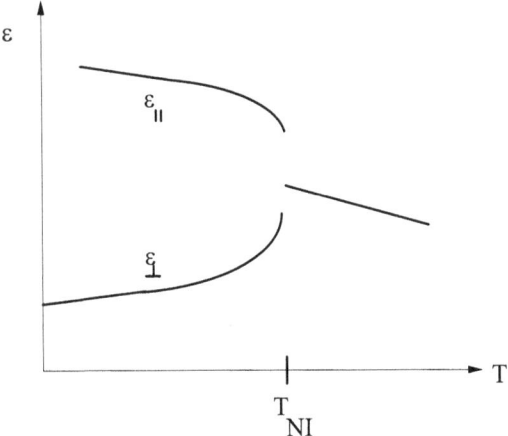

Figure 4.5. Typical behavior of dielectric constant versus temperature for a liquid crystal. At the lower temperature end, $\Delta\varepsilon$ is typically of the order of 1 to 10.

If the cell is then filled with a nematic, and the thickness is not too small, a magnetic field can be used to align **n** parallel (perpendicular) to the plates to determine $\varepsilon_\parallel (\varepsilon_\perp)$, and the dielectric anisotropy $\Delta\varepsilon$. *This method works in principle only.* In practice, when a dc voltage is applied to the cell, charges will soon accumulate on the conducting electrodes. The accumulated charge will reduce the effective voltage appearing across the sample, and chemical reactions may also occur, which will corrode the electrodes. A way to avoid these effects is to use ac voltages of about 1–2 kHz. As an added precaution, one could use glass plates which have a thin dielectric coating (which is called a *passivation layer*) on top of the conducting electrodes. This is a common practice in display applications. An added benefit of the use of a passivation layer is that it will prevent leaching out of other ions (such as Na^+) from glass substrates. Depending upon the particular device, it may sometimes be necessary to avoid the use of soda–lime glass and use, for example, borosilicate glass which is less prone to releasing ions. Lastly, there are free ions in the typical liquid crystal. Thus these capacitors are not ideal and can be modeled as a resistor in parallel with a capacitor.

'Ball-park' capacitance and resistance values

It is instructive to estimate approximate values of the capacitance and resistance, and to discuss the practical aspects of a typical experiment. Our aim is to construct a thin transparent cell, which would allow simultaneous optical observation. Assume that we start with clean glass plates coated with a conductive oxide film. The problem of a conducting dielectric in this geometry is discussed in Ref. 9. Essentially we have a parallel plate capacitor and from electrostatics we can calculate the capacitance as: $C = \varepsilon_0 \varepsilon A/d$, where A is the area of the coated glass plates, d is the spacing of the plates, and $\varepsilon_0 = 8.85 \times 10^{-12}$ F/m. Using $A = 10^{-4}$ m^2 (plates of area 1 cm $\times$ 1 cm), and $\varepsilon = 5$, we find

$$C = 443 \text{ pF for } d = 10 \ \mu\text{m} \qquad \text{('thin cell')},$$

$$C = 44 \text{ pF for } d = 100 \ \mu\text{m} \qquad \text{('thick cell')}.$$

The capacitance of course will have different value for different alignments of **n**. Using $\Delta\varepsilon = 1$, we find that the capacitance difference between the homeotropic and homogeneous alignment is

$$\Delta C = 88 \text{ pF} \qquad \text{for the 10 } \mu\text{m cell and}$$

$$\Delta C = 9 \text{ pF} \qquad \text{for the 100 } \mu\text{m cell.}$$

The resistivity of the liquid crystal material at frequency $f = 1$ kHz is typically of the order of 10^6 Ωm. The resistance of our cell is given by

$$R = \rho \frac{d}{A}.$$

Thus, in the examples used above, this translates to: $R = 1$ MΩ for the 100 μm cell and $R = 0.1$ MΩ for the 10 μm cell. For the above estimates we assumed for simplicity that the resistances parallel and perpendicular to **n** are the same.

To characterize the cell one must determine both R and C. There are a number of techniques suitable for doing this. There is a large specialized literature in this area [10] and those who need measurements at frequencies above a few hundred kHz should consult this literature. Furthermore one expects that there will be a frequency dependence to the dielectric constant as well.

We now present one simple low frequency method for determining R and C. This method uses a voltage divider, ac excitation and phase sensitive detection (a lock-in amplifier) to determine R and C as shown in Fig. 4.6.

While one can use a standard lock-in for this measurement it is even easier to use a 'vector' lock-in which gives both the absolute value of the signal voltage and its phase relative to the reference voltage. Analysis of this circuit leads to the following results:

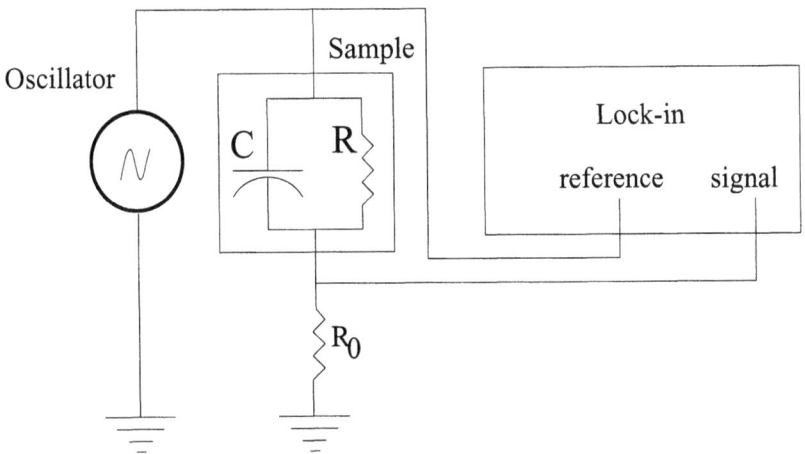

Figure 4.6. A block diagram of the lock-in amplifier technique for measuring the capacitance, C, and resistance, R, of a liquid crystal sample. (See Ref. 11.)

$$\tan\theta = \frac{\omega R^2 C}{R + R_0 + \omega^2 R_0 R^2 C^2}$$

$$\frac{V_{sig}}{V_{source}} = \frac{\left(R^2_0 + (\omega R R_0 C)^2\right)^{\frac{1}{2}}}{\left((R + R_0)^2 + (\omega R R_0 C)^2\right)^{\frac{1}{2}}}$$

where $\omega = 2\pi f$, and V_{sig} is the measured signal voltage and V_{source} is the oscillator (source) voltage. Using the above two equations, we can determine R and C by measuring V_{sig} and the angle θ. For the 100 μm cell, with $R_0 = 1$ MΩ, we find that at 1 kHz, $\theta = 8°$. Using $\Delta\varepsilon = 1$, we find that the phase angle changes by about 2° between the homeotropic and homogeneous alignment. This is a small effect, yet it can be easily measured using a lock-in amplifier, which would read V_{sig} and θ. This method can also be used for measurements versus frequency up to 100 kHz, which is the typical upper limit of the operating range for lock-in amplifiers. Application of this technique to liquid crystals is described in Ref. 11.

Experimental considerations

Defining the cell thickness To define the thickness for thin cells one could use monodisperse polystyrene spheres (e.g., from Polysciences, Inc.) or precision glass fibers (e.g., from E. Merck) as spacers. Teflon or Mylar film can be used for thicker cells. The thickness of the cell should be determined *before* filling the cell. For thick cells one can use a microscope to focus on the inner surfaces of the top and bottom glass plates. The thickness of the empty cell can be determined from the readings of the micrometer of the vertical translation stage of the microscope. For thin cells, one can use interference of light. The interference method can also be used to check the uniformity of the thickness, using monochromatic light. The experimental set-ups for the interference techniques are described in detail in Refs. 12 and 13.

Attaching the leads and sealing First, coated glass plates are commercially available (e.g., Delta Technologies), so there is no need to construct these. There are several options for attaching the leads to the conducting oxide film. One can use indium shot (Johnson Matthey) and a conventional soldering iron. In doing so, one should clean the soldering iron very well after use, otherwise, because indium forms an amalgam with copper, it will not wet the regular solder material. Alternatively, one could start with a new soldering tip, and not use it with any other types of solder. Conductive epoxy (e.g., BIPAX by TRA-CON Inc.) or conductive paint (Fulham Inc.) can also be used for attaching the leads. The expoxy should be mixed fresh, and works better if the leads are clamped on the glass. The conductive paint should be stirred thoroughly, and the contacts should be checked for electrical continuity after the paint dries. The cell can be sealed using epoxy (Epotek, Epoxy Technology Inc. expands very little) or optical adhesives (e.g., UV curable adhesive NO-65 from Norland Products Inc.).

Stray capacitance and similar experimental difficulties The capacitance of the cables may not be negligible and may not even be constant. Assume that the total length of the leads is 1 m, and they are simply two pieces of wire. From electrostatics the capacitance of two parallel wires, C_W, is: $C_W = 2\varepsilon_0/\cosh^{-1}(1 - D^2/2r^2)$, where r is the radius of the wires and D is the distance between wires. If during the experiment the distance between wires is changed from 0.5 cm to 1 cm, then the stray capacitance due to the wires will change from 18 pF to 12 pF. Thus, the difference in capacitance between the homogeneous and homeotropic alignment for the 'thick' cell is comparable to the change in the capacitance of the circuit produced by a rearrangement of the wires! This problem can be eliminated by using coaxial cables throughout and, as an additional precaution, by making sure that all connections in the circuit are firm, and all the contacts are clean. The stray capacitance due to the coaxial cables can be determined by a measurement of the cell before filling-in the sample. From the above estimates it is obvious that the measurements on thick cells are more sensitive to the stray effects. Thin cells on the other hand are more difficult to construct, and require higher aligning fields. The aligning fields needed to produce well aligned samples may be estimated from the Frederiks transition fields discussed in Section 4.5. We use the critical field values, derived in Eq. (4.18) of Section 4.5. For typical values of the elastic coefficients and diamagnetic anisotropy one finds a minimum field of 1.1 T and 0.1 T for the 10 μm and 100 μm cells, respectively. We note that a 5 μm cell will require more than 2.2 T, which is near the limit of what can be attained by common electromagnets (typically of the order of 2.5 T).

Spontaneous polarization of ferroelectrics

Of particular interest in ferroelectric liquid crystals is the measurement of the spontaneous polarization or polarization density, **P**. In the ferroelectric smectic-C, or the SmC* phase, the ferroelectric behavior derives from permanent dipoles that are approximately transverse to the molecular axis.

The Sawyer–Tower circuit

The classical method of detecting ferroelectric response is through the Sawyer–Tower circuit [14], shown in Fig. 4.7. The circuit is essentially a voltage divider. The oscilloscope displays the hysteresis loop of polarization versus applied field. The spontaneous polarization is obtained by extrapolating the saturation part to zero applied field. A refinement of the method can be found in Ref. 15.

Measurement of polarization current

A technique which has been used extensively with ferroelectric smectic liquid crystals to measure the polarization current is the following. The dipole current density is defined as

$$\mathbf{J}_p = \frac{\partial \mathbf{P}}{\partial t}.$$

In these experiments the liquid crystal sample is a thin film (5–30 μm) between two conducting glass plates. The sample is aligned with the molecules in the plane of the conducting electrodes, thus the polarization is perpendicular to the electrodes. Typically the cell has an area of approximately 1 cm × 1 cm. Switching the voltage applied to the electrodes will result in a reversal of the polarization direction, and thus a polarization current, which derives from the rotation of the molecules about their long axis. The polarization current is transient, i.e., it will cease to flow once the

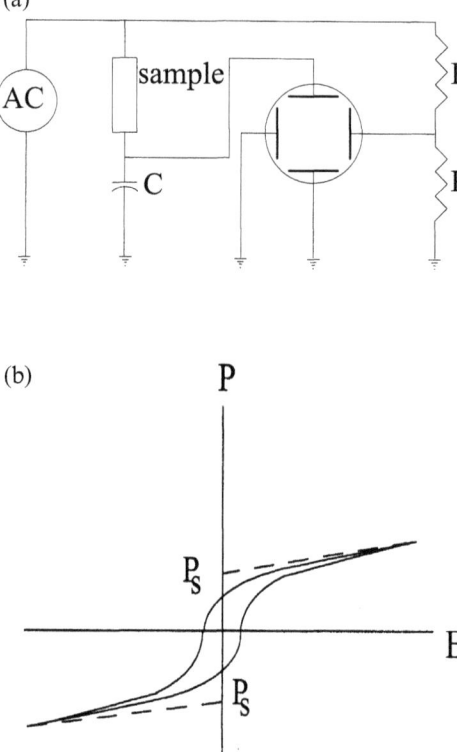

Figure 4.7.   (a) The Sawyer–Tower circuit, used to detect ferroelectric response of a sample. (b) A typical hysteresis loop. The polarization versus applied field is observed on the oscilloscope screen. P_s, the spontaneous polarization, is determined by extrapolation as shown.

reorientation is complete. In addition to the polarization current, there is also a resistive current, due to the conductivity of the sample. This current will persist, as long as the voltage is applied. To determine the spontaneous polarization, one needs to separate the two contributions. The polarization will follow the direction of the applied voltage, if the frequency of the applied voltage is low compared to the relaxation frequency of the molecules about their long axis. The latter is typically in the order of a few hundred hertz. In practice, the applied frequencies are in the order of a few

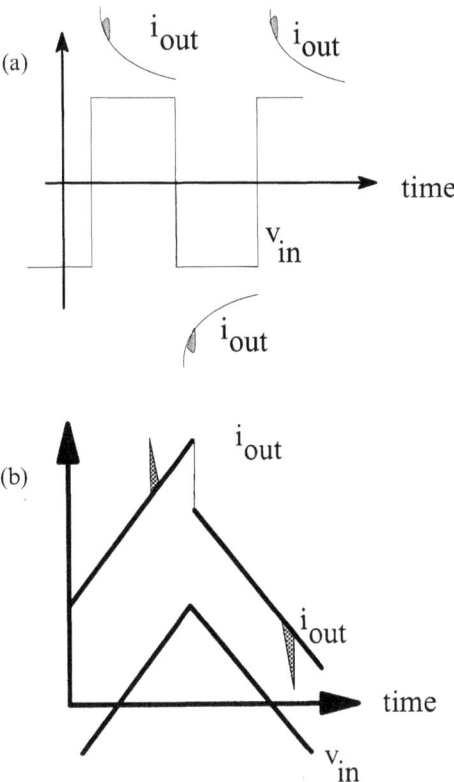

Figure 4.8.   Measurement of spontaneous polarization using a square wave and a triangle wave. This figure shows input voltages and the corresponding output currents for the square wave (a) and triangle wave (b) methods of determining spontaneous polarization in ferroelectric liquid crystals. The polarization current appears as a hump over the *RC* decay curve of the circuit.

hertz, in the form of a square or a triangular wave. The current through the sample (typically of order a microampere) can be measured by connecting the sample to a series resistor (typically approximately 1 MΩ). The resulting waveforms for the current are shown in Fig. 4.8 for square and triangular voltages.

Calculation of the spontaneous polarization

The polarization is obtained from the shaded areas shown in the figure, which correspond to the integral of the polarization current. As the amplitude of the applied voltage is increased, the area under the peak will increase up to a saturation value which corresponds to the flat part of the

P vs. *E* curve in Fig. 4.7. The spontaneous polarization in C/m² may be obtained from the integral of **J**, or its area on an oscillogram, divided by the surface area of the sample.

The details of the square wave method can be found in Ref. 16 for ferroelectric crystalline materials, and Ref. 17 for ferroelectric smectics. The triangular wave method is described in Refs. 18 and 19. The methods of sample alignment, and an extensive discussion on the structure and properties of ferroelectric liquid crystals can be found in Ref. 20. Patel and Goodby [19] observed a dependence of the measured spontaneous polarization on the thickness. The measured polarization increases with sample thickness, and levels off for thicknesses larger than 20 μm. The dependence on the thickness was attributed to the ionic impurities, which screen the applied voltage.

4.3 Optical properties

Introduction

The study of the index of refraction is of prime importance for understanding the fascinating optical properties of liquid crystals. The colorful textures seen when observing a liquid crystal sample under polarized light microscopy result from the dependence of the speed of light on the wavelength, and on the direction of the plane of polarization relative to the director. The former is known as dispersion, the latter as birefringence. For a transparent isotropic material at optical frequencies (of the order of 10^{15} Hz) the square of the index of refraction is equal to the high-frequency dielectric permittivity of the material:

$$n_{\parallel}^2 = \varepsilon_{\infty\parallel} \text{ and } n_{\perp}^2 = \varepsilon_{\infty\perp},$$

where $\perp$ and $\parallel$ stand for perpendicular and parallel to the director. The interaction of the electromagnetic wave with the optical medium is through the dielectric tensor. It is the direction of the electric field of the electromagnetic wave, i.e., the direction of polarization, that matters, rather than the direction of propagation of the wave. Thus we have the following definitions:

Polarizations perpendicular to the optic axis (the director) 'see' the so-called *ordinary refractive index* $n_o = n_{\perp}$.

Polarizations parallel to the optic axis 'see' the so-called *extraordinary refractive index* $n_e = n_{\parallel}$.

This distinction is significant, and should be kept in mind.

Birefringence

Consider a nematic film of thickness d, with a monochromatic polarized beam incident perpendicular to the film. If **n** is normal to the film, then the beam sees the same refractive index, n_o, for any polarization of the beam. The beam in this case is propagating parallel to **n**, which defines the *optic axis* of the nematic phase. What is important, though, is that in this geometry the polarization (i.e., the electric field **E** of the wave) is always perpendicular to **n**. If an analyzer is placed at 90° to the incident polarization, the beam will not propagate past the analyzer. This is what happens in the *homeotropic* or *pseudoisotropic* texture observed under a polarizing microscope. A more interesting situation arises when **n** is in the plane of the film. This results in the so-called *homogeneous* or *uniform planar texture*. If **E** ⊥ **n**, then the beam propagates with the ordinary refractive index, n_o. If **E** ∥ **n**, the beam sees the extraordinary refractive index, n_e. In either case the beam will not propagate past an analyzer at 90° to the incident polarization. These two directions correspond to the extinction positions observed with polarizing microscopy, and are separated by 90° rotations of the microscope stage. For the intermediate positions of the stage, **E** forms an intermediate angle with **n** and the sample is bright or colorful, for any angle of the analyzer.

The birefringence is defined as the difference between the extraordinary, n_e, and ordinary, n_o, indices:

$$\Delta n = n_e - n_o. \tag{4.10}$$

For thermotropic nematics the magnitude of $\Delta n \sim 0.1$ to 0.01. Experimentally [3, 4] one finds that $n_e^2 - n_o^2 \sim \rho S$, where ρ is the density and S is the nematic order parameter.

Transmission of polarized light

The propagation of light in anisotropic media is described in several well known texts [21]. We will briefly review the mathematics which describe the state of polarization of a wave transmitted through a sample in the homogeneous alignment when **E** and **n** form an arbitrary angle φ. For simplicity, we take **n** in the x-direction, and the incident beam along the z-axis, as shown in Fig. 4.9. The wave enters the sample at $z = 0$. We can analyze the electric field vector into two orthogonal components:

$$\mathbf{E}(z = 0) = \mathbf{E}e^{i\omega t} = \mathbf{i}E\cos\phi e^{i\omega t} + \mathbf{j}E\sin\phi e^{i\omega t},$$

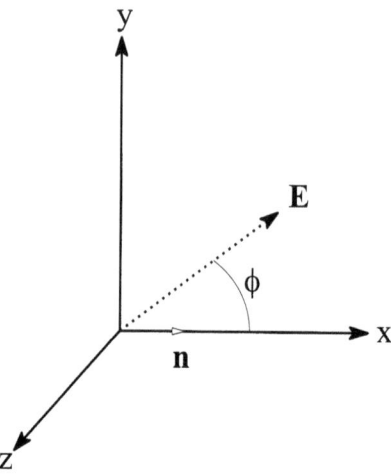

Figure 4.9.   The electric field vector and director to be used in the analysis of the transmission of polarized light. The electric field vector is in the x–y plane and makes an angle of ϕ with the director. The director is parallel to the x-axis and the beam is incident along the z-axis.

where $\mathbf{i}$ and $\mathbf{j}$ are the unit vectors in the x and y directions. E_x propagates in the sample with refractive index n_e and E_y propagates with refractive index n_o. Upon exiting the sample, at $z = d$, the phase of the components has increased by $2\pi n_e d/\lambda$ and $2\pi n_o d/\lambda$ respectively thus:

$$\mathbf{E}(z = d) = \mathbf{i}E\cos\phi\, e^{i(\omega t - 2\pi n_e d/\lambda)} + \mathbf{j}E\sin\phi\, e^{i(\omega t - 2\pi n_o d/\lambda)}$$
$$= E(\mathbf{i}\cos\phi + \mathbf{j}\sin\phi\, e^{i2\pi\Delta n d/\lambda})e^{i\omega t}e^{-i2\pi n_e d/\lambda}.$$

We see that the two components of the polarization exit the sample with a phase difference, δ:

$$\delta = 2\pi\,\Delta n\, d/\lambda. \tag{4.11}$$

The phase difference is proportional to the difference in the optical path, $d\cdot\Delta n$, and inversely proportional to λ. Also, n_o and n_e may obey different dispersion relations, and thus introduce additional dependence of the phase shift δ on λ. The latter dependence is the origin of the colors observed under polarized light microscopy. We can rewrite the above equation as

$$\mathbf{E}(z = d) = E(\mathbf{i}\cos\phi + \mathbf{j}\sin\phi\, e^{i\delta})e^{i\omega t}e^{-2\pi i n_e d/\lambda}. \tag{4.12}$$

The expression for the wave as it propagates past the sample can be obtained by multiplying Eq. (4.12) by $e^{-i2\pi z/\lambda}$. The state of polarization of the emerging wave is determined by the term in parentheses in Eq. (4.12). This allows one to describe retardation plates, as well as right and left handed circularly polarized light. It is interesting to note that linearly polarized light can be decomposed into two circularly polarized waves: a right and a left handed component of equal amplitudes [21, 22]. This decomposition into circular modes will be useful in our discussion of the optical properties of cholesterics. So far, we have considered the light wave emerging from the liquid crystal. But what happens in the liquid crystal film? This question is important both for theoretical analysis and for practical design. Consider for example an aligned nematic film, where the director is in the yz-plane and forms an angle θ with the z-axis. The wave is incident in the z-direction. A beam polarized along the x-axis sees a refractive index equal to n_o, since the polarization is perpendicular to $\mathbf{n}$. But a polarization along the y-axis is neither perpendicular nor parallel to the optical axis. What is the refractive index seen in the y-polarization? To answer this question we need to construct the *index ellipsoid* (which in our case is an ellipse!). This is described in the above texts. Using these one finds

$$n(\theta) = n_o n_e \, (n_e^2 \cos^2 \theta + n_o^2 \sin^2 \theta)^{-1/2}. \tag{4.13}$$

Optical properties of cholesterics

In cholesteric materials the direction of $\mathbf{n}$ is not constant, but describes a helix. If we take the axis of the helix parallel to the z-direction, the components of $\mathbf{n}$ are given by

$$n_x = \cos(2\pi z/P),$$
$$n_y = \sin(2\pi z/P),$$
$$n_z = 0.$$

Here we have taken $\mathbf{n}$ parallel to the x-axis at $z = 0$. Thus, the variation of $\mathbf{n}$ is periodic in z, and the period or pitch of the helix is P. The magnitude of P depends on the material and on the temperature, and can vary over several orders of magnitude. Note that the directions $\mathbf{n}$ and $-\mathbf{n}$ are equivalent, thus the periodicity of $\mathbf{n}$ and other material properties is actually $P/2$. The analysis of the beam propagation in a twisted nematic or a cholesteric liquid crystal requires the solution of Maxwell's equations, and is given in detail in Refs. 2 and 23. Analysis and numerical results for oblique incidence can be found in Refs 24 and 25. Here we will describe the main features of the results for normal incidence. An essential assumption for our discussion is that the cholesteric is locally uniaxial. This assumption is well justified by the experimental results.

Rotation of the plane of polarization

When $P(n_e - n_o) \ll \lambda$ the medium transmits both right and left circularly (more correctly nearly circularly) polarized light, each travelling with a different speed [2]. Plane polarized light may be viewed as the resultant of right and left circularly polarized waves of equal amplitudes. Thus, when a plane polarized wave is incident on a cholesteric liquid crystal, it may be analyzed into its two circular components which will propagate through the sample at different speeds. These circularly polarized components will emerge with equal amplitudes, but *phase shifted* relative to each other. The emerging electric field is the sum of these two circular components, and consequently it is plane polarized, generally with a different plane of polarization than the incident beam. Thus the sample acts as an optically active material. This is similar to the situation in isotropic optically active materials. The only difference is that the rotatory power can be much larger than common chiral materials.

The Mauguin limit

In the other limit, when the pitch is much larger than the wavelength, we have the so-called *Mauguin limit*, which can be stated as

$$\lambda \ll (n_e - n_o) \, P.$$

Under these conditions the sample can *transmit* plane polarized light. When the incident polarization is perpendicular to one of the two principal axes at the point of incidence, the polarization rotates as it travels along the helix, and emerges perpendicular to the direction of the corresponding principal axis at the exit point. The polarization of the emerging wave is rotated by an angle $2\pi d/P$ with respect to the original direction, where d is the thickness of the sample. The sample thus acts as a 'wave-guide'. If the polarization of the incident beam is not perpendicular to either principal axis at the point of incidence, then the emerging beam is elliptically polarized. The principle of the twisted nematic cell is based on this wave-guide mechanism.

Bragg reflection

For conditions intermediate between the two limiting conditions discussed above, the sample will transmit left and right elliptically polarized light, with one exception. As the wave propagates in the cholesteric material, it sees an essentially periodic medium. If the round-trip distance between successive equivalent orientations of **n**, i.e., points separated by $P/2$, is equal to one-half the wavelength of light *in the medium*, then we have reflections which are in phase. This occurs at every point throughout the sample. Under these conditions, the sample would totally reflect circularly polarized light of the same handedness as the helix. Circularly polarized light of opposite handedness to that of the helix would be transmitted through the sample. Thus, if the handedness of the light matches the helix, we have a *Bragg reflection of first order*:

$$P = \lambda / \langle n \rangle,$$

where $\langle n \rangle$ is average index of refraction of the cholesteric,

$$\langle n \rangle = \left(\frac{n_0^2 + n_e^2}{2} \right)^{\frac{1}{2}}.$$

Higher order reflections are observed only for oblique incidence. The Bragg reflection occurs in the range

$$n_o < \lambda/P < n_e,$$

where λ is the wavelength (in vacuum) of the incident beam. For wavelengths outside this range, the intensity of the Bragg reflection falls off rapidly, and the transmission of elliptical or nearly circular light resumes.

Measurement of the index of refraction
Angle of minimum deviation of a prism

A simple way to determine the index of refraction is by measuring the angle of minimum deflection for a ray passing through a prism. In the case of a liquid crystal, the material can be filled in a thin wedge formed by two glass slides or a commercially available prism-shaped cell (e.g., Starna Inc.). This technique is discussed in Ref. 26.

Thin wedge

For a thin wedge, the angles between the plates and the deflection angle are small, and using Snell's law one finds $\theta = (n - 1)\alpha$, where θ is the angular displacement caused by the insertion of the cell. The deflection angle is independent of the angle of incidence for near normal incidence.

With proper alignment of the sample, these methods allow the measurement of the ordinary and extraordinary indices with an accuracy of 1 to 0.1%.

Abbe refractometer

A standard commercial instrument is the Abbe refractometer. The operation of this instrument is based on total internal reflection. The refractometer is not suitable for alignment by magnetic fields. However, the
ordinary or extraordinary ray can be selected using a polarization filter in
front of the eyepiece of the instrument. We should note at this point that
total reflection occurs only if the refractive index to be measured (ordinary
or extraordinary) is less than the refractive index of the glass prisms of the
refractometer, thus special prism materials may be necessary.

Conoscopic observation and the sign of birefringence

The sign of the birefringence refers to the difference $n_e - n_o$. Thus, in uniaxially *positive* materials the ordinary ray travels faster than the extraordinary, and vice versa for *negative* materials. The sign of the birefringence for
a nematic can be determined by *conoscopic* observation under polarized
light microscopy. This amounts to looking through the sample, as opposed
to observing the image of the sample. The simplest way to achieve this is by
simply removing the eyepiece of the microscope. For homeotropic alignment of the nematic sample under crossed polarizers an interference
pattern will appear. The pattern divides the field of view into four quadrants. This pattern is referred to as 'Maltese cross'. An alternative method
is to introduce a lens (the so-called Bertrand lens) above the analyzer. In
either case we are observing the image formed by a strongly converging
beam of light, which samples a wide range of angles of incidence. The next
step is to insert a gypsum plate, which causes a full wave retardation in the
yellow. The slow axis of this plate is conventionally aligned parallel to the
short side of its mount. Upon introducing the gypsum plate, the quadrants
become blue and yellow. In a uniaxially positive crystal, the direction of the
slow axis bisects the blue quadrants, and for a negative sample, the slow axis
bisects the yellow quadrants. Details of this technique and alternative
methods are given in Ref. 27.

Birefringence measurements

Birefringence measurements are used in many studies of liquid crystals. To
understand these measurements consider an aligned nematic film of thickness d, and a polarized incident beam of wavelength λ, as shown in Fig.
4.10. In a typical experiment, the incoming beam is polarized at 45° to the

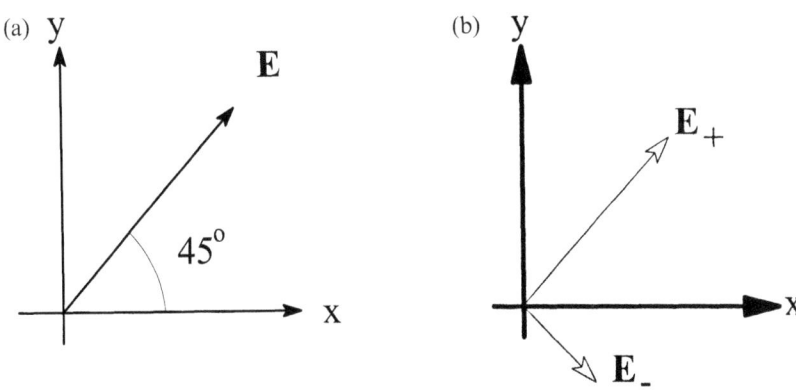

Figure 4.10.   The (a) incoming and (b) outgoing electric field vectors for an experiment which measures birefringence. The incident beam is along the z-axis. The
director is in the yz-plane and forms an angle θ with the z-axis. The beam is polarized at 45° to the x-axis. Thus the x-component of the polarization travels with
refractive index n_o and the y-component with $n(\theta)$ given by Eq. (4.13). (b) Shows the
polarization component of the emerging beam.

x-axis, thus the two components of polarization (x and y) of the incident
beam are equal. As discussed above, the x-component of the polarization
propagates with ordinary index n_o, and the y-component propagates with
$n(\theta)$ given by Eq. (4.13). The difference in *optical path length* between the
x- and y-components of the polarization of the emerging beam is:

$$\Delta l = [n_o - n(\theta)]d,$$

which corresponds to a phase shift of

$$\delta = 2\pi[n_o - n(\theta)]d/\lambda.$$

The beam emerges elliptically polarized, and measurement of Δl and d
would determine $n_o - n(\theta)$. We now discuss three methods for measuring
$n_o - n(\theta)$.

Compensator methods

These devices are used to determine the retardation between the two components of polarization in terms of the wavelength λ, or a distance or angle
which can be related to phase shift. These devices typically consist of bire-
fringent crystals and are commercially available. Soleil–Babinet compensa-
tors are available (e.g. Karl Lambrecht Corp.) with retardation ranges from
1 to 6 wavelengths, depending on the angle of the wedges.

Polarization beam splitter

Beam splitters are devices that separate an incoming light beam into its two orthogonal polarizations. The x and y polarization components of a beam emerging from a liquid crystal sample are

$$E_x = E_0 \cos(\omega t + \delta/2),$$
$$E_y = E_0 \cos(\omega t - \delta/2).$$

Assume (by construction of the apparatus) that the beam splitter selects the polarizations at $\pm 45°$ to the x-axis, as shown in Fig. 4.10. The x and y components of the beam emerging from the beam splitter are

$$E_+ = (E_x + E_y) \cos 45° = 2E_0 \cos \omega t \cos \delta/2 \cos 45°,$$
$$E_- = (E_x - E_y) \cos 45° = -2E_0 \sin \omega t \sin \delta/2 \sin 45°.$$

The intensity of the beams emerging from the beam splitter can be measured using two photodiodes. The measured intensities are proportional to the time average of the square of the corresponding electric field. Noting that

$$\langle \cos^2 \omega t \rangle = \langle \sin^2 \omega t \rangle$$

we find the ratio of the two intensities:

$$\frac{I_-}{I_+} = \frac{\langle E_-^2 \rangle}{\langle E_+^2 \rangle} = \tan^2\left(\frac{\delta}{2}\right) \text{ or } \delta = 2 \tan^{-1}\left(\frac{I_-}{I_+}\right)^{\frac{1}{2}}.$$

Polarization beam splitters are available in various designs. In the cube design, one of the emerging beams is parallel to the incident beam direction and the other is at $90°$ to the direction of the incident beam. In the Rochon prism, the second beam emerges at a small angle to the incident direction (about $6°$ for calcite). A larger separation angle is achieved by the Wollaston prism; in this case, however, neither of the emerging beams is in the incident direction. The description of polarization beam splitters can be found in Guenther's book [21, Chapter 13] or Ref. 26, Chapter 7. Although the method does not provide the sign of δ, it is very useful in practice and is used extensively in monitoring the director realignment, as will be described in Sections 4.5 and 4.6. An application of this method is to the study of the Frederiks transition (see Section 4.5). Error considerations can be found in Refs. 28 and 29.

Modulated beam method

In the geometry of Fig. 4.10, the intensity at $90°$ to the incident polarization is

$$I_- = I_o \sin^2(\delta/2).$$

This intensity can be measured by using an analyzer at $-45°$ to the x-axis, and a single photodiode, after proper correction for reflection losses. An application of this simple method to the study of the Frederiks transition can be found in Ref. 30. The quality of the results depends on the polarization purity of the incoming beam and the quality of the analyzer. For small δ the quadratic dependence of I_- on δ, and the inherent difficulties in measuring low light intensities cause the resolution to decrease with decreasing δ. Improved results can be obtained by superimposing a variable phase shift on the beam emerging from the sample. This can be achieved, for example, using a Pockels cell. The Pockels cell (typically a KH_2PO_4 or $NH_4H_2PO_4$ crystal) when properly biased induces a phase shift proportional to the applied voltage. In this technique, a combination of a high dc (few kV) and low ac (about 100 V) is applied to the cell. As a result, the phase shift of the beam becomes

$$\delta - \delta_o - \delta_m \sin \omega_m t,$$

where δ_o is the phase shift produced by the dc voltage, $\delta_m \sin \omega_m t$ is the phase shift produced by the ac component which has frequency $\omega_m/2\pi$. If the bias is chosen so that the total phase shift is small, we can write

$$I_- = (1/4)I_o (\delta - \delta_o - \delta_m \sin\omega_m t)^2.$$

Upon expansion, the above expression yields a constant term, a term proportional to $\sin^2 \omega_m t$, which oscillates at *twice* the modulation frequency, and the linear term of angular frequency ω_m:

$$I_m = -(1/2) I_o (\delta - \delta_o) \delta_m \sin \omega_m t.$$

The amplitude of this term can be measured directly using a lock-in amplifier. A description of measurements using this set-up, which can achieve resolution of Δn in the order of 10^{-9} to 10^{-10}, can be found in Refs. 31 and 32. The principle of operation of the Pockels cell is described in Ref. 21a, Chapter 14.

4.4 Diamagnetic properties

Liquid crystals, like most organic materials are diamagnetic. This means that an applied magnetic field induces a magnetic moment on the molecules, which points opposite to the applied field. Since the molecules in a liquid crystal material are anisotropic, the magnitude of the induced moment depends on the relative orientation of the applied field with respect to the molecular axes. The diamagnetic anisotropy of hydrocarbon chains is about two orders of magnitude lower than the anisotropy of aromatic rings. Thus, whenever benzene rings are present in the molecular structure, they dominate the diamagnetic behavior of the molecules.

Electric versus magnetic fields

Magnetic interactions are orders of magnitude weaker than electric interactions. Thus, they are totally ignored when discussing the propagation of electromagnetic waves. In addition, since the interaction between induced magnetic dipole moments is weak in diamagnetic materials, the internal magnetic field is negligible; thus, the interpretation of the material properties in terms of the molecular susceptibility is considerably simplified in comparison to the dielectric properties. Also, in contrast to electric fields, static magnetic fields do not cause accumulation or chemical reactions at electrodes. Thus, in experimental and theoretical studies the use of magnetic fields offers significant advantages over electric fields. On the other hand, in view of the field strengths required, magnetic fields are not practical for switching in device applications.

The volume susceptibility

The relation between the susceptibility of the sample and the properties of its constituent molecules is derived in the literature [1–4]. We briefly outline the main ideas. The interactions between magnetic moments are so small that to a good approximation the induced magnetic moment is given by: $\mathbf{m} = \mu_o^{-1}\chi\mathbf{B}$, where $\mathbf{m}$ is the magnetic moment, μ_o is the permeability of vacuum, χ is the magnetic susceptibility tensor per unit volume, and $\mathbf{B}$ is the applied magnetic induction. The average diamagnetic susceptibility, $\bar{\chi}$, is

$$\bar{\chi} = \tfrac{1}{3}(\chi_{\parallel} + 2\chi_{\perp}),$$

and the diamagnetic anisotropy is

$$\Delta\chi = \chi_{\parallel} - \chi_{\perp},$$

where $\chi_{\parallel}$ is the magnetic susceptibility parallel to the director and $\chi_{\perp}$ is the magnetic susceptibility perpendicular to the director.

Mass and molar susceptibilities

The above susceptibilities are volume susceptibilities. It is often more convenient to introduce mass and molar susceptibilities. These are defined as follows: χ^m, the mass susceptibility in units of m^3/kg, $\chi^m \equiv \chi/\rho$; and χ^M, the molar susceptibility in $m^3/mole$, $\chi^M \equiv \chi^m M$ where ρ is the mass density of the liquid crystal, and M is the molecular weight of the liquid crystal. Most liquid crystals are diamagnetic and both $\chi_{\parallel}$ and $\chi_{\perp}$ are of order 10^{-5} SI units. Furthermore, $\Delta\chi$ is generally positive.

Relationship between microscopic and macroscopic variables

The relationship between $\Delta\chi$, the nematic order parameter, and molecular parameters is complex. Following the standard texts [1–4] we define the molecular polarizability tensor, κ. This too can be diagonalized. Letting c be the number of molecules per unit volume one can show that

$$\Delta\chi = \chi_{\parallel} - \chi_{\perp} = \left(\frac{c}{\mu_0}\right)(\kappa_{\parallel} - \kappa_{\perp})S. \tag{4.14}$$

The difference $\kappa_{\parallel} - \kappa_{\perp}$ depends on the detailed structure and rigidity of the molecules, and can be derived from magnetic resonance measurements.

The experimental methods, which will be described below, determine the susceptibility per unit volume. Moreover, as these measurements are carried out in a strong magnetic field, the measurements will always yield only one component of the volume diamagnetic susceptibility. For example, if the diamagnetic anisotropy is positive, the director will align parallel to the field during the experiment, and the measurement will yield $\chi_{\parallel}$, but not $\chi_{\perp}$. To get $\chi_{\perp}$ and the anisotropy we can use the fact that the value of the average mass susceptibility does not change with temperature, and thus is the same in the nematic and isotropic phases. Suppose that the measurement yields $\chi_{\parallel}$. One then calculates $\chi_{\parallel}^m (= \chi_{\parallel}/\rho)$. A separate measurement in the isotropic phase yields $\bar{\chi}$ and by calculation $\bar{\chi}^m (= \bar{\chi}/\rho)$. From the definitions above we obtain $\chi_{\perp}^m = \frac{1}{2}(3\bar{\chi}^m - \chi_{\parallel}^m)$ and $\Delta\chi^m = \frac{3}{2}(\chi_{\parallel}^m - \bar{\chi}^m)$ which upon multiplication by the density at the corresponding temperature would yield $\chi_{\perp}$ and $\Delta\chi$, respectively.

The benzene rings in thermotropic nematics result in positive $\Delta\chi^m$ in the order of 10^{-9} m^3/kg. Since the susceptibility of diamagnetic materials is negative, $\Delta\chi^m < 0$ means that in terms of absolute values $\chi_{\perp}^m$ is larger than $\chi_{\parallel}^m$. Negative anisotropies result when cyclohexanes are substituted for the benzene rings in the molecular structure, and in some micellar nematics. In the latter case, $\Delta\chi$ is also two orders of magnitude smaller than normal thermotropics.

As indicated by Eq. (4.14), $\Delta\chi^m$ is directly proportional to the order parameter S. Thus the temperature dependence of $\Delta\chi^m$ follows that of Fig. 4.2. To appreciate the relative magnitudes involved in the variation of ρ and $\Delta\chi$, note that for a temperature change of 10 K, the density varies by 1% and $\Delta\chi$ by 10%.

Measurements of the susceptibility

The magnetic susceptibility can be measured by adopting standard methods. A detailed description of these methods can be found in Bates [33]. Here we will describe the Gouy, the Faraday, the oscillating sample methods, and a SQUID-based apparatus.

Gouy method

Consider a nematic liquid in a long (10–20 cm) glass sample tube suspended vertically in a magnetic field, as shown in Fig. 4.11(a). Precision NMR tubes are very suitable for this experiment. The vertical force on a unit volume dV and height dz is

$$dF = \frac{(\chi - \chi^{air})}{2\mu_0}\left(\frac{dB_x^2}{dz} + \frac{dB_y^2}{dz} + \frac{dB_z^2}{dz}\right)dV.$$

If the nematic has positive diamagnetic anisotropy then in the above equation $\chi = \chi_\parallel$, i.e., the volume susceptibility along the director. In this experiment, one needs a uniform high field in the gap, and nearly zero field outside. Cylindrical pole faces with a small gap (1–2 cm) work very well. The field is assumed horizontal, in the y-direction, and the volume element $dV = A\,dz$, where A is the constant cross-section area of the sample. The total force on the sample is calculated by integrating dF over the entire sample. In this experiment the top of the sample is outside the gap, in zero field, and the bottom roughly at the midpoint of the pole faces, in field B. Integrating we find

$$F = \frac{(\chi - \chi^{air})}{2\mu_0}A\int_0^B \frac{dB_y^2}{dz}\,dz = \frac{(\chi - \chi^{air})}{2\mu_0}AB^2.$$

For most thermotropics, χ^{air} corresponds to a small percentage correction. The main contribution to it comes from the oxygen in air, which is

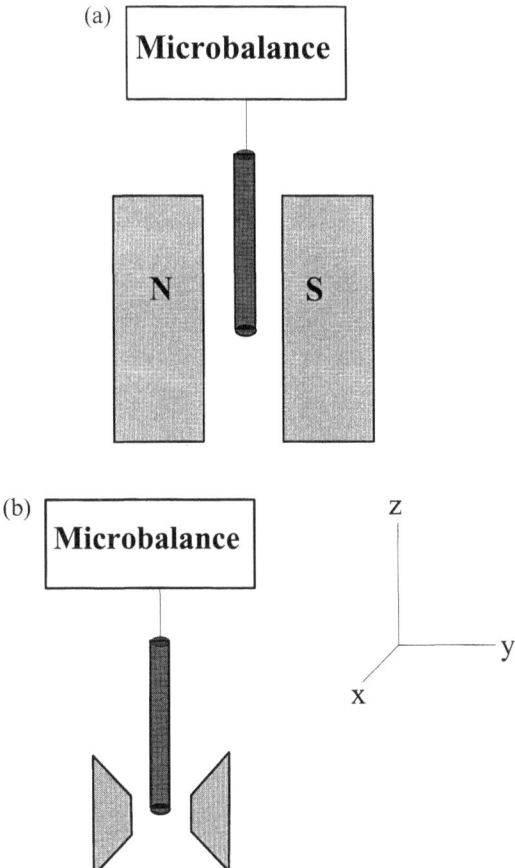

Figure 4.11.   (a) A schematic diagram of the Gouy method of determining magnetic susceptibility. In this apparatus, the magnetic field is uniform in the gap and zero outside. (b) A schematic diagram of the Faraday method of determining magnetic susceptibility. The magnetic field is symmetrical about the vertical mid-plane, and thus has a gradient along z.

paramagnetic, and thus makes a contribution which is temperature dependent. The contribution of oxygen is particularly significant in systems without aromatic rings, e.g., lyotropic micellar liquid crystals. The effects of oxygen can be eliminated by carrying the experiment in a helium or nitrogen atmosphere. This amounts to allowing gas, at slow flow rate, through the volume of the thermostated chamber containing the sample. To measure the magnetic force, one weighs the sample with the field on and with the field off. The difference between the two readings yields the net magnetic force. This entails suspending the sample from a microbalance, securely positioned above the magnet, and properly shielded. The parts used in the suspension of the sample tube must be diamagnetic as well, and preferably non-porous, to avoid accumulation of moisture when air humidity can be a problem. Also, the selection of the sample tube and suspension parts should take into account the fact that the capacity of microbalances is typically of the order of 10 to 20 grams. A separate measurement with the sample tube empty is also necessary, to account for the contribution of the sample tube. A detailed description of this method is given in Ref. 34.

Faraday method

This method uses a strong inhomogeneous magnetic field. Typically the design uses tapered pole caps, and a gap width comparable to the diameter of the pole caps. Figure 4.11(b) illustrates a commonly used arrangement. The sample is suspended vertically through the middle plane of the gap. The field gradient is perpendicular to the field lines. Due to the symmetry, the field gradient seen by the sample is along the z-axis. Thus the force on the unit volume is

$$dF = \frac{(\chi - \chi^{\text{air}})}{2\mu_0} \frac{dB_y^2}{dz} dV.$$

If the nematic has positive diamagnetic anisotropy then in the above equation $\chi = \chi_\parallel$, i.e., the volume susceptibility along the director. Integration over the entire volume of the sample yields the total magnetic force:

$$F = \frac{(\chi - \chi^{\text{air}})}{2\mu_0} \int_{\text{sample vol}} \frac{dB_y^2}{dz} dV = \frac{(\chi - \chi^{\text{air}})}{\mu_0} \int_{\text{sample vol}} B_y \frac{dB_y}{dz} dV.$$

Note that the above expression involves integration of $B_y(dB_y/dz)$ over the volume occupied by the sample. This integral is independent of the material measured, and depends only on the magnet and the geometry of the experiment. Thus, to evaluate the integral we could measure the field and the vertical field gradient dB_y/dz on the plane midway between pole caps. Alternatively, as the integral depends on the geometry of the field and of the sample, we could use a fluid of known χ, to calibrate the set-up. In doing so, the sample shape and position must be reproduced accurately. This is achieved by using the same volume of sample, in the same or identical sample tube, and suspending the tube at the same position in the field. To measure the total magnetic force, the sample is suspended from a microbalance. The magnetic force is the difference between the readings with the field on and the field off. A separate measurement with the sample tube empty yields the force on the tube, which must be subtracted from the total force to yield the net magnetic force on the material. The corrections regarding χ^{air} and precautions stated for the Gouy method apply to the Fardaday method as well. A detailed description of the method is given in Ref. 35.

Oscillating sample method

This method uses a sample of volume V, filled in a cylindrical tube and suspended vertically from a thin wire. If the sample is slightly rotated about the vertical and then allowed to oscillate, the angular frequency for small amplitudes of oscillation will be

$$\omega_0^2 = D/I.$$

Here D is the torsion constant of the wire, which provides the restoring torque for the oscillation, and I is the moment of inertia of the filled tube about the center axis. If the sample is well aligned horizontally, and a magnetic field is initially applied parallel to the director, then as the sample oscillates, and the director forms an angle θ with the field, the sample will experience a magnetic torque. Taking the field in the z-axis and the director in the yz-plane, we can write Eq. (4.6) for the magnetic susceptibility in the lab frame. We find that the resulting torque per unit volume is along the vertical axis and can be written as

$$\mathbf{N} = \Delta\chi B^2 \sin\theta\cos\theta.$$

This torque will tend to realign $\mathbf{n}$ along $\mathbf{B}$, if $\Delta\chi$ is positive (as assumed here) and will act as restoring torque in addition to the torsion of the wire. This will cause a change in the angular frequency of oscillation from ω_0 to ω. If τ, the time required for the realignment of the director by the field, is large compared to the period of oscillation, and V is the volume of liquid crystal then

$$\Delta\chi = \frac{\mu_0 D}{VB^2}\left(\frac{\omega^2}{\omega_0^2} - 1\right).$$

Thus, by measuring the period of oscillation with a known field and at zero field, one determines $\Delta\chi$ directly. The details of this method, as applied to a micellar liquid crystal, are given in Ref. 36.

Note that the method described here is applicable to materials with long reorientation times. For faster reorientations, the rotating field method would be more suitable. This method is described in detail in Ref. 37.

The Superconducting Quantum Interference Device, or SQUID, is a super-conducting loop which in its simplest form consists of a pair of Josephson junctions connected in parallel. Under ideal conditions the device can detect changes in magnetic flux as low as about 10^{-21} Wb. This is by far the most sensitive device, although its high sensitivity makes it susceptible to noise from stray fields as well. Proper isolation is the main challenge in the design of these experiments, and entails not only magnetic shielding but also thermal insulation between the sample chamber and the low tempera-ture environment of the superconducting circuitry. Again, a separate meas-urement is required to determine the magnetic moment of the sample cell. The principles and practical uses of the various SQUID devices and their operating modes can be found in Ref. 38. The details of the SQUID magnetometer are given in Ref. 39 and a measurement on liquid crystals in Ref. 40.

4.5 Elastic properties

The influence of an external force, or even thermally induced fluctuations in a nematic liquid crystal, cause changes in the relative orientation of the molecules. These changes in orientation cause an increase in energy. There are three fundamental deformations and three elastic constants. In this section we will briefly define these deformations, the corresponding elastic constants, discuss the energy associated with these deformations, and present a simple technique for measuring these elastic constants.

Deformations of the director: splay, twist and bend

Consider a monodomain thin nematic film between two parallel glass plates in the *homeotropic* alignment. The director is strongly anchored per-pendicular to the surface at both plates. Now assume that we tilt the top plate slightly. If the director is to remain anchored at both surfaces, the orientation within the film must change in order to accommodate the normal anchoring at each surface, as shown in Fig. 4.12(a). Assume that the variation of the director within the film is smooth. In the deformed state the director remains in the *yz*-plane. The director field in the undeformed state is parallel to the *z*-axis. The deformation corresponds to bending the lines of the director field. We shall call this type of deformation a bend. Now consider the nematic film in the uniform planar alignment. The direc-

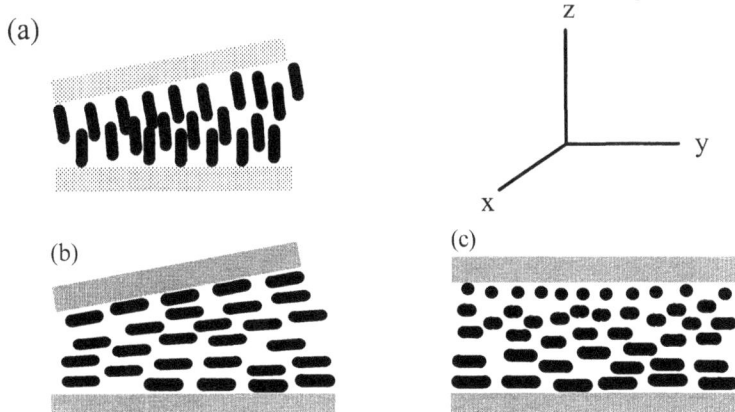

Figure 4.12. The (a) bend, (b) splay, and (c) twist deformations of the nematic director.

tor field in the undeformed state is a set of lines parallel to the y-axis. Tilting the top plate will result in a different type of deformation, known as the splay deformation, shown in Fig. 4.12(b). Here the director field is radial, in the yz-plane. A third type of deformation is achieved if the top plate in the uniform planar alignment is rotated slightly about the z-axis. Here the director remains in the xy-plane, however the direction twists around the z-axis. This is the twist deformation, and is shown in Fig. 4.12(c).

The continuum model

In the above examples, the director in the deformed state varies with position. By assuming that the deformations are negligible over lengths comparable to molecular dimensions we can treat the liquid crystal as a *continuum*. This approach allows us to define the director as a smooth function of position, $\mathbf{n(r)}$, and use the spatial derivatives of $\mathbf{n(r)}$ as a measure of the deformation. In this case the elastic energy in the deformed state is a quadratic expression in the derivatives of $\mathbf{n(r)}$. The formulation of the continuum theory for nematics is described in detail by de Gennes and Prost [2].

The elastic constants

The expression for *elastic energy per unit volume* (or free energy density) is

$$F_\mathrm{d} = \tfrac{1}{2} K_1 (\nabla \cdot \mathbf{n})^2 + \tfrac{1}{2} K_2 (\mathbf{n} \cdot \nabla \times \mathbf{n})^2 + \tfrac{1}{2} K_3 (\mathbf{n} \times \nabla \times \mathbf{n})^2 \qquad (4.15)$$

where K_1 is the *splay elastic constant*, K_2 is the *twist elastic constant*, and K_3 is the *bend elastic constant*.

These three elastic constants are usually referred to as the Frank or Oseen–Frank elastic constants, and have dimensions of force. The elastic constants decrease with increasing temperature, and are typically in the order of 10^{-11} N. It is often convenient in theoretical discussions to assume that the three constants are equal. This is the so-called *one-constant approximation*. In practice, the deformations of the director may involve more than one type or mode of deformation. In addition to the deformations produced by tilting or rotating one of the plates defining the nematic film, one could simply move the plates parallel to each other. This would produce a *shear deformation*. Nematics, like normal liquids, *do not support static shear deformations*. If the nematic film is sheared between two plates, the molecules will simply slide past each other, at the expense of some energy to overcome the viscosity of the liquid. This situation is discussed in Section 4.6.

Application to the Frederiks transitions

We will assume that a deformation from an initially uniform director is caused by an external field and calculate the field necessary for a major change in alignment to occur. This transition is called the Frederiks transition and is a useful technique for measuring the elastic constants. It is important to realize that both deformations and external magnetic or electric fields contribute to the free energy.

Deformation free energy

Assume that the components of the director are $\mathbf{n} = (0, \sin \theta(z), \cos \theta(z))$. Remembering that $\mathbf{n}$ is a unit vector, a straightforward calculation shows that

$$\nabla \cdot \mathbf{n} = -\sin \theta \, (d\theta/dz), \qquad \nabla \times \mathbf{n} = -\mathbf{i} \cos \theta \, (d\theta/dz), \qquad \mathbf{n} \cdot \nabla \times \mathbf{n} = 0,$$

$$\nabla \times \nabla \times \mathbf{n} = (-\mathbf{j} \cos^2 \theta + \mathbf{k} \sin \theta \cos \theta)(d\theta/dz).$$

Here $\mathbf{i}$, $\mathbf{j}$ and $\mathbf{k}$ denote the unit vectors in the lab coordinate system. The quantity $d\theta/dz$ is not known at this point. Substituting the above derivatives into Eq. (4.15) we obtain the deformation energy per unit volume:

$$F_d = \tfrac{1}{2}(K_1 \sin^2 \theta + K_3 \cos^2 \theta)(d\theta/dz)^2. \tag{4.16}$$

Note that $F_d = 0$ for $(d\theta/dz) = 0$, that is, a uniform director. Furthermore, the deformation involves *both* the splay and bend elastic constants.

Magnetic free energy

We will now consider the effect of an external magnetic field on this sample. We will assume that θ is small throughout the film and analyze this limiting case to derive an expression for the critical field, B_c, at which a major change in alignment occurs. To do so, we need to examine the total energy, which is the sum of F_d and the magnetic energy F_m. From magnetostatics we know that for an induced magnetization $\mu_0^{-1} \chi_L \cdot B$, the interaction energy with the applied field $\mathbf{B}$ is (per unit volume)

$$F_m = -\frac{1}{2\mu_0}(\chi_L \cdot \mathbf{B}) \cdot \mathbf{B},$$

here we will assume $\mathbf{B} = (0, B, 0)$ and the subscript in χ_L is the reminder that we need to evaluate the volume susceptibility in the lab coordinate system. We can use Eq. (4.6) to write the magnetic susceptibility in the lab coordinate system, and evaluate the matrix products to find

$$F_m = -\frac{1}{2\mu_0}(\chi_L \cdot \mathbf{B}) \cdot \mathbf{B} = -\frac{1}{2\mu_0}B^2(\chi_\perp \cos^2\theta + \chi_\| \sin^2\theta)$$

$$= -\frac{1}{2\mu_0}\chi_\perp B^2 - \frac{1}{2\mu_0}\Delta\chi B^2 \sin^2\theta. \tag{4.17}$$

In the last line of the above equation, the energy density has two terms. The first term is the magnetic energy in the undeformed state ($\mathbf{n}$ parallel to the z-axis). The second term is the magnetic energy due to the deformation.

The free energy difference between deformed and undeformed states

The free energy difference between deformed and undeformed states per unit volume is

$$\Delta F = \frac{1}{2}(K_1 \sin^2\theta + K_3 \cos^2\theta)\left(\frac{d\theta}{dz}\right)^2 - \frac{1}{2\mu_0}\Delta\chi B^2 \sin^2\theta.$$

The above equation is obtained by subtracting F_m in the undeformed state (i.e., Eq. (4.17) at $\theta = 0$) from the sum of Eqs. (4.16) and (4.17). Note that our discussion at this point assumes equilibrium states only, and our equations do not involve time. We are discussing the energetics of the deformation.

Stability analysis for small deformations

In the limit of *small* θ, i.e., near B_c, we can write: $\theta(z) = \theta_{max}\cos(\pi z/d)$. Then using series expansions for $\sin\theta$ and $\cos q$ and neglecting terms in θ_{max}^4 we find

$$\Delta F_A = \frac{d}{4}\,\theta_{max}^2 \left[K_3 \left(\frac{\pi}{d}\right)^2 - \frac{1}{\mu_0}\Delta\chi B^2 \right].$$

Note that the only material parameters which this expression depends on are K_3 and $\Delta\chi$, thus one anticipates being able to determine K_3.

Critical field in the bend geometry

As long as $\Delta F_A > 0$ the undeformed state is the more stable state. When **B** is increased sufficiently, ΔF_A becomes negative, and the director switches to the deformed state. The transition occurs when the quantity in the square brackets in the above equation becomes zero. This condition determines the critical field $\mathbf{B}_c$ in the geometry of the experiment:

$$\mathbf{B}_c = (\pi/d)\,(\mu_0 K_3/\Delta\chi)^{1/2}. \qquad (4.18)$$

Note that the result involves K_3 only, hence the designation *bend geometry*. Note also that for given K_3 and $\Delta\chi$, in other words for the *same material at the same temperature*

$$\mathbf{B}_c d = \text{constant}.$$

This result was experimentally established in the original work by Frederiks [41].

Magnetic coherence length

The surface anchoring imposes the alignment of **n** at the boundaries of the film. In fact, a basic assumption in our discussion is that the glass surfaces will not let go of the director. This is the so-called strong anchoring condition, which is achieved by proper treatment of the glass. A measure of the film thickness that is effectively aligned by the surface is given by the quantity

$$\xi_3^M = (1/B)(\mu_0 K_3/\Delta\chi)^{1/2}, \qquad (4.19)$$

which is known as the *magnetic coherence length*. Obviously, the transition to the deformed state cannot occur if the sample thickness $d < 2\xi_3^M$. The factor 2 accounts for the fact that we have two glass surfaces. In fact, combining Eqs. (4.18) and (4.19) shows that *the transition will occur if $d \geq \pi\xi_3^M$*.

Measurement of K_3

As the critical field in the bend geometry depends only on the elastic constant K_3, a measurement of B_c can be used to determine the ratio $K_3/\Delta\chi$. The onset of the deformation can be established by conoscopic observation under polarized light microscopy (see Section 4.3). In the undeformed state, under crossed polarizers we would observe a cross centered on a pattern of bright circular rings. At the transition, the rings start deforming [42]. Below the critical field θ_{max} is zero. As the field is increased above the critical value θ_{max} increases continuously, rapidly at first, and levels off at higher fields. This is termed a *second order transition* because θ_{max} increases from zero continuously. Near the critical field, the equilibrium state is achieved very slowly. The derivation of B_c was done in the limit of small deformations, and assuming $\theta(z) = \theta_{max} \cos(\pi z/d)$, which is true only near the transition. More generally one may derive the relation between θ and z away from B_c. This problem can be solved by applying the methods of *calculus of variations* (see, for example, Chapter 9 of Ref. 6 or Chapter 17 of Ref. 7) and is also discussed in the liquid crystal texts [1–4].

Tilted fields

In the preceding discussion we used **B** perpendicular to **n** in the undeformed state. This was the assumption in deriving Eq. (4.17), and the results that followed. We also assumed that all the molecules turn in the same sense, e.g., clockwise. However, as the molecules do not carry permanent magnetic moments, the rotation is independent of the polarity of **B**, and both the clockwise and counterclockwise rotations are completely equivalent. This means that, in practice, parts of the sample may turn clockwise and other parts may turn counterclockwise. When the field **B** is switched off, the sample may not return to the initial homeotropic alignment. Instead, it could build *defects*. To avoid creating defects we could apply the field in the *yz*-plane, but at a small angle to the *y*-axis. The analysis for the oblique field follows the procedure outlined above, and can be found in Ref. 43.

Frederiks transitions

The sample in our example was homeotropically aligned in the *z*-direction and the field was applied in the *y*-direction, i.e., the *bend geometry*. We could have started with a homogeneous planar alignment, e.g., **n** parallel to **j** and **B** perpendicular to **n**. In fact there are two more such possibilities:

  n parallel to **j** and **B** parallel to **k** (the splay geometry) and
  n parallel to **j** and **B** parallel to **i** (the twist geometry).

Here **i**, **j**, **k** are the unit vectors parallel to the *x*-axis, the *y*-axis and the *z*-axis respectively in the lab frame. Similar transitions are observed in these geometries, involving different elastic constants.

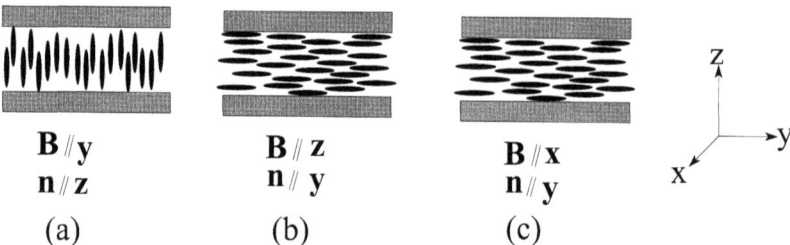

$\mathbf{B} \parallel \mathbf{y}$ $\mathbf{B} \parallel \mathbf{z}$ $\mathbf{B} \parallel \mathbf{x}$

$\mathbf{n} \parallel \mathbf{z}$ $\mathbf{n} \parallel \mathbf{y}$ $\mathbf{n} \parallel \mathbf{y}$

(a)   (b)   (c)

Figure 4.13. Geometries of the Frederiks transitions for positive diamagnetic anisotropy. (a) Bend, (b) splay, and (c) twist geometries are shown.

Figure 4.13 summarizes the results for the splay, twist and bend geometries. The following table lists the expressions for the critical field and coherence length in each case. The term *Frederiks transition* is applied generically to the three geometries, and magnetic coherence lengths can be introduced for each geometry (or elastic constant) by analogy to Eq. (4.19).

Bend geometry	Splay geometry	Twist geometry
$B_{c\,\text{bend}} = \dfrac{\pi}{d}(\mu_0 K_3/\Delta\chi)^{\frac{1}{2}}$	$B_{c\,\text{splay}} = \dfrac{\pi}{d}(\mu_0 K_1/\Delta\chi)^{\frac{1}{2}}$	$B_{c\,\text{twist}} = \dfrac{\pi}{d}(\mu_0 K_2/\Delta\chi)^{\frac{1}{2}}$
$\xi_3^{\text{M}} = \dfrac{1}{B}(\mu_0 K_3/\Delta\chi)^{\frac{1}{2}}$	$\xi_1^{\text{M}} = \dfrac{1}{B}(\mu_0 K_1/\Delta\chi)^{\frac{1}{2}}$	$\xi_2^{\text{M}} = \dfrac{1}{B}(\mu_0 K_2/\Delta\chi)^{\frac{1}{2}}$

Using electric fields

We have discussed the deformation caused by a magnetic field. However, the analysis is valid for electric fields as well. We could transcribe the equations for the critical field and the coherence length, which in the bend geometry, for example, read:

$$E_c = (\pi/d)\,(K_3/\varepsilon_0\Delta\varepsilon)^{\frac{1}{2}},$$

$$\xi_3^{\text{E}} = (1/E)\,(K_3/\varepsilon_0\Delta\varepsilon)^{\frac{1}{2}},$$

where $\Delta\varepsilon$ is the dielectric anisotropy at the frequency of the applied field. The use of electric fields is more in tune with the electro-optic applications of liquid crystals. Nevertheless, measurements of the elastic constants using magnetic fields are preferable in that they are not affected by charge impurities, which are invariably present in liquid crystal samples. In addition, electric fields can not be used to deform conducting samples, such as lyotropic micellar liquid crystals.

Field effects in cholesterics

Similar transitions are observed with cholesteric liquid crystals and twisted nematics. The latter are of particular interest to the operation of the twisted nematic display cell. Consider a cholesteric sample, of positive diamagnetic anisotropy. If the sample is thick enough, so that the restrictions imposed by the boundaries are negligible, the sample will align with the axis of the helix perpendicular to the applied field. As the applied field along the y-axis is increased from zero, the molecules will favor the alignment along the y direction. This is already the case for z values in the region of $P/4$, and odd multiples of $P/4$, and these regions will grow as the field is increased. The twist in these regions will decrease. On the other hand molecules above and below $z = P/2$ will tend to rotate in opposite directions, and consequently the twist in these regions will increase. If the diamagnetic anisotropy were negative, then the energy of the system would be a minimum when the axis of the helix is parallel to the applied field and the helix would remain undistorted.

Unwinding the cholesteric helix

A sufficiently high field will cause a field induced cholesteric to nematic transition, that is, the helix unwinds. A complete treatment can be found in Refs. 44–46. In the case where $n_x = \cos\theta(z)$, $n_y = \cos\theta(z)$, and $n_z = 0$ this *field induced cholesteric to nematic transition* occurs at a critical field value:

$$B_c = \frac{\pi^2}{P_0}\left(\frac{\mu_0 K_2}{\Delta\chi}\right)^{\frac{1}{2}},$$

where P_0 is the unperturbed pitch. It is interesting to note that as the field is decreased, the transition back to the cholesteric occurs at a field value lower than B_c. Thus, there is hysteresis associated with the transition. For an externally applied electric field the situation is rather similar to the magnetic case. If the sample has positive dielectric anisotropy, then the field induced transition will occur at

$$E_c = \frac{\pi^2}{P_0}(K_2/\varepsilon_0\Delta\varepsilon)^{\frac{1}{2}}.$$

Twisted nematic cell

For thin samples, the effects at the boundary become important. Under these conditions a sample of positive diamagnetic anisotropy is not free to align its helix perpendicular to the applied field. In this case it is possible to align the field parallel to the axis of the helix. A case of particular interest is the twisted nematic cell. Consider a nematic sample of positive diamagnetic anisotropy, between two parallel glass plates. The alignment at the surfaces is planar, in the xy-plane, however the director turns by 90° from the bottom ($z = 0$) to the top ($z = d$) plate. Thus the director describes one quarter of a helix. When a field is applied perpendicular to the glass plates, the axis of the helix does not reorient perpendicular to the applied field, due to the anchoring of the director at the surfaces. The situation is analogous to the Frederiks transition in the splay geometry, except that the director is twisted. For low applied fields, the configuration of quarter helix remains undisturbed. As the field is increased above a critical value, the director away from the glass surfaces moves out of the xy-plane, and as the field is further increased the director finally aligns parallel to the applied field. The critical field values for the magnetic and electric cases are [47, 48]

$$B_c = \frac{\pi}{d} \left(\mu_0 \frac{K_1 + (K_3 - 2K_2)/4}{\Delta \chi} \right)^{\frac{1}{2}}$$

and

$$E_c = \frac{\pi}{d} \left(\frac{K_1 + (K_3 - 2K_2)/4}{\varepsilon_0 \Delta \varepsilon} \right)^{\frac{1}{2}},$$

respectively. We note that the expressions for the critical fields involve all three elastic constants. For the electric case, a very useful formula provides the critical voltage

$$V_c = \pi \left(\frac{K_1 + (K_3 - 2K_2)/4}{\varepsilon_0 \Delta \varepsilon} \right)^{\frac{1}{2}}.$$

The term in the numerator is in the order of 10^{-11} N, $\varepsilon_0 = 8.85 \times 10^{-12}$ F/m and $\Delta \varepsilon = 0.1$ is a typical value. Using these values in the above equation we find V_c in the order of 5 V. To achieve the twisted nematic configuration, the cell must be very thin, otherwise the quarter helix will become unstable, which leads to defects in the alignment of the director. When the sample thickness is large enough so that the Mauguin limit is applicable, i.e.,

$$\lambda \ll (n_e - n_0)P = (n_e - n_0)4d,$$

then the polarization of a normally incident beam will emerge rotated by 90°. Thus, when the cell is placed between two crossed polarizers, the transmission of light will be maximum when the applied field is zero. If the applied field is well above the critical value, the beam is essentially polarized perpendicular to the optic axis, and travels with the ordinary refractive index, without change in the polarization direction. The intensity of the transmitted light in this case will be minimum. Typically, d is in the order of 10 μm, and the magnitude of $(n_e - n_0)$ is in the order of 0.1, which insures the applicability of the Mauguin limit, while keeping the sample thickness small.

Other methods for determining elastic constants

Light scattering measurements provide a powerful method for determining the elastic constants of liquid crystals. The principles of light scattering and the experimental techniques are described in detail in Chapter 6. Martins and coworkers proposed a method to measure the ratio K_3/K_1 in nematic polymers using an NMR technique [49] and Grupp [50] used a torsion balance to determine K_2 directly.

4.6 Viscous properties

In our discussion of the elastic constants we have assumed that the distortions did not result in *flow* of molecules; we will now relax this condition. Consider a nematic film between two parallel glass plates. If the plates slide parallel to each other, the molecules will flow, just as an ordinary fluid. In our discussion the lower plate will be held stationary and the upper plate will be moved with constant velocity **v** parallel to the y-axis. If **v** is small, then, for Newtonian fluids, the force per unit area of the plates (or stress) σ is

$$\sigma = \eta \, (dv/dz),$$

where η is the viscosity coefficient having dimensions of $ML^{-1}T^{-1}$ and measured in Pa · s and dv/dz the velocity gradient, which in what follows is assumed constant. For an isotropic liquid, the general behavior of η with temperature follows the relation

$$\eta \sim \exp(E/K_B T),$$

where E is an 'activation' energy, which measures the energy required to slide molecules past each other.

The Miesowicz coefficients

For a nematic liquid crystal, the stress σ should depend on the director **n** as well. Consider the three limiting geometries shown in Fig. 4.14. In Fig. 4.14

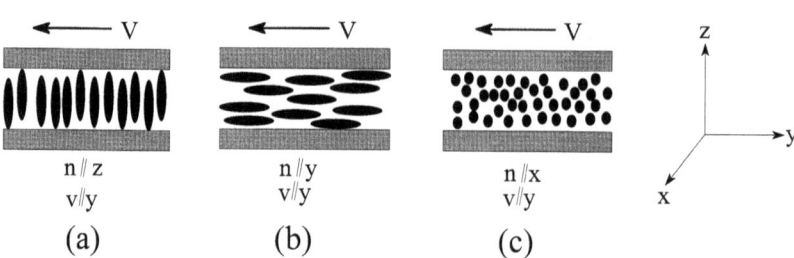

Figure 4.14. The director and velocity directions for the three Miesowicz coefficients η_1 (a), η_2 (b), η_3 (c). The top plate is moving parallel to the y-axis, and the velocity gradient is parallel to the z-axis. y–z is the shear plane.

we assume **v** along the y-axis, and the velocity gradient along the z-axis. The plane defined by **v** and the velocity gradient is the shear plane, here the yz-plane. An experiment using the geometry of Fig. 4.14(a) where **n** is by proper means (e.g., a strong magnetic field) maintained parallel to the velocity gradient, would measure a viscosity coefficient η_1. In Fig. 4.14(b), where the director is parallel to the velocity, the experiment would measure a different viscosity coefficient η_2. Finally, the geometry of Fig. 4.14(c), where the director is normal to the shear plane, measures a viscosity coefficient η_3. In practice, the values of the three coefficients differ substantially. For example, for MBBA, at 25 °C the values reported by Gahwiller [51] are

$$\eta_1 = 103 \times 10^{-3} \text{ Pa} \cdot \text{s}, \ \eta_2 = 24 \times 10^{-3} \text{ Pa} \cdot \text{s, and } \eta_3 = 42 \times 10^{-3} \text{ Pa} \cdot \text{s.}$$

Obviously, the viscosity coefficients depend on the order parameter S, and should become equal at the transition to the isotropic phase. As the order parameter depends on the temperature, dependence of the η_i ($i = 1$, 2, 3) on S should contribute to the temperature dependence of η_i as well. However, the temperature behavior of η_i is dominated by the activation energy behavior in a manner analogous to ordinary fluids. The three viscosity coefficients η_i are known as the Miesowicz coefficients. Note that different authors adopt different labels for the η_i. Here we follow the notation of de Jeu's book [3, Chapter 7]. At this point it is worth examining whether we could use the Miesowicz coefficients to express the viscosity measured for an arbitrary geometry. Consider the geometry shown in Fig. 4.15, which is intermediate between Figs. 14(a) and 14(b). In this figure, the director forms an angle θ with the velocity, and ϕ is the angle between **n** and the velocity gradient. The measured viscosity is $\eta(\phi, \theta)$. Obviously, $\eta(0, \pi/2) = \eta_1$ and $\eta(0,0) = \eta_2$ and $\eta(\pi/2, \pi/2) = \eta_3$. The question is: *can we express* $\eta(\phi, \theta)$ *in terms of* η_1, η_2 *and* η_3 *alone?* The answer is no! In this more general case one finds

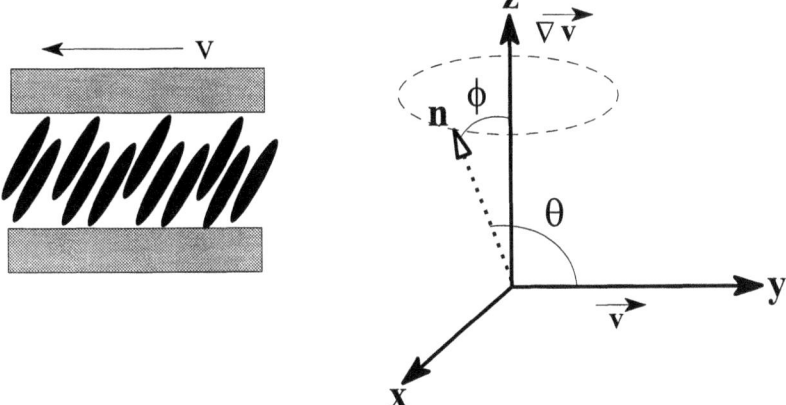

Figure 4.15.   The general case when the director makes an angle θ with the direction of flow, **v**, and an angle ϕ with the velocity gradient. In this case **v** is parallel to the y-axis, and the velocity gradient is along the z-axis.

$$\eta_1(\theta,\phi) = \eta_1 \sin^2 \theta \cos^2 \phi + \eta_2 \cos^2 \theta + \eta_3 \sin^2 \theta \sin^2 \phi + \\ \eta_{12} \cos^2 \theta \sin^2 \theta \cos^2 \phi.$$

The η_{12} term is new, and can be visualized as a 'pull' from the ends of the molecules. This contribution vanishes in the three limiting cases, and is maximum when $\theta = 45°$ and $\phi = 0$.

A set of five independent viscosity coefficients

From our discussion of tensor properties (Section 4.1) such as the magnetic susceptibility, we recall that the respective property of the uniaxial nematic at any orientation depended on two principal values (parallel and perpendicular to the director) and the angles. Here we have four such values so far! What is the difference? Looking back at the definition relating stress and velocity gradient we note that most generally the 'gradient' (dv/dz) refers to the gradient of the vector **v**. So the 'gradient' is part of a tensor with nine components. The other components will come from dv/dx and dv/dy. The stress is also a tensor with nine components. As for the viscosity, it is in fact a tensor of rank four. Fortunately, its components are not independent, in fact, it turns out that for our uniaxial nematic there are *five independent viscosity coefficients*. Thus we are looking for a *fifth coefficient*. This fifth coefficient is the *rotational viscosity coefficient*, which relates the viscous torque per unit volume, N_v, to the time rate of change of the alignment angle θ:

$$N_v \equiv -\gamma_1(d\theta/dt) \tag{4.20}$$

where γ_1 is the *rotational viscosity coefficient*, which along with η_1, η_2, η_3, and η_{12} forms a *complete set of viscosity coefficients for the uniaxial nematic.*

The Leslie coefficients

Although the set of viscosities introduced above is easy to visualize, theoretical treatments of the viscous properties use different sets of coefficients. The most common is a set introduced by Leslie [52] ($\{\alpha_i\}$), which are related to η_i, η_{12} and γ_1 as follows:

$$\alpha_1 = \eta_{12}$$
$$\alpha_2 = (-\eta_1 + \eta_2 - \gamma_1)/2$$
$$\alpha_3 = (-\eta_1 + \eta_2 + \gamma_1)/2$$
$$\alpha_4 = 2\eta_3$$
$$\alpha_5 = (2\eta_1 + \eta_2 - 3\eta_3 - \gamma_1)/2.$$

The development of the theory can be found in Ref. 3, Chapter 7, Ref. 4, Chapter 8, and Ref. 2, Chapter 5. A good starting point would be Landau and Lifshitz [53].

Measurement of viscosity: the flow alignment angle

Standard methods for measuring the viscosity of ordinary fluids include the falling ball method, the Poiseuille flow cell, and the Couette viscometer. The falling ball method was in fact applied to nematics [54], however, the interpretation of the results in terms of the five viscosity coefficients has not been established. In the Poiseuille cell, the liquid flows in a layer, with a controlled pressure differential at the two ends. The Couette viscometer consists of two rotating coaxial cylinders, with the sample introduced in the gap between the two cylinders. In a typical arrangement the flow may be distorting the direction of alignment, unless special precautions are taken to guarantee the integrity of the alignment. The experimental set-ups use strong magnetic fields to align the director, and low shear rate. A straightforward realization of the geometries indicated in Fig. 4.14 can be used to determine the viscosity coefficients η_i. This method could also determine η_{12}. This was done in the original work by Miesowicz who used a slow-moving plate to induce a simple shear [55].

The Poiseuille cell

More recent measurements are on laminar flow through a Poiseuille cell. Essentially, the method measures the volumetric flow rate, Q (volume of sample flowing through the cell per unit time), under the influence of an applied pressure difference, Δp, across the cell. The flow rate and the pressure differential are related by

$$Q = c\Delta p/\eta,$$

where c is a constant. Here η is the viscosity coefficient for the particular alignment geometry of the experiment. The alignment is achieved by placing the sample in a magnetic field. The cell can be calibrated with oils of known viscosity. Demountable glass flow cells of various thicknesses are commercially available (e.g. Hellma, or Starna) with a long dimension of about 3 cm [56]. A cell with an effective length of 85.5 cm was designed by Kneppe and Schneider [57] by interleaving 20 brass plates.

The transverse effect

A very interesting situation arises when the director alignment is maintained out of the shear plane, e.g., in a Poiseuille cell when the field is applied in a plane parallel to the plates at an angle to the x-axis, and the shear is in the yz-plane. In this case a pressure differential develops in the direction transverse to the flow, i.e., parallel to the x-axis, and the flow lines are deflected towards **n**. The analysis and experimental set-up to study this effect are given in Ref. 58.

Attenuation of ultrasonic waves

Shear waves do not propagate in fluids. Consequently, when a shear (transverse) wave propagating in a solid meets a solid–liquid interface, it decays very rapidly (exponentially) within the liquid, and is reflected back into the solid. The characteristic length of this decay, the so-called penetration depth, is given by [53, Chapter 2]

$$\delta = (\eta/\pi\rho f)^{1/2},$$

where η is the viscosity of the fluid, ρ is the density and f is the frequency of the shear wave. This technique has been extensively used by Martinoty and Candau [59] to determine viscosities of nematics. The velocity gradient is not constant in this experiment. The device can be calibrated, and the liquid crystal sample can be measured in the three geometries of Fig. 4.14.

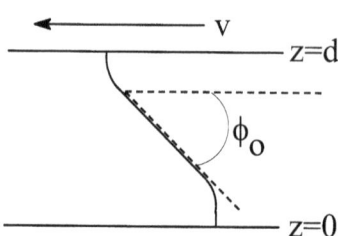

Figure 4.16. The geometry and director as a function of thickness, z, for flow alignment experiment to determine α_3/α_2. The angle between the local director and the flow is ϕ_0.

The viscosities determined in this way are not the same as those determined in flow experiments. These concerns as well as the relation of the effective viscosities measured by the ultrasonic method to the set of viscosity coefficients introduced above can be found in Chapter 5 of de Gennes and Prost [2].

Measurement of the flow alignment angle

To determine the fifth viscosity coefficient, γ_1, we need an experiment that involves the viscous torque. This is achieved by measuring the flow alignment angle, ϕ_0, or by a direct measurement of γ_1. Consider the Poiseuille cell shown in Fig. 4.16. By proper treatment, the director at the glass surfaces is aligned parallel to the z-axis, which is normal to the direction of flow. The flow alignment angle, as measured from the y-axis, is given by

$$\tan^2 \phi_0 = \alpha_3/\alpha_2,$$

where the Leslie coefficients are related to the set of viscosity coefficients used here as shown above. The derivation of this equation can be found in Chapter 7 of de Jeu's book [3]. If the cell is sufficiently thick, the flow alignment prevails over most of the film, except the boundary layers. The flow alignment angle can be measured optically. A beam along the z-axis is incident normal to the film, and is polarized at 45° to the y-axis. The x-component of the polarization is perpendicular to $\mathbf{n}$, and thus sees the ordinary refractive index. The y-component of the polarization is at an angle ϕ_0 to $\mathbf{n}$, and thus sees the refractive index $n(\phi_0)$. To get $n(\phi_0)$, we substitute $\theta = \pi/2 - \phi_0$ in Eq. (4.13). By using the birefringence methods described earlier, we can determine ϕ_0, and thus the ratio α_3/α_2. Note that ϕ_0 exists only if α_3/α_2 is positive. A description of this experiment can be found in Ref. 60. Meiboom and Hewitt measured ϕ_0 in a specially constructed

Couette viscometer, which was also used to determine γ_1. The description of the experiment can be found in Ref. 61.

Measurement of γ_1

Several methods have been used to measure γ_1. When a field is applied at an angle to the director, the magnetic torque is given by $N = \mu_0^{-1} \Delta\chi B^2 \sin\theta \cos\theta$, and the frictional torque associated with the reorientation of the director is given by Eq. (4.20). We can ignore the inertial term, and write for the balance of torques:

$$\gamma_1 \frac{d\theta}{dt} + \mu_0^{-1} \Delta\chi B^2 \sin\theta \cos\theta = 0.$$

This equation can be solved for θ to give: $\theta(t) = \theta(0) \tan^{-1}(t/\tau)$, where $\theta(0)$ is the angle between $\mathbf{n}$ and $\mathbf{B}$ at $t = 0$, and τ is the *relaxation time*, defined as $\tau = \mu_0 \gamma_1 / \Delta\chi B^2$.

The common methods to measure γ_1 rely on measuring the viscous torque directly, or monitoring the realignment with time. In the latter method, time dependent data on θ are fit and τ is derived as a parameter of best fit. Knowledge of $\Delta\chi$ and $\mathbf{B}$ yields γ_1. The main concern in both methods is to avoid creation of defects in the alignment. Experiments used to derive γ_1 and τ are described in what follows.

Rotating magnetic field method

Consider a sample of positive diamagnetic anisotropy in a rotating magnetic field $\mathbf{B}$. The magnetic field rotates with an angular velocity ω. The sample is in a cylindrical tube, suspended by a thin fiber, e.g., quartz. The rotation is very slow, typically less than 1 turn per second. As a result of the viscous torque, the director will not be parallel to $\mathbf{B}$, but will lag the field by a certain angle. In the steady state the lag angle is constant, and the director rotates with the angular velocity, ω, of the field. Opposing the rotation of the sample tube is the torsion of the suspending fiber, which is equal to the torsion angle of the fiber, α, times the torsion constant, D. The angle α can be measured by attaching a small reflector at the top of the sample tube, and D can be determined by measuring the period of oscillation of the suspended tube with the field turned off. One finds

$$\gamma_1 = D\alpha/\omega V,$$

where ω is the angular velocity of the rotating field and V the volume of the liquid crystal sample. A modification of this experiment is to use a fixed field and rotate the sample tube, using a stepping motor. Details on the rotating field method can be found in Ref. 37 and the rotating sample method in Ref. 62.

Other measurements of τ

Other methods have also been used to monitor the reorientation of the director. Essentially they assume the time dependence, $\theta(t)$, given above, and determine the relaxation constant, τ. Examples are the work of Martins *et al.* [49], Reizlein and Hoffmann [63] using NMR techniques, Saupe's group [64] using electrical conductance measurements, the SQUID magnetometer method [40] described above, and the measurement of Meiboom and Hewitt [61] using capacitance measurements to monitor the alignment of the director in a Couette cell.

Dynamics of the Frederiks transition

A very interesting situation occurs when the sample is in the form of a thin film, and a field is switched on or off. This is essentially the case of the Frederiks transition. When the applied field is above the critical value of the transition for the corresponding geometry (see Fig. 4.13) the director reorients. The angle of the director can be measured as a function of time (e.g., using the birefringence method). For small deformations, i.e., when the field is applied obliquely at a small angle to the direction of the undeformed **n**, the variation of the phase shift with time is essentially $\delta(t) = \delta_{max} \exp(-2t/\tau_0)$, as the field is switched off, and $\delta(t) = \delta_{max} [1 - \exp(-2t/\tau_H)]$, as the field is switched on. The factor 2 in the exponent comes from the fact that the phase shift is proportional to $n_o - n(\theta)$, which for small angles is proportional to θ^2, as can be verified using Eq. (4.13). Here the two relaxation times are

$$\tau_0 = d^2 \gamma_1^* / \pi^2 K_3$$

and

$$\frac{1}{\tau_H} = \frac{1}{\tau_0}\left[\left(\frac{B}{B_c}\right)^2 - 1\right]$$

where B_c is the critical field, B the applied field, d the thickness of the sample and K_3 the bend elastic constant. γ_1^* is an effective rotational viscosity, which is actually smaller than γ_1. Note that the above equations assume small deformation, and apply to the slowest mode of the reorientation. What that means is that when fitting the experimental data we have to rely more on the trailing part of the fit. The theory of the so-called dynamics of the Frederiks transition is given in Ref. 65 and examples of the application of the method can be found in Refs. 30 and 66.

Finally, as in the case of elastic constants, the viscosity can be studied using light scattering techniques, which are described in Chapter 6 of this book.

4.7 Electrical conductance

The basic equation describing the transport of charge through a medium under the influence of an applied electric field $\mathbf{E}$ is

$$\mathbf{j} = \sigma \cdot \mathbf{E},$$

where $\mathbf{j}$ is the current density and σ is the conductivity tensor. For conductors, which are characterized by free charge carriers, this equation is essentially a statement of Ohm's law. For a uniaxial liquid crystal, σ is a second rank tensor with two principal values: $\sigma_{\parallel}$ parallel to the director and $\sigma_{\perp}$ perpendicular to the director. The values of σ can vary over many orders of magnitude depending on the nature of the material. Thus, lyotropic liquid crystals can have high ionic concentrations, and as a result high conductance. On the other hand, most thermotropic liquid crystals are dielectrics, i.e., they do not have free charge carriers, and consequently one would expect a vanishing conductivity. In practice ionic charge carriers are present even in highly purified compounds. Besides the charged impurities, weak dissociation of the compound and charges injected from the metal electrodes can contribute to the conductance. For highly polar molecules there is also a contribution from the polarization current density:

$$\mathbf{j}_p = d\mathbf{P}/dt$$

In many studies, the samples were doped with controlled amounts of ionic species. Nevertheless, in comparison to ionic compounds, the conductance of thermotropics is generally low. The measurements in both cases are carried out using ac voltages, typically in the order of 1 kHz, for several reasons: to avoid polarization of the electrodes, chemical reactions, and electrolysis in the case of aqueous micellar liquid crystals. The design of the conductivity cells depends on the magnitude of the conductance to be measured. For ionic materials the separation of the electrodes is large, typically in the order of 1 cm. For low conductance materials, the spacing of the electrodes is small (1 mm or less) and the cross-section of the cell is large, typically in the order of a few cm^2.

Conductance measurements

The conductance of the cell can be measured using impedance bridges, or by comparing the cell to a standard resistor. The principal values of the conductivity are then determined by proper alignment of the director by surface treatment or by magnetic fields. The description of conductance mechanisms and associated electrochemical processes in liquid crystal cells can be found in Ref. 67. The interpretation of the Williams and chevron domains can be found in Refs. 2 and 67. Examples of conductance measurements on thermotropics can be found in Ref. 68 and for lyotropics in Refs. 64 and 69.

Williams domains and dynamic scattering

When the electric field within the dielectric liquid crystal is high (10 V in a film thickness of 10 μm would produce a field of 10^4 V/cm) several interesting effects may occur. At low frequencies (below 100Hz) the so-called Williams domains appear above a frequency-dependent critical voltage, which can be in the order of 5 V. These domains are periodic lines, with spacing comparable to the sample thickness, and can be viewed with light polarized perpendicular to the domain walls. The pattern reflects the periodic deformation of the director. As the voltage is further increased, the Williams domains become fluid, the pattern becomes turbid, and diffuse light scattering (polarization independent) occurs. This is the so-called dynamic light scattering regime. As the frequency is increased above 100 Hz, the critical voltage increases markedly to about 100 V or so, and the observed pattern consists of parallel striations with spacing much smaller than the sample thickness. These are the so-called *chevron domains*. The transition from the Williams domains to the chevrons occurs at a frequency which is inversely proportional to the reorientation time of the molecules, and proportional to the conductivity of the sample. Conductance measurements are usually carried out at about 1 kHz. The director will remain undisturbed as long as the applied voltage is below the threshold voltage for the chevrons, which is rather high at this frequency.

4.8 Thermal conductance

Heat diffusion is described by the equation

$$J = - \kappa^t \cdot \nabla T,$$

where J is the heat flux (i.e., energy per unit area and time), T is the temperature (which is a function of time and position) and κ^t is the thermal conductivity tensor (see Chapter 5 of Ref. 70). Here we assume that heat transfer due to convection and radiation are negligible. The equation describing the temperature distribution in space and time is

$$\rho c\, \partial T/\partial t = \nabla \cdot [\kappa^t \cdot \nabla T],$$

which is a second-order partial differential equation. The above equation is further complicated by the fact that the density, ρ, and the specific heat, c, may be functions of temperature as well. Even if ρ and c are constant, the solution of the diffusion equation involves non-elementary functions [71 and 72]. The situation is greatly simplified when the temperature distribution is constant in time. This is the so-called steady state. In what follows we will describe an experimental method based on steady-state conditions, and a dynamical experiment measuring the diffusion of a heat pulse with time. We will also describe the photoacoustic experiment, which is essentially a dynamical experiment as well.

Steady-state method

In the steady state, $\partial T/\partial t = 0$. For thin samples, e.g., a liquid crystal film between two plates, we can assume that the temperature distribution is constant on the yz-plane, which is taken parallel to the plates. In addition, if the director is well aligned over the film, e.g., parallel to the x-axis, then the temperature gradient dT/dx is constant, and equals $\Delta T/d$ where ΔT is the temperature difference between the two plates, and d is the film thickness. Under these assumptions the heat diffusion equation becomes

$$J = -\kappa^t_\parallel \Delta T/d.$$

Using the fact that $J = AQ$, where A is the area of the film and Q is the heat flowing from the warmer plate into the sample per unit time, we find

$$\kappa^t_\parallel = \frac{dJ}{A\Delta T}.$$

Analogous equations apply to the homogeneous alignment, which is used to measure $\kappa^t_\perp$. A steady-state experimental set-up to measure the thermal conductivity coefficients is described in Ref. 73. The cell was constructed of two copper plates. The temperature of one of the plates was kept constant. The other plate was electrically heated at a constant rate. The plates were held together by nylon screws, and the sample area was sealed by a rubber gasket. The thickness of the sample was about 0.75 mm. ΔT was measured as the temperature difference between the two copper plates. The heat flow per unit time, Q, was determined by the electrical power dissipated at the heated plate. The sample thickness was high enough to minimize the surface alignment effects. The principal values parallel and perpendicular to the director were measured by aligning the director in the proper geometry using a magnetic field. The results for MBBA and PAA show that the conductance is larger parallel to the director.

Heat pulse method

Consider a thin film between two plates. As in the steady-state experiment, we will assume that the temperature distribution is constant on the yz-plane, which is taken parallel to the plates, and that the director is well aligned over the film, e.g., parallel to the x-axis. A heat pulse of flux J and duration τ enters one side of the film (x = 0) at time t = 0. Solving the diffusion equation under these conditions [71] we can find the temperature rise at the other side of the film (x = d), as a function of time. In the experiment described by Akahane et al. [74] a 30 μm film of 8CB was aligned between two plates. A 5 mm by 400 μm aluminum strip was evaporated onto one of the glass plates and served as a heater. The temperature rise at x = d was monitored by measuring the resistance change in a 5 mm by 40 μm aluminum strip evaporated onto the second glass surface. The heat pulses were 5 ms in duration. The data were taken in the isotropic, nematic and smectic-A phases. The results show that the conduction parallel to the director is higher than that perpendicular to the director.

Photoacoustic method

The photoacoustic effect (not to be confused with the acousto-optic effect, used to induce optical gratings for beam deflection) was discovered by A. G. Bell, over a century ago. A chopped beam is incident on the sample. As a result of the absorption of radiation, the sample is periodically heated, which in turn heats the boundary air-layer above the sample. The periodic heating creates a pressure wave in the sampler chamber, which can be detected using a microphone and a lock-in amplifier. The sound wave has the frequency of the chopped beam, thus the frequency of the chopper (few Hz to few kHz) can be used as reference for the lock-in amplifier. The measurement provides the amplitude and phase of the photoacoustic signal, which can be used to determine the thermal conductance and the specific heat of the sample. The theory of the photoacoustic effect is described in Ref. 75. An application to liquid crystals, and details of the design of the experiment, which uses a magnetic field to align the director, are given in Ref. 76. Additional discussion on the specific heats can be found in Chapter 7 of this book.

4.9 Density measurements

The density is a scalar property, yet the alignment of the molecules in liquid crystal phases affects the packing, and thus the density of the phase. This becomes obvious at the nematic-to-isotropic phase transition. The more efficient packing of the molecules in the nematic phase is generally manifested as a jump in the density. As discussed above, the density versus temperature data are essential to the study of diamagnetic properties of the material. We will describe three methods of measurement, namely a buoyancy method, the vibrating U-tube apparatus, and the dilatometric method.

The buoyancy method

This method is very simple in principle, although the conduct of the experiment is by no means trivial. The density of the liquid crystal is measured by determining the buoyancy, which is proportional to the weight of the displaced fluid. The sample is introduced in a thermostated container. A solid object, e.g., a glass sphere, is weighed while totally submerged in the sample. In terms of mass, the reading of the balance, R_{LC}, equals:

$$R_{LC} = m - \rho_{LC}(T) \, V(T) + m_{wire},$$

where m is the mass of the sphere, $\rho_{LC}(T)$ is the density of the liquid crystal, $V(T)$ is the volume of the sphere, and m_{wire} accounts for the suspension wire. The calibration of the experimental set-up consists of determining m and $V(T)$. The mass, m, can be found by measuring the sphere in air and in distilled water at ambient temperature. $V(T)$ is determined by a separate measurement with the sample submerged in water or a suitable liquid. This measurement should be taken over the temperature range of interest. The absolute accuracy of the measurement depends on the calibration of the balance, which is accomplished by measuring standard weights. This set-up can readily achieve resolution 10^{-4} g/cm^3. The method requires relatively large amounts of liquid crystal sample, in the order of 5 g. As the sample is essentially in an open container, additional precautions may be necessary to preserve the composition of the sample over the length of the experiment [34]. Several manufacturers (e.g., Cahn) offer balances with readout suitable for computer controlled data acquisition.

Vibrating U-tube densitometer

In this technique the natural period of oscillation of a capped U-tube containing the fluid of interest is measured. The period of natural oscillation, T, is related to the density, ρ, of the liquid as follows:

$$T = (A\rho + B)^{\frac{1}{2}},$$

where A and B are apparatus parameters, which depend on temperature. The instrument counts the number of oscillations in a selectable time interval. As the parameters A and B are temperature dependent, a liquid of known density must be used to calibrate the apparatus over the temperature range of interest. Instruments of this type are commercially available (e.g., Anton Paar Density Meter) in several designs. The instrument resolution ranges from four to six decimal places, depending on the model. The U-tube is housed in a double-walled glass container which is detachable in some units. The instrument allows fluid circulation for thermostating purposes, with space for a temperature sensor, and a digital readout interface. The capacity of the U-tube ranges from milliliters to microliters, with plugs at both ends, which makes the instrument suitable for hard-to-find samples. Care must be exercised in cleaning the U-tube (the instrument comes with a pump to blow out cleaning solvents) and to avoid bubbles in the sample. An application of the vibrating U-tube method can be found in Refs. 77 and 78. A discussion of errors in this type of measurements when very precise measurements are made is found in Ref. 79.

Dilatometry

Dilatometry is essentially a technique to measure thermal expansion. In its simplest form, a liquid sample is introduced into a U-tube, sealed at one end. An amount of mercury is then introduced into the tube, with the sample liquid floating over the mercury. The other end of the U-tube is jointed into a glass capillary. As the temperature of the sample is changed, the height of the mercury column in the capillary changes proportionally.

The change in the column height is amplified by a factor proportional to the ratio of the sample tube cross-section to the cross-section of the capillary, which could be rather high depending on the specifics of the design. In the dilatometer described by Armitage and Price [80] where the sample volume was about 2 cm^3 and the diameter of the capillary 0.5 mm the amplification factor was about 10^5. The column height can be measured using a cathetometer. However, in this experimental set-up, the height of the column was monitored more conveniently by forming a coaxial capacitor between the mercury column and a conductor surrounding the capillary. The dilatometer was introduced as a branch of a bridge circuit, which was designed to produce a signal which is linear with the column height. The signal was measured by a lock-in amplifier. The set-up was used to measure the expansion of cholesteryl nonanoate [80], MBBA and PAA [81].

Acknowledgments

I am most thankful to my longtime mentor A. Saupe for teaching me about liquid crystals. Also, I had the good fortune of working with people who generously shared their knowledge and time with me: I am thankful to A. Figueiredo Neto, E. Andreoli de Oliveira, C. Rosenblatt, M. Stefanov, C. Frank, S. Lotz and last but not least my brother Demetri, for many illuminating discussions. The editor and I are grateful to Mike Fisch for his help.

References

1. S. Chandrasekhar, *Liquid Crystals*, 2nd Edn., Cambridge University Press, Cambridge (1992).
2. P. G. de Gennes and J. Prost, *The Physics of Liquid Crystals*, 2nd Edn., Clarendon Press, Oxford (1993) Chapter 2.
3. W. H. de Jeu, *Physical Properties of Liquid Crystalline Materials*, Academic Press, New York (1980).
4. G. Vertogen and W. H. de Jeu, *Thermotropic Liquid Crystals, Fundamentals*, Springer-Verlag, Berlin, Heidelberg (1988).
5. P. J. Collings and J. S. Patel (Eds.), *Handbook of Liquid Crystal Research*, Oxford University Press, New York, Oxford (1997).
6. M. L. Boas, *Mathematical Methods in the Physical Sciences*, 2nd Edn., John Wiley and Sons, New York (1983) Chapter 10.
7. G. Arfken, *Mathematical Methods for Physicists*, 3rd Edn., Academic Press, Orlando (1985) Chapter 4.
8. C. Kittel, *Introduction to Solid State Physics*, 7th Edn., John Wiley and Sons, New York (1996) Chapter 13.
9. J. R. Reitz, F. J. Milford and R. W. Christy, *Foundations of Electromagnetic Theory*, Addison-Wesley, Reading (1980), p. 157.

10. See for instance, J. P. Runt and J. J. Fitzgerald (Eds.), *Dielectric Spectroscopy of Polymeric Materials*, ACS Press, Washington, DC (1997); J. R. Macdonald (Ed.), *Impedance Spectroscopy*, John Wiley and Sons, New York (1987); M. Pfeiffer, S. Hiller, S. Wrobel and W. Hasse, *Ferroelectrics* **147**, 419 (1993).

11. D. Meyerhofer, *J. Appl. Phys.* **46**, 5084 (1975).

12. M. Francon, N. Krauzman, J. P. Mathieu and M. May, *Experiments in Physical Optics*, Gordon and Breach, New York (1970) Chapter 3.

13. R. Wood, *Physical Optics*, Dover Publications, New York (1961) Chapter 6.

14. C. B. Sawyer and C. H. Tower, *Phys. Rev.* **35**, 269 (1930).

15. M. M. Berkens and Th. Kwaaitaal, *J. Phys. E* **16**, 516 (1983).

16. W. J. Merz, *J. Appl. Phys.* **27**, 938 (1956).

17. Ph. Martinot-Lagarde, *J. Phys.* **38**, L17 (1977).

18. K. Miyasato, S. Abe, H. Takazoe, A. Fukuda and E. Kuze, *Jpn. J. Appl. Phys.* **22**, 661 (1983).

19. J. S. Patel and J. W. Goodby, *Chem. Phys. Lett.* **137**, 91 (1987).

20. J. W. Goodby, *Ferroelectricity and Related Phenomena*, Vol. 7, *Ferroelectric Liquid Crystals*, Chapter 9, G. W. Taylor (Ed.), Gordon and Breach, Philadelphia (1991).

21. See for instance, R. Guenther, *Modern Optics*, John Wiley and Sons, New York (1990); M. Born and E. Wolf, *Principles of Optics*, 5th Edn., Pergamon Press, New York (1975); M. V. Klein and T. E. Futak, *Optics*, 2nd Edn. John Wiley and Sons, New York (1986).

22. D. S. Klinger, J. W. Lewis and C. E. Randall, *Polarized Light in Optics and Spectroscopy*, Academic Press, San Diego (1990) Chapter 2.

23. E. B. Priestley, 'Introduction to the Optical Properties of Cholesteric and Chiral Nematic Liquid Crystals', in *Introduction to Liquid Crystals*, Plenum Press, New York (1975) E. B. Priestley, P. J. Wojtowicz and P. Sheng (Eds.), Chapter 11.

24. D. W. Berreman and T. J. Scheffer, *Mol. Cryst. Liq. Cryst.* **11**, 395 (1970).

25. R. Dreher and G. Meier, *Phys. Rev. A* **8**, 1616 (1973).

26. O. S. Heavens and R. W. Ditchburn, *Insight into Optics*, John Wiley and Sons, Chichester (1991) Chapter 2, Section 5.

27. N. H. Hartshorne and A. Stuart, *Practical Optical Crystallography*, American Elsevier, New York (1969) Chapter 5.

28. T. Haven, D. Armitage and A. Saupe, *J. Chem. Phys.*, **75**, 352 (1981).

29. P. Boonbrahm, Ph.D. Dissertation, Kent, Ohio (1983).

30. W. Rupp, H. P. Grossmann and B. Stoll, *Liq. Cryst.* **3**, 583 (1988).

31. G. Maret and K. Dransfield, *Physica* **86B**, 1077 (1977); G. Maret and G. Weill, *Biopolymers* **22**, 2727 (1983).

32. C. Rosenblatt, F. Torres de Araujo and R. B. Frankel, *Biophys. J.* **40**, 83 (1982).

33. L. F. Bates, *Modern Magnetism*, Cambridge University Press, London (1961).

34. M. Stefanov, Ph.D. Dissertation, Kent, Ohio (1983).

35. W. H. de Jeu and A. P. Classen, *J. Chem. Phys.* **68**, 102 (1978).

36. S. Plumley, Y. K. Zhu, Y. W. Hui and A. Saupe, *Mol. Cryst. Liq. Cryst.* **182B**, 215 (1990).

37. J. Prost and H. Gasparoux, *Phys. Lett A* **36**, 245 (1971); H. Gasparoux and J. Prost, *J. Physique* **32**, 953 (1971).

38. J. C. Gallop, *SQUIDs, the Josephson Effects and Superconducting Electronics*, Adam Higler, Bristol (1991).

39. J. S. Philo and W. M. Fairbank, *Ber. Sci. Instr.* **48**, 1529 (1977); R. E. Sarwinski, *Cryogenics* **17**, 671 (1977).
40. Th. Dries, K. Fuhrmann, E. W. Fischer and M. Ballauff, *J. Appl. Phys.* **69**, 7539 (1991).
41. V. Frederiks and A. Repiewa, *Z. Phys.* **42**, 532 (1927); V. Frederiks and V. Zolina, *Trans. Faraday Soc.* **29**, 919 (1933).
42. J. Prost and H. Gasparoux, *C. R. Acad. Sci. Paris* **C273**, 355 (1972).
43. H. J. Deuling, M. Gabay, E. Guyon and P. Pieranski, *J. Phys.* **36**, 689 (1975).
44. P. G. de Gennes, *Solid State Commun.* **6**, 163 (1968).
45. R. B. Meyer, *Appl. Phys. Lett.* **14**, 208 (1969).
46. G. Durand, L. Leger and F. Rondelez, *Phys. Rev. Lett.* **22**, 227 (1969).
47. M. Schadt and W. Helfrich, *Appl. Phys. Lett.* **18**, 127 (1971).
48. F. M. Leslie, *Mol. Cryst. Liq. Cryst.* **12**, 57 (1970).
49. A. F. Martins, P. Esnault and F. Volino, *Phys. Rev. Lett.* **57**, 1745 (1986).
50. J. Grupp, *Phys. Lett.* **99**, 373 (1983); *Rev. Sci. Instrum.* **54**, 754 (1983).
51. Ch. Gahwiller, *Phys. Lett A* **36**, 311 (1971).
52. F. M. Leslie, *Q. J. Mech. Appl. Math.* **19**, 367 (1966); *Arch. Ration. Mech. Anal.* **28**, 265 (1968).
53. L. D. Landau and E. M. Lifshitz, *Fluid Mechanics*, Pergamon Press, Oxford (1959).
54. M. Roy, J. P. McClymer and P. H. Keyes, *Phys. Rev A* **30**, 3156 (1984).
55. M. Miesowicz, *Nature* **27**, 158 (1946); see also reference [52].
56. W. W. Beens and W. H. de Jeu, *J. Phys.* **44**, 129 (1983).
57. H. Kneppe and F. Schneider, *Mol. Cryst. Liq. Cryst.* **65**, 23 (1981).
58. P. Pieranski and E. Guyon, *Phys. Lett A* **49**, 237 (1974); E. Guyon and P. Pieranski, *J. Physique Coll.* **C1**, 36, 203 (1975).
59. P. Martinoty and S. Candau, *Mol. Cryst. Liq. Cryst.* **14**, 243 (1971).
60. Ch. Gahwiller, *Phys. Rev. Lett.* **28**, 1554 (1972).
61. S. Meiboom and R. C. Hewitt, *Phys. Rev. Lett.* **30**, 261 (1973).
62. H. Kenppe and F. Schneider, *J. Phys. E.* **16**, 512 (1983).
63. K. Reizlein and H. Hoffmann, *Progr. Colloid Polym. Sci.* **69**, 83 (1984).
64. P. J. Photinos and A. Saupe, *J. Chem. Phys.* **85**, 7467 (1986); A. Saupe, S. Y. Xu, S. Plumley, Y. K. Zhu and P. Photinos, *Physica* **174A**, 195 (1991).
65. P. Pieranski, F. Brochard and E. Guyon, *J. Phys. (Paris)* **34**, 35 (1973).
66. E. Zhou, M. Stefanov and A. Saupe, *J. Chem. Phys.* **88**, 5137 (1988).
67. L. M. Blinov, *Electro-Optical and Magneto-Optical Properties of Liquid Crystals*, John Wiley Interscience, Chichester (1983) Chapter 5.
68. M. Bertolotti, F. Scudieri, D. Sette and R. Bartolino, *J. Appl. Phys.* **43**, 3914 (1972); M. Schadt and C. von Planta, *J. Chem. Phys.* **63**, 4379 (1975).
69. N. Boden, S. A. Corne and K. W. Jolley, *Chem. Phys. Lett.* **105**, 99 (1984).
70. L. D. Landau and E. M. Lifshitz, *Theory of Elasticity*, 2nd Edn., Pergamon Press, Oxford (1970).
71. H. S. Carslaw and J. C. Jeager, *Conduction of Heat in Solids*, 2nd Edn., Clarendon Press, Oxford (1959).
72. J. Crank, *The Mathematics of Diffusion*, 2nd Edn., Oxford University Press, Oxford (1975).
73. R. Vilanove, E. Guyon, C. Mitescu and P. Pieranski, *J. Phys. Paris* **35**, 153 (1974).
74. T. Akahane, M. Kondoh, K. Hashimoto and M. Nakagawa, *Jpn. J. Appl. Phys.* **26**, L1000 (1987).
75. A. Rosencwaig and A. Gersho, *J. Appl. Phys.* **47**, 64 (1976); see also A. Rosencwaig, *Phys. Today* **28**(9), 23 (1975).

76. J. Thoen, C. Glorieux, E. Schoubs and W. Lauriks, *Mol. Cryst. Liq. Cryst.* **191**, 29 (1990).
77. A. A. Barbosa and A. V. A. Pinto, *J. Chem. Phys.* **98**, 8345 (1993).
78. W. H. de Jeu and A. P. Classen, *J. Chem. Phys.* **68**, 102 (1978).
79. A. Zywocinski, *J. Chem. Phys.* **103**, 3087 (1999).
80. D. Armitage and F. P. Price, *J. Appl. Phys.* **47**, 2735 (1976).
81. D. Armitage and F. P. Price, *Phys. Rev. A* **15**, 2496 (1977).

5

NMR studies of orientational order

PAUL UKLEJA

Department of Physics, University of Massachusetts at Dartmouth, North Dartmouth, MA 02747, USA

AND

DANIELE FINOTELLO

Department of Physics, Kent State University, Kent, OH 44242, USA

5.1 Introduction

The study of liquid crystal orientational order by nuclear magnetic resonance (NMR) has inherent difficulties that do not usually arise in solids or liquids. On one hand, molecules in liquid crystal phases are not fixed in position and orientation because interactions between molecules are not as static as in solids. On the other hand, their reorientations are not sufficiently rapid and random to time average most interactions to simple values, as happens in liquids. It is just this intermediate behavior which, although it introduces difficulties, creates a wealth of possibilities to probe the orientation and dynamics of molecules in liquid crystal phases.

NMR is a non-invasive technique that probes the local environment of specific nuclei in the sample. Through interactions of a nucleus with its surroundings, NMR becomes a tool that is sensitive to molecular structure, orientation, configuration, and dynamics. These molecular properties are related to the bulk properties such as order parameters, viscosities, and elastic constants as well as the molecules' response to external influences such as electric and magnetic fields, shear stresses, surface interactions, and so forth.

A list of the type of information that can be obtained using NMR includes the following:

- molecular structures: using standard solvents or using liquid crystal solvents to align the subject molecules,
- phase identification of new compounds or mixtures,
- molecular alignments and conformations in various phases,
- translational, rotational, and conformational dynamics of molecules, segments of molecules, and aggregates of molecules over a wide range

155

of time and length scales; from this, physical properties such as diffusion coefficients and viscosities may be inferred,

• phase diagrams and the kinetics of phase changes and phase separations.

This chapter starts with a discussion of the basic physics that describes the interaction of nuclear spins with a static magnetic field. This interaction is then described from both the classical and the quantum mechanical points of view. The next section is devoted to a discussion of the more subtle interactions of spins with their environments, and how these lead to information on the alignment and the dynamics of molecules in liquid crystal phases. Then comes a discussion of the typical hardware of an NMR spectrometer, followed by a brief presentation of practical matters specific to working with a liquid crystalline materials. Selected examples of NMR experiments are included in Section 5.6, and the chapter concludes with a short but detailed bibliography.

5.2 Nuclear spins in a magnetic field

The phenomenon of magnetic resonance [1–9] arises from the interaction between a static magnetic field and the magnetic moment of a charged, rotating object, specifically the nucleus of an atom. This interaction leads to a precession of the axis of rotation about the field. That is, the axis itself moves at constant speed around the surface of a cone centered on the static field, much the same as the axis of a spinning top may precess about the vertical. In turn, this precession will induce an ac signal in a coil located in the vicinity. The frequency of this motion and of the induced signal is called the Larmor frequency, f_L. It is independent of the angle between the field and the rotational axis while it is proportional to the field. It also depends on the way the charge and mass of the object are distributed with the result that, in a given magnetic field, each nuclear isotope resonates at a characteristic frequency allowing the experimenter to distinguish between different nuclei. (See Table 5.1.) For nuclear spins, the spectrum and the dynamical behavior (e.g., decay time) of the induced signal depend on details of various interactions between the spins and their surroundings. They also depend on details of the way in which the spins are manipulated by the experimenter. It is in this manner that the NMR signal yields so much detailed and useful information. Thus, one NMR experiment may be used to determine some particulars of the molecular structure, another to study the orientational order, and yet another to study the translational diffusion of one or more types of molecules in the sample.

Before discussing the equations of motion, let us consider the relation-ship between the angular momentum and magnetic moment of an object. A rotating, rigid body with an electric charge acts like a simple current loop or a compass needle in that it: (1) produces a magnetic field, and (2) expe-riences a torque when placed in an external magnetic field. This behavior is due to a property called the magnetic moment of the object. A flat loop of current i around an area A has a magnetic moment of $\boldsymbol{\mu} = i\mathbf{A}$, where the direction of $\mathbf{A}$ is the normal (perpendicular) to the plane of the loop. A simple example is that of a uniformly charged hoop rotating about an axis through its center and perpendicular to the plane of the hoop. If the hoop has radius r, charge q, and rotates with a period T, it effectively behaves as a current loop with $i = q/T$ and area πr^2. This results in a magnetic moment of magnitude $\mu = q\pi r^2/T$ or $q\omega r^2/2$, when expressed in terms of its angular velocity $\omega \equiv 2\pi/T$. For positive charges, the magnetic moment and angular velocity vectors would be parallel. The same would be true for a particle with mass m and charge q moving in a circular orbit with the same period. Since the hoop or particle would have an angular momentum $L = mr^2\omega$, the two vectors are proportional, or

$$\boldsymbol{\mu} = \frac{q}{2m}\mathbf{L}. \tag{5.1}$$

It is often convenient to work with the dimensionless quantity $\mathbf{I} \equiv \mathbf{L}/\hbar$, the angular momentum measured in units of the Planck constant $\hbar$, in which case the above result becomes

$$\boldsymbol{\mu} = \frac{q\hbar}{2m}\mathbf{I}. \tag{5.2}$$

It is customary to define the following units of magnetic moment for the electron and proton:

$$\mu_B = (e\hbar/2m_e) = 9.27402 \times 10^{-24} \text{ JT}^{-1} \quad \text{(Bohr magneton)}$$
$$\mu_N = (e\hbar/2m_p) = 5.05095 \times 10^{-27} \text{ JT}^{-1} \quad \text{(nuclear magneton)}.$$

These are defined for convenience as energy changes are proportional to them. For any particle, the magnetic moment and angular momentum are proportional to each other, but the proportionality constant, called the magnetogyric ratio and denoted by γ, depends on the structure of the par-ticle, so that Eqs. (5.1) and (5.2) become

$$\boldsymbol{\mu} = \gamma\mathbf{L} = \gamma\hbar\mathbf{I}. \tag{5.3}$$

Typical values of γ are included in Table 5.1. Also note that the quantity I appearing in the second column of Table 5.1 is the quantum number asso-ciated with $\mathbf{I}$ and will be discussed later.

Table 5.1 *Properties of selected isotopes.*

Isotope	I	Natural abundance (percent)	γ (10^7 rad T^{-1} s^{-1})	Larmor frequency in a 1 T field (MHz)
^{1}H	$\frac{1}{2}$	99.985	26.7519	42.5770
^{2}H	1	0.015	4.1066	6.5360
^{7}Li	$\frac{3}{2}$	92.58	10.3975	16.5482
^{13}C	$\frac{1}{2}$	1.108	6.7283	10.706
^{19}F	$\frac{1}{2}$	100	25.181	40.062
^{17}O	$\frac{5}{2}$	0.037	-2.2407	5.7739
^{31}P	$\frac{1}{2}$	100	10.841	17.235

Classical description

The energy U that describes the interaction between a magnetic moment $\boldsymbol{\mu}$ and a magnetic field $\mathbf{B}$ is $U = -\boldsymbol{\mu}\cdot\mathbf{B} = -\mu B \cos\theta$, where θ is the angle between the field and the magnetic moment. The torque on such a magnetic moment is $\boldsymbol{\mu}\times\mathbf{B}$, which leads to the following equation of motion:

$$\boldsymbol{\mu}\times\mathbf{B} = \frac{d\mathbf{L}}{dt} = \hbar\frac{d\mathbf{I}}{dt}, \tag{5.4}$$

or utilizing Eq. (5.3)

$$\frac{d\boldsymbol{\mu}}{dt} = \gamma\boldsymbol{\mu}\times\mathbf{B} = \gamma\begin{vmatrix} \hat{\imath} & \hat{\jmath} & \hat{k} \\ \mu_x & \mu_y & \mu_z \\ B_x & B_y & B_z \end{vmatrix}. \tag{5.5}$$

Since the vector resulting from the cross product of two vectors is perpendicular to both of the vectors, Eq. (5.5) implies that changes in $\boldsymbol{\mu}$ are perpendicular to $\boldsymbol{\mu}$ itself, meaning that its length μ is fixed and that the motion is a rotation. In fact, the magnetic field must be the axis of the rotation since changes in $\boldsymbol{\mu}$ are also perpendicular to it. In the presence of a constant magnetic field directed along the z-axis, $\mathbf{B}_0 = B_0\hat{k}$, this can be expressed as

$$\frac{d\mu_x}{dt} = \gamma B_0(\mu_y), \tag{5.6a}$$

$$\frac{d\mu_y}{dt} = -\gamma B_0(\mu_x), \tag{5.6b}$$

$$\frac{d\mu_z}{dt} = 0. \tag{5.6c}$$

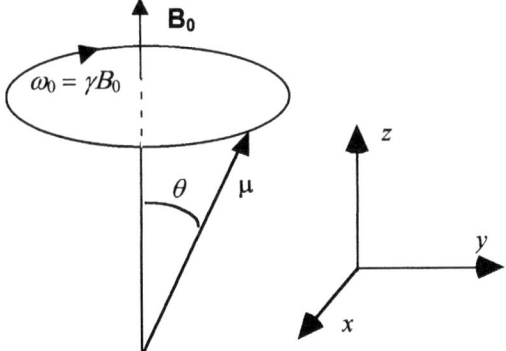

Figure 5.1. Precession of a magnetic moment μ about a constant magnetic field $\mathbf{B}_0$.

One way to 'see' that this set of equations implies a rotation is to take $\dfrac{d}{dt}$ of Eq. (5.6a) and substitute in Eq. (5.6b) to eliminate μ_y; this produces an equation of motion for μ_x alone:

$$\frac{d^2\mu_x}{dt^2} = (\gamma B_0)^2 \mu_x = -(\omega_0)^2 \mu_x. \qquad (5.7a)$$

Similarly,

$$\frac{d^2\mu_y}{dt^2} = (\gamma B_0)^2 \mu_y = -(\omega_0)^2 \mu_y. \qquad (5.7b)$$

The solutions to Eqs. (5.7) are

$$\mu_x = \mu_0 \cos(-\omega_0 t + \phi) \quad \text{and} \quad \mu_y = \mu_0 \sin(-\omega_0 t + \phi), \qquad (5.8a)$$

or in complex notation,

$$\mu_x + i\mu_y = \mu_0 \exp(-i\omega_0 t + i\phi), \qquad (5.8b)$$

where $\omega_0 = \gamma B_0$ is the angular velocity. The tip of the magnetic moment vector thus traces out a circle in the plane perpendicular to $\mathbf{B}_0$ (the x–y plane) at an angular velocity $\omega_0 = \gamma B_0$. The direction of ω_0 is anti-parallel to the magnetic field as shown in Fig. 5.1 for a positive γ. In this case, the energy $U = -\mu B_0 \cos\theta = -\mu_z B_0$ is constant by virtue of Eq. (5.6c) and the angle θ between the moment and the (constant) magnetic field is fixed.

It is instructive to point out the difference between this case and a compass needle which, after all, tends to line up parallel to an external field, and not move at a constant angle to it. In the needle, the microscopic magnetic moments of the iron atoms interact very strongly with each other and thus with the lattice of the material, fixing the net magnetic moment of the material in the macroscopic body of the needle. The magnetic moments are due to electron motions and the electrons only account for a small fraction of the mass of the needle. Thus, the net magnetic moment is not proportional to the angular momentum of the needle due to any macroscopic rotation. In the case previously discussed, the magnetic moments were assumed to interact only with the external field, not with each other or with the lattice. The latter interactions of course exist, but will be introduced later.

To experimentally interact with the magnetic moment it is required to somehow change the angle θ. Given that a constant magnetic field leads to a constant angle, the field must be varied. It is here that we see that a relatively small field, if it varies at or near the Larmor frequency, can have a large effect. For simplicity, imagine that we maintain the static field, B_0 as before, and momentarily turn on an additional small field B_x along the x-axis for a short time (as compared to the period) just as the magnetic moment breaks the y–z plane. During this short time, the analysis made above is still valid, but now the direction of the magnetic field has an x-component; the rotation of the moment has a component along the x-axis that causes the angle θ to increase. Turning off B_x now allows $\boldsymbol{\mu}$ to again precess about the z-axis, but with this increased value of θ. A positive B_x has this effect during any part of the cycle when μ_y is positive. A negative value of B_x has this effect during the other half of the cycle; thus, a $B_x(t)$ which alternates sign *in resonance with the precession* can be used to adjust the angle θ to some particular value, and then turned off. In most NMR experiments, the relevant frequencies are in the radio-frequency (RF) regime, so such pulses are called RF pulses and are applied by feeding an RF current through a coil located near or surrounding the sample.

Rotating frame

In many everyday problems, we are in the habit of working in a non-inertial frame of reference as it allows us to concentrate on a particular interaction. When we steer our car, we generally forget that we are attached to a planet which rotates once a day and orbits the sun, motions which actually move the car much faster than the speedometer indicates. Of course, it is the car's motion relative to the earth's surface that we are usually interested in, so a reference frame attached to that surface is very helpful to us. In NMR, a useful point of view to adopt is a frame of reference that rotates about the static field (z-axis) with an angular velocity at or close to ω_0. Not only is this frame of reference useful for thinking about various experiments, but the signal most experimenters work with corresponds to the projection of the magnetic moment along a fixed direction in the x–y plane of such a rotating frame. When the applied magnetic field is constant, the magnetic moment is stationary in such a rotating frame, resulting in a considerable simplification!

Those who have studied rotations may have already noticed the similarity between Eq. (5.5) and the equation expressing the rate of change of a vector viewed from a rotating frame. Formally, for any vector quantity $\mathbf{A}$, the time derivative, $\partial \mathbf{A}/\partial t$, computed in a frame rotating with angular velocity ω, is related to the derivative $d\mathbf{A}/dt$ in the lab frame through the equation:

$$\frac{\partial A}{\partial t} = \frac{dA}{dt} + \mathbf{A} \times \omega. \tag{5.9}$$

Applying Eq. (5.9) to our equation of motion for μ, we find that

$$\frac{\partial \mu}{\partial t} = \gamma \mu \times \mathbf{B} + \mu \times \omega = \gamma \mu \times \mathbf{B}_e, \tag{5.10}$$

where $\mathbf{B}_e \equiv \mathbf{B} + \dfrac{\omega}{\gamma}$ is called the 'effective field'. This is in the same form as the equation of motion Eq. (5.5), but with an effective field, $\mathbf{B}_e$, about which the magnetic moment rotates in the new frame of reference. When 'on resonance', the case referred to above, $\omega = \omega_0 = -\gamma \mathbf{B}_0$, the effective field is zero, and the magnetic moment is stationary in this frame.

The effect of an RF pulse can be seen more simply in the rotating frame. Assume that it is possible to produce an RF field, $\mathbf{B}_1$, which rotates in the x–y plane. Such a field, if it rotates in the same sense, will appear fixed in the rotating frame and on resonance, $\mathbf{B}_e \equiv \mathbf{B}_1$, that is, the effective field is just the applied RF field that is fixed in the rotating frame. In this frame, the magnetic moment precesses about $\mathbf{B}_1$ at an angular velocity $\omega_1 = \gamma \mathbf{B}_1$. Although it is possible to create such a rotating field, in practice, the actual RF field is usually applied along an axis that is fixed in the lab frame; in reality, such a field is a superposition of two fields rotating in opposite sense. One of these fields yields the effect just discussed while the other can be neglected since in the rotating frame it appears to be rotating at twice the resonant frequency and gives zero average effect. To see this, let $\mathbf{A} = \cos(\omega t)\hat{i} + \sin(\omega t)\hat{j}$ and $\mathbf{B} = \cos(-\omega t)\hat{i} + \sin(-\omega t)\hat{j} = \cos(\omega t)\hat{i} - \sin(\omega t)\hat{j}$ be two unit vectors rotating in opposite senses in the x–y plane. The sum, $\mathbf{A} + \mathbf{B} = 2 \cos(\omega t)\hat{i}$, is a vector oscillating along the x-axis.

The RF pulses used in various experiments are generally labeled according to the effects they produce on the magnetization. Thus, a pulse that rotates the magnetic moment 90° clockwise about the rotating frame x-axis could be labeled a $90°_x$ or a $(\pi/2)_x$ pulse. The clockwise sense corresponds to a positive magnetogyric ratio ($\gamma > 0$). The definition of the rotating frame axes corresponds to a choice of the phase of the receiver pulse with respect to a reference oscillator that is also used by the receiver to be described later.

Free induction decay, relaxation, and the Bloch equations

We can inject a bit of the real world into this simplistic picture by considering how an NMR signal arises. The precession of the magnetic moment of any given nucleus produces a fluctuating magnetic field, which in turn induces an ac electrical signal in a nearby coil. According to what we have considered so far, as long as we do not add an RF pulse into this picture, this signal would continue indefinitely! Of course, if you actually placed, for instance, a sample of water in the coil of an NMR probe placed in a magnetic field, and observed the signal coming from the ^{1}H nuclei after a 90° pulse, you would find that:

1. You must wait for some time between repetitions of the experiment, say, 20 seconds for the ^{1}H resonance, to obtain the greatest magnitude of the induced signal. You need to allow the nuclear spins to recover their magnetization so a time delay equivalent to several relaxation time constants is introduced between repetitions.
2. The signal decays with a time constant that can be much shorter than the initial decay and can last for as long as several seconds, depending, among other things, upon the uniformity of the magnetic field and the purity of the water. This signal is called the free induction decay, or FID.

Broadly speaking, we find two time constants, referred to as T_1 and T_2, that characterize the approach of the sample's magnetization to its equilibrium state. The first time constant, T_1, is characteristic of the growth of the longitudinal or z-component of the sample's magnetization to its equilibrium value. It is a function of the sample composition, temperature, and field. The second time constant, T_2, corresponds to the decay of the transverse (x–y plane) components. Generally, any interaction that contributes to the first process necessarily speeds up the second but not the other way around; thus, T_2 is necessarily shorter than T_1. These times T_1 and T_2 are often known as the *longitudinal*, or *spin-lattice relaxation* and the *transverse*, or *spin-spin relaxation time* respectively, in 'honor' of interactions that are important contributors to the relaxation processes.

It is possible to make some sense of the second process. If we take into account the fact that any macroscopic sample contains a large number of nuclei, the resulting signal is the sum of signals arising from the individual nuclei. Any slight difference in the precession rate from nucleus to nucleus means that, with time, the spins will get out of phase with each other and individual signals will cancel. Alternatively, one can first add the magnetic moments together and consider the signal as arising from the precession of the net magnetic moment, or work with the magnetization (magnetic moment per volume), **M**, of the sample. In some cases, such as that of a sample that extends outside the coil, one may need to take the size and shape of the sample and the coil into account. For purposes of signal detection and simple manipulations of the spins by applied pulses, one may separately consider the moments arising from different nuclear species (e.g., ^1H, ^2H, or ^{13}C), since they precess at vastly different frequencies and can be manipulated and detected separately.

The phenomenological *Bloch equations* model this situation by introducing the two time constants, T_1 and T_2, that become important parameters for discussing any NMR experiments, even in cases where such a simple behavior is not observed. Values of T_1 commonly range from microseconds to tens of seconds and, for a given nucleus, depend on the resonant frequency, the nature of its neighbors, and on the motions and orientations of the molecules in the sample. Thus, this time constant varies dramatically from sample to sample and even from location to location in the same molecule, so that not all the samples can be characterized by a single value of T_1. The same can be said about T_2 with the addition that it also includes effects of sample shape and size, and the uniformity of the external field.

Making the simplest assumption and now working with the magnetization instead of an individual nuclear magnetic moment, one expects the z-component of the magnetization to decay towards some equilibrium value, M_0 from an initial value, $M_z(0)$ according to the formula:

$$[M_0 - M_z(t)] = [M_0 - M_z(0)]\exp(-t/T_1), \qquad (5.11)$$

that can be obtained by adding $(\mu_0 - \mu_z(t) (\mu_0 - \mu_z(t))/T_1$ to the right-hand side of Eq. (5.6c), a term which drives μ_z towards its equilibrium value. Perturbations in the environment of the nucleus that are at the Larmor frequency and its low multiples are what give rise to these effects.

Since the equilibrium value of the x- and y-components of the magnetic moment is zero, the corresponding terms for the right-hand sides of Eqs. (5.6a) and (5.6b) are just $-\mu_x/T_2$ and $-\mu_y/T_2$. This relaxation is affected by processes that relax the z-component, so that T_1 is necessarily an upper bound on T_2. The relaxation can also occur by processes only affecting the phase of the precession (angle between the x-axis and the projection of magnetization in the x–y plane) from one spin to the next. In such cases T_2 is often orders of magnitude smaller than T_1. Even static effects such as variation of the magnetic field across the sample cause this relaxation. Introducing variations in the NMR experiments, such as a rapid rotation of the sample, it is often possible to reduce some of these interactions and unravel those arising from various internal and external causes.

Adding the terms described above to the components of the equation of motion, Eq. (5.5), and when applied to the net magnetization, gives rise to the phenomenological Bloch equations:

$$\frac{\partial M_x}{\partial t} = \gamma(\mathbf{M} \times \mathbf{B}_0)_x - \frac{M_x}{T_2}, \tag{5.12a}$$

$$\frac{\partial M_y}{\partial t} = \gamma(\mathbf{M} \times \mathbf{B}_0)_y - \frac{M_y}{T_2}, \tag{5.12b}$$

$$\frac{\partial M_z}{\partial t} = \gamma(\mathbf{M} \times \mathbf{B}_0)_z - \frac{(M_0 - M_z)}{T_1}, \tag{5.12c}$$

which are the usual starting points to discuss NMR phenomena from a classical point of view.

Simple one- and two-pulse experiments

Before turning to a necessary, quantum mechanical description of NMR, it seems worthwhile to describe a few simple experiments in classical language. This will serve the purpose of applying the concepts already discussed to concrete experiments. We begin with one- and two-pulse experiments to determine T_1 and T_2 and end the section with a discussion of the spin echo.

Let us first assume that our equipment is set up to record and display a signal proportional to M_y', the component of the net magnetization along the y'-axis in the rotating frame at resonance. For simplicity, consider first the case of a liquid such as benzene in a uniform magnetic field, in which case all the hydrogen nuclei are chemically equivalent, and resonate at precisely the same frequency having the same relaxation times.

At equilibrium, the magnetization lies fully along the z-axis and our signal consists only of the noise picked up or generated by our equipment. If we then apply a $90°_x$ pulse, the magnetization, viewed in the rotating frame, is simply rotated to the y'-axis and stops there, giving a signal proportional to M_0. Any time we apply a $90°_x$ pulse, the signal immediately

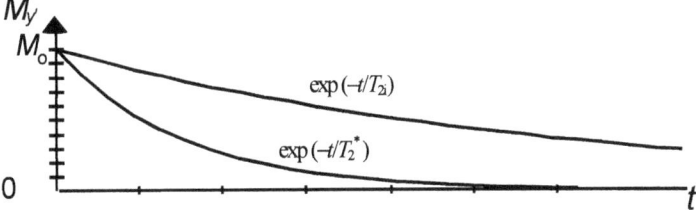

Figure 5.2.   The lower curve shows the decay of the magnetization after a 90°
pulse. The top curve shows the decay that would contribute if only internal
processes were important.

after the pulse gives us a measurement of M_z just before the pulse. (There
are, of course, some practical difficulties that complicate this simple picture,
but we will attend to them later.)

After a duration of T_2, the signal has decreased to M_0/e and is described
in the lab frame by

$$M_y(t) = M_0 \exp(-t/T_2^*) \cos(\omega_0 t), \qquad (5.13a)$$

and in the rotating frame (on resonance) by

$$M_{y'}(t) = M_0 \exp(-t/T_2^*). \qquad (5.13b)$$

Here, T_2^* denotes the experimentally observed decay time, which arises
from the combination of internal processes (spin-spin interactions) and
external causes such as a significant variation of the applied field across the
sample. In this straightforward case, we can write

$$M_{y'}(t) = M_0 [\exp(-t/T_{2i})][\exp(-t/T_{2e})]. \qquad (5.13c)$$

Another way to express this is to say that the internal and external processes
act in parallel and their decay *rates* add, resulting in the expression

$$1/T_2^* = 1/T_{2i} + 1/T_{2e}. \qquad (5.14)$$

This situation is presented in Figure 5.2.

One can determine the time to decay to M_0/e directly from the signal
displayed. Of course, it is simple to take many points from the curve and
fit them to the exponential decay, Eq. (5.13b), or to fit $\ln(M)$ to a straight
line of slope T_2^*. Most spectrometers will take the FID and Fourier transform the signal; in such case the line width (full width at half-maximum
or *fwhm*) is $1/\pi T_2^*$. In practice, the third term in Eq. (5.13c) is a more
complicated function that reflects the sample shape and the variations of
the applied field across it, even for the simple, isotropic liquid we have been

discussing. For liquid crystals, the situation is much more 'interesting'. In any case, this simple one-pulse experiment can provide lots of information!

To determine T_1, at least two pulses are required: one (at $t = 0$) to move the magnetization away from equilibrium and a second ($90°_x$) at time t to determine $M_z(t)$. If both pulses are $90°_x$, the first pulse makes $M_z(0) = 0$ and, according to the decay Eq. (5.11), the difference $M_0 - M(z)$ decreases from M_0 to zero with a time constant, T_1. To introduce a larger change in the M_z values, the first pulse can be a $180°$ pulse; in such a case $M_z(0) = -M_0$ and the difference $M_0 - M(z)$ decreases exponentially from a starting value of $2M_0$. If the signal is weak, this helps.

Spin echoes

Two RF pulses separated by some time τ will generally produce, in addition to the FIDs we have discussed and that start immediately after each pulse, a third signal that reaches a peak at a time τ after the second pulse. In the case discussed here ($90°_x - \tau\,180°_y$), the echo is called the 'Hahn' echo after its discoverer, E. L. Hahn. In brief, various processes, internal and external, cause the FID after the first RF pulse to decay. The second pulse reverses some of this and 'refocuses' the magnetization. This is similar to the effect of a distant building or cliff face, which reverses the direction of a sound wave and causes it to retrace its path to the experimentalist (or tourist, as the case may be). For the spin echo, the height of the signal that forms depends on which of the processes were reversed by the second pulse. In short, it is the remarkable ability to 'tune' the NMR experiment to different interactions that gives the NMR technique its extraordinary utility in many applications, for instance in medical imaging.

To the extent that the external causes of relaxation are constant in time (e.g., each atom 'sees' a field which is constant during the experiment), this part of the decay is reversed by the action of a $180°_y$ pulse, giving rise to the picture shown in Fig. 5.3. In this figure, the shape shown for the FID and the echo is based on a Gaussian curve and would result from a Gaussian distribution of magnetic field values throughout the sample volume that is large enough to dominate compared to the internal interactions responsible for T_{2i}.

In this ideal picture, no signal appears right after the second pulse, which rotates each magnetic moment $180°$ about the y-axis. In a sense, some information about the spins has been retrieved through the use of the second pulse. In this simple case, the echo height allows us to determine T_{2i}! To

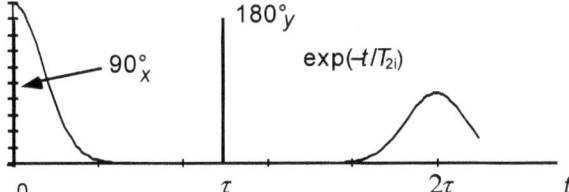

Figure 5.3.   When a second pulse is added at $t = \tau$, an echo appears at 2τ.

consider an even more realistic case, imagine that the molecules diffuse during the time of the experiment and that the applied field varies linearly in one direction (e.g., along the x-axis) so that the static field is given by

$$B_z(x) = B_0 + Gx, \tag{5.15}$$

where G is called the magnetic field gradient. During any time interval Δt, each molecule moves (diffuses) some distance. In this case, only the displacement in the x-direction affects the field, thus, the resonant frequency. The mean value of Δx is zero unless the sample is flowing (in which case the phase of the echo is affected), but the mean-squared displacement along the x-axis, at least in simple cases, is not. It is this mean-squared displacement that increases with time, at a rate depending upon the temperature and composition of a sample. For a simple, isotropic medium, the relation is given by

$$(\Delta x^2) = 2D(\Delta t), \tag{5.16}$$

where D is called the diffusion coefficient. The diffusion process causes an irreversible decrease in the echo height, leading to the expression for the echo height:

$$M_y(2\tau) = M_0 \exp(-2\tau/T_{2i}) \exp(-2DG^2\gamma^2\tau^3/3). \tag{5.17}$$

In conclusion, from this simple experiment one can determine both T_{2i} and D for the specific material. Those wishing further discussion of spin echoes from this semi-classical viewpoint may wish to consult some of the references provided in the bibliography. In particular, Stejskal and Tanner [20] present a straightforward derivation of the diffusion terms, a more general form of Eq. (5.16), and apply their derivation to the cases of several time-dependent gradients.

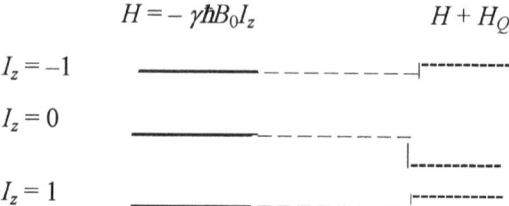

Figure 5.4.   Energy levels for a nucleus of spin $I=1$ in a magnetic field B_0. The equally spaced Zeeman levels are shown on the left while the perturbing effect of the quadrupolar interaction shifting the energy levels is shown on the right.

Quantum mechanical description. Spin interactions that complicate and illuminate

The basic interaction in any NMR experiment is the interaction between the magnetic dipole moment $\boldsymbol{\mu}$ of a nucleus and the external magnetic field $\mathbf{B}_0$. Through this well-known Zeeman interaction, the applied field splits the degeneracy existing among the different components of the nuclear spin. The relevant Hamiltonian is

$$H = -\boldsymbol{\mu} \cdot \mathbf{B}_0 = -\gamma \hbar B_0 I_z. \tag{5.18}$$

If the field is applied along the z-direction, I_z is the z-component of the dimensionless nuclear spin angular momentum operator. Although other interactions are also relevant to NMR (see next section), the Zeeman interaction is by far the strongest; all others can be viewed as perturbations to the Zeeman energy levels.

For a nucleus like deuterium with $I=1$, in a magnetic field B_0, the Zeeman energy levels that correspond to the Hamiltonian given in Eq. (5.18) will have energy eigenvalues of 0, $\pm\gamma\hbar B_0$. These are sketched in Fig. 5.4. Moreover, in the presence of a perturbing quadrupolar Hamiltonian H_Q (cf. Eq. (5.20)), the energy levels are shifted. The so corrected energy levels (also given later in Eq. (5.23)) are shown in the right column of Fig. 5.4.

It is the interactions of the spins with their environments that give us information about bonding, ordering, and dynamics of molecules. The electrons adjust their orbits slightly due to the external magnetic field, shifting the nuclear magnetic resonances by small amounts that depend upon the location of the nuclei within the molecule. Each magnetic moment itself creates a small magnetic field at the sites of its neighbors, the electric quadrupole moments of the nuclei with spin $\frac{1}{2}$ interact with the local electric field gradient. These interactions serve to change the observed spectrum and relaxation times in ways dependent upon the locations of the nuclei and the detailed manner in which the orientations and spacings of molecules are modulated in space and time. Although these interactions complicate the picture considered so far, namely, that of a spin interacting only with an external magnetic field, they illuminate our picture to reveal the beauty of molecular interactions in a manner unique among physical tools.

In the specific case of liquid crystal materials, there are three interactions relevant to NMR: the previously mentioned electric quadrupole coupling, chemical shift, and direct magnetic dipole coupling. The latter describes the coupling of nuclear magnetic dipole moments with each other while the first two involve interactions between the nucleus of interest and electrons in the neighborhood. These NMR interactions are motionally averaged yielding important information on liquid crystals properties.

For the three types of interaction it is found that their Hamiltonian has the same general form:

$$H_\lambda = C_\lambda \sum_l \sum_{m=-l}^{l} (-1)^m \langle R_{l,-m} \rangle T_{l,m}. \tag{5.19}$$

In Eq. (5.19) the subscript λ denotes the spin interaction of interest with C_λ the appropriate constant for each interaction. $R_{l,m}$ and $T_{l,m}$ are known as irreducible tensor operators representing the spatial and spin parts of the Hamiltonian; the $\langle\,\rangle$ brackets stand for the time average of the spatial part. We give their values corresponding to each spin interaction in Tables 5.2 and 5.3.

Note that sometimes the tensors R and T are given a superscript H or L (or P) to remind us that they are referred to a laboratory frame (or to a principal axis frame fixed with respect to the liquid crystal molecular directors) whose z-axis is determined by the direction of the magnetic field. For instance, in Eq. (5.20) below, we will use R^H to emphasize that we are working in the laboratory frame.

H_λ given in Eq. (5.19) is observed as a perturbation to the Zeeman interaction described in Eq. (5.18), which, except for the case of the quadrupole interaction for some nuclear spins like ^{14}N, is often orders of magnitude larger. Thus, when treated as perturbative terms to the Zeeman Hamiltonian, only the $m = 0$ terms of Eq. (5.19) need to be considered.

We refer the interested reader to the articles by Chidichimo and Golemme [12] and Doane [15] to find a nice description of the laboratory and principal axis frames and how they transform. Below, we proceed with a more detailed description of the spin interactions.

Table 5.2 *Constants C_λ and irreducible standardized basis sets $T_{l,m}$ for the three spin interactions. Q is the quadrupole moment of the nucleus and $I_\pm = (I_x \pm iI_y)$. The indices j and k refer to the jth and kth nuclei respectively, and e is the elementary charge.*

λ (interaction)	C_λ	$T_{0,0}$	$T_{2,0}$	$T_{2,\pm 2}$
Quadrupole (Q)	$\dfrac{eQ}{2I(2I-1)}$	$(\mathbf{I})^2$	$\dfrac{1}{\sqrt{6}}[3(I_z)^2 - (\mathbf{I})^2]$	$\frac{1}{2}(I_\pm)^2$
Chemical shift (CS)	$\gamma\hbar$	$I_z B_z$	$\sqrt{\dfrac{2}{3}}I_z B_z$	0
Dipole (D)	$-2\gamma_j\gamma_k\hbar^2$	$\mathbf{I}_j \cdot \mathbf{I}_k$	$\dfrac{1}{\sqrt{6}}[3I_{jz}I_{kz}) - \mathbf{I}_j \cdot \mathbf{I}_k]$	$\frac{1}{2}I_{j\pm}I_{k\pm}$

Table 5.3 *Irreducible spherical tensor operators $R_{l,m}$ for the three spin interactions in a principal axis system. η represents the asymmetry in the electric field gradient.*

λ	$R_{0,0}$	$R_{2,0}$	$R_{2,\pm 2}$
Q	0	$\sqrt{\dfrac{3}{2}}eq$	$\dfrac{1}{2}eq\eta$
CS	$\dfrac{\sigma}{3} = \left(\dfrac{\sigma_{xx} + \sigma_{yy} + \sigma_{zz}}{3}\right)$	$\sqrt{\dfrac{2}{3}}\dfrac{1}{2}(3\sigma_{zz} - \sigma)$	$\dfrac{1}{2}(\sigma_{xx} - \sigma_{yy})$
D	0	$\sqrt{\dfrac{3}{2}}r_{jk}^{-3}$	0

Quadrupole interaction

An important NMR interaction in liquid crystals is the coupling of electric field gradients with the quadrupole moment of nuclei with $I \geq 1$. The electric field gradients have their origin in the asymmetrical average distribution of electrons around the nucleus, and, in most cases, they are associated with the position of bond electrons. The interactions are of considerable significance in deuterium NMR; they do not exist for protons due to symmetry as the quadrupole moment of a spin $\frac{1}{2}$ nucleus is zero.

The quadrupolar spins that are abundant in liquid crystal materials are

of spin 1; these include nitrogen and deuterium. For this case, the perturbing quadrupolar Hamiltonian reduces to

$$H_Q = \frac{eQ}{2I(2I-1)} \frac{1}{\sqrt{6}} \langle R_{2,0}^H \rangle [3I_z^2 - I^2]. \tag{5.20}$$

For the case of a uniaxial phase, namely, a phase that has three-fold or
higher rotational symmetry, the Hamiltonian can be recast in the form

$$H_Q = \frac{eQV_{zz}}{4I(2I-1)} [3I_z^2 - I(I+1)] \frac{3\cos^2\alpha_0 - 1}{2}, \tag{5.21}$$

where V_{zz} is the component of the electric field gradient tensor averaged by
conformational and overall molecular motions. On the other hand, for a
biaxial phase, a phase with less than three-fold symmetry to rotations, the
Hamiltonian is further changed taking the form

$$H_Q = \frac{eQV_{zz}}{4I(2I-1)} [3I_z^2 - I(I+1)] \left[\frac{3\cos^2\alpha_0 - 1}{2} + \frac{\eta \sin^2\alpha_0 \cos 2\beta_0}{2} \right], \tag{5.22}$$

with α and β the Euler angles (again, we recommend the article by Doane
[15]).

For an $I = 1$ nucleus, the Zeeman energy levels corrected by the quadrupolar interaction are

$$E_1 = -\gamma\hbar B_0 + \tfrac{1}{4}\varepsilon_Q eQV_{zz},$$

$$E_0 = -\tfrac{1}{2}\varepsilon_Q, \tag{5.23}$$

$$E_{-1} = \gamma\hbar B_0 + \tfrac{1}{4}\varepsilon_Q eQV_{zz},$$

where ε_Q is given by the angular part of Eq. (5.22). These energy levels,
$E_{0,\pm1}$, were shown earlier in Fig. 5.4. Now, a magnetic field oscillating in a
direction perpendicular to B_0 will induce transitions between adjacent
energy levels, at frequencies

$$\omega = \gamma B_0 \pm \pi\delta\nu\varepsilon_Q, \tag{5.24}$$

with

$$\delta\nu = \frac{3}{2} \frac{eQV_{zz}}{h} \tag{5.25}$$

the quadrupole splitting. This quadrupole splitting is sample and temperature dependent and, for liquid crystals, is directly related to the orientational order. We defer additional discussion of quadrupolar NMR

including the typical NMR spectra, to the liquid crystals examples presented in Section 5.6.

Chemical shift

The term chemical shift expresses the magnetic coupling of the nucleus to magnetic fields produced by the motion of electric charges. In condensed matter, nuclei are surrounded by atomic or molecular electron clouds interacting with the nuclear spin angular momentum. From the orbital motion of the electrons, at the nucleus there is an additional magnetic field $\mathbf{B}'$, which is experimentally found proportional to the applied external field $\mathbf{B}_0$. The Hamiltonian term associated with the chemical shift for a nucleus can be written in the form:

$$H_{CS} = \gamma_i \hbar \mathbf{I}_i \cdot \overset{\leftrightarrow}{\sigma} \cdot \mathbf{B}_0 \qquad (5.26)$$

where $\overset{\leftrightarrow}{\sigma}$ is the dimensionless chemical shift tensor containing the proportionality between the induced field $\mathbf{B}'$ and the external field $\mathbf{B}_0$ and is dependent on the local electronic environment:

$$\mathbf{B}' = - \overset{\leftrightarrow}{\sigma} \cdot \mathbf{B}_0. \qquad (5.27)$$

Effectively, electron orbitals are perturbed by the applied magnetic field and mask or magnetically shield that field at the site of the nucleus. This chemical shift effect causes precession (Larmor) frequencies to be slightly displaced in a manner characteristic of the chemical environment. In ordered environments like solids or liquid crystals, molecules may possess rotational anisotropy and thus the shielding effect of the electron cloud, namely σ, has a tensorial character.

The fact that the chemical shift depends on the local chemical environment means that it is a fingerprint of the material. Therefore, chemical shift NMR is an essential tool in organic chemistry, particularly in matters pertaining to structural determinations. For the reader interested in this aspect of the NMR technique, a number of appropriate references are listed in the bibliography section.

Magnetic dipole–dipole interaction

The final interaction that we briefly wish to discuss here is the magnetic dipole–dipole coupling. The dipole–dipole interaction describes the coupling of nuclear magnetic moments with each other. As opposed to the previous two single spin interactions, this is a multiple spin interaction, and therefore quite complicated. Although the separation of atomic nuclei in condensed matter is rather large, the intrinsic magnetic moment associated with each nuclear spin dipole is able to exert a large influence on its neighbors. This occurs because the magnetic field produced by the dipole acts on the dipole moments of remote spins. Although complex, if not too many spins are involved, it can yield more information on molecular conformation and order than the previous two interactions.

The Hamiltonian for N interacting nuclei is given by

$$H_D = \sum_{j<k}^{N} \left[\frac{\boldsymbol{\mu}_j \cdot \boldsymbol{\mu}_k}{r_{jk}^3} - \frac{3(\boldsymbol{\mu}_j \cdot \boldsymbol{r}_{jk})(\boldsymbol{\mu}_k \cdot \boldsymbol{r}_{jk})}{r_{jk}^5} \right]. \tag{5.28}$$

Here $\boldsymbol{r}_{jk}$ is the radius vector between nuclei j and k and $\boldsymbol{\mu}_i = \gamma_i \hbar \boldsymbol{I}_i$ is the magnetic dipole moment of the ith nucleus.

In solids, since the inter-nuclear vectors have fixed orientations, the line shape is dominated by dipolar interactions that have cylindrical symmetry. In contrast, for liquids, molecular motion causes H_D to fluctuate. If the molecular tumbling motion corresponds to a greater angular frequency (or is 'faster') than the dipolar interaction strength (true for all except the most viscous liquids), the dipolar Hamiltonian of Eq. (5.28) is averaged to zero. Essentially, interactions between spins belonging to different molecules are normally averaged to zero because of the translational diffusion that takes place in many of the liquid crystal phases. As before, several references that expand on these aspects are provided in the bibliography section.

5.3 Experimental apparatus

In this section, we provide a brief discussion on the typical NMR spectrometer hardware. It could be argued that the heart of the spectrometer is the computer as it controls all of the components of the spectrometer. In a typical NMR setup, the RF components under control of the computer are the RF frequency source and pulse sequence. The source produces a sine wave of the desired frequency. The programmer sets the pulse width, the time delays, and, in some cases, the shape of the RF pulses. The RF amplifier increases the pulse power from milliwatts to hundreds of watts; the amplified signal is then fed to the receiver. The computer contains the software to perform Fourier analysis of the signals. In Fig. 5.5 we sketch a block diagram for the NMR set-up, as used in our laboratory, with electronic equipment from Tecmag Inc.

Since a thorough discussion of each element composing the NMR spectrometer is beyond the scope of this chapter, we will only focus on a few of the important elements. For the most part, the discussion will follow that found in the book by Fukushima and Roeder [4]; in addition, we briefly

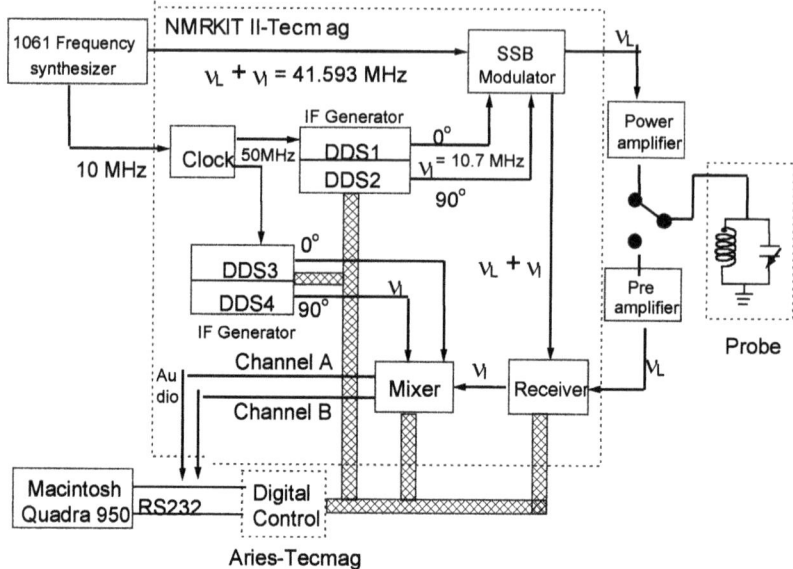

Figure 5.5.   Schematic block diagram of the NMR experimental set-up used for quadrupole splitting line shape studies of deuteratedliquid crystals.

comment on some of the main features of the digital electronics that have being added to the spectrometer.

Transmitter The transmitter is composed of a frequency source (synthesizer) and amplifiers to increase the RF signals to the appropriate level. For pulsed NMR, the RF irradiation must be modulated into pulses. For this, one employs an oscillator that can be turned on and off. Specifically, one generates NMR pulses at low power by gating a continuous RF signal on and off, effectively, a gated pulse generator. This signal is subsequently fed to the amplifiers to obtain the desired signal level. Typically, the power required ranges from 100 to 1000 W. Such a high output power must be delivered in short pulses, typically less than $10\,\mu$s. The pulses should have very short rise and fall times compared to their widths.

In the set-up shown in Fig. 5.5, the RF frequency is obtained from the General Radio 1061 frequency synthesizer. This is highly stable, about 1 part in 10^9, and accurate from 0.4 to 160 MHz at an output level of 0 to $+20$ dB. The gated RF is amplified using a Bruker system, which includes selective switchable amplification stages of wide bandwidth. It has a maximum output of 1 kW at 30.6 MHz.

Receiver NMR receivers consist of amplifier stages to enhance the signal, followed by a detector whose job is to remove the RF carrier so that only the signal envelope, which contains the desired information, is left. The remaining or detected signal, also known as the audio signal, is amplified further so that a digitization process can occur. The amplification within the receiver is crucial because of the small size of the nuclear magnetic moments. In addition, the receiver should not introduce distortions or electrical noise. This can be accomplished by making sure that the receiver can discriminate unwanted signals. This is achieved by making a frequency window (narrow-banding) that only allows the desired signal and part of the noise spectrum to pass through it, i.e., filtering the signal. The expert reader will be able to appreciate that having such a narrow window is the main reason why NMR has an easier time 'seeing' narrow lines than broad lines.

The narrow window can be achieved using the superheterodyne technique, which involves shifting the frequencies to a more convenient one before additional signal processing is performed. In a sense, this is how lock-in amplifiers work. This is accomplished using mixers, or phase sensitive detectors which form the sum or difference of the input frequencies. The net result is that the amplifier is always tuned to the correct frequency (ν_1 in Fig. 5.5) regardless of the Larmor or carrier frequency. In fact, we have shown above how the addition and subtraction of frequencies takes place in different parts of the experimental set-up.

Finally, there should be perfect isolation between the receiver and the transmitter. The power delivered to the NMR probe must be quickly dissipated after the RF pulse is turned off. Such time is typically on the order of $20\,\mu s$ and is known as the receiver dead time.

Digital commercial equipment Recently (circa 1995) we upgraded the control electronics by purchasing from Tecmag Inc. an ARIES acquisition unit with an NMR kit II containing the transmitter and receiver. The system is controlled through the ARIES interface by a Macintosh Quadra 950 computer. The spectrometer has a resolution of 100 ns with a maximum acquisition speed of $1\,\mu s$ per complex point, which corresponds to a sweep width of ± 500 kHz. The control logic is provided with five independent, 16-bit loop counters, $2K \times 78$ bits of memory, and a 78-bit micro-control word length giving the system the versatility to work with and operate at any pulse sequence. The signals are first sampled using an analog module that simultaneously samples two channels; such a signal is then fed to a digital module with 32K of 32-bit memory and a 32-bit arithmetic unit coupled to a 32-bit digital multiplexer for signal averaging.

Magnet. There are different kinds of magnets, the preferred one these days is the superconducting magnet that offers the desirable combination of stability and field strength. In fact, fields greater than 7 T are only possible through superconducting magnet technology. Standard superconducting magnets achieve reasonably good homogeneities of approximately 1 part in 10^6 over a large sample area. Should higher homogeneity be required, room temperature shim (or correction) coils are available. Something to keep in mind is that the larger the sample space, the smaller the maximum field strength at a given cost.

Specifically referring back to Fig. 5.5, the magnetic field is produced by a Nalorac 4″ room temperature superconducting magnet set at 4.685 T, corresponding to a frequency of 30.6 MHz for deuterium nuclei. Its homogeneity is adjusted with room temperature shim coils. Typically, the magnet drifts by about 1 Hz per day, which is a particularly good stability. For this magnet, the magnetic coherence length (cf. Eq. (5.34) and discussion in Section 5.6), which decreases with increasing magnetic field, is approximately 1.5 μm. This is an important consideration when performing NMR studies in confined liquid crystals (again see Section 5.6). If one is to learn about how bounding surfaces promote a specific molecular alignment, the magnetic coherence length must be larger than the confining size. Under such conditions, the magnetic field will not be able to align the liquid crystal molecules and surface effects will be singled out.

Probe The probe head is what effectively isolates the transmitter from the receiver. The RF transmitter and the receiver are both tuned to the Larmor frequency of the nucleus under investigation (in our lab, 30.6 MHz for deuterons). To maximize the efficiency of the coupling of the RF energy to the nuclear spins of the sample, a single solenoidal coil is used for both transmitting and receiving. The coil forms a part of the resonant circuit (its inductance L) completed with the addition of variable, high voltage capacitors that serve to tune the 'tank' circuit to the resonant frequency. The coil, extending for our magnet over a 2.5 cm distance, is made from a few turns (6 to 10) of either copper or silver wire (about 1 mm cross-section) and is oriented with its axis perpendicular to the magnetic field. The coil cross-sectional area, effectively the sample space, is typically 5–6 mm diameter. Changing the number of turns changes the inductance thus requiring additional tuning. We stress that proper tuning is essential to maximize the power transfer. This should be verified for every sample and checked at different temperatures.

Finally, the sample area within the coil is such that it permits a 360 degrees sample rotation about the coil axis. Changing the angle is accomplished at room temperature. As we shall see later, the ability of changing a confined sample orientation with respect to the NMR field is vital in understanding which liquid crystal configurations are produced.

5.4 Practical details

Removing impurities of various sorts is standard practice in most NMR experiments. For example, in high-resolution NMR or T_1 measurements in isotropic liquids, oxygen dissolved in the sample broadens lines and decreases the relaxation times. In the case of liquid crystals, the effects of impurities may be even more serious, modifying the phase transition temperatures, giving rise to two-phase regions, and so forth. Even after purifying and sealing a sample, it may degrade with time, so the 'history' of a sample may need to be recorded and transition temperatures checked from time to time.

Purity The synthesis of the materials making up liquid crystalline samples generally leaves residues of solvents and, even more important, precursors of the final product, which must be removed. These may greatly affect the properties of the sample. Thus the source of the materials must be carefully checked and, in most cases, samples should be studied under the microscope to observe the phase transitions and the optical 'textures' formed.

Freeze–thaw cycling For thermotropic samples, a standard technique is to alternately freeze then thaw a sample while pumping away any gases that are driven off at the transitions. This cycling removes much of the dissolved oxygen and the solvent used in the synthesis of the liquid crystalline material. It is especially important to remove the oxygen, as it tends to react with many liquid crystals and break them down with time. The amount of gas given off at the first melting can be quite substantial and care must be taken to avoid having much of the sample pumped away. A more gentle approach is to slowly feed nitrogen gas into the sample to flush out the gases that are driven off at the transitions. Typically, one wants to cool the sample down to the crystalline phase; this can be done quickly with a liquid nitrogen bath. Instead of doing this the first time on a very expensive sample, it is probably wise to practice on something such as PAA, which is very inexpensive and easy to obtain. It is useful to use a standard such as PAA to check the system operation. For example, it can be used to see that the temperature control system does not overshoot and that the transition temperatures are where they should be.

Since lyotropic samples generally include liquids such as water as one of the components, the vacuum treatment is not as appropriate. Some of the lyotropic phases are particularly 'fragile' and extremely sensitive to just about any sort of impurity. In general, it is best to include a trap between the sample and the vacuum pump. For one thing, you do not want your valuable sample to end up inside the pump itself. For another, you probably do not want to be breathing the solvents that are being driven off. Always make sure that evacuated material does not obstruct the vacuum lines.

Sample alignment Sample alignment is strongly dependent upon the sample's physical parameters, its 'history', boundary conditions, and the strength of the magnetic field. Since many NMR parameters (line widths, splittings, couplings) will depend on the distribution of molecular orientations, knowledge and control of the sample's alignment is critical. Even in the isotropic phase, large magnetic fields and/or surface interactions can give rise to pretransitional ordering.

Nematics The typical 2 to 9 T magnetic field available in most laboratories is sufficient to align most nematics, but the time required to fully align a sample can range from milliseconds to hours or days. It depends on viscosity, elastic constants, field strength, diamagnetic anisotropy (dc), the initial alignment, and the sample container. Surface interactions and viscosity in a cylindrical sample tube, for instance, become more important as you decrease the tube diameter.

Nematics having a positive magnetic anisotropy will align with their directors parallel to the applied field. In a cylindrical sample tube placed in an electromagnet, with the field perpendicular to the cylinder's axis, one can typically observe this phenomenon in progress by observing the FID: rotating the sample quickly through an angle of 30–50°, and watching the FID recover its original shape. This is in fact one way to identify a nematic phase using NMR. It is not an unambiguous test since the time may be quite considerable in some nematic samples (high viscosity, small order parameter, small sample diameter) and smectics will sometimes align in the presence of a strong field. For small rotations, the director field stays uniform throughout the sample. The sample director rotates uniformly with the container and then the magnetic torque rotates it back to the equilibrium position. The angle, θ, between the field and the director decays (roughly) exponentially with a time constant proportional to the rotational viscosity, γ_1. Since the dipolar and quadrupolar splittings that give rise to characteristic features of the spectrum scale with the Legendre polynomial of second degree, $P_2 (\cos \theta)$, it is sometimes possible to follow this decay by recording the FID, or by simply measuring the time interval for a particular feature (a local minimum, for instance) to move from one time to another on the FID. This is, in fact, one way to measure γ_1.

For a slowly rotating sample, the competition between the viscous torque at the container–sample interface and the magnetic torque gives rise to a sample having its director field uniform but at an angle with respect to the field. Varying the rotational velocity allows one to control the angle. In this way, one can perform experiments in a nematic in which the angle θ can be varied, typically up to 30 or 40°. When a nematic sample is rotated quickly compared to the reorientation time, it tends to align in a planar powder pattern with local directors in the plane perpendicular to the axis of rotation. See, for example, Ukleja, Pirs, and Doane [18], where these effects were used to study the angular dependence of T_1 in PAA.

For those nematics with a negative magnetic anisotropy, the interaction with the magnetic field leads to an equilibrium with the local sample directors perpendicular to the magnetic field, but not, of course, necessarily (or even likely) parallel. Cooling such a material from the isotropic to the nematic phase will usually produce a sample with local directors uniformly distributed in the plane perpendicular to the field. The obtained spectra thus represent a 'planar' powder pattern. These effects can be over-ridden by surface interactions, by, for instance, inserting into the sample many parallel glass plates which may induce an alignment along the common normal to the plates.

Smectics In addition to orientational order as in nematics, smectic materials possess translational order, namely, a layered structure. Several different smectics exist depending, among other things, on the molecular orientation with respect to the layers themselves. For smectics with a nematic phase at higher temperatures, one can take advantage of the alignment in the nematic phase (if $\Delta\chi > 0$) and cool into the smectic phase with the directors uniformly aligned. For a smectic-A liquid crystal, this usually gives a very well aligned phase. In electromagnets, one can then rotate the smectic-A sample so that the director makes any desired angle with the field. Rotating to the magic angle (where $P_2 (\cos \theta) = 0$), the dipole–dipole and electric quadrupole interactions scale to zero and a liquid-like NMR signal is obtained. For example, one can use a $\pi/2$-τ-π sequence and get an echo even with values of τ in the millisecond range for ^{1}H; this enables one to study diffusion in such samples. The alignment can be judged by measuring the sharpness of this effect, for instance, setting up an echo experiment and seeing how the echo height depends on angle near the magic angle position. Often, such an echo disappears within a degree or two of the magic angle. If the field is large enough or the sample is relatively 'soft', the uniform director pattern may be distorted or the sample may even re-align. Finally, for some smectics, the magnetic field alone will align the sample.

The scenario is more complex for a tilted phase, such as a smectic-C. Cooling into a smectic-C phase from the nematic, one starts with the directors all parallel to the field but with a random distribution of the planes normal. The director in many of the planes can follow the field to some extent if the sample is rotated with respect to the field. Alternatively, if one starts with an aligned smectic-A phase and cools into the smectic-C, the smectic planes may survive intact, but the directors move away from the field.

Surface alignment Especially noticeable in lyotropic samples, the interaction between liquid crystals and a clean glass surface often orders the sample director locally along the normal to the surface. This interaction often slowly orders a lyotropic sample, particularly in capillary tubes of small diameter such as those used in x-ray apparatus, or in the rectangular cross-section tubes employed in microscope studies. In the latter case, one can 'grow' the alignment in a sample so it is fairly uniform and perpendicular to the wider sides of the rectangular cross-section. For a cylindrical tube, the likely configuration is one with most of the molecules aligned radially. This can be observed by placing the sample between crossed polarizers and seeing the uniformity of the optical activity. Often, it only takes a small physical disturbance to disrupt the order in such a sample, quite visible to the eye in this case. For many smectics, one can also create an aligned sample between two glass plates (e.g., cover slips) by moving the glass surfaces back and forth until the sample appears clear.

If the surface interaction in such a surface-aligned sample is strong enough compared to the magnetic interaction, the sample would have its director at an angle of 90° with respect to the magnetic field with the usual geometry in a superconducting magnet, in which the long axis of the sample is generally placed parallel to the magnetic field. This, for instance, gives a dipole–dipole or electric quadrupole splitting one-half the full value (if aligned at 0°). In contrast, if the long axis is perpendicular to the field, as is usually the case in an electromagnet, one can get a planar powder pattern (a spectrum from a sample in which all directors in a plane are equally represented) in a cylindrical tube, or can vary the field-director angle with a rectangular cross-section tube.

A lesson? The important lesson in all this is that the alignment of one's sample may change with time and, unless care is taken, from one repetition of an experiment to the next. In most experiments, the sample alignment is a critical variable that must be controlled. Usually in nematics, the director is uniformly parallel to (or, if $\Delta\chi < 0$, perpendicular to) the magnetic field. For smectics, one may prefer to work with an unaligned sample. This can be done in many cases by, while outside the magnet, rapidly cooling a sample from the isotropic to the smectic phase, or to a lower lying crystalline phase, and then inserting the sample into the field. This does not always work if the sample flows during the process, or if the surfaces align the sample, or if the field is simply too large. In such cases, the distribution of sample directors may vary widely from experiment to experiment.

Phase separation For mixtures of any sort, which means *always* for lyotropic samples, the boundaries between different phases include two-phase regions (or worse!). That is, the sample does not merely change from entirely phase A to entirely phase B at a single temperature, but there will be a range of temperatures for which the two phases coexist in various proportions at equilibrium. This leads to a separation of the sample into domains having different compositions and physical properties. In many such cases, like the isotropic+nematic regions for lyotropic samples, the densities of the different phases are different enough and the viscosity low enough that the sample 'de-mixes' with the lighter phase rising to the top in a matter of seconds or even less. Different domains will have different transition temperatures to the adjacent phases, and if the resulting domains are large enough and the viscosity relatively high, merely changing temperature to what should be a single-phase region if the sample were well-mixed may not produce a uniform sample. Since the resulting mosaic of domains is sensitive to the initial uniformity of the sample and the times spent at different temperatures and the rates of change of temperature during the experiment, much care must be taken. The NMR spectra obtained in such cases would exhibit effects such as the following:

- spectrum changes with time at fixed temperature (sample is de-mixing or re-mixing by diffusion),
- spectrum changes from one repetition to the next,

- spectra are not symmetric (regions with different chemical shifts and line-splittings),
- spectrum depends on radius of sample container (small diameter tubes restrict flow),
- transitions are broadened, and depend upon time and thermal history,
- lines are broadened, and depend upon time and thermal history.

An example on this aspect of phase separation will be presented in Section 5.5.

Typical values for relaxation times, line widths, spectral widths, signal/noise
Values for most NMR parameters vary by several orders of magnitude from sample to sample and, even for one sample, can vary greatly with temperature or even site to site on a molecule. Thus, it is wise to keep several well-known and stable samples to check the operation of the spectrometer, data acquisition parameters, and so forth, recording standard spectra, relaxation times, transition temperatures, signal/noise, 90° pulse widths, and so on. You must be able to tell whether your spectrometer is actually working! You may, for example, want to record the appearance of the FID of a standard (e.g., water) at various stages of your instrument, starting with the signal coming out of the pre-amp. Always check your spectra using different repetition times to make sure all features are fully relaxed. For example, in DNMR spectra from samples with many sites deuterated, you may often find that the inner lines from, say, methyl groups, have T_1 values on the order of seconds while those on the rings relax fully in 100 ms. Repeating too fast (for instance, every 100 ms) will give a spectrum in which the methyl lines appear too small in comparison to the lines from the ring deuterons. Depending upon the pulse sequence, of course, other distortions are possible.

One final caution concerning the great variations in, for example, line splittings, line widths, and relaxation times, is to take great care to keep in mind the time scales relevant in performing and interpreting the results of a given experiment. For instance, in measuring a translational diffusion coefficient with a two-pulse echo experiment, you have some control of the diffusion time, but you are typically studying mean-squared molecular displacements over a range of 10 ms to 1 s. When comparing your results to those from other experiments, be aware of the different time and length scales involved. For instance, for a water molecule having a diffusion coefficient of $2 \times 10^{-9} \, \text{m}^2/\text{s}$, the r.m.s. displacement of water molecules in 0.1 s is about 20 μm.

5.5 Selected examples

In this section, we focus the NMR discussion onto liquid crystal aspects by discussing three different examples. As a review, we recall that in an NMR experiment, the nuclei with magnetic moments different from zero serve as the probe. Their interaction with the NMR external magnetic field results in equidistant Zeeman energy levels. The application of a perpendicular rotating magnetic field induces transitions among the energy levels when the frequency of the rotating field corresponds to the Larmor frequency of the nucleus. For typical magnets and deuterium nuclei, we recall that the frequency is in the megahertz range: 6.5 MHz for a 1 T magnet, 30 MHz for a 4.7 T magnet.

For deuterons, $I = 1$, the dominant interaction is that between the electric quadrupole of the nuclei and the electric field gradient which causes a non-uniform shift of the Zeeman energy levels. Consequently the NMR line is either broadened or split into several lines. In fact, from deuterons in rigid molecules, the NMR spectrum consists of two lines separated in frequency by an amount (frequency splitting)

$$\Delta\nu_q = \frac{3}{2}\frac{e^2qQ}{h}\frac{1}{2}(3\cos^2\Theta_0 - 1). \tag{5.29}$$

This is a generalization of Eq. (5.25) with V_{zz} defined as qe, where q is the field gradient. Thus, e^2qQ/h represents the static quadrupole coupling constant: it is roughly 165 kHz for deuterons in the alkyl chain, while it is 185 kHz for aromatic deuterons. Θ_0 is the angle between the electric field gradient tensor symmetry axis, which in liquid crystals is typically parallel to the C–D bond, and the external magnetic field.

Specifically for nematic liquid crystals, the angular factor is averaged over molecular motions. Principally, these are conformational changes of the molecule, its rotation about its long axis and fluctuations of such axis about a local preferred direction. The latter is referred to as the local director **n**. Then, for a uniaxial system, the splitting of the NMR line is given by

$$\Delta\nu_q = \frac{3}{2}\frac{e^2qQ}{h}\frac{1}{2}(3\cos^2\Theta_B - 1)$$

$$\left[S_{zz}\left\langle\frac{1}{2}(\cos^2\beta - 1)\right\rangle + \frac{1}{2}(S_{xx} - S_{yy})\langle\sin^2\beta\cos 2\alpha\rangle\right]. \tag{5.30}$$

α and β denote the azimuthal and polar angles that the electric field gradient tensor symmetry axis forms with the long molecular axis. The quantity S is the well-known scalar orientational order parameter:

$$S = \left\langle \frac{1}{2}(3 \cos^2 \Theta - 1) \right\rangle, \tag{5.31}$$

where Θ is the angle between the instantaneous direction of the long molecular axis and $\mathbf{n}$, while Θ_B is the angle between the director and the magnetic field. The quantity $(S_{xx} - S_{yy})$ is the molecular biaxiality order parameter and describes the deviation of the molecular shape from an exact cylindrical symmetry. It is usually almost one order of magnitude smaller than S. For present purposes, what is important is that the frequency splitting of the NMR line is directly proportional to the orientational order parameter. For cylindrically symmetric molecules, the splitting is a maximum when the director is parallel to the magnetic field; it decreases to half of the maximum value when the molecules are oriented perpendicular to the applied field.

When the nematic liquid crystal temperature is raised, the kilohertz frequency splitting decreases. Once the temperature is further increased above the nematic to isotropic phase transition, the NMR lines collapse into a single line whose width, typically 100 Hz or less, is several orders of magnitude below the nematic splitting. It should be stressed that the width of the isotropic line is the result of some experimental limitations, particularly non-homogeneities in the NMR magnet, and consequently it is not a measure of any intrinsic spin interaction. In Fig. 5.6, for 4'-n-pentyl-4-cyanobiphenyl (or 5CB) liquid crystal deuterated at the alpha position (the first position in the hydrocarbon chain), we show a typical NMR spectrum for the bulk nematic completely aligned by the magnetic field, and also after raising the temperature into the isotropic phase where orientational order is non-existent. Also included in Fig. 5.6 is the temperature dependence of the quadrupole splitting of nematic 5CB. Notice that with increasing temperature the nematic splitting decreases in size, vanishing discontinuously at the transition to the isotropic phase. The nematic to isotropic phase transition is a weakly first order phase transition.

We have just learned what the typical NMR spectrum for a nematic looks like, and we have acquired the knowledge of the mathematical expression for the quadrupolar splitting. Together they provide an excellent starting point to appreciate the usefulness of the NMR technique in studies of liquid crystal materials. With them in mind, we proceed to discuss three representative examples in some detail: the first two describe liquid crystals under confinement and the last one is concerned with phase separation.

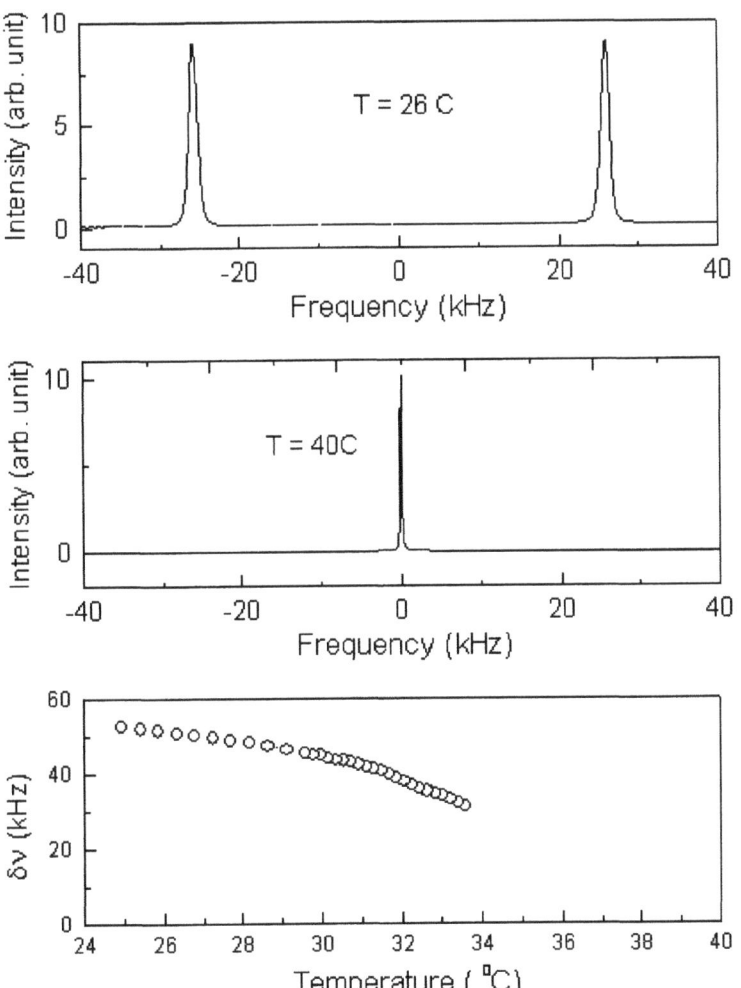

Figure 5.6.   NMR spectra for 5CB in the nematic (top) and isotropic (middle) phases. The quadrupole splitting dependence on temperature is shown at the bottom.

Cylindrically confined nematics. NMR determination of director fields

When a liquid crystal is confined to sub-micrometer dimensions, the director field $\mathbf{n}(\mathbf{r})$ may exhibit spatial as well as time dependence. Further, the orientational order scalar parameter S may also vary with position within the sample, $S = S(\mathbf{r})$. Under these conditions, the NMR spectrum for a nematic will be given by

$$\Delta\nu_q(\mathbf{r}) = \Delta\nu_{q0}1/2(3\cos^2\Theta_B(\mathbf{r}) - 1)S(\mathbf{r}). \qquad (5.32)$$

In Eq. (5.32), $\Delta\nu_{q0}$ represents the quadrupole splitting of a perfectly aligned nematic along the direction of the magnetic field ($S = 1$). The NMR spectrum is then a sum of contributions from all parts of the sample. The frequency distribution represents the complete director field as long as the lineshape is not distorted. Distortions could arise from the spatial dependence of S, from the magnetic field, or from molecular self-diffusion; however, these can be neglected if certain conditions are met as discussed below.

The molecules' translational diffusion may partly smear the spectral distribution if, within the time scale of the NMR measurement, the molecules diffuse to regions of the sample possessing different director orientations. The NMR time scale is

$$t_{NMR} = (2\pi\Delta\nu_{q0})^{-1}. \qquad (5.33)$$

For a quadrupole splitting of 60 kHz (as for 5CB at low temperatures) t_{NMR} is 2.7 μs. Using the bulk 5CB diffusivity at room temperature of 4×10^{-11} m²/s, the diffusion length is estimated to be $\langle r^2 \rangle^{1/2} \sim (6Dt_{NMR})^{1/2} \sim 26$ nm. Since, in general, changes in the nematic director orientation take place over much larger distances, except in small cavities of 100 nm size or less, the translational self-diffusion can be neglected.

When the nematic material is confined to large pores, the alignment induced by the bounding surface, which presumably propagates throughout the sample, could be distorted by the NMR magnetic field. The extent of the surface influence can be estimated by comparison to the magnetic coherence length, namely, the length scale above which there would be magnetic field induced alignment. From Landau–de Gennes elastic type of arguments, it can be shown that the magnetic coherence length is expressed in terms of

$$\xi_B = \sqrt{\frac{K}{\Delta\chi}\frac{1}{B}}, \qquad (5.34)$$

where B is the strength of the magnetic field, $\Delta\chi$ is the diamagnetic susceptibility anisotropy, and K is one of the liquid crystal elastic constants. At room temperature, using typical liquid crystal values, the magnetic coherence length is ~1.7 μm (8 μm) for a 4.7 T (1 T) magnet. Therefore, only when the liquid crystal is confined to pores larger (or at distances from the surface farther) than the magnetic coherence length does one expect distortions in the director field due to the magnetic field.

The final effect that could lead to distortions in the director field is the possible dependence of S on location within the sample. In general, its variation is only important for very small cavities (10 nm or less), for temperatures near the phase transition, or close to structural defects. For most experiments involving confined liquid crystals, S = constant is a reasonable approximation.

Therefore, under confining conditions that make the above-mentioned distortions negligible, the NMR spectrum is particularly useful in determining the surface-induced liquid crystal alignment within a pore. In pores where the director is everywhere parallel to the magnetic field, the NMR spectra will consist of absorption lines, e.g., the quadrupole splitting $\Delta\nu_q$. This would be the case for cylindrical confinement where the pore axes are oriented parallel to the field and the liquid crystal molecules are axially (parallel) oriented. If, on the other hand, the director is everywhere perpendicular to the magnetic field (a radial director alignment in a cylinder), the spectrum will also consist of two lines, but now separated by an amount $\frac{1}{2}\Delta\nu_q$. If the director orientation is randomly distributed in space, the NMR spectrum is a Pake-type powder pattern with two singularities at $\pm\frac{1}{2}\Delta\nu_q$ and two shoulders at $\pm\Delta\nu_q$. The shoulders are due to those molecules that are aligned parallel to the magnetic field; the singularities arise from regions where the liquid crystal director is perpendicular to the magnetic field. Finally, it should be noted that in cylindrical pores, other, more complex configurations are possible, each of them yielding a unique NMR spectral pattern.

To exemplify one of the many virtues of the NMR technique, a few of the liquid crystal configurations that are possible in cylindrical symmetry are illustrated in Fig. 5.7. Each of them produces unique NMR spectral patterns, which are presented in Fig. 5.8.

In this type of cylindrical confinement, it is possible to produce structural changes such that one goes from one configuration to another. As discussed in the next example, this can be achieved by changing the pore size (its radius) or by changing the interaction between the surface and the confined liquid crystal material. The latter is obtained by treating the host porous material with a chemical surfactant prior to introducing the liquid crystal (see Crawford and collaborators [10, 13, 14]). Configurational transitions have also been found in random geometries by Zeng and Finotello [19]. In short, NMR is a powerful tool to investigate surface-induced alignment in confined geometries.

Measuring the K_{24} surface elastic constant in a confined nematic

The second example is concerned with the analysis of the NMR spectral patterns and how it led to the determination of an 'elusive' elastic constant. The equilibrium state of nematic liquid crystals is characterized by the

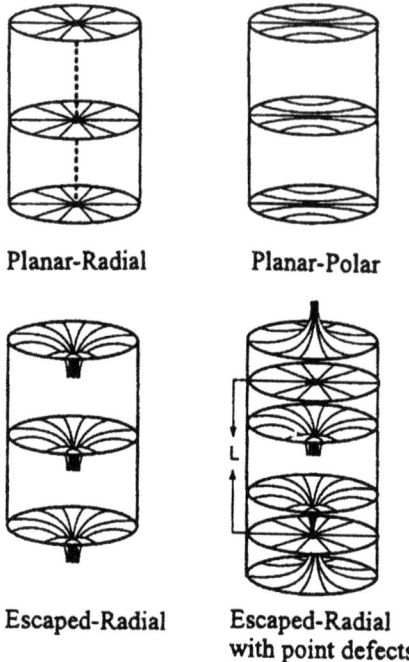

Planar-Radial **Planar-Polar**

Escaped-Radial **Escaped-Radial with point defects**

Figure 5.7. A few of the liquid crystal configurations that can be found in cylindrical geometry. Configurational transitions from one configuration to another can be stimulated by, for instance, changing the surface-liquid crystal interaction or by varying the confining size, or through changes in theelastic constants.

parallel alignment of all local directors. However, when surfaces, an external field, or thermal fluctuations are present, the system can be elastically distorted. Nematics are described by the elastic continuum theory which starts by considering the equilibrium state or state of minimum free energy. A widely used form for the elastic free energy that describes the bulk and surface elastic characteristics that are shown by liquid crystal materials is

$$F = \int d^3r \frac{1}{2} \left\{ K_{11}(\nabla \cdot \hat{n})^2 + K_{22}(\hat{n} \cdot (\nabla \times \hat{n}))^2 + K_{33}(\hat{n} \times \nabla \times \hat{n})^2 - \right.$$

$$\left. K_{24}\nabla \cdot (\hat{n} \times \nabla \times \hat{n} + \hat{n}\nabla \cdot \hat{n}) \right\} + \frac{1}{2} \int d^2r W_0 \sin^2 \phi. \qquad (5.35)$$

Here K_{11}, K_{22}, and K_{33} are the phenomenological bulk elastic constants describing the splay, twist and bend deformations respectively, and are material properties. The term in the second integral represents the interaction between the nematic liquid crystal and a solid surface with ϕ the angle

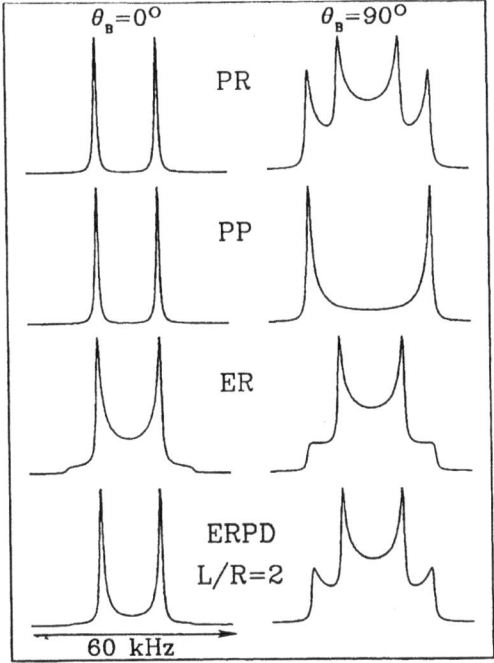

Frequency (kHz)

Figure 5.8. NMR spectral pattern associated with the liquid crystal configurations shown in Fig. 5.7. θ_B is the angle between the cylindrical pore axes and the external NMR field. Note that angular dependence studies are required to distinguish between the different configurations as some of them yield the same pattern at $\theta_B = 0$. The legends are PR for planar-radial, PP for planar-polar, ER for escaped-radial and ERPD for escaped-radial with point defects.

between $\hat{n}$ and the alignment induced by the surface, and W_0 denotes the strength of the interaction. Finally, K_{24} is usually referred to as the saddle-splay surface elastic constant. In the above expression, the contribution due to another elastic constant K_{13} has been neglected.

The free energy given by Eq. (5.35) predicts a variety of stable configurations. In particular, when a liquid crystal is confined to a cylindrical geometry of radius R, there is a competition between the bulk and the surface parts of the free energy if the surface favors a radial liquid crystal alignment. For instance, as previously seen in Fig. 5.7, some of the configurations are known as planar-radial (PR), planar-polar (PP), escaped-radial (ER), and escaped-radial with point defects (ERPD). The first three configurations correspond to a minimum in the free energy. By changing the radius, R, of the cylinder, or the anchoring strength, or the elastic constants, it is possible to promote transitions from one type of configuration to another. More specifically, K_{24} and W_0 participate in the configurational transition from the PP to the ER structure.

It should be emphasized that W_0, the anchoring strength, strongly depends on the type of surface and surface treatment used. In the so-called strong anchoring limit, $W_0 \to \infty$, and since the nematic director becomes perpendicular to the surface, the surface energy term is replaced by the boundary condition $\phi = 0$. The saddle-splay elastic constant K_{24}, on the other hand, is a material parameter and is independent of any short range interactions occurring at the surface. Consequently, the determination of W_0, and more so K_{24}, was quite difficult until recently when Allender, Crawford, and Doane [10] (ACD) cleverly utilized the power of NMR technique to study confined nematic liquid crystals and simultaneously measure both W_0 and K_{24}.

The effect of diminishing anchoring strength becomes important if the liquid crystal is confined to small pore sizes. The curvature imposed by the confining walls competes with the anchoring strength, effectively tilting the molecules away from the preferred perpendicular anchoring alignment. Under these conditions, the molecular anchoring of the director at the surface is given by the dimensionless surface parameter

$$\sigma = \frac{RW_0}{K_{11}} + \frac{K_{24}}{K_{11}} - 1, \tag{5.36}$$

indicating the linear dependence of σ on pore size R.

The method suggested and used by ACD to determine σ and, from that, W_0 and K_{24} consisted of probing nematic director configurations that are stable in appropriate size cylindrical pores. Confining deuterated nematic 5CB liquid crystal to the sub-micrometer size pores found in polycarbonate Nuclepore membranes, NMR was able to confirm the stable configurations predicted by the free energy expression. Consequently, the NMR spectra were analyzed using σ as the only fitting parameter.

Specifically, NMR spectra were obtained at temperatures well into the nematic phase for 5CB deuterated at the β position (the second position in the alkyl chain) and confined to Nuclepore membranes. The size of the Nuclepore pores was varied from $0.3\,\mu m$ down to $0.05\,\mu m$. The spectra were recorded at different sample orientations, in particular, Fig. 5.9, with the pore axes parallel to the field ($\theta_B = 0°$) and with the pore axes perpendicular to the field ($\theta_B = 90°$).

Best fits were numerically calculated for each pore size and a correspond-

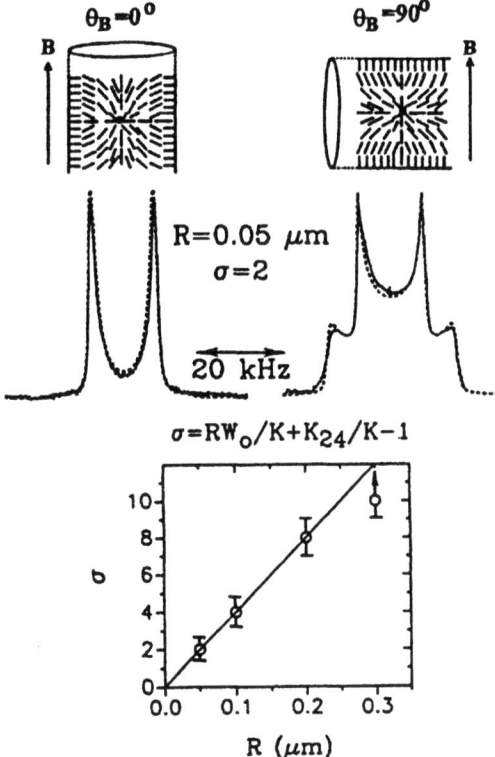

Figure 5.9. Experimental deuterium NMR spectra (solid line) of 5CB-βd₂ confined to the 0.05 μm pores of Nuclepore membranes at room temperature. The theoretical fit (dashed line) used the fitting parameters $\sigma = 2$ and $L/R = 1.5$. Shown at the bottom are the values of σ from equivalent studies as a function of pore size. From the linear relationship Eq. (5.36), W_0 and K_{24} normalized by K are determined from the slope and ordinate intercept, respectively. (From Crawford and Zumer [14].)

ing value for σ obtained. The 0.3 μm sample is in the strong anchoring regime and thus provides only a lower bound of $\sigma \geq 8$. For the smaller pore sizes, 0.05, 0.1, and 0.2 μm, it was found that $\sigma = 1.7 \pm 0.2$, 2.8 ± 0.4, and 7.0 ± 1.0 respectively. Plotting σ as a function of the pore size as shown in Fig. 5.9, yields W_0 and K_{24}. The fit gives values of $W_0/K = 3.5 \times 10^7$ m⁻¹ and $K_{24}/K \approx 1$. The magnitude of the saddle-splay surface elastic constant was found comparable to that of the bulk elastic constants of 5CB.

This measurement of K_{24} was further confirmed by treating the Nuclepore pores with a surfactant. The surfactant chosen, lecithin, was prepared in a 2 % solution which was then evaporated leaving a coating of lecithin in the pores. This weakened the anchoring strength but preserved the homeotropic or perpendicular anchoring at the pores' surface. The deuterium NMR patterns for pore sizes less than 0.4 μm were consistent with the liquid crystal molecules aligned in a planar-polar configuration. The configuration changed into an escaped-radial with a defect when the radius of the pore reached 0.5 μm. This result was particularly useful since the PP configuration is independent of K_{24} and thus sensitive to RW_0/K, while the escaped-radial with defects depends on σ and consequently K_{24}. The outcome of the analysis of these studies is that $K_{24}/K \approx 1.1$, which is consistent with the result from the untreated pores.

In summary, the combination of a powerful technique like NMR to study the 'ideal' system and the cleverness of the experimenters led to the determination of a quite elusive yet important elastic constant. The reader interested in a more complete discussion on this subject is referred to the original articles by Allender, Crawford, and Doane [10] and to the review article by Crawford and Zumer [14], listed in the bibliography section.

Phase separation

Although phase separation was mentioned earlier as a cause of truly horrendous difficulties, it is in general a process of immense importance in the manufacturing of many important materials ranging from steel to polymers, where a microscopic examination reveals domains having differing concentrations of the constituent materials. The use of phase separation in thermotropic mixtures of liquid crystals has taken on significant technological importance in the creation of polymer-dispersed liquid crystal displays. Lyotropics are of course inherently mixtures. On some length scales, phases such as the micellar phase could be considered as self-organized phase-separated materials.

NMR is a useful tool to map out two-phase regions and to study the dynamics of domain formation. Examples of its use in accurately mapping out phase diagrams for lyotropics can be found in the article by Boden, Jolley, and Ukleja [11]. One case in which NMR was able to unambiguously determine that a particular area of a phase diagram was indeed a two-phase region and not a new smectic phase occurs in the study of a mixture of 65% DB70CN and 35% 80CB in which the 80CB had been deuterated in the α position. An incommensurate SmA phase had been reported to exist between the SmA_d and SmA_2 regions for concentrations of 80CB between 25% and about 42%. However, subsequent x-ray experiments suggested strongly that this was a biphasic region.

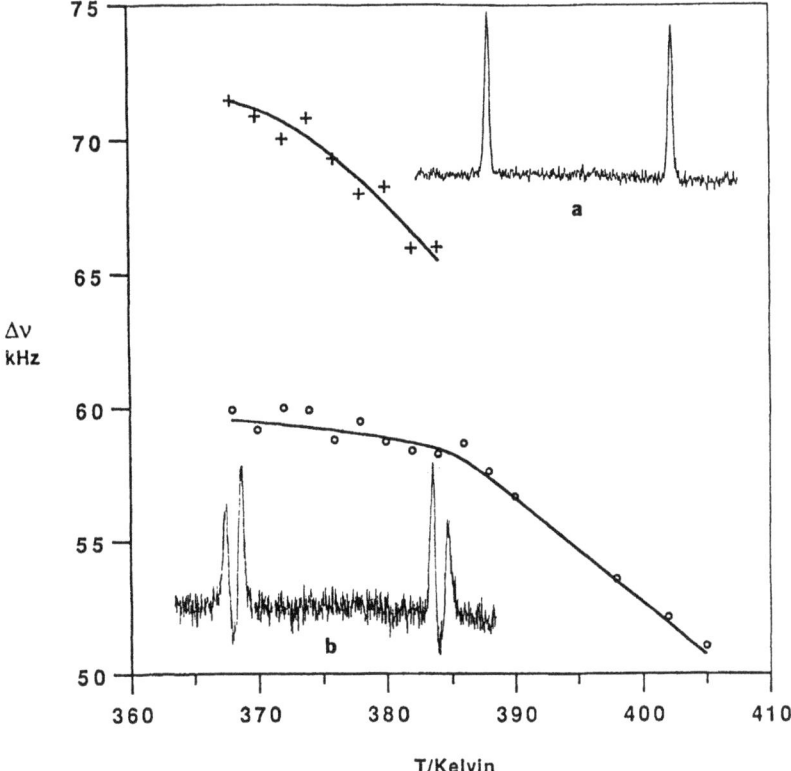

Figure 5.10. Deuterium NMR spectra and quadrupolar splittings versus temperature for the mixture 65 mole% DB7OCN + 35 mole% 8OCB-d₂. Inset (a) shows the spectrum upon cooling an aligned sample to 390 K. Inset (b) shows the spectrum after additional cooling to 380 K. (From Patel, Kumar, and Ukleja [16].)

In the NMR experiment, summarized in Fig. 5.10, the sample was aligned in a 7 T magnetic field on heating from room temperature to the smectic phases, and its spectrum was a single doublet (splitting) in the single-phase regions. On cooling into the region of the phase diagram thought earlier to contain an incommensurate smectic-A phase, a new doublet grew at the expense of the old, which is consistent with an interpretation of this region as a two-phase region. The existence of two sets of doublets implies that 80CB molecules apparently do *not* sample both molecular environments during the evolution of the NMR signal, several hundred microseconds. This implies domains on the order of at least tenths of micrometers (assuming a diffusion coefficient around $10^{-11}\,\mathrm{m}^2/\mathrm{s}$), or very strong barriers to diffusion at boundaries between the regions. On the other hand, the sample spectrum very quickly reverted to the single pair of doublets when heated back into the SmA_d phase, which argues that bulk demixing does not occur over a period of several hours. Thus the domains cannot be smaller than tens of micrometers, and do not coalesce into larger domains over times on the order of hours.

When jumps in temperature were made down into the two-phase region, the growth of one sort of domain at the expense of the other had a half-time of roughly one hour. This could be seen from the relative areas under the peaks corresponding to the two phases. In addition, after 12 to 14 hours at a temperature of 372 K, the sample abruptly changed into a lower lying phase, indicating that the sample was metastable and in a supercooled state. Interestingly, the line splitting was much lower in this phase. The article by Patel, Kumar, and Ukleja [16] thoroughly discusses more of this investigation.

Acknowledgment

One of us (D. F.) would like to acknowledge support from the NSF ALCOM Grant No. 89-20147 while this review chapter was written.

Bibliography

There are several NMR textbooks at many levels of sophistication, thus suitable for readers with various backgrounds. A few of them are listed below.

Introduction to NMR

1. A. Abragam, *Principles of Nuclear Magnetism*, Oxford University Press (1961). A classic, now also available in paperback.
2. P. T. Callaghan, *Principles of Nuclear Magnetic Resonance Microscopy*, Oxford Science Publications (1991). Although this text is aimed at NMR imaging, the introductory chapter includes concise and practical explanations of the NMR basics, including spin echoes and the basic interactions.
3. J. N. S. Evans, *Biomolecular NMR Spectroscopy*, Oxford University Press (1995).
4. E. Fukushima and S. B. W. Roeder, *Experimental Pulse NMR. A Nuts and Bolts Approach*, Addison-Wesley, Reading, MA (1981). Nice explanations and many practical details including instrumentation.
5. R. K. Harris, *Nuclear Magnetic Resonance Spectroscopy. A Physicochemical View*, Pitman, New York (1986).
6. C. P. Slichter, *Principles of Magnetic Resonance*, Springer-Verlag, Berlin (1989). A classic with detailed explanations. In addition, a thorough bibliography guiding the reader to specific aspects of NMR is included.

Specialized books

7. R. Y. Dong, *Nuclear Magnetic Resonance of Liquid Crystals*, Springer-Verlag, Berlin (1994). A discussion of NMR and molecular theory and experiments, in particular as relevant to liquid crystal work.
8. J. W. Emsley (Ed.), *Nuclear Magnetic Resonance of Liquid Crystals*, D. Reidel, Dordrecht (1985). Proceedings of the NATO Advanced Study Institute, San Miniato, Italy, 1983. NATO ASI series C, vol. 141.
9. G. R. Luckhurst and C. A. Veracini, *The Molecular Dynamics of Liquid Crystals*, Kluwer Academic, New York (1994). Proceedings of the NATO Advanced Study Institute, Barga, Italy, 1989, NATO ASI series C, vol. 143.

Selected articles

10. D. W. Allender, G. P. Crawford, and J. W. Doane, Determination of the Liquid-Crystal Surface Elastic Constant K_{24}, *Phys. Rev. Lett.* **67**, 1442 (1991). As the title suggests, this is the first NMR paper discussing how to determine the saddle-splay elastic constant.
11. N. Boden, K. W. Jolley, and P. Ukleja, NMR of Liquid Crystals, an overview, in *Condensed Matter News*, **1** (2) (1991).
12. G. Chidichimo and A. Golemme, Nuclear Magnetic Resonance Spectroscopy of Liquid Crystals, in *Physics of Liquid Crystalline Materials*, Khoo, I. C., and Simoni, F. (Eds.), Chapter 6, Gordon and Breach Science Publishers, Philadelphia, PA (1991). A simplified discussion of the basic spin interactions, motional averaging and molecular dynamics.
13. G. P. Crawford and J. W. Doane, Ordering and Ordering Transitions in Confined Liquid Crystals, *Mod. Phys. Lett. B* **28**, 1785 (1993). A review of the behavior of liquid crystal materials confined to different porous media including a brief discussion of the determination of the saddle-splay elastic constant.
14. G. P. Crawford and S. Zumer, Saddle-Splay Elasticity in Nematic Liquid Crystals, *Int. J. Mod. Phys. B* **18 & 19**, 2469 (1995). A review of NMR measurements in the cylindrical pores on Nuclepore membranes with emphasis on the determination of K_{24}.
15. J. W. Doane, NMR of Liquid Crystals, in *Magnetic Resonance of Phase Transitions*, Owen, F. J., *et al.* (Eds.), Chapter 4, Academic Press, New York (1979). A thorough discussion of NMR in liquid crystals including uniaxial and biaxial phases, Euler angles, orientational order, and spin-lattice relaxation.
16. P. Patel, S. Kumar, and P. Ukleja, The Case of Missing Incommensurate Smectic A Phases, *Liq. Cryst.* **16**, 35 (1994). For those interested in learning more about phase separation and particularly about incommensurate phases.
17. M. Vilfan and N. Vrbancic-Kopac, Nuclear Magnetic Resonance of Liquid Crystals with an Embedded Polymer Network in *Liquid Crystals in Complex Geometries Formed by Polymer and Porous Networks*, G. P. Crawford and S. Zumer (Eds.), Chapter 7, Taylor & Francis, Bristol, PA (1996). An excellent review of deuterium NMR studies on confined liquid crystals.
18. P. Ukleja, J. Pirs, and J. W. Doane, Theory for Spin-Lattice Relaxation in Nematic Liquid Crystals, *Phys. Rev. A* **14**, 414 (1976). Relaxation measurements in nematics as a function of temperature, frequency, order parameter, alignment of sample in several patterns, and the angle between field and director were performed to test a model for the molecular motions.

19. H. Zeng and D. Finotello, Anisotropy Induced Liquid-Crystal Configurational Transitions, *Phys. Rev. Lett.* **81**, 2703 (1998). Very recent work showing how even in a random interconnected network of cavities, well-defined liquid crystal configurations are established. For this particular host medium, its anisotropy induces configurational transitions.
20. O. Stejskal and B. Tanner, *J. Chem. Phys.* **42**, 288 (1965). This article presents a detailed discussion of the spin echoes from the semi-classical view point.

6

Light scattering and quasielastic spectroscopy

JOHN T. HO

Department of Physics, State University of New York at Buffalo, Buffalo,
NY 14260, USA

6.1 Introduction

It is well known that the discovery of liquid crystals was prompted at least in part by the striking turbidity of nematic liquid crystals caused by light scattering, and this dramatic property has been exploited to elucidate valuable information about liquid crystals ever since. Even before its application to liquid crystals, light scattering was generally recognized as a powerful experimental technique in condensed matter physics. It has been widely used in the study of conventional fluids [1]. The phenomenon is based on the scattering of light by inhomogeneities in the dielectric constant (or polarizability) of the medium through which light is passed. The dielectric inhomogeneities can be either static or dynamic, and originate from intermolecular or intramolecular fluctuations in the fluid. The intensity of the scattered light thus contains information about the magnitude of the fluctuations, while the frequency shift or time-correlation function of the scattered light contains information about their dynamics. Depending on the direction of the scattered light relative to that of the incident light, one can study fluctuations on a length scale ranging from a fraction of the wavelength to an experimental limit of several hundred times the wavelength of the incident light.

Light scattering has been used to probe fluctuations in condensed matter that originate from the hydrodynamic modes of the system, which exist from either conservation laws or symmetry. They are characterized by decay times that are proportional to some power of the wavelength. Among these are the 'Goldstone modes', which are long-lived collective excitations associated with a symmetry that can be continuously broken and thus have restoring forces that vanish in the limit of infinite length scale [2]. In a conventional isotropic fluid, for example, the conservation of mass, energy, and momentum leads to hydrodynamic behavior in the fluctuations of the density, temperature, and the three components of the momentum. The solution of the equations of motion of the system leads to two propagating modes (longitudinal sound waves) and three diffusive modes (two transverse shear waves and one mode associated with temperature or entropy fluctuations). These modes have typical frequencies in the megahertz to gigahertz range, and are typically probed using the technique of Brillouin scattering. In addition, another class of fluctuations that can result in the scattering of light originates from the internal degrees of freedom of the molecules. These intramolecular modes lead to Raman scattering which has frequency shifts of more than 10^9 Hz. Raman scattering in fluids is generally a local probe that provides little information about the collective behavior of the system.

The early light scattering experiments in nematic liquid crystals by Chatelain [3] showed that the scattering was intense and that the scattered light was largely depolarized. It was de Gennes [4] who first pointed out that the reason for the strong light scattering by nematics lies in spontaneous fluctuations of the molecular orientations. The molecules in the nematic phase in equilibrium are aligned along an average direction. Thermal fluctuations, however, cause the molecules to deviate from this direction temporally as well as spatially. These spontaneous orientational fluctuations cause fluctuations in the dielectric tensor which is associated with the molecular anisotropy, causing light passing through the liquid crystal to be scattered. It should be mentioned that the scattering of light by a liquid crystal can also result from the usual hydrodynamic modes, such as density fluctuations, as mentioned above in a conventional isotropic fluid, as well as from intramolecular modes, but such scattering is considerably weaker (by a factor $\sim 10^{-6}$) than that caused by orientational fluctuations. The dominant scattered light in a liquid crystal can thus be used to obtain information about the magnitude of the orientational fluctuations as well as their dynamics. Light scattering is particularly powerful in elucidating the properties of liquid crystals near phase transitions, since these fluctuations are expected to exhibit pronounced pretransitional behavior. As a result of techniques that have been developed over the last 25 years, light scattering has become an important laboratory tool in liquid crystal research. This chapter is devoted primarily to describing the use of light scattering to probe the hydrodynamic orientational fluctuations in liquid crystals. Other aspects of light scattering, such as diffraction, Brillouin and Raman scattering, will be mentioned only briefly. For additional material on the subject of light scattering in liquid crystals, the reader is referred to standard monographs [5, 6] and review articles [7, 8].

6.2 Principles of light scattering

Consider the incident light to be a plane wave with electric field $\mathbf{E}_i$ in the form

$$\mathbf{E}_i(\mathbf{r},t) = \mathbf{i} E_0 \exp[i(\mathbf{k}_i \cdot \mathbf{r} - \omega t)], \tag{6.1}$$

where E_0 is the field amplitude, $\mathbf{i}$ is a unit vector in the direction of the incident polarization, $\mathbf{k}_i$ is the incident wavevector, and ω is the angular frequency. According to classical electromagnetic theory, this field polarizes the molecules of the medium and the resulting oscillating dipoles emit electromagnetic radiation. The scattered light in a given direction is the sum of the emitted radiation in that direction from the entire irradiated volume of the medium, taking into account interference due to phase differences among the radiated waves. If the polarizability, or dielectric constant, is uniform throughout the medium, then it can be shown that this summation results in only a transmitted beam in the forward direction and no scattered light. Thus light scattering originates from the presence of inhomogeneities in the dielectric constant of the medium. The intensity of the scattered light depends on the magnitude of the dielectric fluctuations, while its frequency is affected by their temporal behavior.

If the local dielectric tensor $\varepsilon(\mathbf{r},t)$ of the medium is divided into an average part $\bar{\varepsilon}$ and a fluctuating part $\delta\varepsilon(\mathbf{r},t)$:

$$\varepsilon(\mathbf{r},t) = \bar{\varepsilon} + \delta\varepsilon(\mathbf{r},t), \tag{6.2}$$

then it can be shown [9] that the magnitude of the scattered electric field E_s at a large distance R from the scattering volume V with wavevector $\mathbf{k}_f$ and polarization along the unit vector $\mathbf{f}$ is given by

$$E_s(t) = \frac{E_0}{4\pi R} \exp[i(k_f R - \omega t)] \int_V \exp(i\mathbf{q}\cdot\mathbf{r})[\mathbf{f}\cdot[\mathbf{k}_f \times \mathbf{k}_f \times (\delta\varepsilon(\mathbf{r},t)]\} dr, \tag{6.3}$$

where $\mathbf{q}$ is the scattering wavevector defined by

$$\mathbf{q} \equiv \mathbf{k}_i - \mathbf{k}_f. \tag{6.4}$$

A typical scattering geometry is depicted in Fig. 6.1(a). The significance of the scattering wavevector $\mathbf{q}$ is illustrated in Fig. 6.1(b). By considering the path difference between the incident rays at two points separated by a distance $\mathbf{r}$ and the path difference between the scattered rays from these two

(a)

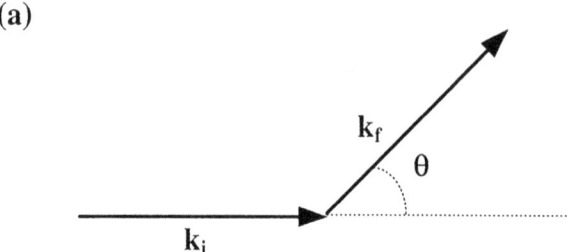

(b)

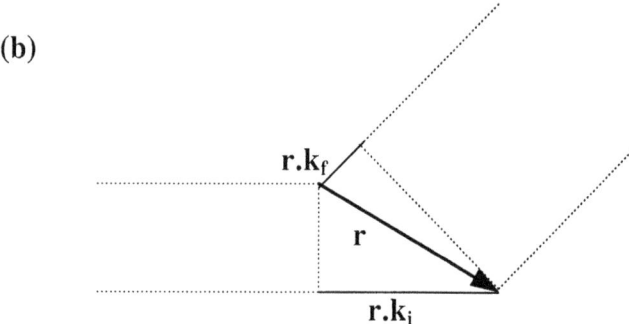

Figure 6.1. (a) Typical light scattering geometry showing incident wavevector $\mathbf{k}_i$ and final wavevector $\mathbf{k}_f$. (b) Phase difference between two points separated by $\mathbf{r}$.

points, it can be seen that the net phase difference Δ between the two scattered rays is

$$\Delta = \mathbf{r} \cdot \mathbf{k}_i - \mathbf{r} \cdot \mathbf{k}_f = \mathbf{r} \cdot (\mathbf{k}_i - \mathbf{k}_f) = \mathbf{r} \cdot \mathbf{q}. \tag{6.5}$$

Equation (6.3) represents the sum of the phase-shifted waves emitted from different parts across the volume V of the scattering medium.

In Eq. (6.3), the angular frequency of the scattered light is assumed to be basically unchanged from that of the incident light. This is a reasonable assumption, since the dominant fluctuations in liquid crystals scatter light quasielastically. In discussing light scattering in isotropic liquids, it usually follows that the magnitudes of the incident and scattered wavevectors are also essentially identical, resulting in the useful results $k_f \cong k_i = k$ and

$$q = 2k \sin(\theta/2), \tag{6.6}$$

where θ is the angle between $\mathbf{k}_i$ and $\mathbf{k}_f$. However, it is important to note that such simplifications are generally not valid in liquid crystals. Because of the optical anisotropy of liquid crystals, the refractive index depends on the direction of the polarization and the wavevector relative to optic axis. Since the magnitude of wavevector is proportional to the refractive index, the wavevectors $\mathbf{k}_i$ and $\mathbf{k}_f$ typically have different magnitudes, and it is necessary to evaluate the scattering wavevector $\mathbf{q}$ in each specific case in accordance with Eq. (6.4).

Introducing the spatial Fourier transform of the dielectric fluctuations

$$\delta\varepsilon\,(\mathbf{q},t) \equiv \int_V \delta\varepsilon\,(\mathbf{r},t)\exp(i\mathbf{q}\cdot\mathbf{r})d\mathbf{r}, \tag{6.7}$$

Eq. (6.3) can be simplified to

$$E_s(t) = \frac{k_f^2 E_o}{4\pi R}\exp[i(k_f R - \omega t)\delta\varepsilon_{if}\,(\mathbf{q},t), \tag{6.8}$$

where

$$\delta\varepsilon_{if}\,(\mathbf{q},t) \equiv \mathbf{f}\cdot\delta\varepsilon\,(\mathbf{q},t)\cdot\mathbf{i} \tag{6.9}$$

is the component of the dielectric fluctuation tensor along the initial and final polarization directions.

An important quantity in both static and dynamic light scattering measurements is the time-correlation function $C_E(\mathbf{q},\tau)$ of the scattered electric field:

$$C_E(\mathbf{q},\tau) \equiv \langle E_s^*(t)E_S(t+\tau)\rangle, \tag{6.10}$$

where τ represents a delay time. From Eqs. (6.8) and (6.10), one obtains

$$C_E(\mathbf{q},\tau) = \frac{k_f^4 I_o}{16\pi^2 R^2}\exp(-i\omega\tau)\langle\delta\varepsilon_{if}\,(\mathbf{q},t)\delta\varepsilon_{if}\,(\mathbf{q},t+\tau)\rangle, \tag{6.11}$$

where $I_0 = |E_0|^2$ is the incident light intensity.

Depending on the experimental design, the quantities that can be measured in light scattering include the intensity of the scattered light and its time-correlation function [9, 10]. The average intensity $I_s(\mathbf{q})$ is obtained by setting τ to zero in Eq. (6.11).

$$I_s(\mathbf{q}) = \langle E_s^*(t)E_s(t)\rangle = \frac{k_f^4 I_o}{16\pi^2 R^2}\langle\delta\varepsilon_{if}^2(\mathbf{q})\rangle. \tag{6.12}$$

It can be measured directly by any suitable photodetector, such as a photomultiplier tube. The photomultiplier tube is an example of a square-law detector, in which the instantaneous output is proportional to the square of the electric field (and hence proportional to the intensity) of the light at the photocathode. Equation (6.12) shows that the average scattered intensity $I_s(\mathbf{q})$ is a direct measure of the mean-square amplitude of the dielectric fluctuations $\langle \delta\varepsilon_{if}^2(\mathbf{q}) \rangle$.

The measurement of the time-correlation function of the scattered light intensity to obtain information about the dynamics of orientational fluctuations in liquid crystals represents an example of a special class of optical spectroscopy, commonly referred to as 'quasielastic', 'light-beating' or 'photon-correlation', which becomes preferable when the processes to be examined have relatively slow relaxation times of the order of 10^{-6} s or longer. These techniques are the optical analog of the beating techniques developed in radio-frequency spectroscopy [11], and do not involve any filtering optics. In contrast, optical spectroscopy to examine processes faster than $\sim 10^{-6}$ s would require the use of frequency-filtering techniques based on either a diffraction grating (for Raman scattering) or a Fabry–Perot interferometer (for Brillouin scattering).

Photon-correlation spectroscopy in liquid crystals is usually conducted in the *homodyne*, or 'self-beating', mode. In this approach, only the scattered light impinges on the photocathode, and the measured intensity is recorded and manipulated electronically to compute its time-correlation function $C_I(\mathbf{q},\tau)$, which is defined as

$$C_I(\mathbf{q},\tau) \equiv \langle I_s(t)I_s(t+\tau) \rangle = \langle E_s^*(t)E_s(t)E_s^*(t+\tau)E_s(t+\tau) \rangle, \qquad (6.13)$$

where $I_s(t)$ is the instantaneous value of the scattered intensity at time t. If the scattered optical field obeys Gaussian statistics, which is usually true in experiments involving liquid crystals, the time-correlation functions $C_E(\mathbf{q},\tau)$ and $C_I(\mathbf{q},\tau)$ of the scattered field and the scattered intensity, respectively, are related to each other. The relation is most easily expressed in terms of normalized first- and second-order autocorrelation functions defined by

$$g^{(1)}(\tau) \equiv \frac{\langle E_s^*(t)E_s(t+\tau) \rangle}{\langle E_s^*(t)E_s(t) \rangle} = \frac{C_E(\mathbf{q},\tau)}{I_s(\mathbf{q})} \qquad (6.14)$$

and

$$g^{(2)}(\tau) \equiv \frac{\langle E_s^*(t)E_s(t)E_s^*(t+\tau)E_s(t+\tau) \rangle}{\langle E_s^*(t)E_s(t) \rangle^2} = \frac{C_I(\mathbf{q},\tau)}{I_s^2(\mathbf{q})}. \qquad (6.15)$$

These autocorrelation functions obey the Siegert relation [10, 12],

$$g^{(2)}(\tau) = 1 + |g^{(1)}(\tau)|^2. \qquad (6.16)$$

Combining Eqs. (6.10) to (6.16), one obtains

$$C_I(\mathbf{q},\tau) = I_s^2(\mathbf{q}) + |C_E(\mathbf{q},\tau)|^2 = I_s^2(\mathbf{q})\left[1 + \frac{\langle\delta\varepsilon_{if}(\mathbf{q},t)\delta\varepsilon_{if}(\mathbf{q},t+\tau)\rangle^2}{\langle\delta\varepsilon_{if}^2(\mathbf{q})\rangle^2}\right].(6.17)$$

Thus the scattered intensity correlation function $C_I(\mathbf{q},\tau)$ contains two parts, a constant or 'dc' term which depends on the average intensity and a time-dependent term. The latter depends on the dielectric correlation function

$$C_\varepsilon(\mathbf{q},\tau) \equiv \langle\delta\varepsilon_{if}(\mathbf{q},t)\delta\varepsilon_{if}(\mathbf{q},t+\tau)\rangle. \tag{6.18}$$

A common form of $C_\varepsilon(\mathbf{q},t)$ for scattering from diffusive modes is an exponential decay

$$C_\varepsilon(\mathbf{q},\tau) = \langle\delta\varepsilon_{if}^2(\mathbf{q})\rangle\exp(-\Gamma\tau), \tag{6.19}$$

where Γ is the relaxation rate which reflects the dynamics of the dielectric fluctuations. The reciprocal of Γ is usually referred to as the correlation time. The intensity correlation function $(C_I(\mathbf{q},\tau)$ will then have the simple time dependence

$$C_I(\mathbf{q},\tau) = I_s^2(\mathbf{q})[1 + \exp(-2\Gamma\tau)]. \tag{6.20}$$

It should be noted that, if the surface of the photocathode being illuminated contains more than one coherence area of the scattered light, the magnitude of the time-dependent term in Eq. (6.17) will be reduced relative to that of the dc term, but its functional form will not be affected [13].

Either by design or by accident, photon-correlation spectroscopy is sometimes conducted in the *heterodyne* mode, in which a local oscillator is mixed with the dynamically scattered light on the photocathode. The local oscillator is usually derived from light statically scattered by defects in the sample or from a small portion of the unscattered laser beam. Depending on the intensity I_{LO} of the local oscillator relative to that of the dynamically scattered light, the correlation function $C_I(\mathbf{q},\tau)$ of the total measured intensity can take on a very complex form [10]. In the limit of $I_{LO} \gg I_s(\mathbf{q})$, it can be shown that [9]

$$C_I(\mathbf{q},\tau) \cong I_{LO}^2 + 2I_{LO}\,\text{Re}[C_E(\mathbf{q},\tau)]. \tag{6.21}$$

In this case, the time dependence of $C_I(\mathbf{q},\tau)$ will be identical to that of $C_E(\mathbf{q},\tau)$, which in turn is proportional to $C_\varepsilon(\mathbf{q},\tau)$.

6.3 Scattering in the isotropic phase

The primary interest in conducting light scattering in the isotropic phase of a liquid crystal material lies in studying the pretransitional behavior of short range orientational order in the neighborhood of a phase transition to a liquid crystal phase, most commonly a nematic phase. Based on symmetry considerations, the orientational order can be described by a symmetric traceless tensor $Q_{\alpha\beta}$, which in general has five independent components. If we ignore biaxiality, we can write $Q_{\alpha\beta}$ as

$$Q_{\alpha\beta} \equiv \tfrac{1}{2} S(3n_\alpha n_\beta - \delta_{\alpha\beta}), \tag{6.22}$$

where S is the scalar nematic order parameter and n_α and n_β are the Cartesian components of a unit vector $\mathbf{n}$, called the director, which defines the direction of the optic axis. The dielectric tensor $\varepsilon_{\alpha\beta}$ is related to $Q_{\alpha\beta}$ by

$$\varepsilon_{\alpha\beta} = \bar{\varepsilon}\delta_{\alpha\beta} + \tfrac{2}{3} \varepsilon_a^0 \, Q_{\alpha\beta}, \tag{6.23}$$

where $\bar{\varepsilon}$ is the average dielectric constant and ε_a^0 is the dielectric anisotropy in the completely ordered state ($S = 1$).

In the isotropic phase, using the Landau–de Gennes mean-field model, the free energy F_1 per unit volume in the vicinity of the isotropic–nematic transition can be expanded in a power series [14, 15]:

$$F_1 = F_0 + \tfrac{1}{2} A Q_{\alpha\beta} \, Q_{\beta\alpha} + \tfrac{1}{3} \tilde{B} Q_{\alpha\beta} \, Q_{\beta\gamma} \, Q_{\gamma\alpha} + \tfrac{1}{4} C(Q_{\alpha\beta} \, Q_{\beta\alpha})^2$$
$$+ \tfrac{1}{2} L_1 \partial_\alpha \, Q_{\beta\gamma} \, \partial_\alpha \, Q_{\beta\gamma} + \tfrac{1}{2} L_2 \partial_\alpha \, Q_{\alpha\gamma} \partial_\beta \, Q_{\beta\gamma} + O(Q^5), \tag{6.24}$$

where ∂_α denotes $\partial/\partial\chi_\alpha$ and summation over repeated indices is implied. The terms in Eq. (6.24) are chosen to be invariant with respect to rotation. In the mean-field approximation,

$$A(T) \propto T - T_{\text{IN}}^*, \tag{6.25}$$

where T_{IN}^* is an apparent isotropic–nematic transition temperature, and the other higher-order coefficients $\tilde{B}, C, L_1$ and L_2 are essentially temperature-independent. The presence of the coefficient $\tilde{B}$ results in the isotropic–nematic transition being first order and occurring at an actual transition temperature T_{IN} which is somewhat above T_{IN}^*.

To evaluate the change in the total free energy associated with order-parameter fluctuations, we expand $Q_{\alpha\beta}$ in terms of its spatial Fourier components:

$$Q_{\alpha\beta}(\mathbf{r}) = V^{-1} \sum_q Q_{\alpha\beta} \, (\mathbf{q}) \exp(-i\mathbf{q} \cdot \mathbf{r}) \tag{6.26}$$

and integrate F_I over the sample volume V. Keeping only the second-order term in Eq. (6.24) and applying the equipartition theorem, it can be shown that [8, 15]

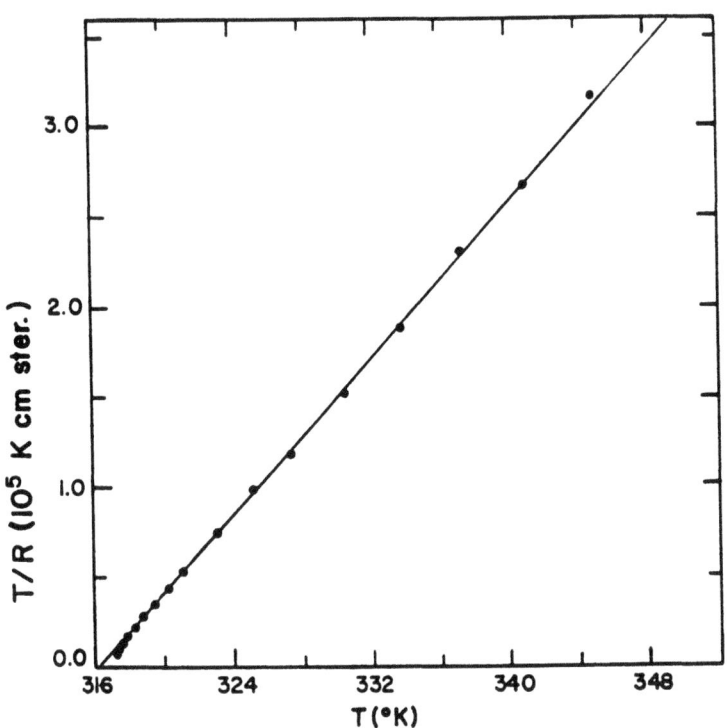

Figure 6.2. Temperature dependence of the reciprocal of the Rayleigh ratio R (which is proportional to the light scattering intensity) above the isotropic–nematic transition in MBBA. The straight line is the fit to Eq. (6.28). (Reprinted with permission from Ref. 16.)

$$\langle Q^2_{\alpha\beta}(\mathbf{q})\rangle \propto k_B T/A \propto (T - T^*_{\mathrm{IN}})^{-1}. \qquad (6.27)$$

From Eq. (6.23), one sees that $\langle Q^2_{\alpha\beta}(\mathbf{q})\rangle$ is directly related to the mean-square amplitude of the dielectric fluctuations $\langle \delta\varepsilon^2_{ij}(\mathbf{q})\rangle$, which is in turn proportional to the average light scattering intensity $I_s(\mathbf{q})$ in accordance with Eq. (6.12). Thus the reciprocal of the intensity is expected to be proportional to T:

$$I_s^{-1}(\mathbf{q}) \propto T - T^*_{\mathrm{IN}}. \qquad (6.28)$$

The behavior predicted by Eq. (6.28) has been borne out experimentally [16]. The results for N-p-methoxybenzylidene-p'-n-butylaniline (MBBA) are illustrated in Fig. 6.2.

The angular dependence of the light scattering intensity can be obtained from Eq. (6.24) by including the gradient terms which depend on L_1 and L_2. By choosing the $\mathbf{x}_3$-axis to be along $\mathbf{q}$, it can be shown, for example, that [8]

$$\langle Q_{33}^2(\mathbf{q}) \rangle = \frac{2k_B T}{3A(1 + \xi_1^2 q^2 + \frac{2}{3}\xi_2^2 q^2)}, \tag{6.29}$$

where the two correlation lengths ξ_1 and ξ_2 are defined by

$$\xi_1 \equiv \sqrt{L_1/A} \propto (T - T_{IN}^*)^{-0.5} \tag{6.30}$$

and

$$\xi_2 \equiv \sqrt{L_2/A} \propto (T - T_{IN}^*)^{-0.5}. \tag{6.31}$$

In the isotropic phase, the value of q is related to the scattering angle θ by the simple expression in Eq. (6.6). Thus Eq. (6.29) suggests a decreasing light scattering intensity as one moves away from the forward direction. In practice, the typical values of ξ_1 and ξ_2 are of the order of 100 Å. With the wavelength λ of light in the visible range, Eq. (6.29) predicts a variation in the light scattering intensity with angle of the order of only about 1%, making its experimental verification somewhat difficult. Nevertheless, the results of careful measurements are in general agreement with the typical angular dependence of the intensity described in Eq. (6.29) and the temperature dependence of the correlation lengths described in Eqs. (6.30) and (6.31) [17].

Simple theoretical argument suggests that the dynamics of short range orientational order in the isotropic phase of a nematic liquid crystal should also show pretransitional behavior, similar to the critical slowing down found near other phase transitions [18]. As a result, the time-correlation function of the scattered light intensity is expected to exhibit the exponential decay depicted in Eq. (6.20), with the relaxation rate Γ proportional to

$$\Gamma \propto (T - T_{IN}^*)/\eta, \tag{6.32}$$

where η is a transport coefficient which does not show pretransitional behavior, but has an Arrhenius temperature dependence with an activation energy similar to that of the shear viscosity. This behavior has been observed experimentally [19].

If the liquid crystal phase to which the isotropic phase transforms upon cooling is chiral in nature (typically cholesteric), there is additional complexity and richness in the pretransitional behavior. The general orientational order parameter $Q_{\alpha\beta}$ has five independent elements. In a chiral system, it is appropriate to represent $Q_{\alpha\beta}$ as a linear combination of five basis tensors, each of which describes a different 'structural mode' [20, 21]. The fluctuations of the various modes are expected to diverge near the isotropic–cholesteric transition similar to Eq. (6.27), but the five apparent transition temperatures of the modes are slightly different from each other and depend on the intrinsic chirality q_c and the wavevector q of the modes. By properly choosing the experimental geometry, these modes can be studied individually with light scattering using circularly polarized light [21, 22]. The role of the chirality can be examined by using binary mixtures with different proportions of a cholesteric and a nematic liquid crystal to vary the value of q_c.

6.4 Scattering by nematic fluctuations

The nematic phase is ordered orientationally but not translationally. An ideal nematic liquid crystal is characterized by a uniform equilibrium director $\mathbf{n}_0$, so that the molecules are on the average aligned along $\pm \mathbf{n}_0$. To describe thermal fluctuations in the alignment, as well as deformations resulting from boundary constraints or external fields, it is convenient to use a continuum theory, in which the detailed order on a molecular scale is ignored and the local director $\mathbf{n}$ is treated as a continuous function of the position $\mathbf{r}$ [5]. One can write

$$\mathbf{n}(\mathbf{r}) = \mathbf{n}_0 + \delta\mathbf{n}(\mathbf{r}), \tag{6.33}$$

where $\delta\mathbf{n}(\mathbf{r})$ is the deviation of $\mathbf{n}(\mathbf{r})$ from equilibrium. Since the fluctuations are small and $\mathbf{n} \cdot \mathbf{n} = 1$, we have, to first approximation,

$$\delta\mathbf{n} \cdot \mathbf{n}_0 = 0. \tag{6.34}$$

Suppose we would like to consider the $\mathbf{q}$-th Fourier component of the director fluctuations. If we choose the $\mathbf{x}_3$-axis to be along $\mathbf{n}_0$ and take the plane containing $\mathbf{q}$ and $\mathbf{n}_0$ as the $\mathbf{x}_1 - \mathbf{x}_3$ plane, as shown in Fig. 6.3, so that $\mathbf{q}$ is perpendicular to $\mathbf{x}_2$, then Eq. (6.34) suggests that $\delta\mathbf{n}$ has only two components δn_1 and δn_2. The components of $\mathbf{q}$ are

$$\mathbf{q} = (q_1, 0, q_3) = (q_\perp, 0, q_\parallel). \tag{6.35}$$

It follows from Eq. (6.22) that the orientational order parameter $Q_{\alpha\beta}(\mathbf{r})$ also varies spatially in accordance with the behavior of $\mathbf{n}(\mathbf{r})$. From Eqs. (6.22) and (6.23), the dielectric tensor can be written as

$$\varepsilon_{\alpha\beta}(\mathbf{r}) = \varepsilon_\perp \delta_{\alpha\beta} + (\varepsilon_\parallel - \varepsilon_\perp) n_\alpha(\mathbf{r}) n_\beta(\mathbf{r}), \tag{6.36}$$

where $\varepsilon_\perp$ and $\varepsilon_\parallel$ are the dielectric constants perpendicular and parallel to the optic axis, respectively. The dielectric anisotropy ε_a is defined as

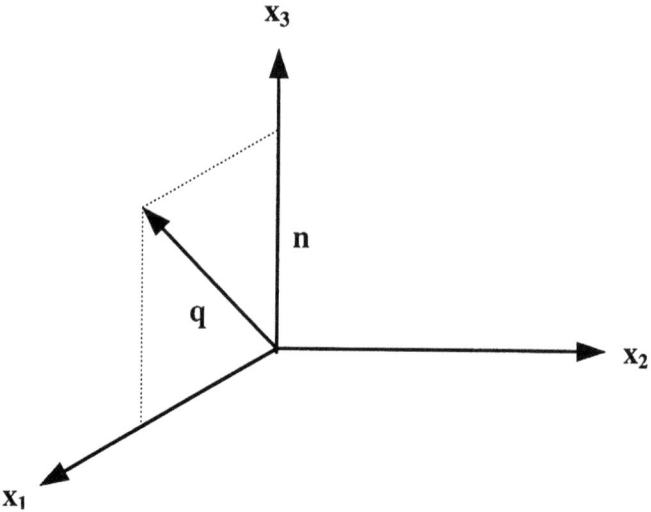

Figure 6.3.   Cartesian coordinates in relation to the director **n** and scattering
wavevector **q** to describe scattering in the nematic and smectic-A phases.

$$\varepsilon_a \equiv \varepsilon_\parallel - \varepsilon_\perp. \tag{6.37}$$

From Eqs. (6.36) and (6.37), we can express the fluctuations in the dielectric tensor as

$$\delta\varepsilon_{\alpha\beta} = \varepsilon_a(\delta n_\alpha n_{0\beta} + n_{0\alpha}\delta n_\beta). \tag{6.38}$$

If we take the spatial Fourier transform of $\delta\varepsilon$ as defined by Eq. (6.7) and
consider the component with respect to the initial and final polarization
directions **i** and **f** as described in Eq. (6.9), then the mean-square amplitude
of the dielectric fluctuations as a function of the wavevector **q** can be
expressed as

$$\langle \delta\varepsilon_{\mathrm{if}}^2(\mathbf{q}) \rangle = \varepsilon_a^2 \sum_{\alpha=1,2} (i_\alpha f_3 + i_3 f_\alpha)^2 \langle |\delta n_\alpha(\mathbf{q})|^2 \rangle, \tag{6.39}$$

where i_α and f_α are the projections of **i** and **f** on the Cartesian axes and
$\delta n_\alpha(\mathbf{q})$ are the spatial Fourier transforms of the components of the director
fluctuations defined by

$$\delta n_\alpha(\mathbf{q}) \equiv \int_V \delta n_\alpha(\mathbf{r})\exp(i\mathbf{q}\cdot\mathbf{r})d\mathbf{r}. \tag{6.40}$$

From Eqs. (6.12) and (6.39), the average intensity of the scattered light is
given by

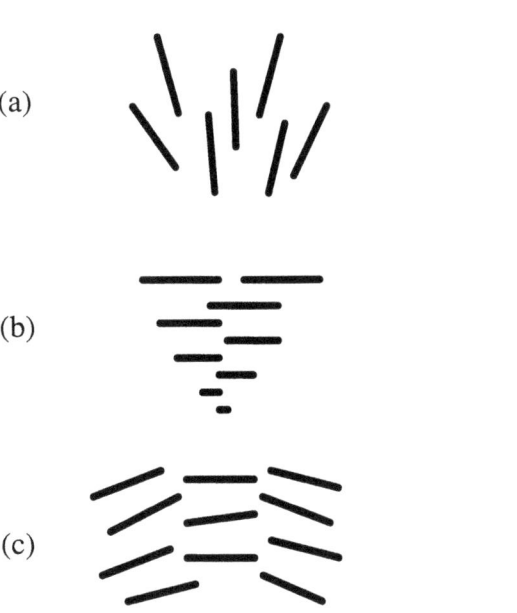

(a)

(b)

(c)

Figure 6.4. Schematic representation of (a) splay, (b) twist and (c) bend deformations. Depicted in (b) are projections on the plane of the diagram of various molecules, some of which are twisted away from the plane.

$$I_s(\mathbf{q}) = \frac{k_f^4 I_0 \varepsilon_a^2}{16\pi^2 R^2} \sum_{\alpha=1,2} (i_\alpha f_3 + i_3 f_\alpha)^2 \langle |\delta n_\alpha(\mathbf{q})|^2 \rangle. \qquad (6.41)$$

It can be seen from Eq. (6.41) that the scattered intensity depends on the mean-square amplitude of the thermal fluctuations of the two uncoupled eigenmodes represented by δn_1 and δn_2, which are the two Goldstone modes associated with the nematic order. The allowed fluctuations are related to the three basic types of orientational deformation that are present in a nematic liquid crystal, namely, splay, twist and bend [5]. These are depicted schematically in Fig. 6.4. To evaluate the amplitudes of these normal modes, we start with the free energy per unit volume, F_N, in a nematic liquid crystal with a spatially varying director $\mathbf{n(r)}$ [23]:

$$F_N = \tfrac{1}{2} K_1 (\nabla \cdot \mathbf{n})^2 + \tfrac{1}{2} K_2 (\mathbf{n} \cdot \nabla \times \mathbf{n})^2 + \tfrac{1}{2} K_3 (\mathbf{n} \times \nabla \times \mathbf{n})^2, \qquad (6.42)$$

where K_1, K_2 and K_3 are the Frank elastic constants for splay, twist and bend, respectively, that describe the energetics of the various types of deformation. With the equilibrium director $\mathbf{n}_0$ along the x_3-axis and recognizing that the director fluctuation $\delta\mathbf{n}$ to first approximation has only two components δn_1 and δn_2, Eq. (6.42) is reduced to

$$F_N = \tfrac{1}{2}\{K_1(\partial_1 n_1 + \partial_2 n_2)^2 + K_2(\partial_2 n_1 - \partial_1 n_2)^2 + K_3[(\partial_3 n_1)^2 + (\partial_3 n_2)^2]\}. \quad (6.42)$$

Each component of the director fluctuations can be expanded in terms of its Fourier series:

$$\delta n_\alpha(\mathbf{r}) = V^{-1} \sum_q \delta n_\alpha(\mathbf{q}) \exp(-i\mathbf{q} \cdot \mathbf{r}). \quad (6.44)$$

Choosing the coordinates for a given $\mathbf{q}$ as prescribed above and integrating over the illuminated volume V, we have

$$\int_V F_N d\mathbf{r} = \tfrac{1}{2} \sum_q (K_1 q_\perp^2 + K_3 q_\parallel^2) \delta n_1^2(\mathbf{q}) + (K_2 q_\perp^2 + K_3 q_\parallel^2) \delta n_2^2(\mathbf{q})]. \quad (6.45)$$

By applying the equipartition theorem to Eq. (6.45), we obtain the mean-square amplitudes of the two normal director modes for the wave vector $\mathbf{q}$:

$$\langle |\delta n_1(\mathbf{q})|^2 \rangle = \frac{k_B T}{K_1 q_\perp^2 + K_3 q_\parallel^2}, \quad (6.46)$$

$$\langle |\delta n_2(\mathbf{q})|^2 \rangle = \frac{k_B T}{K_2 q_\perp^2 + K_3 q_\parallel^2}. \quad (6.47)$$

It can be seen from Eqs. (6.46) and (6.47) that the first mode associated with δn_1 involves a combination of bend and splay distortions, while the second mode associated with δn_2 involves a combination of bend and twist distortions. When the wave vector $\mathbf{q}$ is along the direction of the equilibrium director $\mathbf{n}_0$, the two modes are degenerate and only bend fluctuations are present.

Combining Eqs. (6.41), (6.46) and (6.47), we arrive at a general expression for the average intensity of the scattered light:

$$I_s(\mathbf{q}) = \frac{k_f^4 I_o \varepsilon_a^2 k_B T}{16\pi^2 R^2} \sum_{\alpha=1,2} \frac{(i_\alpha f_3 + i_3 f_\alpha)^2}{K_\alpha q_\perp^2 + K_3 q_\parallel^2}. \quad (6.48)$$

Equation (6.48) suggests that individual Frank elastic constants can be studied by choosing appropriate directions of the scattering wave vector $\mathbf{q}$ and the polarizations $\mathbf{i}$ and $\mathbf{f}$ relative to the optic axis. Examples of specially chosen experimental geometries are shown in Fig. 6.5. In these examples, the incident and scattered polarizations are chosen to be normal to and on the scattering plane, respectively.

In the example in Fig. 6.5(a), the scattering plane is normal to the optic axis, and the magnitudes and directions of $\mathbf{k}_i$ and $\mathbf{k}_f$ are such that $\mathbf{q}$ is perpendicular to $\mathbf{k}_f$. We therefore have

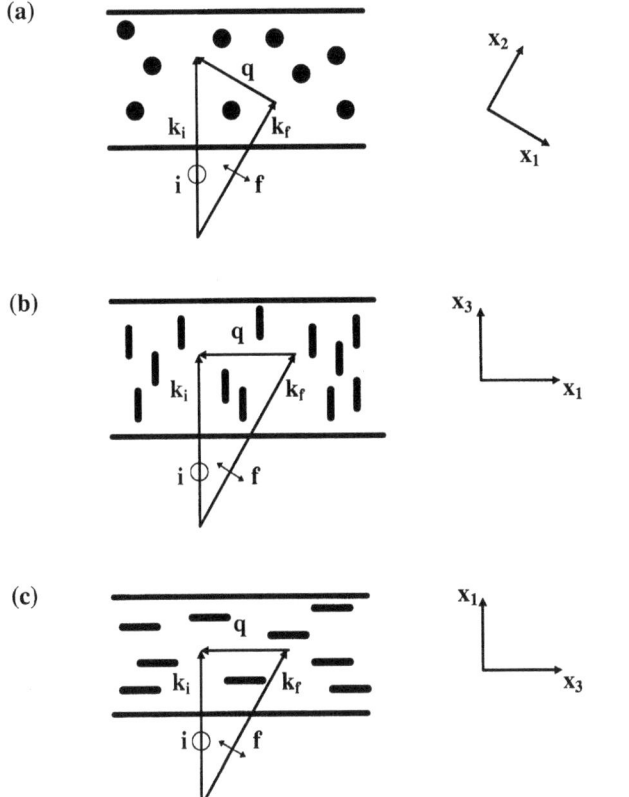

Figure 6.5. Light scattering geometries for the (a) splay, (b) twist and (c) bend modes.

$$i_1 = i_2 = f_2 = f_3 = q_{\parallel} = 0, \quad i_3 = f_1 = 1, \quad q_{\perp} = q. \tag{6.49}$$

The result is a scattered intensity which depends only on the splay elastic constant K_1:

$$I_s(\mathbf{q}) = \frac{k_f^4 I_o \varepsilon_a^2 k_B T}{16\pi^2 R^2 K_1 q^2}. \tag{6.50}$$

In the example in Fig. 6.5(b), the optic axis lies on the scattering plane, and $\mathbf{q}$ is chosen to be normal to both $\mathbf{k}_1$ and the optic axis, resulting in

$$i_1 = i_3 = f_2 = q_{\parallel} = 0, \quad i_2 = 1, \quad q_{\perp} = q. \tag{6.51}$$

The scattered intensity then depends only on the twist elastic constant K_2:

$$I_s(\mathbf{q}) = \frac{k_f^4 I_o \varepsilon_a^2 k_B T f_3^2}{16\pi^2 R^2 K_2 q^2}. \tag{6.52}$$

In Fig. 6.5(b), the incident light beam is parallel to the optic axis. Other twist-mode geometries with the incident light beam not parallel to the optic axis are often advantageous [24].

Finally, in the example in Fig. 6.5(c), the optic axis also lies on the scattering plane, and $\mathbf{q}$ is chosen to be normal to $\mathbf{k}_i$ but parallel to the optic axis. We then have

$$i_1 = i_3 = f_2 = q_\perp = 0, \quad i_2 = 1, \quad q_\| = q. \tag{6.53}$$

The scattered intensity in this case depends only on the bend elastic constant K_3:

$$I_s(\mathbf{q}) = \frac{k_f^4 I_o \varepsilon_a^2 k_B T f_3^2}{16\pi^2 R^2 K_3 q^2}. \tag{6.54}$$

The formalism up to this point pertains only to the amplitudes of the director fluctuation modes, which govern the intensity of the scattered light. To discuss the frequency spectrum of the scattered light or time-correlation function of the scattered intensity, it is necessary to examine the dynamics of the director fluctuations, which are related to viscosity coefficients. In liquid crystals, the translational motions are coupled to the orientational motions of the molecules. In the analysis of liquid crystal dynamics, it is necessary to include several new transport parameters corresponding to the director field $\mathbf{n}(\mathbf{r},t)$ in addition to those corresponding to the velocity field that are used in ordinary fluids. According to the macroscopic approach proposed by Ericksen [25, 26], Leslie [27, 28], and Parodi [29], among the many viscosity coefficients that have been defined for a uniform incompressible nematic liquid crystal, only five are independent. Among the commonly used ones are the Miesowicz viscosities η_a, η_b and η_c, the Leslie coefficients α_1, α_2, α_3, etc., and the twist viscosity γ_1. The equations of motion of the director for the two eigenmodes can be shown to be in the form [5]

$$J(\delta n_\alpha(\mathbf{q},t)) = -\eta_\alpha(\mathbf{q})\frac{\partial}{\partial t}(\delta n_\alpha(\mathbf{q},t)) - K_\alpha(\mathbf{q})\delta n_\alpha(\mathbf{q},t) \quad (\alpha = 1,2), \tag{6.55}$$

where J is the inertial term, $\eta_a(\mathbf{q})$ are effective viscosities defined by

$$\eta_1(\mathbf{q}) \equiv \gamma_1 - \frac{(\alpha_3 q_\perp^2 - \alpha_2 q_\|^2)^2}{\eta_b q_\perp^4 + (\alpha_1 + \alpha_3 + \alpha_4 + \alpha_5)q_\perp^2 q_\|^2 + \eta_c q_\|^4}, \tag{6.56}$$

$$\eta_2(\mathbf{q}) \equiv \gamma_1 - \frac{\alpha_2^2 q_\|^2}{\eta_a q_\perp^2 + \eta_c q_\|^2}, \tag{6.57}$$

and

$$K_\alpha(\mathbf{q}) \equiv K_\alpha q_\perp^2 + K_3 q_\parallel^2 \quad (\alpha = 1,2). \tag{6.58}$$

The first term on the right-hand side of Eq. (6.55) is due to viscous forces while the second term is due to distortion forces. Since the inertial term is generally found to be negligible, we have

$$\frac{\partial}{\partial t}(\delta n_\alpha(\mathbf{q},t)) = -\frac{K_\alpha(\mathbf{q})}{\eta_\alpha(\mathbf{q})}\delta n_\alpha(\mathbf{q},t). \tag{6.59}$$

The solution of Eq. (6.59) is

$$\delta n_\alpha(\mathbf{q},t) = \delta n_\alpha(\mathbf{q})\exp[-\Gamma_\alpha(\mathbf{q})t], \tag{6.60}$$

where the relaxation rates for the two modes are given by

$$\Gamma_1(\mathbf{q}) = \frac{K_1 q_\perp^2 + K_3 q_\parallel^2}{\eta_1(\mathbf{q})} \tag{6.61}$$

and

$$\Gamma_2(\mathbf{q}) = \frac{K_2 q_\perp^2 + K_3 q_\parallel^2}{\eta_2(\mathbf{q})}. \tag{6.62}$$

From Eqs. (6.38) and (6.60), the dielectric correlation function $C_\varepsilon(\mathbf{q},\tau)$ can be expressed as

$$\langle \delta\varepsilon_{ij}^*(\mathbf{q},t)\delta\varepsilon_{ij}(\mathbf{q},t+\tau)\rangle = \varepsilon_a^2 \sum_{\alpha=1,2} (i_\alpha f_3 + i_3 f_\alpha)^2 \langle|\delta n_\alpha(\mathbf{q})|^2\rangle\exp[-\Gamma_\alpha(\mathbf{q})\tau]. \tag{6.63}$$

Since the scattered intensity correlation function $C_I(\mathbf{q},\tau)$ is related to $C_\varepsilon(\mathbf{q},\tau)$ through Eq. (6.17), it can, in general, be quite complicated. In practice, just as in the case of the measurement of the average intensity mentioned above, special experimental geometry can be chosen so that the right-hand side of Eq. (6.63) contains a single exponential decay associated with only one of the two modes. The corresponding intensity correlation function $C_I(\mathbf{q},\tau)$ will then take on the simple exponential form depicted in Eq. (6.20). Furthermore, by properly selecting the directions of the scattering wavevector and the polarizations, it is possible to study the dynamics of a specific Frank elastic distortion. For example, the geometry of Fig. 6.5(a) leads to a single splay-mode relaxation rate

$$\Gamma_{\text{splay}}(\mathbf{q}) = \frac{K_1 q^2}{\gamma_1 - \alpha_3^2/\eta_b}, \tag{6.64}$$

the geometry of Fig. 6.5(b) leads to a single twist-mode relaxation rate

$$\Gamma_{\text{twist}}(\mathbf{q}) = \frac{K_2 q^2}{\gamma_1},$$ (6.65)

and the geometry of Fig. 6.5(c) leads to a single bend-mode relaxation rate

$$\Gamma_{\text{bend}}(\mathbf{q}) = \frac{K_3 q^2}{\gamma_1 - \alpha_2^2/\eta_c}.$$ (6.66)

Thus a quasielastic light scattering measurement in the appropriate geometry can yield the ratio between the relevant Frank elastic constant and a combination of viscosity coefficients [30–32]. Since the behavior of the Frank elastic constant can be obtained from the light scattering intensity, information about the viscosity coefficients can be extracted.

6.5 Scattering in the presence of smectic-A order

The smectic-A (SmA) phase is characterized by the occurrence of one-dimensional positional order in the form of a density wave whose wave-vector $\mathbf{q}_0$ is parallel to the equilibrium director $\mathbf{n}_0$, which is along the $\mathbf{x}_3$-axis. The density ρ in the presence of SmA order can be described by the expression

$$\rho(\mathbf{r}) = \rho_0 \{ 1 + \text{Re}[\Psi_A \exp(iq_0 x_3)] \},$$ (6.67)

where ρ_0 is the average density, and Ψ_A is the SmA order parameter, which is a complex quantity. The phase of Ψ_A governs the position of the layers along the $\mathbf{x}_3$ direction. We can write Ψ_A in the form

$$\Psi_A = |\Psi_A| \exp(iq_0 u),$$ (6.68)

where $u(\mathbf{r})$ is the displacement of the smectic layers in the $\mathbf{x}_3$-direction away from their equilibrium position. The layer spacing d is given by

$$d = 2\pi/q_0.$$ (6.69)

It has been pointed out by de Gennes [33] and McMillan [34] that interesting analogies exist between the SmA state and the superconductor. Apart from the complex nature of the order parameter, an analog to the Meissner effect exists in the SmA liquid crystal. This can be seen by taking the line integral

$$I = \frac{1}{d} \int_A^B \mathbf{n} \cdot d\mathbf{r}$$ (6.70)

between any two points A and B in the liquid crystal. Since the integral I simply counts the number of SmA layers being crossed, its value must be independent of the path taken in a defect-free sample. This implies that

$$\nabla \times \mathbf{n} = 0. \tag{6.71}$$

From Eq. (6.42), it can be seen that the Meissner-analog represented in Eq. (6.71) implies that bend and twist distortions are excluded. Such influence of the presence of the SmA order on the orientational fluctuations would in turn have profound effects on the light scattering from the director modes.

The simplest model to incorporate the presence of SmA order is the Landau–Ginzburg free energy density proposed by de Gennes [33] in the form

$$F_A = F_N + a|\Psi_A|^2 + \tfrac{1}{2} b|\Psi_A|^4 + \tfrac{1}{2} C_{\|}|\partial_3 \Psi_A|^2$$
$$+ \tfrac{1}{2} C_\perp |(\partial_1 - iq_0\delta n_1)\Psi_A|^2 + \tfrac{1}{2} C_\perp |(\partial_2 - iq_0\delta n_2)\Psi_A|^2, \tag{6.72}$$

where F_N is the nematic free energy density defined in Eq. (6.42), $C_{\|}$ and $C_\perp$ are components of a 'mass tensor' parallel and perpendicular to the equilibrium director $\mathbf{n}_0$, and $\partial_\alpha = \partial/\partial x_\alpha$. In the SmA phase, the long range SmA order parameter has the value

$$\langle |\Psi_A| \rangle = \sqrt{-a/b}, \tag{6.73}$$

In the mean-field approximation,

$$a \propto T - T_{NA}, \tag{6.74}$$

where T_{NA} is the nematic-SmA (NA) transition temperature, and b is a constant. The SmA order parameter is thus predicted to have a temperature dependence of

$$\langle |\Psi_A| \rangle \propto (T_{NA} - T)^{0.5} \tag{6.75}$$

if the mean-field approximation is applicable. In practice, most materials do not exhibit mean-field behavior near the NA transition. However, as described below and discussed in Chapters 3 and 7, our understanding of critical phenomena near the NA transition is still not complete.

If we assume that the fluctuations in Ψ_A occur primarily in its phase and not in its magnitude, then from Eqs. (6.68) and (6.72), the free energy in the SmA phase is given by

$$F_A = F_N + \tfrac{1}{2}\{-a^2/b + B(\partial_3 u)^2 + D[(\delta n_1 + \partial_1 u)^2 + (\delta n_2 + \partial_2 u)^2]\}, \tag{6.76}$$

where

$$B \equiv \langle |\Psi_A| \rangle^2 q_0^2 / C_{\parallel} \tag{6.77}$$

and

$$D \equiv \langle |\Psi_A| \rangle^2 q_0^2 / C_{\perp} \tag{6.78}$$

are the elastic constants related to the restoring forces against deviations from uniformity in the layer thickness and against the molecular orientation pointing away from the normal to the layers, respectively. In the mean-field approximation, $C_{\parallel}$ and $C_{\perp}$ are assumed to be independent of temperature. From Eqs. (6.75), (6.77) and (6.78), one then expects the elastic constants B and D to have linear temperature dependences near the NA transition:

$$B \propto T_{NA} - T, \tag{6.79}$$

$$D \propto T_{NA} - T. \tag{6.80}$$

As mentioned above, the NA transition, which is second order in most materials, is accompanied by considerable critical fluctuations, and has to be described by non-mean field theories.

Using scaling arguments which allow for the existence of anisotropic critical exponents, it can be shown that [35, 36]

$$B \propto \xi_{\parallel} / \xi_{\perp}^2 \tag{6.81}$$

and

$$D \propto 1/\xi_{\parallel}, \tag{6.82}$$

where $\xi_{\parallel}$ and $\xi_{\perp}$ are the SmA correlation lengths parallel and perpendicular to the normal to the layers, respectively. From Eq. (6.72), $\xi_{\parallel}$ and $\xi_{\perp}$ on either side of the NA transition are given by

$$\xi_{\parallel} = \sqrt{C_{\parallel} / |a|} \tag{6.83}$$

and

$$\xi_{\perp} = \sqrt{C_{\perp} / |a|}. \tag{6.84}$$

Thus, in the mean-field approximation, both $\xi_{\parallel}$ and $\xi_{\perp}$ are expected to behave as

$$\xi_{\parallel} \propto |T - T_{NA}|^{-0.5} \tag{6.85}$$

and

$$\xi_{\perp} \propto |T - T_{NA}|^{-0.5}. \tag{6.86}$$

To allow for deviations from mean field and for the presence of anisotropy, it is useful to introduce the critical exponents $\nu_\parallel$ and $\nu_\perp$:

$$\xi_\parallel \propto |T - T_{NA}|^{-\nu_\parallel},\tag{6.87}$$

$$\xi_\perp \propto |T - T_{NA}|^{-\nu_\perp}.\tag{6.88}$$

The elastic constants B and D in the SmA phase are therefore expected to have the following temperature dependences:

$$B \propto (T_{NA} - T)^{2\nu_\perp - \nu_\parallel},\tag{6.89}$$

$$D \propto (T_{NA} - T)^{\nu_\parallel}.\tag{6.90}$$

Experimentally, an anisotropy in the critical behavior and an apparent lack of universality among the measured critical exponents have been observed for most materials [37]. As a result, the NA transition has been a subject of considerable theoretical interest over the last two decades [33–36, 38–51]. The free energy density in Eq. (6.72) is analogous to that of the superconductor in the presence of a magnetic field, with $\delta \mathbf{n}$ in place of the magnetic vector potential and the Frank elastic constants in place of the inverse magnetic permittivity. The simplest model would have the asymptotic behavior governed by an isotropic XY fixed point in three dimensions. However, the analogy to the superconductor is not complete, since the free energy density in Eq. (6.72) is not gauge invariant due to the presence of the splay term in F_N. To account for the experimentally observed anisotropy in the critical exponents, searches for an anisotropic fixed point have been conducted using the ε-expansion [39, 40], $1/n$-expansion [42], dislocation-mediated [40, 41, 43–46, 49], and one-loop [50, 51] techniques. The results to date have had partial success in describing some experimental results, but a completely satisfactory theoretical understanding is still lacking.

The effect on light scattering of the presence in the SmA phase of the elastic constants B for layer undulations and D for layer tilting can be exploited experimentally to measure B and D. Using the free energy density in Eq. (6.76), the mean-square amplitude of the director fluctuations can be obtained in a manner similar to that in the nematic phase by taking its Fourier transform and applying the equipartition theorem. Choosing as before the wavevector $\mathbf{q}$ to lie in the $\mathbf{x}_1$–$\mathbf{x}_3$ plane, we have for the two normal modes [7, 52]

$$\langle |\delta n_1(\mathbf{q})|^2 \rangle = \frac{k_B T [1 + (B/D)(q_\parallel/q_\perp)^2]}{K_1 q_\perp^2 + K_3 q_\parallel^2 + (B/D)(D + K_1 q_\perp^2 + K_3 q_\parallel^2)(q_\parallel/q_\perp)^2}\tag{6.91}$$

and

$$\langle |\delta n_2(\mathbf{q})|^2 \rangle = \frac{k_{\mathrm{B}}T}{D + K_2 q_\perp^2 + K_3 q_\parallel^2}. \tag{6.92}$$

The average intensity of the scattered light is then given by Eq. (6.41). Because of the large value of K_3 in the SmA phase, Eqs. (6.91) and (6.92) suggest that light scattering in the SmA phase is significant only near $q_\parallel = 0$.

The dynamics of the director fluctuations can be found by solving the equations of motion in the presence of the SmA order. Using reasonable assumptions, one finds after a fairly complicated calculation that the relaxation rate of the first director mode is given by [7, 52]

$$\Gamma_1(\mathbf{q}) = \frac{Bq_\parallel^2 + (K_1 q_\perp^2 + K_3 q_\parallel^2) q_\perp^2}{\eta_b q^2 + [\alpha_1 + \eta_c(q^2/q_\perp^2)]q_\parallel^2 + [\gamma_1 - 2\alpha_3 + \alpha_2(q_\parallel^2/q_\perp^2)]q_\perp^2}. \tag{6.93}$$

The second director mode is found to be non-hydrodynamic in the SmA phase.

Experimentally, it is sometimes convenient to take the limit of $q_\parallel/q_\perp \ll 1$. Then the following approximate expressions for the mean-square amplitude and relaxation rate of the first director mode are applicable [52]:

$$\langle |\delta n_1(\mathbf{q})|^2 \rangle = \frac{k_{\mathrm{B}}T}{K_1 q_\perp^2 + B(q_\parallel/q_\perp)^2}, \tag{6.94}$$

$$\Gamma_1(\mathbf{q}) = \frac{K_1 q_\perp^2 + B(q_\parallel/q_\perp)^2}{\eta_b + \gamma_1 - 2\alpha_3}. \tag{6.95}$$

Thus, in principle, the elastic constants B and D can be measured relative to the Frank elastic constants using static light scattering and relative to the viscosities using dynamic light scattering. It should be mentioned that, in homeotropic sample alignment, small defects on the glass surface can cause static distortions of the smectic layers that extend appreciably into the sample and scatter light [53, 54]. In performing light scattering from SmA liquid crystals, it is therefore important to eliminate such uninteresting static scattering [55]. Most reported light scattering experiments have been aimed at measuring the layer compressional elastic constant B, especially near the NA transition [56, 59]. The behavior of B measured by this and other techniques [60–62] is to be compared with that depicted in Eq. (6.89), using the values of $\nu_\parallel$ and $\nu_\perp$ predicted theoretically or measured by x-ray scattering. Of particular interest is the possibility that B remains finite at T_{NA} in the fully anisotropic limit of $\nu_\parallel = 2\nu_\perp$ [44]. The experimental results thus far have not presented a consistent picture that sheds convincing light on the theoretical situation. As an example, the results of a light scattering

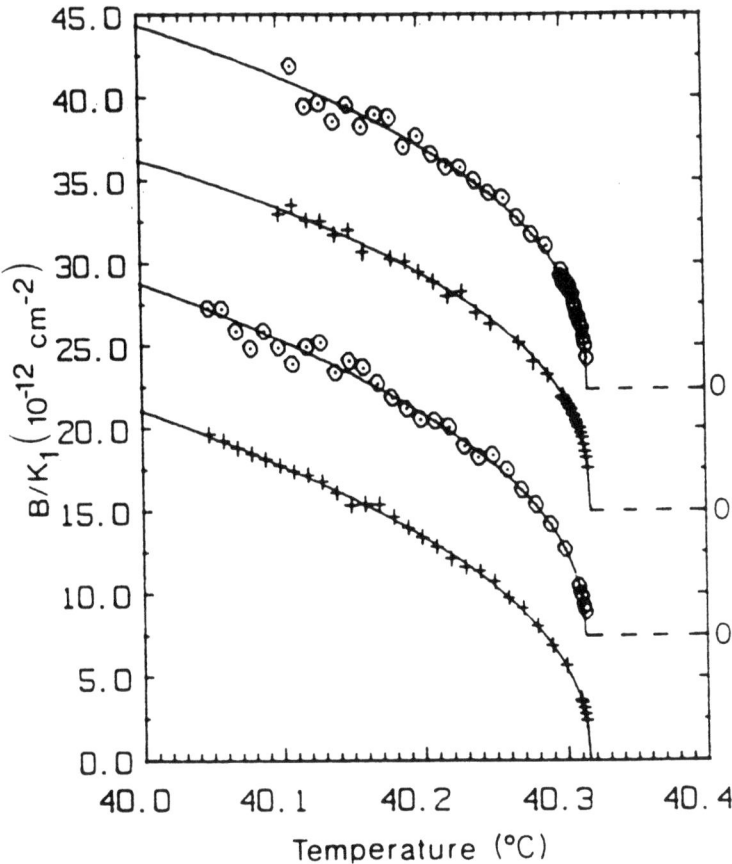

Figure 6.6. Temperature dependence of B/K_1 below the NA transition in 4-*n*-hexyloxybenzoate. The crosses are data from light scattering intensity. The circles are data from light scattering correlation time. The lower two curves are from cooling runs, while the upper two curves are from heating runs. (Reprinted with permission from Ref. 59.)

measurement of B just below the NA transition in 4-*n*-hexyloxybenzoate are shown in Fig. 6.6 [59].

6.6 Scattering above the nematic–smectic-A transition

The exclusion of bend and twist distortions in the SmA phase indicated by Eq. (6.71) suggests that the occurrence of short range SmA order in the nematic phase near the NA transition should also result in a pretransitional increase in the bend and twist elastic constants. The values of K_2 and K_3 are expected to be enhanced above their normal values K_2^0 and K_3^0 in the absence of the NA transition:

$$K_2 = K_2^0 + \delta K_2 \qquad (6.96)$$

$$K_3 = K_3^0 + \delta K_3. \qquad (6.97)$$

The increases in K_2 and K_3 are related to the size of the short range smectic-like clusters in the nematic phase, which grow as the NA transition is approached, and can be shown to be given by [33, 36]

$$\delta K_2 \propto \xi_\perp^2 / \xi_\parallel \propto (T - T_{NA})^{-(2\nu_\perp - \nu_\parallel)} \qquad (6.98)$$

and

$$\delta K_3 \propto \xi_\parallel \propto (T - T_{NA})^{-\nu_\parallel}. \qquad (6.99)$$

The splay elastic constant K_1 is expected to be unaffected by the NA transition. From Eq. (6.48), an increase in K_2 and K_3 would reduce the amplitude of the director fluctuations and result in a decrease in the light scattering intensity near the NA transition. The effect of the NA transition on the dynamics of the director modes, however, is more complicated. As can be seen from Eqs. (6.64)–(6.66), the correlation rates are related not only to the Frank elastic constants, but also to the viscosities, some of which are also expected to increase near the NA transition [36, 63].

Historically, light scattering has been one of the most powerful techniques in studying the critical phenomena associated with the NA transition. The earlier experiments using light scattering intensity confirmed the strong pretransitional increases in the twist and bend elastic constants K_2 and K_3 [64–72]. The intensity measurements have revealed that the critical divergence in K_3 is generally stronger than that in K_2, suggesting an anisotropy in the behavior of the longitudinal and transverse correlation lengths in accordance with Eqs. (6.98) and (6.99), which has subsequently been confirmed by x-ray scattering [73, 74]. Measurement of the critical increase δK_2 of the twist elastic constant K_2 is generally more difficult than that of K_3 because of the relatively large background contribution K_2^0, but quantitative results have been obtained in careful experiments [24, 75], such as those on dihexylazoxybenzene shown in Fig. 6.7. Alternately, information about δK_2 can be obtained from the critical increase in the helical pitch near the cholesteric–SmA transition [76]. In general, the critical exponent for δK_2 has been found to be around 0.5, while that of δK_3 has been found to be around 0.7, but there are considerable variations among the results obtained from different materials, with the length of the nematic temperature range possibly playing a role. A complete reconciliation of these results

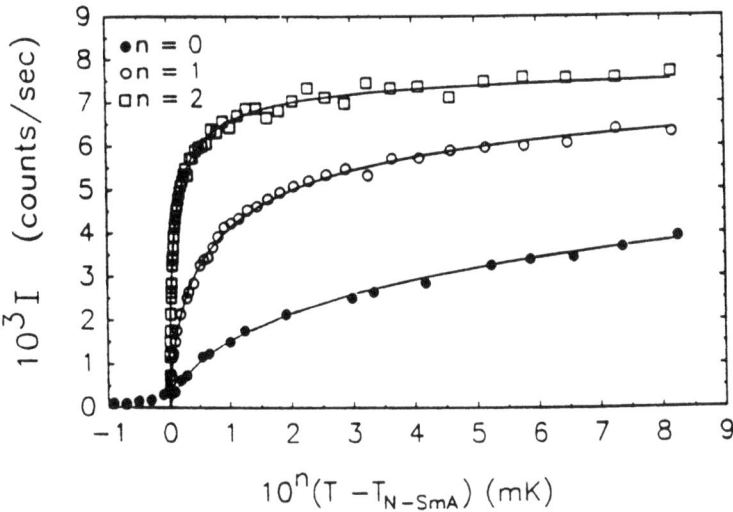

Figure 6.7. Temperature dependence of twist-mode light scattering intensity above the NA transition in dihexylazoxybenzene. The lines are fits to Eq. (6.98). (Reprinted with permission from Ref. 24.)

with our current theoretical understanding [47] is still unavailable. While most of the experiments mentioned above involved the measurement of the light scattering intensity, the behavior of the dynamics near the NA transition was also examined in some experiments [57, 68–71]. The extraction of information about the viscosities from these experiments often involves considerable background subtraction. The quantity that is most commonly measured by dynamic light scattering is the twist viscosity γ_1, which can also be obtained by other techniques, such as dynamic Frederiks deformation [77, 78]. The results of both types of measurements have not fully resolved the question of whether the critical behavior of γ_1 should be mean field [63] or non-mean field [36]. The relatively large temperature dependence in the normal part of γ_1 makes the data analysis rather difficult.

Strictly speaking, the divergence of the Frank elastic constants predicted in Eqs. (6.98) and (6.99) would manifest itself fully only if the size of the measuring probe were large relative to that of the cybotactic clusters. In the case of light scattering, this condition is equivalent to the reciprocal of the scattering wavevector q being much larger than the correlation lengths. Because the longitudinal correlation length $\xi_\parallel$ is the dominant length, the effect is most readily seen in the measurement of the bend elastic constant K_3. As the NA transition is approached and $\xi_\parallel$ increases to the point at which $q\xi_\parallel \approx 1$, one enters the 'non-hydrodynamic' regime in which the growth of the apparent K_3 begins to saturate. This phenomenon represents an experimental complication, but also offers an opportunity to extract absolute values of the correlation lengths from the light scattering data. The details of the non-hydrodynamic effect on light scattering have been calculated using a mode–mode coupling theory [36]. In the experimental geometry in which $q_\perp = 0$, the temperature dependence of the apparent K_3 as measured by the bend-mode light scattering intensity of Eq. (6.54) is expected to be described by the expression

$$K_3 = K_3^0 + \frac{k_B T q_o^2}{8\pi q_\parallel}[(1 + X^{-2})\tan^{-1}X - X^{-1}], \qquad (6.100)$$

where

$$X \equiv \tfrac{1}{2} q_\parallel \xi_\parallel. \qquad (6.101)$$

The non-hydrodynamic effect on the twist elastic constant K_2 is less pronounced and more difficult to measure. Nevertheless, it can be included by considering the amplitude of the second director fluctuation mode. Its amplitude is normally described by the expression in Eq. (6.47). Close to the NA transition, the mode–mode coupling theory [36] predicts that it should be modified to the form

$$\langle|\delta n_2(\mathbf{q})|\rangle^2 = k_B T \left\{ K_2^0 q_\perp^2 + K_3^0 q_\parallel^2 \right.$$

$$\left. + \frac{k_B T q_o^2}{8\pi}\left[q_\parallel^2 + \left(\frac{\xi_\perp}{\xi_\parallel}\right)^2 q_\perp^2\right]^{\frac{1}{2}}[(1 + Y^{-2})\tan^{-1}Y - Y^{-1}]\right\}^{-1} \qquad (6.102)$$

where

$$Y \equiv \tfrac{1}{2}\left[(q_\perp\xi_\perp)^2 + (q_\parallel\xi_\parallel)^2\right]^{\frac{1}{2}}. \qquad (6.103)$$

Experimentally, light scattering in the non-hydrodynamic regime near the NA transition has been used to great advantage for measuring the SmA correlation lengths. Light scattering experiments probing either the bend mode [57, 70, 79] or the second director mode [80] have yielded temperature dependences of the SmA correlation lengths which are in excellent agreement with those measured directly by x-ray diffraction. As an illustration, the temperature dependence of the light scattering intensity above the NA transition due to the pure-bend director mode in octyloxy-*p'*-pentylphenylthiobenzoate (8̄S5) is shown in Fig. 6.8, together with the theoretical fit using Eq. (6.100) [79].

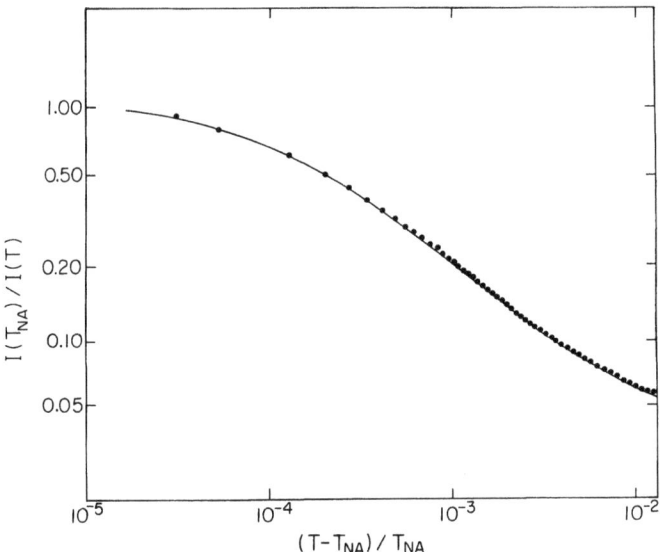

Figure 6.8. Log–log plot of inverse bend-mode light scattering intensity I vs reduced temperature above the NA transition in $\overline{8}S5$. The points are the data and the line is a fit to Eq. (6.100).

6.7 Scattering in or near the smectic-C phase

The smectic-C (SmC) phase is similar to the SmA phase with the exception that the molecules are tilted at a finite angle relative to the normal to the smectic plane. The positions of the centers of mass of the molecules within a layer are fluid-like. In the continuum theory of the SmC phase, the x_3-axis is as usual chosen to be along the normal to the smectic plane. A unit vector $\mathbf{c}$ is introduced to be parallel to the layers and pointing in the tilt direction. The x_1-axis is chosen along the unperturbed direction $\mathbf{c}_0$ of $\mathbf{c}$. The tilt order parameter is usually chosen to be

$$\Psi_c \equiv \theta_t \exp(i\phi), \tag{6.104}$$

where θ_t is the tilt angle and ϕ is the angle between $\mathbf{c}$ and the x_1-axis.

There are two types of degrees of freedom associated with the SmC phase: the displacement $\mathbf{u}(\mathbf{r})$ of the layers along the x_3-direction just as in the SmA phase, and the rotation ϕ of the tilt director $\mathbf{c}$ about the x_3-axis. The free energy density in the SmC phase for non-chiral and non-polar molecules can then be expressed in the form [81, 82]

$$F_C = \frac{1}{2}\left[A_1\left(\frac{\partial^2 u}{\partial x_1^2}\right)^2 + 2A_{12}\left(\frac{\partial^2 u}{\partial x_1^2}\frac{\partial^2 u}{\partial x_2^2}\right) + A_2\left(\frac{\partial^2 u}{\partial x_2^2}\right)^2\right] + \frac{1}{2}\bar{B}\left(\frac{\partial u}{\partial x_3}\right)^2 +$$

$$+\frac{1}{2}\left[B_1\left(\frac{\partial\phi}{\partial x_1}\right)^2 + B_2\left(\frac{\partial\phi}{\partial x_2}\right)^2 + B_3\left(\frac{\partial\phi}{\partial x_3}\right)^2 + 2B_{13}\left(\frac{\partial\phi}{\partial x_1}\frac{\partial\phi}{\partial x_3}\right)\right] \quad (6.105)$$

$$+\left[C_1\left(\frac{\partial^2 u}{\partial x_1\partial x_2}\frac{\partial\phi}{\partial x_1}\right) + C_2\left(\frac{\partial^2 u}{\partial x_2^2}\frac{\partial\phi}{\partial x_2}\right)\right].$$

The A terms describe curvature distortions of the smectic planes, the B terms describe distortions of the tilt director $\mathbf{c}$ and the C terms describe coupling between the tilt and curvature distortions. The A, B, and C coefficients have the same dimensions and probably similar orders of magnitude as the Frank elastic constants. The $\bar{B}$ coefficient is analogous to the layer-compression elastic constant in the SmA phase.

Light scattering in the SmC phase arises primarily from thermal fluctuations in $\mathbf{u}$ and ϕ, with the latter being usually predominant. Applying the equipartition theorem to the Fourier transform of Eq. (6.105), the mean-square amplitude of the fluctuations in ϕ is given by [82]

$$\langle|\delta\phi(\mathbf{q})|^2\rangle = \frac{k_B T}{B_1 q_1^2 + B_2 q_2^2 + B_3 q_3^2 + 2B_{13}q_1 q_3}, \quad (6.106)$$

which is related to the mean-square amplitude of the fluctuations in $\mathbf{n}$ and therefore ultimately to the light-scattering intensity I_s. The dynamics of the director fluctuations in the SmC phase have also been examined theoretically and are necessarily quite complicated [83, 84]. It can be seen in Eq. (6.106) that significant light scattering from the SmC phase occurs for all directions of the scattering wavevector $\mathbf{q}$, resulting in the SmC phase being noticeably more turbid than the SmA phase, where light scattering occurs primarily near $q_3 = 0$.

The main features expected of light scattering in the SmC phase have been observed [85]. By choosing $q_2 = 0$, Eq. (6.106) implies that

$$I_s^{-1} \propto B_1 q_1^2 + B_3 q_3^2 + 2B_{12}q_1 q_3, \quad (6.107)$$

or

$$(B_1\cos^2\theta + B_3\sin^2\theta + 2B_{13}\cos\theta\sin\theta)q^2 I_s \equiv B(\theta)q^2 I_s = \text{constant}, \quad (6.108)$$

where θ is the angle between $\mathbf{q}$ and the smectic plane. Equation (6.108) predicts that a plot of $q_1\sqrt{I_s}$ versus $q_3\sqrt{I_s}$ for various values of θ would have the shape of an ellipse. This prediction has been verified experimentally [85]. The dynamics of the light scattering have also been measured, with the relaxation rate of the tilt fluctuations being found to be describable by the expression

$$\Gamma = \frac{B_1 q_1^2 + B_3 q_3^2 + 2B_{13}q_1 q_3}{\eta(\theta)} \equiv \frac{B(\theta)}{\eta(\theta)} q^2, \tag{6.109}$$

where $\eta(\theta)$ is an effective viscosity which depends on θ.

In a thin SmC film, the fluctuations of the tilt director **c** along the smectic plane offer the opportunity to study the behavior expected in a two-dimensional XY model as proposed by Kosterlitz and Thouless [85, 87]. These studies are usually performed using a ferroelectric smectic-C (SmC*) material, which is composed of chiral molecules exhibiting a spontaneous electric dipole moment $\mathbf{P}_0$ perpendicular to **n** and parallel to the smectic plane [88]. The sample is in the form of a free-standing film of thickness l equivalent to a few molecular layers drawn over a hole in a glass slide [89–91]. Because of the two-dimensional nature of this system, one can concentrate on the rotation ϕ of the tilt director **c** about the $\mathbf{x}_3$-axis. In the presence of an external electric field **E**, it can be shown, starting with a free energy expression similar to Eq. (6.105), that the light scattering intensity is proportional to [89]

$$\langle |\delta\phi(\mathbf{q})|^2 \rangle = \frac{k_B T}{K_b q_1^2 + K_s q_2^2 + 2\pi P_0^2 |q_1| + P_0 E}, \tag{6.110}$$

where K_b and K_s are the two-dimensional bend-like and splay-like elastic constants of the tilt director **c**, respectively, which are related to B_1 and B_2 by

$$K_b = B_1 l, \tag{6.111}$$

and

$$K_s = B_2 l. \tag{6.112}$$

The dynamics of the tilt fluctuations can also be studied. For example, the relaxation rate of the bend-like mode is given by

$$\Gamma = \frac{K_b q_1^2 + 2\pi P_0^2 |q_1| + P_0 E}{\eta_e}, \tag{6.113}$$

where η_e is an effective viscosity. These general features have been verified experimentally [89–91]. There is additional interest in the behavior very close to the SmC–SmA transition in two dimensions. The effect of the critical dynamics on light scattering has been worked out theoretically [92] and observed in the laboratory [93].

In the SmA phase near a SmA–SmC (AC) transition, there are interesting pretransitional properties which can be measured with light scattering.

As the SmC phase is approached, the elastic constant D, which governs the tilting of the molecules away from the normal to the smectic layer, is expected to decrease and eventually vanish at the AC transition temperature T_{AC}. In the mean-field model, the temperature dependence of D is simply

$$D \propto T - T_{AC}. \tag{6.114}$$

The vanishing of D near the AC transition is expected to result in increasing tilt fluctuations in the SmA phase and to have a strong effect on the light scattering from the second director mode. From Eq. (6.92), the mean-square amplitude of this mode can be written as

$$\langle |\delta n_2(\mathbf{q})|^2 \rangle = \frac{k_B T}{D\left(1 + q_\perp^2 \tilde{\xi}_\perp^2 + q_\parallel^2 \tilde{\xi}_\parallel^2\right)}, \tag{6.115}$$

where the tilt correlation lengths $\tilde{\xi}_\perp$ and $\tilde{\xi}_\parallel$ are defined as

$$\tilde{\xi}_\perp \equiv \sqrt{K_2/D} \tag{6.116}$$

and

$$\tilde{\xi}_\parallel \equiv \sqrt{K_3/D}. \tag{6.117}$$

The relaxation rate of this mode would also be similarly affected:

$$\Gamma = \frac{D\left(1 + q_\perp^2 \tilde{\xi}_\perp^2 + q_\parallel^2 \tilde{\xi}_\parallel^2\right)}{\tilde{\eta}}, \tag{6.118}$$

where $\tilde{\eta}$ is an effective viscosity. These effects of the strong tilt fluctuations in the SmA phase near the AC transition on the intensity, angular dependence and dynamics of light scattering have been observed experimentally [94, 95]. The tilt fluctuations are sufficiently strong to actually result in a decrease in the apparent birefringence near the AC transition, which has also been detected [96]. While the earlier light scattering data did not resolve the question about the nature of the AC transition, more recent measurements [79] have confirmed the mean-field description of this transition first suggested by heat-capacity experiments [97]. Experimental data showing the decrease in the intensity of the bend-mode light scattering near the AC transition in $\overline{8}$S5 are illustrated in Fig. 6.9 [79].

When a liquid crystal material which has a nematic–SmC (NC) transition is mixed with another which undergoes both an NA and an AC transition, one obtains an interesting phase diagram containing a nematic–SmA–SmC (NAC) multicritical point [98, 99]. Among the theories [84, 100, 101] proposed to describe the NAC point, the Chen–Lubensky model [100],

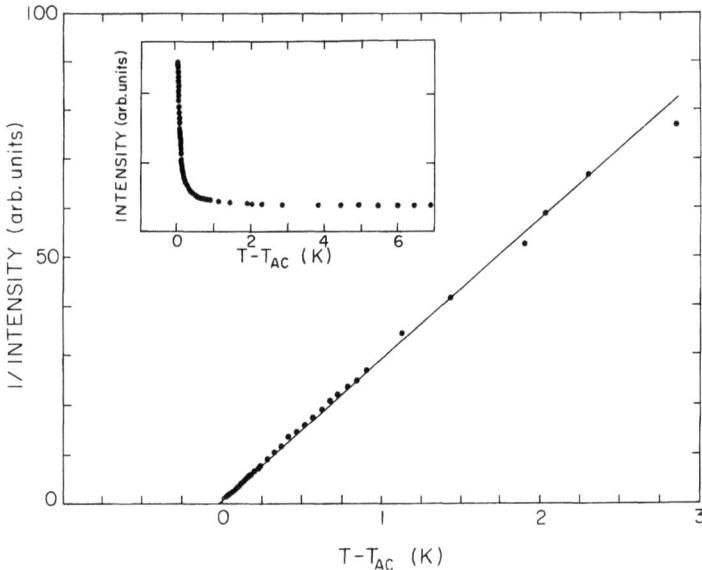

Figure 6.9. Temperature dependence of the inverse bend-mode light scattering intensity above the AC transition in $\overline{8}$S5. The points are the data and the line is a linear fit. The raw data are shown in the insert.

which suggests that the NAC point is analogous to the Lifshitz point originally proposed for magnetic systems, has been the most successful in describing most, but not all, of the experimental findings. In the SmA phase near the NC transition but away from the NAC point, it predicts that the bend elastic constant should increase in the form

$$\delta K_3 \propto \xi_\| \xi_\perp, \tag{6.119}$$

where $\xi_\|$ and $\xi_\perp$ are the longitudinal and transverse mass-density-wave correlation lengths. Since the NC transition is first order, $\xi_\|$ and $\xi_\perp$ are expected to have a mean-field temperature dependence of $(T - T_{NC}^*)^{-0.5}$, where T_{NC}^* is the apparent NC critical temperature. We therefore expect K_3 to increase near the NC transition in the form

$$\delta K_3 \propto (T - T_{NC}^*)^{-1}. \tag{6.120}$$

The linear decrease with temperature of the bend-mode light scattering intensity predicted in Eq. (6.120) has been observed [102]. The data above the NC transition in heptyloxy-p'-pentylphenylthiobenzoate ($\overline{7}$S5) are shown in Fig. 6.10.

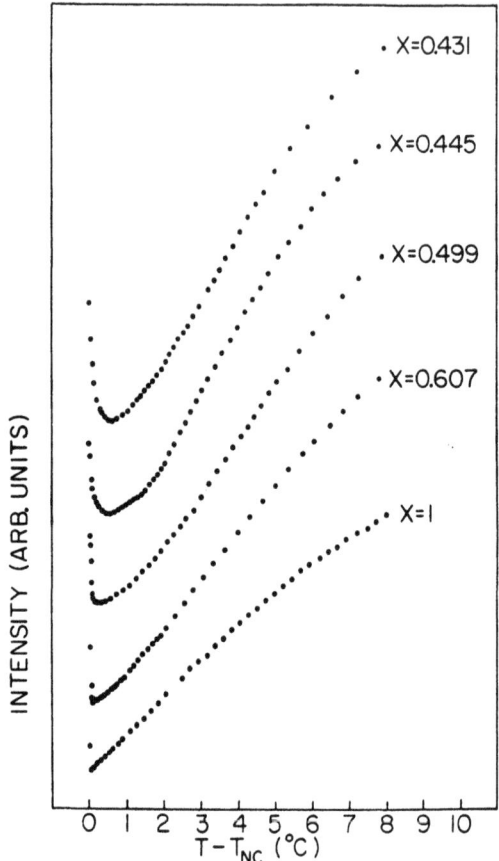

Figure 6.10. Temperature dependence of the bend-mode light scattering intensity above the NC transition in $\overline{7}$S5–$\overline{8}$S5 mixtures of various mole concentrations x of $\overline{7}$S5. The ordinate for each x has been arbitrarily shifted for clarity.

Also shown in Fig. 6.10 are data for $\overline{7}$S5–$\overline{8}$S5 mixtures containing various mole concentrations x of $\overline{7}$S5 [79, 102]. This binary system exhibits an NAC point at $x_{NAC} = 0.42$. For the mixture closest to x_{NAC}, the light scattering intensity upon approaching the NC transition exhibits a steep initial drop, which is followed by a rounding to a minimum value and finally a pronounced increase in the 0.5 K temperature range immediately above the transition. The data for the other mixture with $x > x_{NAC}$ show a gradual evolution of this anomalous temperature dependence in the light scattering intensity to the eventual linear decrease for pure $\overline{7}$S5 ($x = 1$). A non-monotonic temperature dependence on the smectic correlation length $\xi_{\parallel}$ has also been seen in x-ray scattering [103]. It has been suggested [104] that, if the light-scattering intensity is assumed to be inversely proportional to K_3, the light scattering and associated x-ray data are indicative of K_3 and $\xi_\parallel$ being proportional to each other above the NC transition near the NAC point, in apparent agreement with one of the predictions of the Chen–Lubensky model [100]. However, this possibility has been ruled out by a direct measurement of K_3 using Frederiks deformation [105], which showed K_3 increasing monotonically in the nematic phase as the SmC phase was approached even for mixtures close to the NAC point.

The dynamics of director fluctuations above the NC transition near the NAC point have also been studied using quasielastic light scattering [106]. The light scattering due to splay-mode fluctuations has been shown to be quite normal. The intensity correlation function is found to be exponential, as expected in Eq. (6.20) and illustrated in Fig. 6.11(a) for a mixture of $\overline{7}$S5 and octyloxy-cyanobiphenyl (8OCB), and the temperature dependence of the correlation time is consistent with that of the viscosity measured with dynamic Frederiks deformation [107]. The anomalous bend-mode light scattering, on the other hand, has been found to exhibit two correlation times close to the NC transition, as seen in Fig. 6.11(b) for the same mixture. The relative amplitude of the fast fluctuations has a temperature dependence similar to that of the longtitudinal smectic correlation length $\xi_\parallel$, suggesting that the additional light scattering is due to tilt fluctuations that manifest themselves in the non-hydrodynamic regime.

6.8 Scattering in the presence of bond-orientational order

Several smectic phases have been found to exhibit the bond-orientational order that was originally predicted to exist in an intermediate 'hexatic' phase in the theory of defect-mediated melting in two dimensions [108–110]. A hexatic phase does not have long-range translational order, but the orientation of the local molecular arrangement is maintained throughout the sample. X-ray diffraction [111, 112] and electron diffraction [113, 114] have provided the most direct evidence for the existence of bond-orientational order in the hexatic-B (HexB), smectic-I (SmI) and smectic-F (SmF) phases. The molecules are normal to the smectic layer in the HexB phase, but are tilted toward the nearest neighbors and next-nearest neighbors in the SmI and SmF phases, respectively.

Fluctuations of the bond-orientational order can be studied using light scattering in films of the tilted SmI and SmF phases because of its coupling

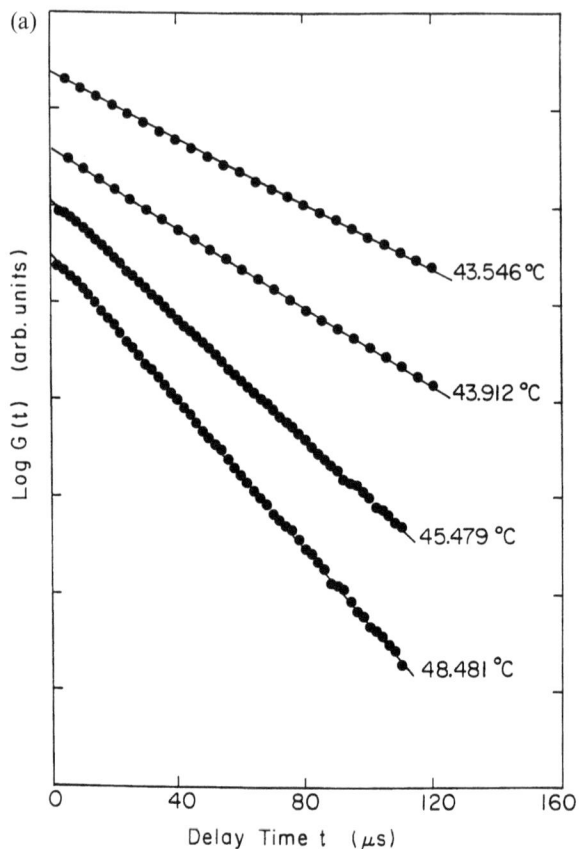

Figure 6.11. Semilog plots of the (a) splay-mode and (b) bend-mode light scattering intensity correlation functions in a $\overline{7}S5_{0.98}8OCB_{0.02}$ mixture. The function
$G(\tau)$ plotted is the correlation function $C_1(\mathbf{q},\tau)$ minus the $\tau=\infty$ baseline. The
ordinates for the different temperatures have been arbitrarily shifted.

with the direction ϕ of the tilt director $\mathbf{c}$. The six-fold bond-orientational
order parameter Ψ_6 can be written as

$$\Psi_6 = |\Psi|_6 2\exp(i6\varphi), \tag{6.121}$$

where φ is the angle between the 'bond' orientation with a nearest neighbor
and some reference axis. A simplified but adequate expression for the elastic
free energy per unit area in a SmI film is [115, 116]

$$F_{\mathrm{I}} = \tfrac{1}{2} K_{\mathrm{b}}|\partial_1\phi|^2 + \tfrac{1}{2} K_{\mathrm{s}}|\partial_2\phi|^2 + \tfrac{1}{2} K_6^{\mathrm{I}}|\partial_1\varphi|^2 + \tfrac{1}{2} K_6^{\mathrm{II}}|\partial_2\varphi|^2$$

$$- H_6\cos[6(\varphi-\phi)] - H_{12}\cos[12(\varphi-\phi)], \tag{6.122}$$

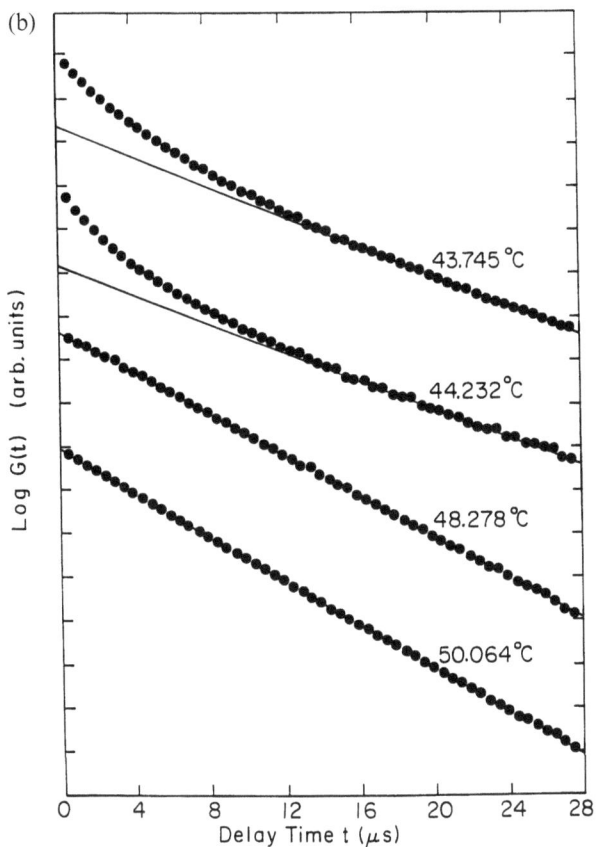

Figure 6.11. (*cont.*)

where K_6^{I} and K_6^{II} are elastic constants for distortions in the bond angle φ, and the coefficients H_6 and H_{12} provide coupling between fluctuations in φ and ϕ. For a SmF film, K_6^{I} and K_6^{II} in Eq. (6.122) should be interchanged and the sign of H_6 reversed. In the two-dimensional reciprocal space, the Fourier transform of Eq. (6.122) can be expressed in terms of the in-plane scattering wave vector $\mathbf{q}_\perp$:

$$F(\mathbf{q}_\perp) = \tfrac{1}{2}[K_6(\chi)q_\perp^2 + H]|\delta\varphi(\mathbf{q}_\perp)|^2 + \tfrac{1}{2}[K(\chi)q_\perp^2 + H]|\delta\phi(\mathbf{q}_\perp)|^2$$

$$\tfrac{1}{2}H[\delta\varphi(\mathbf{q}_\perp)\delta\phi(-\mathbf{q}_\perp) + \text{c.c.}], \tag{6.123}$$

where

$$K(\chi) \equiv K_b\cos^2\chi + K_s\sin^2\chi, \tag{6.124}$$

$$K_6(\chi) \equiv K_6^{\mathrm{I}}\cos^2\chi + K_6^{\mathrm{II}}\sin^2\chi, \tag{6.125}$$

$$H \equiv 36(H_6 + 4H_{12}), \tag{6.126}$$

and χ is the angle between $\mathbf{q}_\perp$ and the tilt director $\mathbf{c}$. The normal modes [117] which diagonalize Eq. (123) consist of an 'optic' mode

$$\theta_- = \varphi - \phi \tag{6.127}$$

and an acoustic mode

$$\theta_+ = \frac{K\phi + K_6\varphi}{K + K_6}. \tag{6.128}$$

Using the equipartition theorem, it can be seen that the mean-square amplitude of the in-plane director fluctuations, which dominate the light scattering from tilted hexatic films, is given by

$$\langle|\delta\phi(\mathbf{q}_\perp)|^2\rangle = \frac{k_{\mathrm{B}}T}{K_+(\chi)q_\perp^2} + \frac{K_-^2(\chi)}{K^2(\chi)}\frac{k_{\mathrm{B}}T}{H + K_-(\chi)q_\perp^2}, \tag{6.129}$$

and

$$K_+ \equiv K + K_6 \tag{6.130}$$

and

$$K_- \equiv KK_6/(K + K_6). \tag{6.131}$$

A complete hydrodynamic model of hexatic smectic phases can be quite complicated [118], but we can expect phenomenologically that the relaxation rates are given by

$$\Gamma_+(\mathbf{q}_\perp) = \frac{K_+(\chi)q_\perp^2}{\eta_+(\chi)} \tag{6.132}$$

and

$$\Gamma_-(\mathbf{q}_\perp) = \frac{H + K_-(\chi)q_\perp^2}{\eta_-(\chi)}, \tag{6.133}$$

where $\eta_+(\chi)$ and $\eta_-(\chi)$ are effective viscosities which depend on χ. Light scattering experiments [117, 119–121] confirming most of the predicted features have been demonstrated in both thick and thin tilted hexatic films. Surprisingly, the magnitude of the bond-tilt coupling is found to be highly material dependent.

6.9 Other forms of light scattering

We shall mention briefly in this section several examples of light scattering in liquid crystals that is not due primarily to director fluctuations.

Raman scattering has been applied extensively to examine the effects of liquid-crystalline order on the various intramolecular vibrational modes [122–124]. Of particular interest is the technique of using the polarization dependence of the Raman scattering intensity to extract information about both the quadrupolar and octupolar nematic order parameters [125, 126].

Brillouin scattering has been used to study the propagating sound waves that exist in a liquid crystal due to density fluctuations [127, 128]. The properties of both the longtitudinal and the transverse sound waves that are predicted [83] to exist in the SmA phase have been verified.

In the bulk SmC* phase, there is a helical structure in which the local director **n** gradually precesses around in space about an axis normal to the smectic layers. In the long-wavelength limit, the spectrum of elementary excitations in the SmC* phase consists of two branches. The 'amplitudon' branch represents variations in the magnitude of the tilt angle. The 'phason' branch represents a collective excitation of the phase of the director field. The dispersion relation of the phason mode has been measured by light scattering [129–131].

Cholesteric blue phases, which exist in a narrow temperature range between the isotropic and cholesteric phases of sufficiently short pitch, form three-dimensional cubic arrays of defects with periodicity of the order of magnitude of the pitch, leading to strong Bragg diffraction in the optical wavelength range. Optical spectroscopy either in the transmission [132] or reflection [133–136] mode has been successfully used to elucidate the structure and polymorphism in blue phases.

6.10 Experimental considerations

Examples of the experimental set-up for measuring either light scattering intensity or correlation function in liquid crystals are shown in Fig. 6.12. It should be noted that there are considerable variations on the optical arrangement governed by the desired scattering geometry relative to the sample orientation. A typical light source is either a He–Ne laser or an Ar^+ ion laser depending on the power requirement. The polarization of the incident beam relative to the scattering plane and the sample director has to be chosen in accordance with the desired scattering mode. In the isotropic

(a)

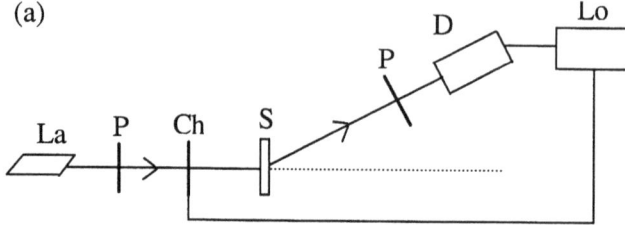

(b)

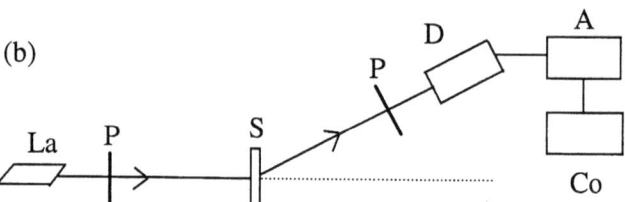

Figure 6.12. Typical experimental set-up for the measurement of light scattering (a) intensity and (b) correlation function. The components indicated are: A, amplifier–discriminator; Ch, light chopper; Co, digital correlator; D, photomultiplier tube; La, laser; Lo, lock-in amplifier; P, polarizer; and S, sample.

phase, because of the relatively weak scattering, samples of thickness in the mm to cm range are usually needed. In the liquid crystal phases, thin aligned samples of thickness in the range of tens of µm between glass slides are typically used. The glass slide are treated with polymer or surfactant coating and, if necessary, are buffed unidirectionally to ensure proper homeotropic or planar alignment of the sample. For experiments on the smectic phases, freely suspended films [89] offer an alternative and often desirable means of obtaining high quality homeotropically aligned samples. Precise temperature control and measurement of the sample is often needed, particularly for studies near phase transitions. Light scattered at a chosen angle and polarization is detected with a photomultiplier tube. The scattered intensity can be measured either by means of the photocurrent or by photon counting using an amplifier–discriminator and a pulse counter. If necessary, stray light and noise can be reduced using a combination of a chopper and a lock-in amplifier. For photon-correlation spectroscopy, the photocounts are stored and analyzed by a digital correlator. Digital correlators with delay times as short as 0.1 µs per channel are available commercially from such manufacturers as ALV, Brookhaven, Langley-Ford and Malvern.

To evaluate the scattering wavevector $\mathbf{q}$ in accordance with Eq. (6.4), it is necessary to consider both the magnitudes and directions of the incident and scattered wavevectors $\mathbf{k}_i$ and $\mathbf{k}_f$. The optical anisotropy of the liquid crystal medium has to be taken into account in this calculation. The magnitude of any wavevector $\mathbf{k}$ is given by

$$k = 2\pi n/\lambda, \tag{6.134}$$

where n is the refractive index relevant to the particular direction and polarization of $\mathbf{k}$ and λ is the wavelength of the light in vacuum. To determine n, a method based on an ellipsoid of refractive indices developed for uniaxial crystals [137] may be used. If the x_3-axis is chosen to be along the director $\mathbf{n}$, and n_0 and n_e are the refractive indices for light polarized perpendicular and parallel to $\mathbf{n}$, respectively, the equation of the ellipsoid is

$$\frac{x_1^2}{n_o^2} + \frac{x_2^2}{n_o^2} + \frac{x_3^2}{n_e^2} = 1. \tag{6.135}$$

Along an arbitrary direction of $\mathbf{k}$ making an angle Ψ with x_3, there are two independent plane-wave propagations that the liquid crystal can support. The two displacement vectors associated with them are in the plane normal to $\mathbf{k}$ through the origin. The intersection of this plane with the ellipsoid of refractive index is an ellipse. One of the two propagations, the 'ordinary' ray, is polarized along the semi-major axis perpendicular to the x_3–$\mathbf{k}$ plane and its refractive index is equal to n_0. The other one, the 'extraordinary' ray, is polarized along the semi-major axis in the x_3–$\mathbf{k}$ plane and its refractive index $n(\Psi)$ is equal to the length of this axis in accordance with the equation

$$\frac{1}{n^2(\psi)} = \frac{\cos^2\psi}{n_o^2} + \frac{\sin^2\psi}{n_e^2}. \tag{6.136}$$

This knowledge of the refractive index is also important to correct for the bending of the light rays at the sample interface when calculating the actual scattering angle θ in the liquid crystal for a given pair of incident and scattered rays external to the sample.

6.11 Conclusion

The anisotropic properties of liquid crystals and associated thermal and critical fluctuations result in interesting phenomena that can be probed using light scattering. The experimental approach is typically very simple. With careful design and execution, many important static or dynamic physical properties have been measured quantitatively. Light scattering has proven to be an extremely powerful laboratory tool for the study of liquid crystals, especially near phase transitions. It is expected that it will continue to be a valuable technique in liquid crystal research, with many novel applications to be explored in the future.

Acknowledgments

Input and materials received from P. P. Crooker, D. L. Johnson, C. Rosenblatt and S. Sprunt in the preparation of this work are gratefully acknowledged. My own involvement in light scattering research was strongly influenced by the mentorship of G. B. Benedek and J. D. Litster and the pioneering work of fellow graduate students D. S. Cannell, N. A. Clark, S. B. Dubin, J. B. Lastovka, P. D. Lazay, J. H. Lunacek and T. W. Stinson a quarter-century ago, and has been enriched by the insights of many graduate students in my laboratory over the years, including N. R. Chen, M. Cheng, S. K. Hark, C. C. Huang, J. Huang, K. C. Lim, R. Pindak, R. Qiu and S. T. Sun. This work was supported in part by the National Science Foundation under Grant No. DMR-9103921.

References

1. P. A. Fleury and J. P. Boon, *Adv. Chem. Phys.* **24**, 1 (1973).
2. J. Goldstone, *Nuovo Cimento* **19**, 154 (1961).
3. P. Chatelain, *Acta Crystallogr.* **1**, 315 (1948).
4. P. G. de Gennes, *C. R. Acad. Sci. Paris* **266**, 15 (1968).
5. P. G. de Gennes and J. Prost, *The Physics of Liquid Crystals*, Oxford University Press, Oxford (1993).
6. S. Chandrasekhar, *Liquid Crystals*, Cambridge University Press, Cambridge (1992).
7. R. Schaetzing and J. D. Litster, in *Advances in Liquid Crystals*, Vol. 4, G. H. Brown (Ed.), Academic Press, New York (1979).
8. S. Sprung and J. D. Litster, in *Phase Transitions in Liquid Crystals*, S. Martellucci and A. N. Chester (Eds.), Plenum Press, New York (1992).
9. B. J. Berne and R. Pecora, *Dynamic Light Scattering*, Wiley, New York (1976).
10. B. Chu, *Laser Light Scattering*, Academic Press, San Diego (1991).
11. A. T. Forrester, *J. Opt. Soc Am.* **51**, 213 (1961).
12. L. Mendel, *Progr. Opt.* **2**, 181 (1963).
13. N. A. Clark, J. H. Lunacek, and G. B. Benedek, *Am. J. Phys.* **38**, 575 (1970).
14. L. D. Landau, in *Collected Papers*, D. Ter Daar (Ed.), Gordon and Breach, New York (1965).
15. P. G. de Gennes, *Mol. Cryst. Liq. Cryst.* **12**, 193 (1971).
16. T. W. Stinson, J. D. Litster, and N. A. Clark, *J. Phys.* (*Paris*) **33**, C1–69 (1972).
17. T. W. Stinson and J. D. Litster, *Phys. Rev. Lett.* **30**, 688 (1973).
18. P. G. de Gennes, *Phys. Lett. A* **30**, 454 (1969).

19. J. D. Litster and T. W. Stinson, *J. Appl. Phys.* **41**, 996 (1970).
20. R. M. Hornreich and S. Shtrikman, *Phys. Rev. A* **28**, 1791 (1983).
21. J. E. Wyse and P. J. Collings, *Phys. Rev. A* **45**, 2449 (1992).
22. H. Zink and W. Van Dael, *Liq. Cryst.* **5**, 899 (1989).
23. F. C. Frank, *Disc. Faraday Soc.* **25**, 19 (1958).
24. H. K. M. Vithana, V. Surendranath, M. Lewis, A. Baldwin, K. Eidner, R. Mahmood, and D. L. Johnson, *Phys. Rev. A* **41**, 2031 (1990).
25. J. L. Ericksen, *Arch. Ration. Mech. Anal.* **4**, 231 (1960).
26. J. L. Ericksen, *Phys. Fluids* **9**, 1205 (1966).
27. F. M. Leslie, *Quart. J. Mech. Appl. Math.* **19**, 357 (1966).
28. F. M. Leslie, *Arch. Ration. Mech. Anal.* **28**, 265 (1968).
29. O. J. Parodi, *J. Phys. (Paris)* **31**, 581 (1970).
30. Group d'Etude des Cristaux Liquides, *J. Chem. Phys.* **51**, 816 (1969).
31. I. Haller and J. D. Litster, *Phys. Rev. Lett.* **25**, 1550 (1970).
32. H. Fellner, W. Franklin, and S. Christensen, *Phys. Rev. A* **11**, 1440 (1975).
33. P. G. de Gennes, *Solid State Commun.* **10**, 753 (1972).
34. W. L. McMillan, *Phys. Rev. A* **4** 1238 (1971).
35. F. Brochard, *J. Phys. (Paris)* **34**, 411 (1973).
36. F. Jähnig and F. Brochard, *J. Phys. (Paris)* **35**, 301 (1974).
37. C. W. Garland and G. Nounesis, *Phys. Rev. E* **49**, 2964 (1994).
38. B. A. Heberman, D. M. Lubkin, and S. Doniach, *Solid State Commun.* **17**, 485 (1975).
39. T. C. Lubensky and J. H. Chen, *Phys. Rev. A* **17**, 366 (1978).
40. J. H. Chen, T. C. Lubensky, and D. R. Nelson, *Phys. Rev. B* **17**, 4274 (1978).
41. W. J. Helfrich, *J. Phys. (Paris)* **39**, 1199 (1978).
42. S. G. Dunn and T. C. Lubensky, *J. Phys. (Paris)* **42**, 1201 (1981).
43. C. Dasgupta and B. I. Halperin, *Phys. Rev. Lett.* **47**, 1556 (1981).
44. D. R. Nelson and J. Toner, *Phys. Rev. B* **24**, 363 (1981).
45. J. Toner, *Phys. Rev. B* **26**, 462 (1982).
46. A. R. Day, T. C. Lubensky, and A. J. McKane, *Phys. Rev. A* **27**, 1461 (1983).
47. T. C. Lubensky, *J. Chim. Phys.* **80**, 31 (1983).
48. J. Prost, *Adv. Phys.* **33**, 1 (1984).
49. C. Dasgupta, *J. Phys. (Paris)* **48**, 957 (1987).
50. B. R. Patton and B. S. Andereck, *Phys. Rev. Lett.* **69**, 1556 (1992).
51. B. S. Andereck and B. R. Patton, *Phys. Rev. E* **49**, 1393 (1994).
52. H. Birecki, Ph.D. Thesis, Massachusetts Institute of Technology (1975).
53. R. Ribotta, G. Durand, and J. D. Litster, *Solid State Commun.* **12**, 27 (1973).
54. N. A. Clark and P. S. Pershan, *Phys. Rev. Lett.* **30**, 3 (1973).
55. R. Ribotta, D. Salin, and G. Durand, *Phys. Rev. Lett.* **32**, 6 (1974).
56. H. Birecki, R. Schaetzing, F. Rondelez, and J. D. Litster, *Phys. Rev. Lett.* **36**, 1376 (1976).
57. H. von Känel and J. D. Litster, *Phys. Rev. A* **23**, 3251 (1981).
58. H. J. Fromm, *J. Phys. (Paris)* **48**, 647 (1987).
59. M. E. Lewis, I. Khan, H. Vithana, A. Baldwin, and D. L. Johnson, *Phys. Rev. A* **38**, 3702 (1988).
60. L. Ricard and J. Prost, *J. Phys. (Paris)* **42**, 861 (1981).
61. M. R. Fisch, P. S. Pershan, and L. B. Sorensen, *Phys. Rev. A* **29**, 2741 (1984).
62. M. Benzekri, T. Claverie, J. P. Marcerou, and J. C. Rouillon, *Phys. Rev. Lett.* **68**, 2480 (1992).
63. W. L. McMillan, *Phys. Rev. A* **9**, 1720 (1974).
64. L. Cheung, R. B. Meyer, and H. Gruler, *Phys. Rev. Lett.* **31**, 349 (1973).
65. M. Delaye, R. Ribotta, and G. Durand, *Phys. Rev. Lett.* **31**, 443 (1973).

66. P. E. Cladis, *Phys. Rev. Lett.* **31**, 1200 (1973).
67. L. Léger, *Phys. Lett. A* **44**, 535 (1973).
68. D. Salin, I. W. Smith, and G. Durand, *J. Phys. (Paris)* **35**, L-165 (1974).
69. K. C. Chu and W. L. McMillan, *Phys Rev. A* **11**, 1059 (1975).
70. M. Delaye, *J. Phys. (Paris)* **37**, C3-99 (1976).
71. H. Birecki and J. D. Litster, *Mol. Cryst. Liq. Cryst.* **42**, 33 (1977).
72. C. C. Huang, R. S. Pindak, and J. T. Ho, *Solid State Commun.* **25**, 1015 (1978).
73. J. Als-Nielsen, R. J. Birgeneau, M. Kaplan, J. D. Litster, and C. R. Safinya, *Phys. Rev. Lett.* **39**, 352 (1977).
74. J. D. Litster, R. J. Birgeneau, M. Kaplan, C. R. Safinya, and J. Als-Nielsen, in *Ordering in Strongly Fluctuating Condensed Matter Systems*, T. Riste (Ed.), Plenum, New York (1980).
75. R. Mahmood, D. Brisbin, I. Khan, C. Gooden, A. Baldwin, D. L. Johnson and M. E. Neubert, *Phys. Rev. Lett.* **54**, 1031 (1985).
76. R. S. Pindak, C. C. Huang, and J. T. Ho, *Phys. Rev. Lett.* **32**, 43 (1974).
77. C. C. Huang, R. S. Pindak, P. J. Flanders, and J. T. Ho, *Phys. Rev. Lett.* **33**, 400 (1974).
78. L. Léger and A. Martinet, *J. Phys. (Paris)* **37**, C3-89 (1976).
79. J. Huang and J. T. Ho, *Phys. Rev. A* **38**, 400 (1988).
80. S. Sprunt, L. Solomon, and J. D. Litster, *Phys. Rev. Lett.* **53**, 1923 (1984).
81. A. Saupe, *Mol. Cryst. Liq. Cryst.* **7**, 59 (1969).
82. Orsay Group on Liquid Crystals, *Solid State Commun.* **9**, 653 (1971).
83. P. C. Martin, O. Parodi, and P. S. Pershan, *Phys. Rev. A* **6**, 2401 (1972).
84. P. G. de Gennes, *Mol. Cryst. Liq. Cryst.* **21**, 49 (1973).
85. Y. Galerne, J. L. Martinand, G. Durand, and M. Veyssie, *Phys. Rev. Lett.* **29**, 562 (1972).
86. J. M. Kosterlitz and D. J. Thouless, *J. Phys. C* **5**, 1124 (1972).
87. J. M. Kosterlitz and D. J. Thouless, *J. Phys. C* **6**, 1181 (1973).
88. R. B. Meyer, L. Liebert, L. Strzelecki, and P. Keller, *J. Phys. (Paris) Lett.* **36**, L69 (1975).
89. C. Y. Young, R. Pindak, N. A. Clark, and R. B. Meyer, *Phys. Rev. Lett.* **40**, 773 (1978).
90. C. Rosenblatt, R. Pindak, N. A. Clark, and R. B. Meyer, *Phys. Rev. Lett.* **42**, 1220 (1979).
91. C. Rosenblatt, R. B. Meyer, R. Pindak, and N. A. Clark, *Phys. Rev. A* **21**, 140 (1980).
92. S. W. Heinekamp and R. A. Pelcovits, *Phys. Rev. A* **32**, 2506 (1985).
93. S. M. Amador and P. S. Pershan, *Phys. Rev. A* **41**, 4326 (1990).
94. M. Delaye and P. Keller, *Phys. Rev. Lett.* **37**, 1065 (1976).
95. M. Delaye, *J. Phys. (Paris)* **40**, C3-350 (1979).
96. K. C. Lim and J. T. Ho, *Phys. Rev. Lett.* **40**, 1576 (1978).
97. C. C. Huang and J. M. Viner, *Phys. Rev. A* **25**, 3385 (1982).
98. G. Sigaud, F. Hardouin, and M. F. Achard, *Solid State Commun.* **23**, 35 (1977).
99. D. Johnson, D. Allender, D. DeHoff, C. Maze, E. Oppenheim, and R. Reynolds, *Phys. Rev. B* **16**, 470 (1977).
100. J. H. Chen and T. C. Lubensky, *Phys. Rev. A* **14**, 1202 (1976).
101. K. C. Chu and W. L. McMillan, *Phys. Rev. A* **15**, 1181 (1977).
102. S. Witanachchi, J. Huang, and J. T. Ho, *Phys. Rev. Lett.* **50**, 594 (1983).
103. L. J. Martínez-Miranda, A. R. Kortan, and R. J. Birgeneau, *Phys. Rev. A* **36**, 2372 (1987).

104. L. Solomon and J. D. Litster, *Phys. Rev. Lett.* **56**, 2268 (1986).
105. J. Huang and J. T. Ho, *Phys. Rev. Lett.* **58**, 2239 (1987).
106. R. Qiu and J. T. Ho, *Phys. Rev. Lett.* **64**, 1122 (1990).
107. J. Huang and J. T. Ho, *Phys. Rev. A* **42**, 2449 (1990).
108. B. I. Halperin and D. R. Nelson, *Phys. Rev. Lett.* **41**, 121 (1978).
109. A. P. Young, *Phys. Rev. B* **19**, 1855 (1979).
110. D. R. Nelson and B. I. Halperin, *Phys. Rev. B* **19**, 2457 (1979).
111. R. Pindak, D. E. Moncton, J. W. Goodby, and S. C. Davey, *Phys. Rev. Lett.* **46**, 1135 (1981); S. Kumar, *J. Phys. (Paris)* **44**, 123 (1983).
112. J. D. Brock, A. Aharony, R. J. Birgeneau, K. W. Evans-Lutterodt, J. D. Litster, P. M. Horn, G. B. Stephenson, and A. R. Tajbakhsh, *Phys. Rev. Lett.* **57**, 98 (1986); E. Gorecka, L. Chen, W. Pyzuk, A. Krowczynski, and S. Kumar, *Phys. Rev. E* **50**, 2863 (1994).
113. M. Cheng, J. T. Ho, S. W. Hui, and R. Pindak, *Phys. Rev. Lett.* **59**, 1112 (1987).
114. M. Cheng, J. T. Ho, S. W. Hui, and R. Pindak, *Phys. Rev. Lett.* **61**, 550 (1988).
115. D. R. Nelson and B. I. Halperin, *Phys. Rev. B* **21**, 5312 (1980).
116. R. Bruinsma and D. R. Nelson, *Phys. Rev. B* **23**, 402 (1981).
117. S. Sprunt, M. S. Spector, and J. D. Litster, *Phys. Rev. A* **45**, 7355 (1992).
118. H. Pleiner and H. R. Brand, *Phys. Rev. A* **29**, 911 (1984).
119. S. B. Dierker and R. Pindak, *Phys. Rev. Lett.* **59**, 1002 (1987).
120. S. Sprunt and J. D. Litster, *Phys. Rev. Lett.* **59**, 2682 (1987).
121. M. S. Spector, S. Sprunt, and J. D. Litster, *Phys. Rev. E* **47**, 1101 (1993).
122. B. J. Bulkin, in *Advances in Liquid Crystals*, Vol. 2, G. H. Brown (Ed.), Academic Press, New York (1976).
123. C. Destrade, F. Guillon, and H. Gasparoux, *Mol. Cryst. Liq. Cryst.* **36**, 115 (1976).
124. S. K. Hark and J. T. Ho, *Mol. Cryst. Liq. Cryst. Lett.* **56**, 99 (1979).
125. S. Jen, N. A. Clark, P. S. Pershan, and E. B. Priestley, *J. Chem. Phys.* **66**, 4635 (1977).
126. K. Miyano, *Phys. Lett. A* **63**, 37 (1977).
127. Y. Liao, N. A. Clark, and P. S. Pershan, *Phys. Rev. Lett.* **30**, 639 (1973).
128. G. W. Bradberry and J. M. Vaughan, *Phys. Lett. A* **62**, 225 (1977).
129. I. Musevic, R. Blinc, B. Zeks, C. Filipic, M. Copic, A. Seppen, P. Wyder, and A. Levanyuk, *Phys. Rev. Lett.* **60**, 1530 (1988).
130. I. Drevensek, I. Musevic, and M. Copic, *Phys. Rev. A* **41**, 923 (1990).
131. I. Musevic, B. Zeks, R. Blinc, H. A. Wierenga, and Th. Rasing, *Phys. Rev. Lett.* **68**, 1850 (1992).
132. S. Meiboon, and M. Sammon, *Phys. Rev. Lett.* **44**, 882 (1980).
133. D. L. Johnson, J. H. Flack, and P. P. Crooker, *Phys. Rev. Lett.* **45**, 641 (1980).
134. J. H. Flack, and P. P. Crooker, *Phys. Lett. A* **82**, 247 (1981).
135. J. Her, B. B. Rao, and J. T. Ho, *Phys. Rev. A* **24**, 3272 (1981).
136. A. J. Nicastro and P. H. Keyes, *Phys. Rev. A* **27**, 431 (1983).
137. A. Yariv, *Quantum Electronics*, Wiley, New York (1989).

7

Calorimetric studies

CARL W. GARLAND

Department of Chemistry and Center for Materials Science and Engineering,
M. I. T., Cambridge, MA 02139, USA

As discussed in Chapter 1, liquid crystalline materials exhibit a rich variety of phases and phase transitions. These transitions are mostly of the order–disorder type, and calorimetric studies provide information on energy effects that complements the structural information obtained from x-ray studies. This chapter contains four sections. Section 7.1 is an introduction to the role of calorimetry in describing and characterizing liquid crystal phase behavior. Section 7.2 describes several high-resolution experimental techniques and gives the necessary phenomenological theory for extracting heat capacity C_p or enthalpy H values. Section 7.3 reviews the theoretical analysis of C_p data and covers both mean-field Landau theory and critical fluctuation results as given by renormalization group theory. Section 7.4 deals with the results obtained for a variety of phase transitions, mostly in thermotropic smectic systems. Major emphasis will be on nematic (N)–isotropic (I), blue phase III (BP$_{III}$)–isotropic (I), chiral nematic (cholesteric, N*)–blue phase (BP), nematic (N)–smectic-A (SmA), smectic-A (SmA)–smectic-C (SmC), polymorphic SmA–SmA, and smectic–hexatic transitions.

7.1 Introduction

Measurement of the temperature dependence of the heat capacity C_p and/or the enthalpy H determines the thermal behavior of liquid crystal phases and helps to characterize the phase transitions which occur. These transitions can be first order or second order, and critical fluctuations often play an important role. Thus, high-resolution techniques are essential to elucidate the nature of liquid crystal transitions.

In a first-order transition, there are discontinuous jumps in the first derivatives of the free energy and two distinct phases α and β coexist at the transition temperature T_1: $\Delta V = V_\beta - V_\alpha \neq 0$ and $\Delta H = H_\beta - H_\alpha \neq 0$. If the transition is first order, one wants to know the magnitude of the latent heat $\Delta H \equiv L$ and the presence or absence of pretransitional heat capacity effects. If the transition is second order, there are no discontinuities in the volume V or enthalpy H but the heat capacity C_p, given by

$$C_p = \left(\frac{\partial H}{\partial T}\right)_p = T\left(\frac{\partial S}{\partial T}\right)_p = -T\left(\frac{\partial^2 G}{\partial T^2}\right)_{p'} \tag{7.1}$$

will exhibit either a discontinuous jump (mean-field behavior) or a divergence (critical fluctuation behavior).

The possible behaviors for H as a function of temperature near a phase transition are shown schematically in Fig. 7.1. For a *strongly first-order* transition (Fig. 1(a)), the variation of H with T is close to linear both above and below T_1 and all the enthalpy change associated with the transition is represented by the latent heat ΔH. For a *weakly first-order* transition (Fig. 1(b)), there is still a discontinuous latent heat, $\Delta H \neq 0$, but the enthalpy shows substantial pretransitional effects. In this case (which is observed for $N - I$ transitions in liquid crystals), the total enthalpy change is the sum of two parts, $\Delta H + \delta H$, where

$$\delta H = \int \Delta C_p dT \tag{7.2}$$

and $\Delta C_p \equiv C_p - C_p(\text{background})$ is the excess heat capacity due to the change in ordering that is associated with the transition. If the transition is second order, $\Delta H = 0$ and there is no two-phase coexistence but the second derivative $\partial^2 G/\partial T^2$ is singular. A *mean-field second-order* transition (Fig. 1(c)) shows a rapid variation in H below T_c (due to changes in the long-range order with T) but a normal linear variation above T_c. Thus C_p exhibits a discontinuous jump. A *critical fluctuation second-order* transition (Fig. 1(d)) exhibits a divergent singularity in the derivative $C_p = (\partial H/\partial T)_p$ at T_c and pretransitional effects occur both above and below T_c. In both the mean-field and fluctuation cases, the enthalpy change associated with the transition is δH, as indicated in Figs. 1(c) and 1(d). For a *supercritical* evolution (Fig. 1(e)), no thermodynamic transition occurs. In this case the high and low temperature 'phases' must have the same point group symmetry so that order can occur in a smooth continuous manner. A familiar example would be the liquid–gas transition in simple fluids, where a line of first-order transitions terminates at an isolated critical point. For supercritical pressures, $p > p_c$, the variation of H with T looks like Fig. 1(e) and C_p goes through a broad, rounded (non-singular) maximum at T_{max}. In liquid crystals, analogous behavior occurs for $SmA_d - SmA_2$ transitions in binary

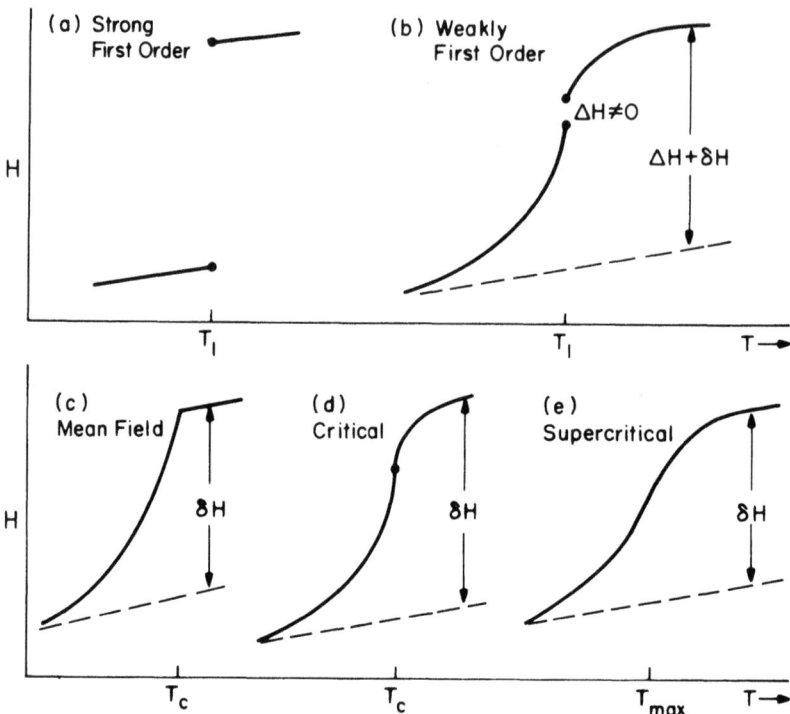

Figure 7.1.   Schematic variation of enthalpy H with temperature for transitions that are (a) strongly first order, (b) weakly first order with pretransitional fluctuation behavior, (c) mean-field second order, (d) critical fluctuation second order, and (e) supercritical. T_1 is the first-order transition temperature; T_c is the Landau second-order transition temperature in case (c) and the second-order critical temperature in case (d). In the supercritical case, there is no true thermodynamic transition but only a continuous evolution of order and a finite maximum in
$$C_p = (\partial H/\partial T)_p \text{ at } T_{\max}.$$

mixtures with $X > X_c$. The magnitude of δH decreases monotonically as the composition (mole fraction X) of the supercritical system moves away from the critical composition value X_c.

The heat capacity behaviors for weakly first-order, mean-field second-order, and critical fluctuation second-order transitions are shown in Fig. 7.2. The excess enthalpy δH associated with a transition and given by Eq. (7.2) is equal to the shaded area. A more extensive discussion of Fig. 7.2 will be given in Section 7.3.

In the period from 1978 to 1998, an impressive variety of calorimetric results were obtained on liquid crystal systems. Several extensive reviews

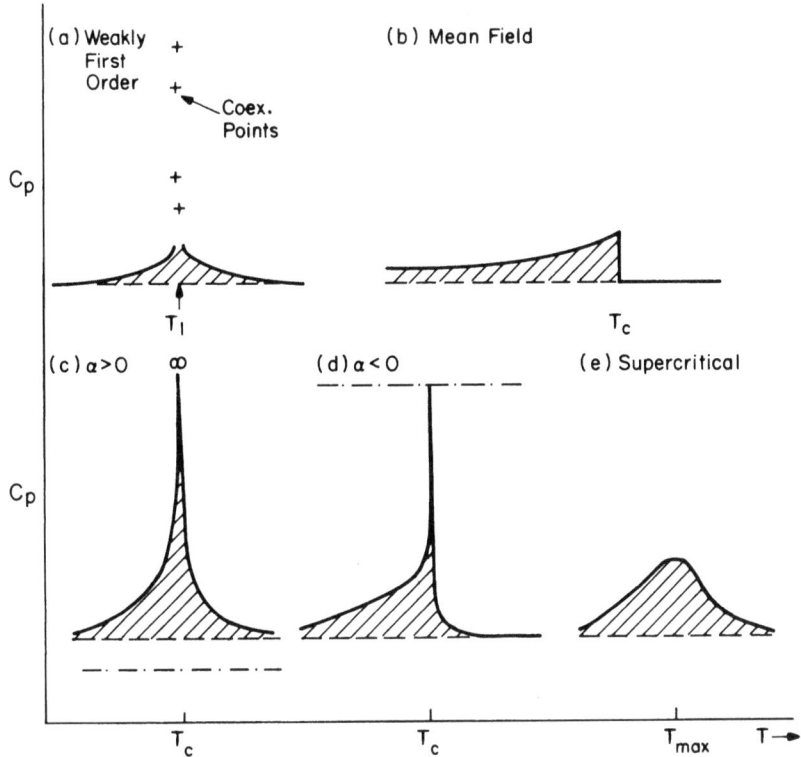

Figure 7.2. Schematic variation of heat capacity C_p with temperature for transitions that are (a) weakly first order, (b) mean-field second order, (c) fluctuation second order with critical exponent $\alpha < 0$, and (d) fluctuation second order with critical exponent $\alpha < 0$. Also shown is the C_p behavior for a supercritical evolution of order (e). In all cases, the dashed line represents C_p (background) and δH is equal to the shaded area. The significance of the dot-dash line is discussed in Section 7.3.

have appeared [1–5], and these complement the present chapter. During this same period, there have been significant theoretical advances (especially the renormalization group approach) in dealing with phase transitions in general and liquid crystal transitions in particular. In the latter category is the de Gennes and Prost revision of de Gennes' classic book [6] and a book by Chaikin and Lubensky [7], as well as several review articles [8, 9]. Throughout this chapter, liquid crystal compounds will be referred to by their commonly used abbreviated names rather than their proper chemical names. These conventional names are defined in terms of the full chemical nomenclature and formulae in Table 7.1.

Table 7.1 *Listing of various liquid crystal compounds by common name, chemical name, and structural formula. The order is alphabetic with respect to the capital letters in the common name. ϕ denotes* —⟨ ⟩— *and an asterisk denotes a chiral center.*

Common name	Chemical name and formula
Cn	4-(3-methyl-2-chloropentanoyloxy)-4'-alkyloxybiphenyl $C_nH_{2n+1}O-\phi-\phi-OOC-{*}CHCl-{*}CH(CH_3)-C_2H_5$
Cn stilbene	4-cyanobenzoyloxy-4'-alkylstilbene $C_nH_{2n+1}-\phi-CH=CH-\phi-OOC-\phi-CN$
nCB	4-alkyl-4'-cyanobiphenyl $C_2H_{2n+1}-\phi-\phi-CN$
CBOOA	4-cyanobenzylidene-octyloxyaniline $C_8H_{17}-\phi-N=CH-\phi-CN$
CN	cholesteryl nonanoate $C_8H_{17}COOC_{27}H_{45}$
CsPFO	cesium perfluoro-octanoate $C_7F_{15}COOCs$
DB$_8$ClCN	octylphenyl-2-chloro-4(p-cyanobenzoyloxy) benzoate $C_8H_{17}-\phi-OOC-\phi(Cl)-OOC-\phi-CN$
DB$_n$CN	4-(alkylphenylbenzoyloxy)-4-cyanobenzoate $C_nH_{2n+1}-\phi-OOC-\phi-OOC-\phi-CN$
DB$_n$ONO$_2$	4-alkyloxyphenyl-4'-(4''-nitrobenzoyloxy) benzoate $C_nH_{2n+1}O-\phi-OOC-\phi-OOC-\phi-NO_2$
MBBPC	4''-(methylbutylphenyl)-4'(methylbutyl)-4 biphenylcarboxylate $C_2H_5-C{*}H(CH_3)-CH_2-\phi-\phi-COO-\phi-CH_2-C{*}H(CH_3)-C_2H_5$
MHPOBC	4-(1-methylheptyloxycarbonyl)phenyl-4'-octyloxy-biphenyl-4-carboxylate $C_{18}H_{17}O-\phi-\phi-COO-\phi-COO-{*}CH(CH_3)C_6H_{13}$
2M4nOBC	2-methylbutyl-4'-n-alkyloxybiphenyl-4-carboxylate $C_nH_{2n+1}O-\phi-\phi-COO-CH_2-{*}CH(CH_3)-C_2H_5$
nO.m	N-(4-n-alkyloxybenzylidene)-4'-n-alkylaniline $C_nH_{2n+1}O-\phi-CH=N-\phi-C_mH_{2m+1}$
$\overline{n}O\overline{m}$	4-n-alkyloxyphenyl-4'-n-alkyloxybenzoate $C_nH_{2n+1}O-\phi-COO-\phi-OC_mH_{2m+1}$
nmOBC	n-alkyl-4'-n-alkyloxybiphenyl-4-carboxylate $C_nH_{2n+1}O-\phi-\phi-COO-C_mH_{2m+1}$
nOBC	4-alkoxy-4'-cyanobiphenyl $C_nH_{2n+1}O-\phi-\phi-CN$
nOPCBOB	4-n-alkyloxyphenyl-4'-(4''-cyanobenzyloxy)-benzoate $C_nH_{2n+1}O-\phi-OOC-\phi-O-CH_2-\phi-CN$
nOSI	4-(2'-methylbutyl) phenyl-4'-n-alkyloxybiphenyl-4-carboxylate $C_nH_{2n+1}O-\phi-\phi-COO-\phi-CH_2-{*}CH(CH_3)-C_2H_5$

Table 7.1 (*cont.*)

Common name	Chemical name and formula
$\bar{n}Sm$	4-*n*-alkyl-phenylthiol-4'-*n*-alkyloxybenzoate $C_nH_{2n+1}O$—ϕ—COS—ϕ—C_mH_{2m+1}
*n*SI	4-(2'-methylbutyl)phenyl-4'-*n*-alkylbiphenyl-4-carboxylate C_nH_{2n+1}—ϕ—ϕ—COO—ϕ—CH_2—*$CH(CH_3)$—C_2H_5
TB*n*A	terephthal-bis-(4*n*)-alkylaniline C_nH_{2n+1}—ϕ—N=CH—ϕ—CH=N—ϕ—C_nH_{2n+1}

7.2 Experimental methods

A variety of techniques are discussed in this section, with emphasis on high-resolution methods that allow one to characterize the critical energy fluctuations at second-order transitions and to distinguish small latent heats from pretransitional enthalpy variations in the case of weakly first-order transitions. In several cases, the experimental apparatus and sample cell design are described (with appropriate references being given for further details) and the necessary phenomenological theory is developed. For each high-resolution experimental technique, an example is given to illustrate the sort of data that can be obtained with that method.

7.2.1 Traditional adiabatic calorimetry

The classical method of measuring enthalpy and heat capacity changes is *adiabatic calorimetry* carried out by the stepwise addition of carefully measured small increments of electrical energy [10]. As in most calorimetric techniques, one measures the total heat capacity of the sample and a sample holder or cell: $C_p(\text{obs}) = C_p(\text{sample}) + C_p(\text{cell})$. In adiabatic calorimetry, Eq. (7.1) is approximated by

$$C_p = \frac{Q_p}{\Delta T}, \qquad (7.3)$$

where Q_p is the known heat input that causes a finite temperature rise ΔT. In order to achieve reasonable resolution near a transition the ΔT steps must be small, which makes the method somewhat tedious and slow. In addition, $C_p(\text{cell})$ is usually quite large so that large amounts of sample (say 10 g) are needed. The investigation of liquid crystals with this technique has been primarily carried out by Anisimov and coworkers [5,11]. Since this method is very well known, further details are unnecessary.

7.2.2 Differential scanning calorimetry

By far the most common thermal technique used to study liquid crystals has been differential scanning calorimetry (DSC). This method has the advantages of high sensitivity for detecting enthalpy changes (as small as ~0.01 J/g), very small sample size (~10 mg), rapid and convenient operation procedures using commercial instruments with flexible computer software programs. DSC is in many ways an ideal survey technique, useful for discovering new phase transitions and determining the qualitative magnitude of thermal features. However, DSC is not well suited to making detailed quantitive measurements near liquid crystal phase transitions.

The difficulty with a DSC measurement can be understood by considering Fig. 7.3, where the DSC responses to first- and second-order transitions are compared. In operation, a constant heating (or cooling) rate $dH(\text{ref})/dt$ is imposed on the reference sample, which results in a constant and rapid linear temperature ramp. Scan rates dT/dt of 5–10 K/min are commonly used. Although slower rates can be used, DSC machines work best for fairly rapid scans ($dT/dt \geq 1$ K/min) and require minimum scan rates of ~0.1 K/min. A servosystem forces the sample temperature to follow that of the reference by varying the power input $dH(\text{sample})/dt$. The differential power $dH/dt = dH(\text{sample})/dt - dH(\text{ref})/dt$ is measured, and the integral $\int(dH/dt)dt$ for a DSC peak approximates the enthalpy associated with the corresponding transition.

The DSC response in Fig. 7.3(b) is very similar for a first-order transition with latent heat ΔH and a second-order transition with a comparable integrated enthalpy δH. The broadening of the DSC first-order signal is due to the fact that the latent heat cannot be delivered to the sample instantaneously. The DSC peak will achieve its maximum at a temperature T_m well above T_{tr} for a heating run and well below T_{tr} for a cooling run; $|T_m - T_{tr}|$ depends on the scan rate that is chosen. The DSC response for a second-order transition (dashed curve in Fig. 7.3(b)) is qualitatively the same. It is very difficult to distinguish first- and second- order transitions and impossible to detect the difference between these and a weak first-order transition with a significant pretransitional heat capacity variation. Furthermore, there is the problem that thermodynamic equilibrium for many liquid crystal phase transitions requires scan rates near T_{tr} of $dT/dt \leq 200$ mK/h = 0.0035 K/min, which is 10^2 to 10^3 times slower than

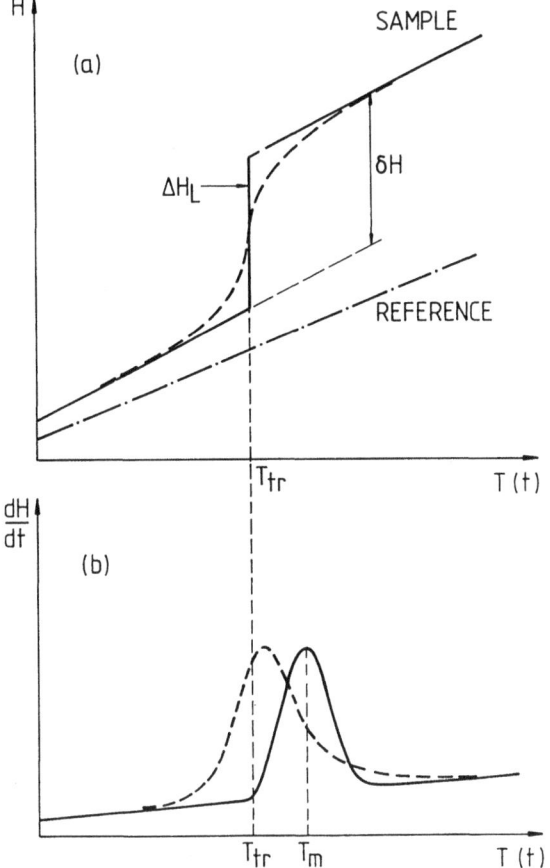

Figure 7.3. Schematic representation of the DSC responses from heating runs through a first-order (solid lines) and second-order (dashed lines) transition. (a) Variation of enthalpy H for the sample as a function of temperature T, which itself varies linearly with time t. The dashed-dot line shows the H variation of a reference material for which $C_p(\text{ref}) \approx$ constant. (b) The differential power input $dH/dt = dH(\text{sample})/dt - dH(\text{ref})/dt$. T_{tr} denotes the transition temperature and T_m is the location of the DSC response maximum. Taken from Ref. 1.

slow DSC scans. The absolute accuracy of latent heats ΔH is typically $\pm 20\%$, and the accuracy of integrated second-order enthalpy changes δH is often $\pm 100\%$ and seldom better than $\pm 50\%$.

There is a way of using DSC to make a qualitative distinction between first- and second-order transitions: make a series of measurements at different scan rates and extrapolate to zero rate. This method exploits a weakness of the DSC technique in detecting continuous enthalpy changes at

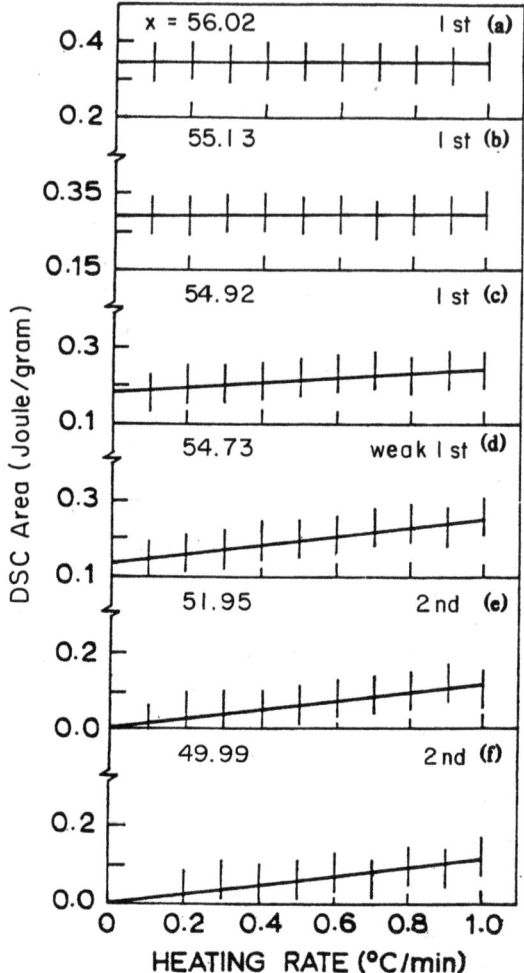

Figure 7.4.   DSC apparent enthalpy, corresponding to the area under DSC peaks such as those shown in Fig. 7.3(b), as a function of scan rate for six mixtures of $DB_8ONO_2 + DB_{10}ONO_2$[12]. The values of $X =$ mole percent $DB_{10}ONO_2$ are given. The transition is strongly first-order for (a) and (b) and second-order for (e) and (f). See text for further details.

very slow scan rates. An example is shown in Fig. 7.4 for mixtures of $DB_8ONO_2 + DB_{10}ONO_2$ [12]. In mixtures (a) and (b) there are strongly first-order SmA_1–SmA_d transitions, while mixtures (e) and (f) undergo second-order N–SmA_1 transitions as confirmed by high-resolution ac calorimetry [13]. The situation is much more complicated for mixtures (c) and (d) [14], and DSC does not provide any indication of the true behavior. One sees that the DSC signals for strongly first-order transitions are independent of scan rate and represent the latent heat ΔH, whereas DSC signals for second-order transitions extrapolate to zero in spite of the fact that δH is comparable to ΔH in magnitude.

7.2.3 High-resolution calorimetry: a simple model

The basic problem with DSC measurements is the intrinsic design requirement of rapid scan rates. The trouble with traditional adiabatic calorimetry is the very long data acquisition period and limited resolution. Discussed below in Section 7.2.4–7.2.7 are a variety of high-resolution scanning techniques well suited to quantitative measurements at liquid crystal (and other) transitions. These are research techniques that in general cannot be implemented with commercial calorimeters. Thus the design and operation will be described in some detail and appropriate references given to sources of more information. As a common background for these techniques, the necessary phenomenological theory for extracting C_p and/or H values is presented here. The simple 'zero-dimensional' model of thermal analysis will suffice for present purposes [15]. This model will be valid if τ_{int}, the internal thermal diffusion time of the sample (depending on its thermal conductivity and geometry), is small and temperature gradients within the sample can be neglected. Discussion of one-dimensional models of thermal analysis, which take such internal gradients into account, are given elsewhere [16].

In addition to assuming that the internal relaxation time is very small compared with the time scale of data acquisition, the zero-dimensional model also assumes that the heat leak to the thermal bath can be represented by a single parameter, the thermal resistance R (reciprocal of the thermal conductance Λ) of the gas and lead/support wires that link the sample to a temperature-controlled bath. The heat input power, $P = dW_e/dt$, usually involves dissipating electrical work in the system although radiant energy can also be used. For constant pressure processes, the conservation of energy yields

$$P = C_p^{\mathrm{sys}}\frac{dT}{dt} + \frac{dL}{dt} + \frac{T - T_{\mathrm{B}}}{R}, \tag{7.4}$$

where $dL = \overline{L}dm$ with $\overline{L}$ being the latent heat per gram associated with a first-order transition and dm the mass of material converted from one coexisting phase to another. T is the sample temperature, and T_{B} is that of the bath. The quantity $C_p^{\mathrm{sys}} \equiv C_p + C_p(\mathrm{cell})$ is the total heat capacity of sample + cell. In the absence of two-phase coexistence,

$$C_p^{\text{sys}} = \frac{dH/dt}{dT/dt} = \frac{P - (T - T_B)/R}{dT/dt} \quad \text{(if no two-phase coexistence)}. \quad (7.5)$$

If two phases coexist at a first-order transition and phase interconversion occurs, one can define an effective heat capacity C_{eff} by

$$C_{\text{eff}}^{\text{sys}} \equiv \frac{P - (T - T_B)/R}{dT/dt} \quad \text{(in presence of two-phase coexistence)}. \quad (7.6)$$

The latent heat L is then given by

$$L = \int_{T_a}^{T_b} \{C_{\text{eff}}^{\text{sys}} - [C_p(\text{coex}) + C_p(\text{cell})]\} \, dT, \qquad (7.7)$$

where $C_p(\text{coex})$ is the heat capacity of the two coexisting phases, $\alpha + \beta$, over a narrow coexistence range from T_a to T_b; i.e., $C_p(\text{coex}) = X_\alpha C_p(\alpha) + X_\beta C_p(\beta)$, where X_α is the mass fraction of phase α (which varies from 1 at T_a to 0 to T_b). Since the coexistence range is almost always small (~ 50–$100\,\text{mK}$), one can take $C_p(\alpha)$ and $C_p(\beta)$ as independent of T and use $C_p(\alpha) = C_p(T_a)$ and $C_p(\beta) = C_p(T_b)$. Equation (7.7) is equivalent to

$$L = \int_{t_1}^{t_2} \left(P - \frac{T - T_B}{R}\right) dt - \int_{T_a}^{T_b} [C_p(\text{coex}) + C_p(\text{cell})] dT, \qquad (7.8)$$

where t_1 and t_2 correspond to the times for the appearance (at T_a) and disappearance (at Tb) of two phases.

7.2.4 Adiabatic scanning calorimeter

In comparison to traditional adiabatic calorimetry, much greater resolution can be achieved with an adiabatic scanning technique. This method, used in the 1970s for the study of liquid–gas [17] and liquid–liquid critical points [18], was first applied to liquid crystals in a beautiful study of 8CB by Thoen [19]. Detailed discussions of this experimental method are given in Refs. 1, 2, and 20. The essential design features are very precise control of the thermal shield 2 in Fig. 7.5 and high-resolution measurements of the sample temperature $T(t)$. Although this apparatus can be operated in several modes (including cooling as well as heating), the most attractive mode involves heating with the power, P, constant and the shield temperature T_2 maintained equal to that of the sample. In this mode, Eq. (7.5) becomes simply $C_{\text{eff}}^{\text{sys}} = P/T$ since $T = T_B$ and Eq. (7.8) becomes

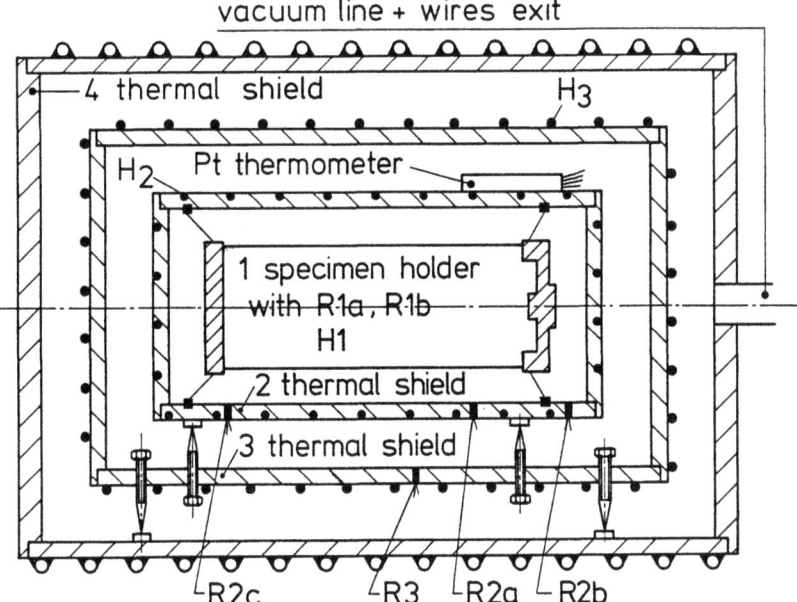

Figure 7.5. Schematic diagram of an adiabatic scanning calorimeter [2, 20]. Electrical heaters and thermistors are denoted by H and R. Details of a sample holder with stirring capabilities are given in Ref. 20.

$$L = H(T_b) - H(T_a) - \int_{T_a}^{T_b} [C_p(\text{coex}) + C_p(\text{cell})] \, dT. \tag{7.9}$$

Typical scan rates are $\dot{T} = 100\text{–}200\,\text{mK/h}$ away from transitions and $\dot{T} = 10\text{–}20\,\text{mK/h}$ near transitions.

An example of liquid crystal heat capacity data obtained with the adiabatic scanning method is given in Fig. 7.6. Data points in the vicinity of the N–I transition represent only the pretransitional C_p wings not latent heat effects in the coexistence region. The enthalpy variation through a typical N–I transition is shown in Fig. 7.7, and other examples are given in Ref. 2. It must be stressed that the direct result obtained with adiabatic scanning is the enthalpy change $H(T) - H(T_s) = P(t - t_s)$ and the associated temperature change $T(t) - T_s(t_s)$, where t_s is the start time for a run that begins with the sample at temperature T_s and enthalpy $H(T_s)$. Heat capacity values, such as those shown in Fig. 7.6, are obtained from numerical differentiation involving a very large number of experimental data points.

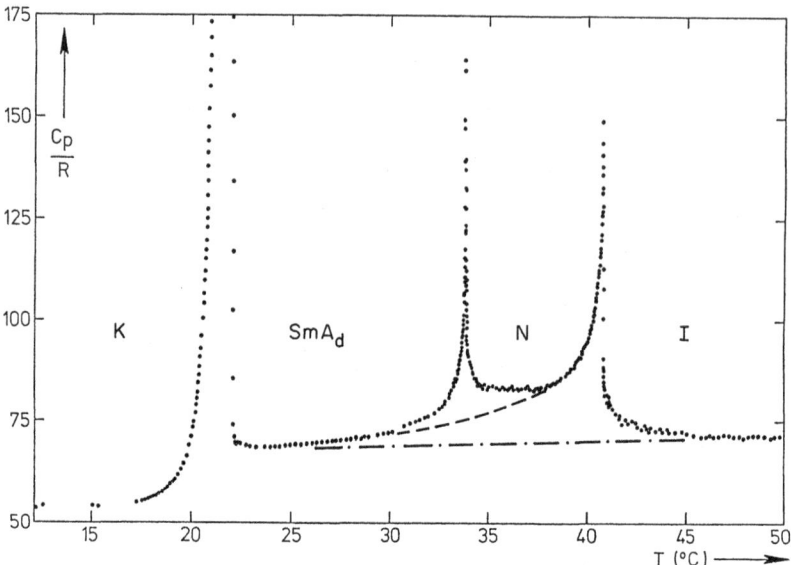

Figure 7.6.   Dimensionless heat capacity C_p/R for octylcyanobiphenyl (8CB) as a function of temperature [19]. K denotes the crystalline phase. C_p in J K^{-1} g^{-1} units equals $0.0285(C_p/R)$. The significance of the dashed and dashed-dot lines is discussed in Section 7.4.

7.2.5 Ac calorimetry

The basic ac method was invented in the 1960s by Kraftmaker [21] and by Sullivan and Seidel [22], and it has been widely used for studying liquid crystals by Johnson [23], Huang [24, 25], Garland [3, 26, 27], Hatta [28] and others. Indeed, much of what is known about the detailed behavior of C_p near liquid crystal transitions comes from high-resolution ac calorimetry, and the technique has even been adapted and extended to high pressures [26, 29] and the study of free-standing smectic films [25].

For ac calorimetry, one uses an oscillating input power, $P = P_0 \cos \omega t + P_0$, and chooses a frequency ω so that $\omega \tau_{int} \ll 1$ in order to avoid temperature gradients inside the sample. In the absence of two-phase coexistence, the term dL/dt disappears from Eq. (7.4) and that equation can be integrated if one makes the reasonable assumption that C_p^{sys} and R are constants over a small time (temperature) interval. In the steady state, the system temperature is given by

$$T(t) = T_{dc} + T_{ac}(t), \qquad (7.10a)$$

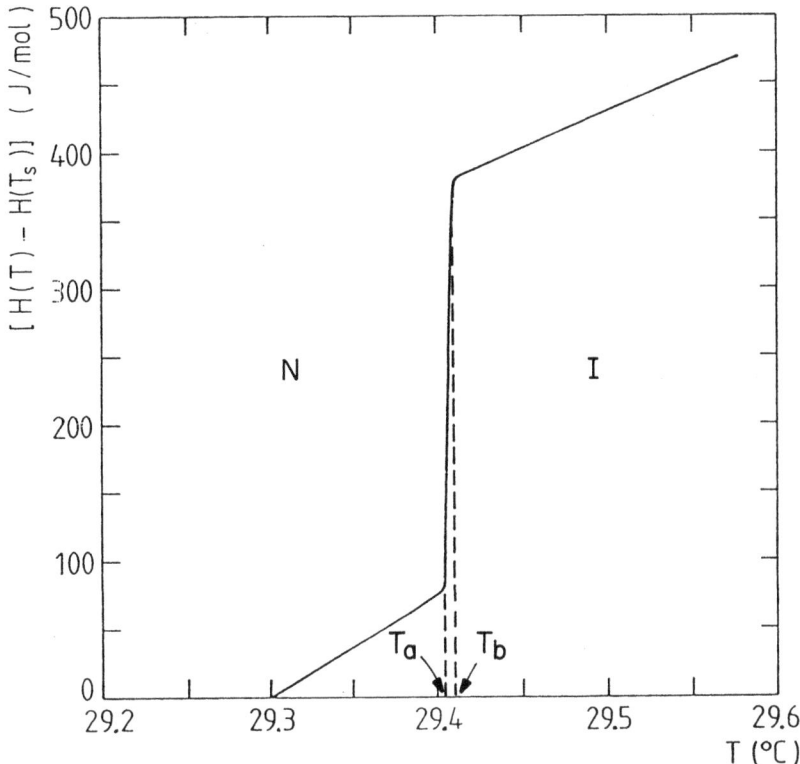

Figure 7.7. Temperature variation of the enthalpy near the N–I transition in 6CB [1, 2]. The temperatures T_a and T_b bracket the N + I coexistence region.

where

$$T_{dc} = T_B + RP_0 \qquad (7.10b)$$

and

$$T_{ac}(t) = \Delta T_{ac} \sin(\omega t + \varphi) = \Delta T_{ac} \cos\left(\omega t + \varphi - \frac{\pi}{2}\right). \qquad (7.10c)$$

The observable quantities ΔT_{ac} and φ are given by

$$\Delta T_{ac} = \frac{P_0}{\omega \, C_p^{sys}} \left(1 + \frac{1}{\omega^2 \tau_{ext}^2}\right)^{-1/2} \qquad (7.11a)$$

$$\tan \varphi = \frac{1}{\omega \tau_{ext}} = \frac{1}{\omega R \, C_p^{ext}} \qquad (7.11b)$$

where $\tau_{\text{ext}} = R\,C_p^{\text{sys}}$ is the external time constant for heat flow from the system to the bath.[†] Eliminating τ_{ext} from the above equations, one finds

$$C_p^{\text{sys}} = \frac{P_0}{\omega \Delta T_{\text{ac}}} \cos \varphi \quad \text{(if no two-phase coexistence).} \quad (7.12)$$

Note that $\Delta \Phi = \varphi - \dfrac{\pi}{2}$ is the phase shift between $T(t)$ and $P(t)$, but it is conventional to call φ the phase shift since $\varphi \approx 0$ when $\omega \tau_{\text{ext}}$ is large ($\omega \tau_{\text{ext}} \gg 1$), which corresponds to the normal operating conditions. In that case, $\cos \varphi \approx 1$ and Eq. (7.12) reduces to the simple expression

$$C_p^{\text{sys}} \simeq \frac{P_0}{\omega \Delta T_{\text{ac}}}. \quad (7.13)$$

This can be used as a good approximation in many cases. If the value of φ is small but not negligible and is not known due to instrumental effects, one can approximate Eq. (7.12) with

$$\left(C_p^{\text{sys}}\right)^2 \simeq \left(\frac{P_0}{\omega \Delta T_{\text{ac}}}\right)^2 - \frac{1}{\omega^2 R^2}, \quad (7.14)$$

where R can be estimated from $T_{\text{dc}} - T_{\text{B}} = R P_0$.

A number of designs for ac calorimeters have been described in the literature [23, 24, 27], and the block diagram for a conventional ac calorimeter used for several years at MIT is given in Fig. 7.8. The sample cell used at MIT is a cold-welded silver cell 1 cm in diameter and 0.05–0.1 cm deep that holds $\sim$50 mg of liquid crystal and a helical coil of fine gold wire to increase the effective thermal conductivity of the sample (i.e., achieve low τ_{int}) [15, 30]. This system operates at a constant frequency $\omega_0 = 2\pi f = 0.1963$, corresponding to a 32-s period for the temperature oscillations. A multiplexer allows the DMM to measure the heater current, the heater voltage, the platinum thermometer (PRT) resistance, and then to measure the resistance of the microbead thermistor every 0.5 s during 5 to 8 periods of oscillation. A typical ΔT_{ac} amplitude of ± 5 mK (see Fig. 7.9) corresponds to a resistance variation of $\Delta R \approx 50\,\Omega$ for a thermistor with $R \approx 60\,k\Omega$. For a DMM with 6.5 decades of resolution, the typical scatter in C_p values is $\pm 0.2\%$. At this low frequency, the ΔT_{ac} value can be obtained with digital lock-in techniques or from a least-squares fit to the $T(t)$ thermistor oscillations. In

[†] A more complete expression for ΔT_{ac} including $\omega \tau_{\text{int}}$ and sample/coupling medium conductances is given in Ref. 22.

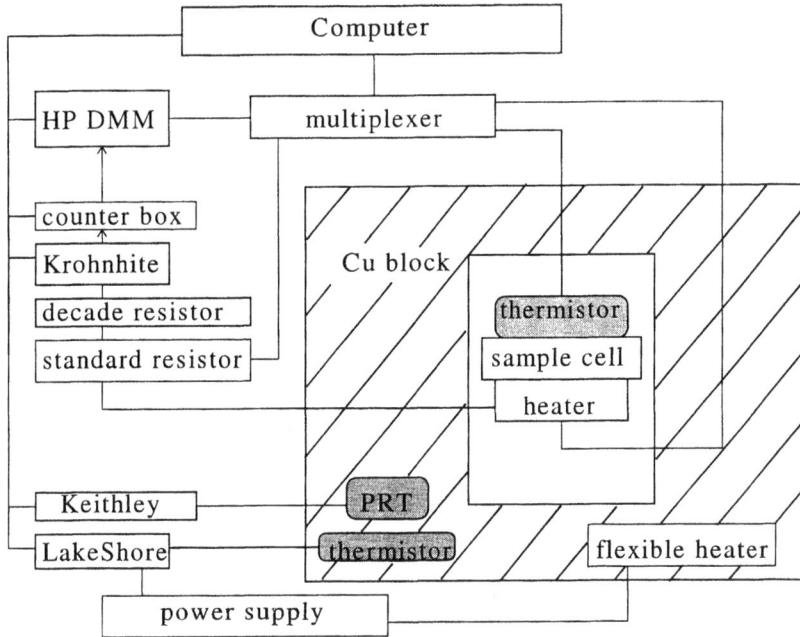

Figure 7.8. Block diagram of an automated ac calorimeter [15, 30]. This instrument can be operated with a personal computer. Krohnhite is a stable frequency synthesizer.

either case, it is also possible to characterize the behavior of tan ϕ. An example of high-resolution C_p data obtained with the ac technique is given in Fig. 7.10. It should be noted that it is feasible to use higher ac frequencies if a much thinner ($\sim 100\,\mu m$) cell is used, in which case $f \simeq 1\,Hz$ is a typical value and analog lock-in detection with a commercial instrument is possible [24].

Advantages of the ac technique are small sample sizes (20–80 mg or much less in the case of very thin cells or free-standing films), ability to carry out either slow heating or cooling scans with equal ease, and excellent resolution in $C_p(T)$ since the oscillatory amplitude can be small, say $\sim 5\,mK$ as shown in Fig. 7.9. A disadvantage is the fact that this method measures C_p directly and cannot quantitatively determine the enthalpy H. Thus, if a first-order transition occurs, one cannot measure the latent heat $\Delta H = L$. However, there is usually a clear qualitative indication of two-phase coexistence in the phase shift variation. If interconversion of two coexisting phases is very sluggish on the time scale of data acquisition (3–4 min per data point), the release of latent heat is so slow that dL/dt in Eq.

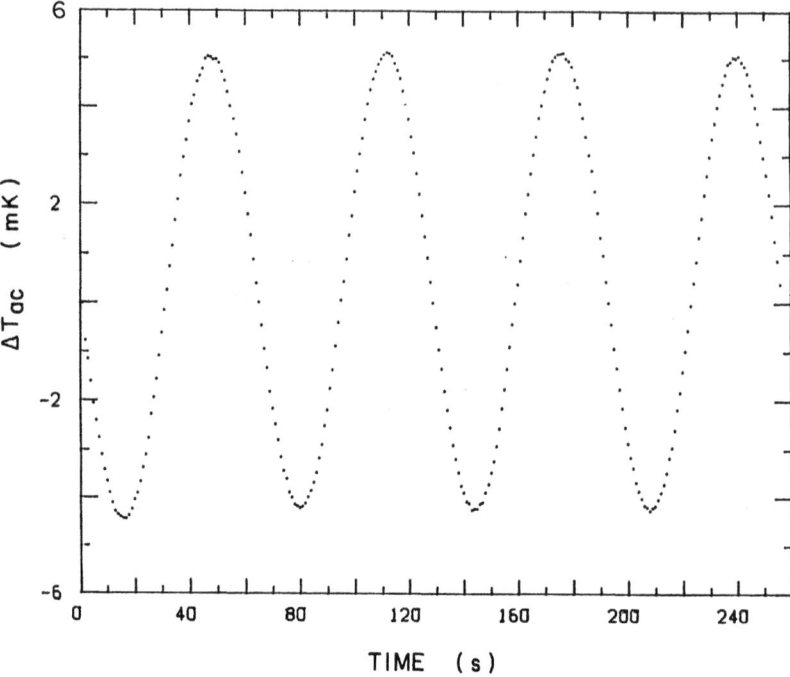

Figure 7.9. Typical T_{ac} sample temperature oscillations during data acquisition [27]. Four periods are shown although 5–8 periods are usually used for a data point.

(7.4) is a smooth, roughly linear, function of time. Thus, latent heat effects are seen only in T_{dc} and Eq. (7.12) yields $C_P^{sys} = C_p(coex) + C_p(cell)$. If inter-conversion is very fast, $(P_0/\omega\Delta T_{ac})\cos\varphi$ yields $C_P^{sys} + dL/dT$, where $dL/dT = \overline{L}\, dm/dT$ with $\overline{L}$ being the latent heat per gram and dm being the mass of phase α converted into phase β. In both of these limiting cases, the sample temperature $T(t)$ does not exhibit any anomalous phase shift. However, the phase conversion rate usually has an intermediate value not low enough or high enough to satisfy either of these limiting cases. In the event, $dL/dt = (dL/dT)(dT/dt)$ has a Fourier component at frequency ω that will act like a power at this frequency and its higher harmonics will also con-tribute. As a result, one observes anomalous variations in φ and artificially high values of the apparent heat capacity. This is illustrated by the data shown in Fig. 7.11 for the first-order SmA–SmC* transition in C7 [15, 32]. Other examples are the $Sm\tilde{C}$–SmC_2 transition in $DB_8ONO_2 + DB_{10}ONO_2$ mixtures (see Fig. 5 in Ref. 33) and the SmC–SmF transition in three TBnA compounds (see Figs. 1 and 2 in Ref. 34). The same type of $\tan\varphi$ anomaly

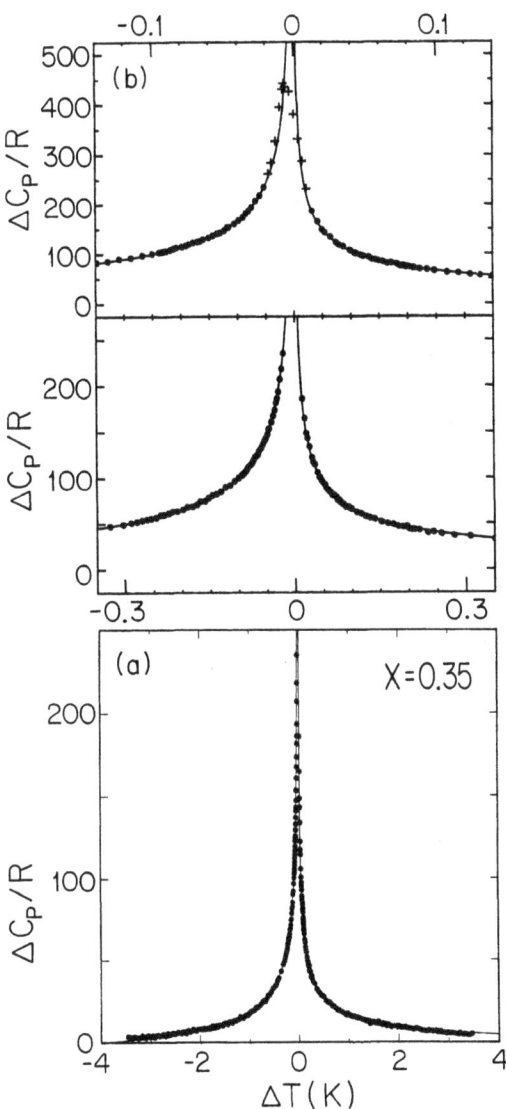

Figure 7.10. Excess heat capacity ΔC_p associated with the N–SmA transition in a 4O.8 + 6O.8 mixture with $X_{6O.8} = 0.35$ [31]. The scale is expanded in (b) to show the detailed behavior very close to T_c. The theoretical curve represents a fit using the tricritical exponent $\alpha = 0.5$.

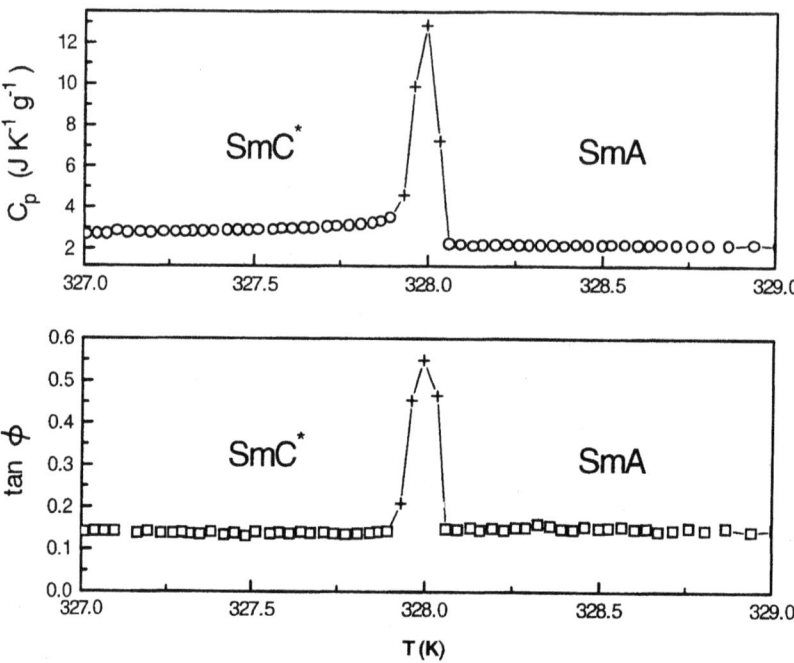

Figure 7.11. C_p and tan φ data for C7 near its SmA–SmC* first-order transition [15]. The anomalous points denoted by + signs are in the two-phase coexistence region.

is also seen at the N–I transition, which is weakly first order with extensive pretransitional C_p wings (see Fig. 3 in Ref. 35).

The work by Huang's group on the heat capacity of free-standing liquid crystal films utilizes a chopped laser beam as the ac power input, a special constant-temperature oven that allows the manipulations required for spreading smectic films, and a thermocouple detector placed near ($\sim$100 μm) but not touching the film. Experimental details are given in Ref. 25. Both thick films ($\sim$100 layers) and films as thin as two smectic layers have been studied, and an example of data for two, three and four layer films is shown in Fig. 7.12.

7.2.6 Relaxation calorimetry

The N–SmA transition in CBOOA was studied with relaxation calorimetry in 1974 [37], and then this method languished until a recent revival by Ema [38, 39]. In conventional relaxation mode operation, the bath temperature T_B is held constant and a dc power is supplied to the cell that is a step

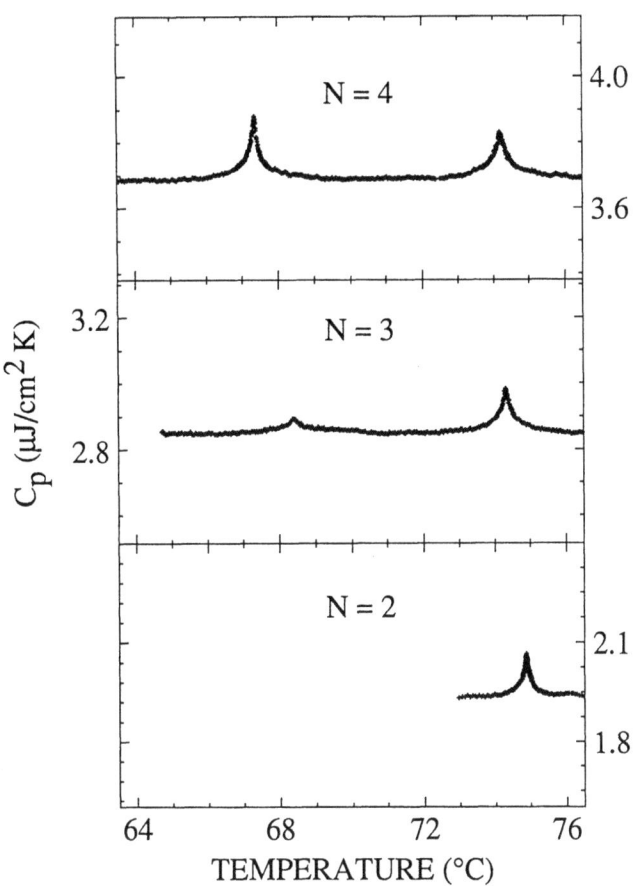

Figure 7.12. Heat capacity of 3(10)OBC films N layers thick near the SmA–HexB phase transition [36]. The C_p peak near 74 °C is associated with the transition in the top and bottom layers, while the peak at ~ 68 °C arises from a transition in the interior layer(s).

function. For a heating run, power is switched from 0 at time $t = 0$ to a constant value P_0; whereas the power is switched from P_0 to 0 for a cooling run. It follows from solving Eq. (7.4) in the absence of latent heat effects that the cell temperature $T(t)$ relaxes exponentially when R and C_p are taken to be constant over the narrow range from $T(0)$ to $T(\infty)$:

$$T(t) = T_B + \Delta T_\infty[1 - \exp(-t/\tau_{ext})] \quad \text{for heat regime,} \quad (7.15a)$$
$$T(t) = T_B + \Delta T_\infty \exp(-t/\tau_{ext})] \quad \quad \text{for cool regime,} \quad (7.15b)$$

where T_B is the constant bath temperature and $\Delta T_\infty = RP_0$. As before, the quantity $\tau_{ext} = RC_p$ is the 'external' time constant for heat flow from the

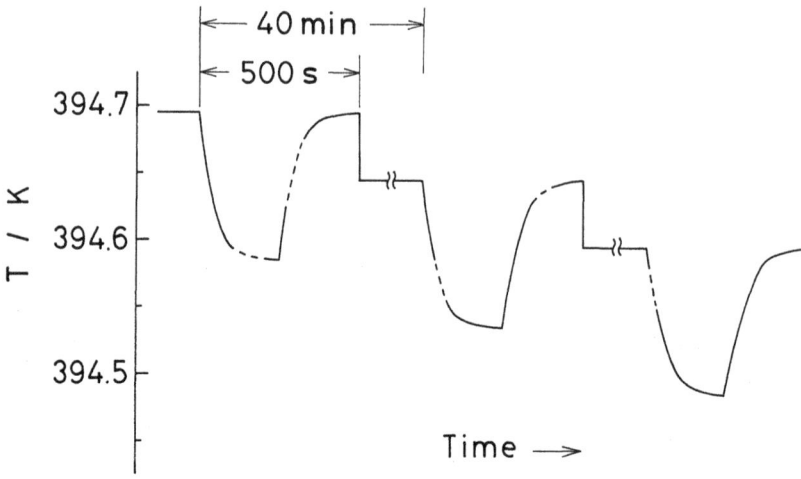

Figure 7.13. Sequence of cool/heat relaxation runs [38]. Periods shown by dashed lines represent regions of non-exponential behavior seen when a first-order transition occurs.

sample to the bath. A fit of the $T(t)$ data with Eq. (7.15a) or (7.15b) will then determine both R and τ_{ext} and therefore will yield $C_p^{sys} = \tau_{ext}/R$. It is usual to carry out a pair of runs (a cooling run followed by a heat run) then reset the bath temperature and make another pair as shown in Fig. 7.13. If a first-order phase conversion occurs between $T(0)$ and $T(\infty)$, there will be an 'anomalous' non-exponential $T(t)$ variation due to latent heat effects. Ema *et al.* [38, 39] have handled this by defining a time-dependent heat capacity $C(t) \equiv \dot{Q}/\dot{T}$, where $\dot{Q} = P_0 - (T - T_B)/R$. This formulation is equivalent to our Eq. (7.6), and their latent heat expression

$$L = \int_0^{\infty} [C(t) - C_0]\dot{T}(t)\, dt \qquad (7.16)$$

is equivalent to Eq. (7.8). A typical example of $C(t)$ behavior associated with a first-order transition having a small latent heat is given in Fig. 7.14.

Recently a new type of relaxation method has been developed [40] in which the heater power is linearly ramped. For a heating run, $P = 0$ for $t < 0$, $P = \dot{P}t$ for $0 \leq t \leq t_1$ where $\dot{P} \equiv dP/dt$ is a constant, and $P = P_0 = \dot{P}t_1$ for $t > t_1$. The initial ($t \leq 0$) sample temperature is T_B and the plateau sample temperature is $T(\infty) = T_B + RP_0$ for $t \gg t_1$. For a cooling run, the power profile is reversed: $P = P_0$ for $t < 0$, $P = P_0 + \dot{P}t$ with $\dot{P}$ negative for $0 \leq t \leq t_1$. In this case the sample temperature is initially $T(\infty)$ and has the

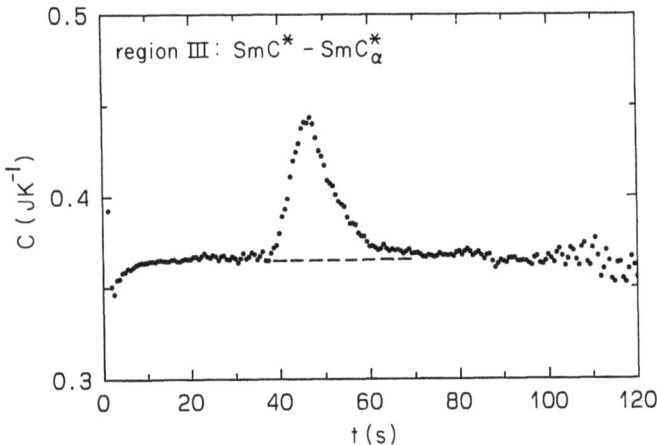

Figure 7.14. Plot of time-dependent heat capacity $C(t)$ in the region of the first-order SmC*–SmC*$_\alpha$ transition in MHPOBC [39]. The latent heat is $12\,\mathrm{J\,mol}^{-1} = 21.5\,\mathrm{mJ\,g}^{-1}$, and this is closely related to the area under the peak and above the dashed baseline; see the text.

final value T_B. The variation of the cell temperature over the time regime $0 \le t \le t_1$ is

$$T(t) = T_B + R\dot{P}(t - \tau_{ext}) + \tau_{ext}\,R\dot{P}\,\exp(-t/\tau_{ext}) \quad \text{heat } (\dot{P} \text{ pos}), \quad (7.17a)$$
$$T(t) = T(\infty) + R\dot{P}(t - \tau_{ext}) + \tau_{ext}\,R\dot{P}\,\exp(-t/\tau_{ext}) \quad \text{cool } (\dot{P} \text{ pos}). \quad (7.17b)$$

The thermal resistance is given by

$$R = (T_\infty - T_B)/P_0, \quad (7.18)$$

and the heat capacity for heating or cooling runs is given by Eqs. (7.5–6), where P is the power at time t' corresponding to sample temperature T lying in the interval T_B to $T(\infty)$, and dT/dt is obtained by fitting $T(t)$ data over a short time interval centered at t'. The advantages of this new ramp relaxation method, which could best be called *non-adiabatic scanning calorimetry*, over the use of a power step function are (a) much better control of the bath temperature T_B at a constant value since step increases or decreases in P cause large transient disturbances in T_B and (b) optimal behavior of dT/dt since T varies almost linearly with t rather than exponentially (except for $t > t_1$ and a brief period just after $t = 0$ and of course regions where the enthalpy H is an unusually rapidly varying function of T due to first-order phase conversion).

The mechanical/thermal and electronic designs of a calorimeter capable of both ac-mode and relaxation-mode operation are given in Figs. 7.15(a)

(a)
outer stainless steel can

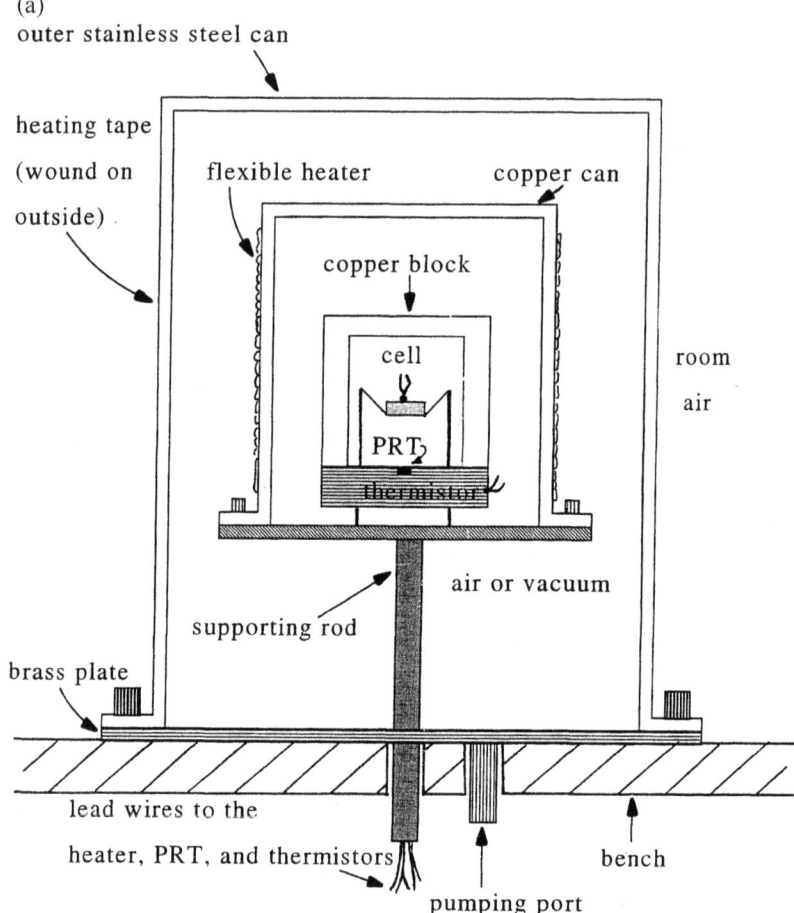

Figure 7.15. Block diagram of a computer-controlled calorimeter capable of operation in the ac mode or in relaxation modes including non-adiabatic scanning
[15, 40]. (a) Mechanical/thermal design.

and (b). The sample is surrounded by a massive copper block which is itself
surrounded by a heated copper can and an outer steel jacket. The 'bath'
temperature is controlled by a thermistor and programmable resistor that
comprise two arms of Wheatstone bridge B. The imbalance signal is fed to
a PID feedback unit that controls the heater attached to the copper can.
The sample thermistor is part of programmable bridge A, whose output is
read with $6\frac{1}{2}$ digit resolution by a DMM. The entire instrument is controlled by a 486SX personal computer. See Ref. 40b for further details.

(b)

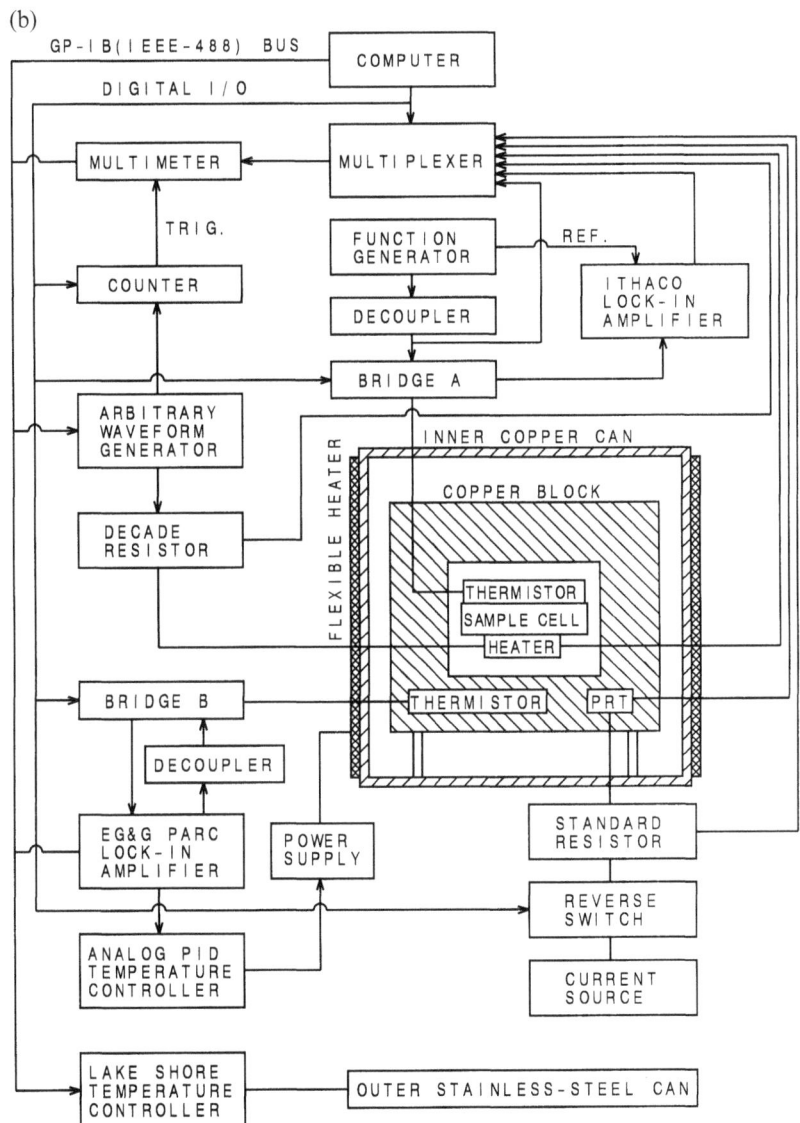

Figure 7.15. (*cont.*) (b) Electronic design.

7.2.7 *Special techniques*

There are a number of other techniques that have been applied to the study of the C_p behavior near liquid crystal transitions. When large quantities of samples are available, as for many lyotropic systems, it is possible to use modified commercial differential calorimeters in which the sample is 'balanced' by a reference cell of comparable heat capacity. Aqueous CsPFO (cesium perfluoro-octanoate) micellar solutions have been studied by this technique with the reference cell containing the solvent [41]. Although the excess ΔC_p peaks at the transitions are quite small, both the N–I and N–neat soap (SmA) transitions were easily observed.

Two higher frequency dynamic techniques that have recently been applied to the study of liquid crystal transitions are photoacoustic and photothermal calorimetry. Photoacoustics is a well-established field [42], but photothermal work, including photopyroelectric detection, is quite new [43]. The photoacoustic technique involves providing heat with a modulated laser beam to a sample that is contained in a sealed gas-filled cell. Coupling between the sample and the gas gives rise to a periodic gas pressure change that is detected by a microphone. In photothermal techniques, pyroelectric detection is employed and this has proved to be a very sensitive method. The major attraction of photoacoustic and photothermal techniques is the possibility of measuring simultaneously the heat capacity C_p and the thermal conductivity κ. The major disadvantage is the complexity of the phenomenological equations relating the amplitude and phase of the detected signal to C_p and κ. The pioneers in using these techniques for liquid crystals are Marinelli *et al.* [44] and Thoen *et al.* [45]. An example of such results is given in Fig. 7.16; however, much remains to be done along these lines.

Another aspect of dynamical behavior for the thermal properties of liquid crystals is the possibility that C_p can be a complex frequency-dependent quantity, i.e., a dynamical response function like the susceptibility. In formal terms, $C_p^*(\omega)$ can be defined by [46]

$$C_p^*(\omega) = C_p(0) + \frac{i\omega}{k_B T^2 \rho V} \int_0^\infty dt \, e^{i\omega t} \langle \delta H(0) \delta H(t) \rangle, \qquad (7.19)$$

where $C_p(0)$ is the usual static specific heat and $\langle \delta H(0) \delta H(t) \rangle$ is a time-dependent correlation function for the total enthalpy where enthalpy fluctuations are given by $\delta H(t) = H(t) - \langle H \rangle$. If the system contains slow degrees of freedom that relax with a time constant τ, $\mathrm{Re}\, C_p(\omega)$ will be frequency dependent (show 'dispersion') and $\mathrm{Im}\, C_p(\omega)$ will exhibit a peak at frequencies near $\omega = 1/\tau$. Very low-frequency critical dynamics – slowing down of cooperative ordering phenomena – has recently been found with ac calorimetry near the SmC–SmI critical point for tilted hexatic ordering in mixtures of 8SI + 8OSI [40a].

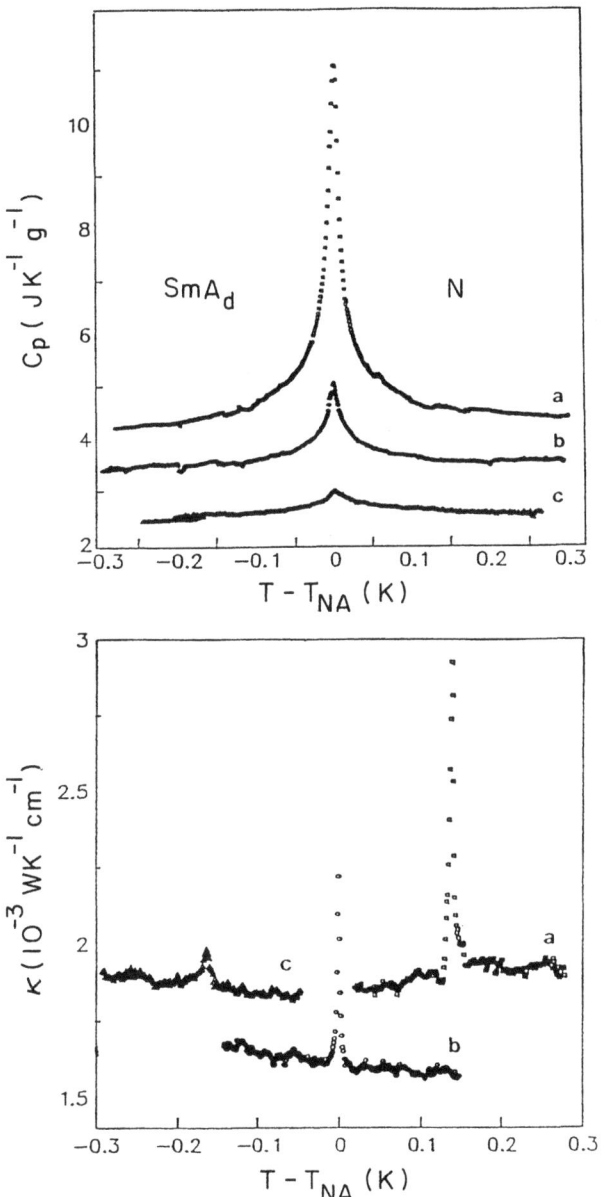

Figure 7.16. Heat capacity C_p and thermal conductivity κ near the N–SmA$_d$ transition in (a) 9CB, (b) 8CB, and (c) 7.76 CB = a mixture of 7CB + 8CB with $X_{8CB} = 0.76$ [44b]. The C_p data for (a) and (b) have been shifted up 2 and 1 J K^{-1} g^{-1} respectively. The κ data for (b) are shifted down by 1 unit and those for (a) and (c) are shifted laterally by 0.15 °C.

7.3 Theoretical analysis

The experimental acquisition of high-quality $C_p(T)$ and/or $H(T)$ data is merely the first step in a calorimetric study of liquid crystals and their phase transitions. The essential second step is a thoughtful theoretical analysis of these data in order to gain insight into the basic physical model that determines the thermal and other properties.

The qualitative determination that a transition is first order rather than second order is interesting and valuable but not sufficient. One generally needs to know the magnitude of the latent heat ΔH relative to the pretransition enthalpy δH defined in Eq. (7.2). This will allow one to judge how close the system is to (a) an isolated critical point beyond which there is supercritical behavior (e.g., the SmA_d–SmA_2 transition [47] or the familiar liquid–gas transition in simple fluids) or (b) a tricritical point beyond which first-order behavior changes into second-order (e.g., the N–SmA transition [1, 2]). Quantitative ΔH values for first-order transitions can also be combined with dilatometric values of ΔV to predict the pressure dependence of transition temperatures via the Clapeyron equation $(dT/dp)_{coex} = T\Delta V/\Delta H)$.

More important than latent heat determinations is the characterization of the pretransitional C_p wings above and below the transition temperature. If the transition is found to be second order (i.e., $\Delta H = 0$), one still needs to know whether critical energy fluctuations occur and what universality class describes them. The highlights of mean-field and critical fluctuation theories of C_p behavior are given below. More detailed treatments are given in Refs. 6–9.

7.3.1 Mean-field analysis

If the interactions responsible for ordering in a system are sufficiently long range, critical fluctuation effects do not occur and the system can be described by the Landau model, which is an attractively simple phenomenological mean-field theory. Of course, fluctuations still play a role since the heat capacity C_p is by definition a measure of the enthalpy fluctuations:

$$C_p = \frac{1}{kT^2}\langle \delta H^2 \rangle, \tag{7.20}$$

where $\delta H = H - \langle H \rangle$ and $\langle \ \rangle$ represents an ensemble average. What is ignored in Landau theory is short range fluctuations in the order. In the Landau model, the Gibbs free energy G has the analytic form

$$G = G_0 + a\psi^2 + b\psi^4 + c\psi^6, \tag{7.21}$$

where G_0 is the free energy of the disordered phase, ψ is the long range order parameter, and $a = a_0 t \equiv a_0(T - T_c)/T_c$. T_c is the critical transition temperature and the dimensionless quantity t is called the reduced temperature, which is used throughout Section 7.3 and 7.4 (whereas t denotes time in Section 7.2). Odd powers in c are omitted in Eq. (7.21) under the assumption that $G(\psi) = G(-\psi)$, a symmetry that clearly exists for ferromagnetic or ferroelectric crystals but may not always hold for liquid crystals.[†] The Landau coefficients a_0 and c are positive, but b can be positive (second order), zero (tricritical), or negative (first order). As in all mean-field models, there is no excess (pretransitional) heat capacity for T above the transition temperature and the disordered phase heat capacity can be well represented by a linear expression $C_p^0(T) = B + E(T - T_{tr})$, which is the background value coming from G_0. Below the transition temperature, the excess heat capacity $\Delta C_p = C_p - C_p^0$ is given by

$$\Delta C_p = A \frac{T}{T_c} \left(\frac{T_k - T_c}{T_k - T} \right)^{1/2}, \quad T < T_c \text{ for } b \quad 0 \tag{7.22}$$

or

$$\Delta C_p = 2A \frac{T}{T_c} \left(\frac{T_k - T_1}{T_k - T} \right)^{1/2}, \quad T < T_1 \text{ for } b < 0 \tag{7.23}$$

where $A = |a_0^2/2bT_c|$ is the mean-field jump in C_p at T_c for a second-order transition and $2A$ is the jump at T_1 for a first-order transition. The metastability limit T_k is given by

$$T_k \equiv T_c + (b^2 T_c/3a_0 c). \tag{7.24}$$

When b 0, a second-order transition occurs at T_c. When $b < 0$, a first-order transition occurs at $T_1 = T_c + (b^2 T_c/4a_0 c) = T_k - (b^2 T_c/12a_0 c)$. When $b = 0$, a Landau or 'classical' tricritical point occurs at $T_k = T_c = T_1$. Since T_c is typically ≥ 300 K and one seldom analyzes C_p data over a range greater than 10 K, the term (T/T_c) varies from ~0.97 to 1.0 and can be set equal to

---

[†] Indeed, the free energy expression for the N–I transition contains a term that is cubic in the scalar nematic orientational order parameter Q since the state $Q = +1$ (all molecules having their long axis parallel to the director **n**) and $Q = -1$ (all molecules having their long axis perpendicular to **n**) are not energetically equivalent. The mean-field theory of the N–I transition is well developed [1, 2, 6], and the Q^3 term causes this transition to be first order. In fact a 'cubic invariant', as it is called, will always cause first-order behavior even in fluctuation dominated cases [6, 7].

unity for fitting purposes. Thus one can rewrite both Eqs. (7.22) and (7.23) in the simpler form

$$\Delta C_{\mathrm{p}} = A^*(T_{\mathrm{k}} - T)^{-1/2}, \tag{7.25}$$

where $A^* = (a_0^3/12c\ T_{\mathrm{c}})^{1/2}$. This form is also valid at the tricritical point where ΔC_p diverges. Schematic representations of Landau heat capacity variations are shown in Fig. 7.17. These sketches include the Landau variation that is obtained in the limiting case where $b > 0$ and $c = 0$. In this case, $\Delta C_p = A(T/T_{\mathrm{c}}) \approx A$ yielding a simple step in C_p.

In mean-field theory the excess heat capacity ΔC_p can be related to the temperature dependence of the *long range* order parameter ψ: $\Delta C_p = -a_0(T/T_{\mathrm{c}})\ d\psi^2/dT$. Since $\psi = 0$ for $T > T_{\mathrm{c}}$, $\Delta C_p = 0$ above the transition. For the simplest second-order Landau model ($b > 0$, $c = 0$) one finds $\psi \sim t^{1/2}$, which leads directly to a step in C_p. For a tricritical point ($b = 0$, $c > 0$), the Landau order parameter dependence becomes $\psi \sim t^{1/4}$, which yields $C_p \sim t^{-1/2}$ as given above.

7.3.2 Critical fluctuation analysis

In reality, C_p is related to the behavior of short range order rather than long range order. For example, in the case of the Ising model the configurational enthalpy (energy) is given by

$$H_{\mathrm{c}} = - N(J/2)\ \langle \sigma_i \sigma_j \rangle, \tag{7.26}$$

where $\sigma_i = \pm 1$ is the spin variable associated with lattice site i and σ_j is that for a nearest neighbor site j, N is the number of spins and J is the Ising coupling constant [7]. Note that the long range order parameter is given by $\psi = \langle \sigma_i \rangle$, and H_{c} is not proportional to ψ^2 as in the mean-field approximation. Since fluctuations in local order cause energy fluctuations, one would expect a pretransitional excess heat capacity ΔC_p above T_{c} as well as below. Critical behavior in well-defined models, such as those with vector order parameters of dimension $n = 1$ (Ising), $n = 2$ (XY) and $n = 3$ (Heisenberg), have been studied with a variety of theoretical techniques. Although Onsager found an analytic solution for the free energy of a two-dimensional Ising ($d = 2$, $n = 1$) model, analytic solutions do not exist for three-dimensional systems. However, sophisticated renormalization-group (RG) analyses have been carried out for three-dimensional n vector models [7, 48] and these results are pertinent to liquid crystal systems [31, 49].

The power-law singularity given by renormalization-group theory for $\Delta C_p = C_p - C_p(\text{background})$ has the form

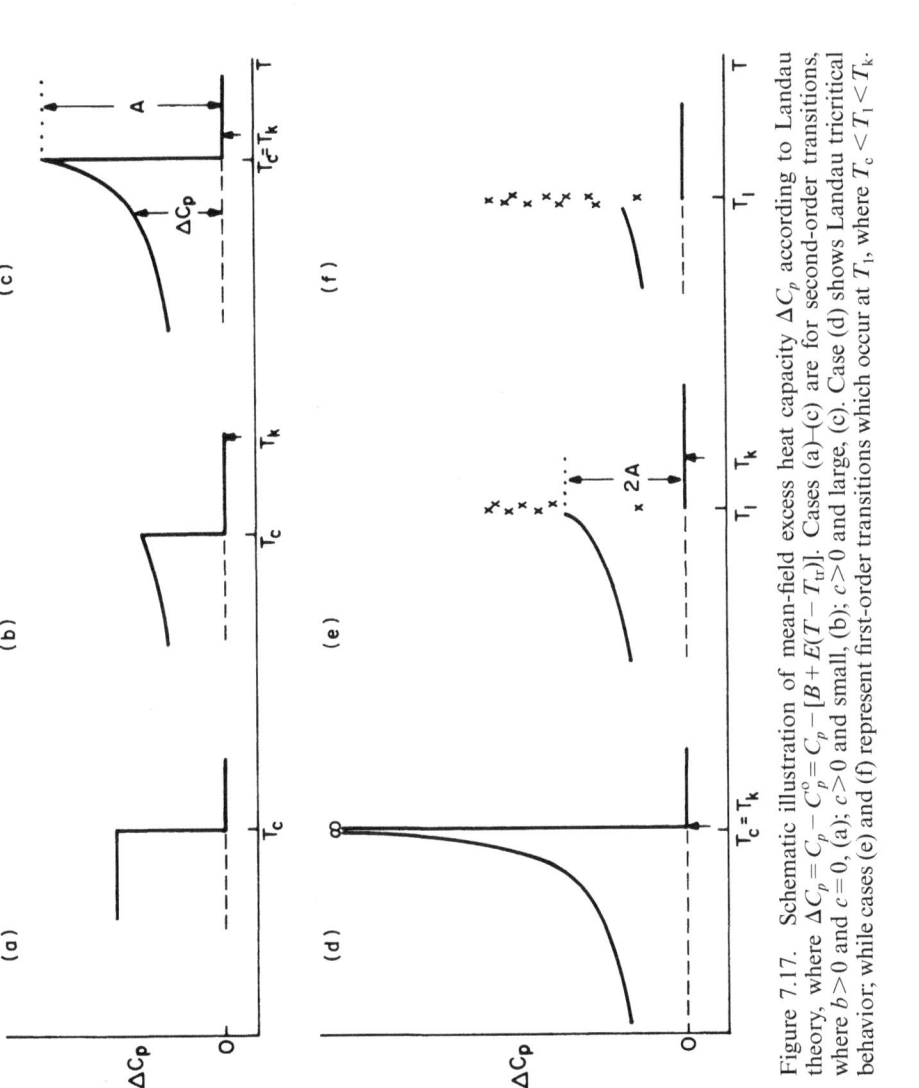

Figure 7.17. Schematic illustration of mean-field excess heat capacity ΔC_p according to Landau theory, where $\Delta C_p = C_p - C_p^o = C_p - [B + E(T - T_{tr})]$. Cases (a)–(c) are for second-order transitions, where $b > 0$ and $c = 0$, (a); $c > 0$ and small, (b); $c > 0$ and large, (c). Case (d) shows Landau tricritical behavior; while cases (e) and (f) represent first-order transitions which occur at T_1, where $T_c < T_1 < T_k$.

$$\Delta C_p = A^{\pm}|t|^{-\alpha}(1 + D_1^{\pm} |t|^{\Delta_1} + D_2^{\pm} |t|^{2\Delta_1} + D_3^{\pm} t + \cdots) + B_c, \quad (7.27)$$

where t is the reduced temperature $(T - T_c)/T_c$ and $\pm$ indicates above and below T_c. The critical exponent for ΔC_p is α, which specifies the character of the leading singularity and determines the shape of the $\Delta C_p(T)$ curve. Note that Eq. (7.27) also includes first- and second-order corrections-to-scaling terms with the exponent $\Delta_1 \approx 0.5$ and the first analytic correction term $D_3^{\pm} t$. Since $2\Delta_1 \approx 1$, the $D_2|t|^{2\Delta_1}$ and $D_3 t$ terms can be effectively merged into a second correction term that is usually small and often neglected in the fitting of experimental data [49, 50]. The critical exponent α has a unique theoretical value for a given universality class ($\alpha = +0.11$ for three-dimensional Ising, $\alpha = -0.007$ for three-dimensional XY, etc.). C_p(background), which represents the non-singular C_p behavior that would have occurred in the absence of the transition, is usually represented by $B_r + E(T - T_c)$, and the quantity B_c is the critical contribution to the regular behavior. The amplitudes $A^{\pm}$, $D_1^{\pm}$, etc. are not universal quantities but the ratios A^-/A^+ and D_1^-/D_1^+ should have fixed values for each universality class. Also it must be stressed that there should be no step discontinuity at T_c for a second-order fluctuation-dominated heat capacity (i.e., $B_c^+ = B_c^- = B_c$) and the same exponent α must be used above and below T_c.

The goal of C_p data analysis for second-order transitions is to obtain the value of the critical exponent α and secondarily the amplitude ratio A^-/A^+ since this will determine the universality class and help to test statistical-mechanical models for the order–disorder transition under consideration. If $\alpha > 0$, Eq. (7.27) shows that C_p should diverge to ∞ at T_c with a temperature-dependent shape determined by α. Also the value of B_c is negative and small in magnitude. For $\alpha < 0$, $C_p(T)$ will exhibit a finite cusp with $\Delta C_p(T_c) = B_c$, where B_c now has a large positive value and both A^+ and A^- have negative values. Sketches of expected $C_p(T) = \Delta C_p + C_p$(background) behavior are shown in Fig. 7.2 for both cases. The dashed line represents C_p(background) $= B_r + E(T - T_c)$ and the dot-dash line represents $B_c + B_r + E(T - T_c)$. The fitting procedure using Eq. (7.27) must be carried out carefully using sophisticated non-linear least-squares techniques since there is considerable coupling between the parameters, especially $A^{\pm}$, α and B_c. Attention should be paid to the presence or absence of systematic deviations $\Delta \equiv \Delta C_p$(obs) $- \Delta C_p$(fit) and to the stability of the parameters on range shrinking (systematic variation of t_{min} or t_{max} in the data set being fit) [19, 49, 50].

A short list of universal critical parameters is given in Table 7.2. The connection between critical C_p data and x-ray data on the order parameter ψ

Table 7.2 *Critical exponents and amplitude ratios for selected universality classes [48].*

	Mean field	Tricritical	Ising $(n=1)$	XY $(n=2)$	Heisenberg $(n=3)$
α	0	0.5	0.110	-0.007	-0.115
A^-/A^+	∞	—	1.85	0.971	
Δ_1	—	0.5	0.496	0.524	0.550
β	0.5	0.25	0.325	0.346	0.365
γ	1	1	1.241	1.316	1.386
ν	0.5	0.5	0.63	0.67	0.70
R_ξ^+	—	—	0.270	0.361	0.435

Note:
The dimensionless two-scale universality quantity R_ξ^+ is defined by $(R_\xi^+)^3 = (\alpha\rho/k_B)A^+\xi_0^3$, where A^+ is in units of $J\ K^{-1}\ g^{-1}$, ξ_0 is in cm, and ρ is the mass density in $g\ cm^{-3}$. Information on other universality classes is given in Refs. 6–9.

below T_c, the susceptibility χ, and the correlation length ξ above T_c should be noted. For second-order and tricritical transitions, one finds $\psi = B|t|^\beta$, $\chi = C^\pm|t|^{-\gamma}$, and $\xi = \xi_0|t|^{-\nu}$ if correction terms are neglected. The Rushbrooke scaling relation $\alpha + 2\beta + \gamma = 2$ and the hyperscaling relation $2 - \alpha = d\nu$ provide consistency tests for critical exponent behavior inferred from independent calorimetric and x-ray experiments. Furthermore, there is a universal relationship called two-scale-factor universality between the non-universal heat capacity amplitude A^+ and the non-universal bare correlation length ξ_0. The constancy of the quantity R_ξ^+ defined in Table 7.2 means that A^+ is proportional to ξ_0^{-3}. Thus, the excess heat capacity ΔC_p can even be too small to detect if the correlation lengths associated with a given transition are very large [51]. However, there are other situations where calorimetry can detect transitions difficult or impossible to observe from microscopic textures. As an example, smectic-A polymorphism was discovered from the thermal behavior near what proved to be a SmA_1–SmA_2 transition in $DB_6CN + TB4A$ mixtures [52].

Thus, heat capacity data provide crucial information for locating phase transitions, establishing the global behavior at a critical point, and testing the consistency of experiment and theory. As an example of the latter point, Fig. 7.18 shows experimental C_p data near the N–SmA_1 transition in 8OPCBOB and a fit to these data with Eq. (7.27) using $\alpha = \alpha_{XY} = -0.007$ and other three-dimensional parameter values [49, 50b]. Another N–SmA example is provided by the data in Fig. 7.10. This 4O.8 + 6O.8 mixture with

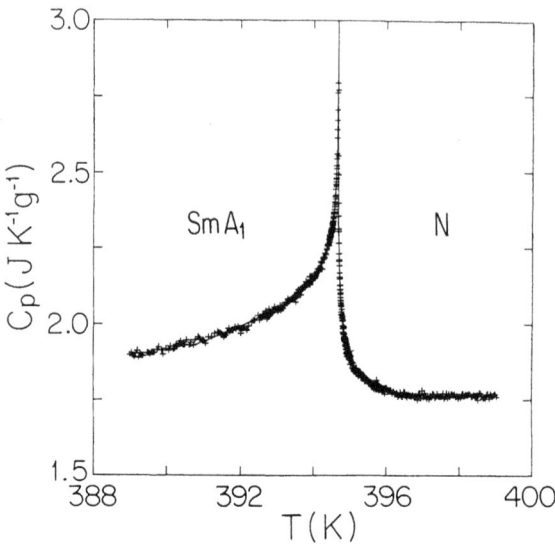

Figure 7.18.   Heat capacity of 8OPCBOB near the N–SmA$_1$ phase transition [49, 50]. The smooth curve represents a fit to these data with Eq. (7.27) based on critical parameters in agreement with the three-dimensional XY model.

$X_{60.8} = 0.35$ corresponds to the tricritical composition, and the smooth curve represents a fit with Eq. (7.27) using $\alpha = \alpha_{TC} = 0.5$.

It is often of interest to study binary mixtures of two liquid crystals so that some special behavior, say a tricritical point [31], can be captured. In such cases, the power-law form for ΔC_p given by Eq. (7.27) should in principle be applied to $\Delta C_{p\phi}$ as a function of t_ϕ, where ϕ indicates a path of constant chemical potential difference. When the mixture consists of homologs or chemically similar molecules, the distinction between $\Delta C_{p\phi}$ and the experimental ΔC_{pX} data obtained at constant composition is not important. However, when the two molecules are sufficiently dissimilar (as in mixtures of polar + non-polar compounds) the slope of the transition line dT_c/dX, where X is the mole fraction of one component in a binary mixture, is large. The magnitude of $(dT_c/dX)^2$ determines the deviation of ΔC_{pX} from ideal behavior, and for large $(dT_c/dX)^2$ Fisher renormalization is observed. Fisher [53] worked out the consequences of approaching a critical point via a constant-composition path and found an exponent renormalization where

$$2 - \alpha_R = \frac{2 - \alpha}{1 - \alpha}, \quad \gamma_R = \frac{\gamma}{1 - \alpha}, \quad \nu_R = \frac{\nu}{1 - \alpha}. \tag{7.28}$$

The critical ΔC_p behavior asymptotically close to T_c is governed by the renormalized value α_R in such cases. Note that Fisher renormalization has a large effect on the heat capacity: $\alpha = +0.11 \to \alpha_R = -0.12$ for Ising, $\alpha = -0.007 \to \alpha_R = +0.007$ for XY, $\alpha = 0.5 \to \alpha_R = -1$ for tricritical. Fisher renormalization has been observed for the Ising-like SmA_1–SmA_2 transition in $DB_6CN + TB4A$ [54] and the tricritical N–SmA transition in $\overline{7}S5 + 8OCB$ [55].

A final theoretical aspect deserves mention – *crossover* between different kinds of critical behavior. An important example of this phenomenon in liquid crystals is the evolution of the N–SmA transition from second order to first order via a tricritical point as the composition of a binary mixture is varied. At the tricritical composition one should observe tricritical exponents (for C_p, $\alpha_{TC} = 0.5$) and sufficiently far from the tricritical point, three-dimensional XY exponents are expected ($\alpha_{XY} = -0.007$). However, in many experimental systems an intermediate crossover behavior is observed with an effective C_p exponent α where $-0.007 < \alpha < 0.5$ [56]. Another type of crossover is from mean-field behavior at large reduced temperatures to critical fluctuation behavior asymptotically close to T_c. The correction terms in Eq. (7.27) are related to this crossover. However, there are cases where critical fluctuation results are expected but mean-field behavior is observed since the width t_{crit} of the critical regime is very small and all the data lie at $t > t_{crit}$, where a mean-field behavior occurs. The Ginzburg criterion [57] for predicting t_{crit} shows that $t_{crit} \propto \xi_0^{-6}$; thus the size of the critical region is very sensitive to the range of correlations in any particular case.

7.4 Calorimetric results

Phase transitions that play an important role in thermotropic smectic liquid crystals are listed in Table 7.3, whch provides a summary of theoretical predictions and presently available experimental results on the character of each transition. Discotic liquid crystals will not be discussed here. Most phase transitions in discotics are first order with little pretransitional fluctuation character, and these systems are discussed elsewhere [58]. Also omitted from Table 7.3 are most of the transitions involving $Sm\tilde{A}$ (the fluid antiphase) and $Sm\tilde{C}$ (the ribbon phase). These 'modulated smectic' phases are discussed in Ref. 4. Presented below are highlights of the best calorimetric results available for six groups of transitions among those listed in Table 7.3. Emphasis will be on work done since 1988. No attempt has been made to give an exhaustive treatment of any of these transitions. Rather the goal is to show what can be accomplished for a typical system

Table 7.3 *Theoretical predictions and available calorimetric results concerning the character of various liquid crystal transitions in smectic materials.*

Transition	Theory [Ref.]	Experiment
N–I	Q model [59] 1st due to cubic invariant	1st, pretrans. C_p wings
BP_{III}–I	1st → CP [60]	1st → CP
BP–BP	1st	1st
N*–BP	1st	1st
SmA–I	1st	1st, weak pretrans. C_p wings
N–SmA_m[a]	XY → TCP → 1st [6]	2nd → TCP → 1st
N–SmA_d	XY → TCP → 1st [6]	2nd → TCP → 1st
N–SmA_1	XY [61, 62]	XY → TCP
N–SmA_2	XY → TCP → 1st [63]	1st and 2nd seen
N–SmC	1st [64]	1st
N–SmA–SmC point	Lifshitz [65]	Lifshitz and TCP
SmA_1–SmA_2	Ising [66]	Fisher RN Ising
SmA_1–Sm$\tilde{A}$	1st [9, 63]	1st, unusual C_p wings
Sm$\tilde{A}$–SmA_2	1st [9, 63]	broad 1st coex.
SmA_d–SmA_2	1st → CP [67]	1st → CP (unusual)
SmA_d–SmA_1	{ 1st → CP [67] { 1st → N_r region [68]	1st → N_r region
SmA_m–SmC_m[a]	XY [6]	MF (Ginzburg criterion)
SmA_2–SmC_2	XY [6]	MF step (Ginzburg criterion)
SmA–HexB	XY [69]	unusual 2nd/weak 1st
SmA–CrB	1st	1st, weak pretrans. C_p wings
SmC–SmI	1st → CP [70]	1st → MF CP

Notes:
1st, first order; 2nd, second order; CP, critical point; TCP, tricritical point; MF, mean field; Ising and XY are $n = 1$ and $n = 2$ vector model second-order transitions. General theoretical references of value are Refs. 6–9. The symbol BP stands for any of the three blue phases and N* denotes a chiral nematic (cholesteric) phase.
[a] The subscript m denotes *monomeric* and is used for non-polar compounds to distinguish these monolayer smectics from SmA_1 and SmC_1 phases of polar 'frustrated smectics'.

with high-resolution calorimetric experiments and careful theoretical analysis. References are provided to the pertinent research literature and to more encyclopedic review articles.

7.4.1 *N–I transitions*

An overview of the N–I heat capacity peak is given in Fig. 7.6 for 8CB. In this case as in many others, there is a second feature – the N–SmA peak –

sitting on the low-temperature wing of the N–I peak. The C_p variation expected in the absence of this N–SmA transition is shown by the dashed line, which also serves as the nematic C_p(background) choice for the determination of ΔC_p(NA). The quantity C_p(background) for the assessment of the excess heat capacity ΔC_p(NI) associated with the N–I transition is the dash-dot line in Fig. 7.6. See Refs. 26 and 51 for a demonstration of the validity of these ideas that can be seen when the nematic range $T_{NI} - T_{NA}$ is increased by applying pressure or by varying the composition of a binary mixture.

Thus for typical materials an excess heat capacity ΔC_p(NI) is observed over a considerable temperature range from roughly 10–15 K above T_{NI} to 20–30 K below T_{NI}. In spite of the presence of significant pretransitional heat capacity near the N–I transition, this transition is expected to be first order due to a cubic invariant in the free energy (see Section 7.3.1). This first-order character is clearly seen experimentally, as indicated by the latent heat shown for 6CB in Fig. 7.7. The almost vertical section of the H vs. T curve yields a latent heat $\Delta H = 293 \pm 3 \, J \, mol^{-1}$ [1, 2]. The N + I two-phase coexistence region $T_b - T_a$, which is induced by small amounts of impurities in the sample, is quite narrow in this case ($\sim 20 \, mK$). Coexistence regions at first-order transitions in liquid crystals are typically 20–100 mK wide. The values of ΔH(NI) in the homologous series nCB range from 300 to 1200 J mol^{-1} [2], and these values are quite small compared to the latent heat of melting the crystalline phase of cyanobiphenyls ($\sim 30 \, kJ \, mol^{-1}$ [19]).

A detailed view of the heat capacity variation near T_{NI} for 6CB, as obtained with adiabatic scanning calorimetry, is given in Fig. 7.19. No data points are displayed in the coexistence region, whose width is indicated by the narrow box around the arrow marking T_{NI}. Although analyses have been given for the N–I transition in terms of mean-field Landau–de Gennes theory [5, 11], the presence of excess heat capacity above T_{NI} clearly indicates that fluctuation effects play a role, as expected since the orientational interactions between liquid crystal molecules are short range. The N–I order parameter is a 3×3 symmetric traceless tensor with $n = 5$ independent components, the so-called Q model of Priest and Lubensky [59].[†]

Attempts to determine a critical exponent α for the N–I transitions have been largely unsuccessful due to the substantial first-order character of this transition. Fits with Eq. (7.27) must be made separately for data above and

---

[†] In the mean field approximation, rotational invariance allows the order parameter to be taken as diagonal and one obtains the Lebwohl-Lasher or Maier–Saupe model.

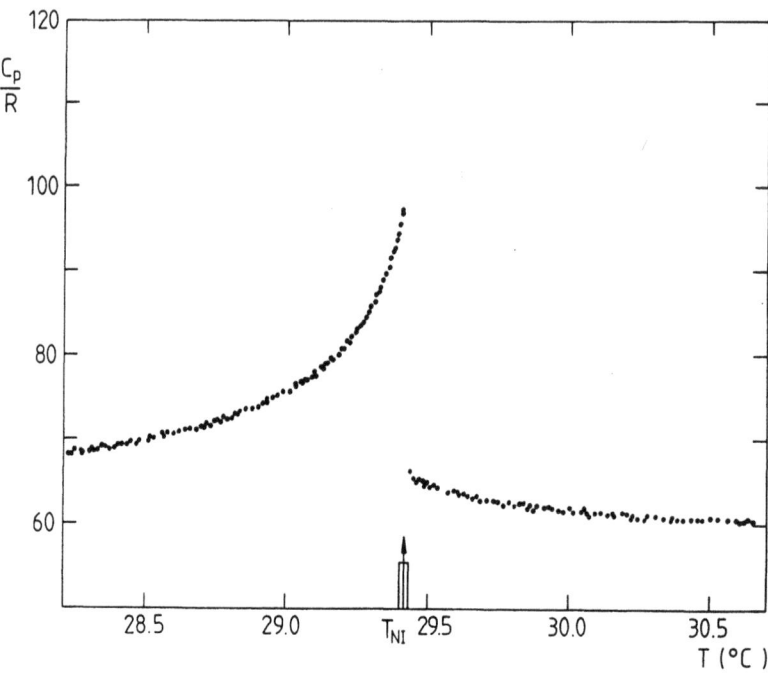

Figure 7.19. Detail near the N–I transition for 6CB [1]. The dimensionless quantity C_p/R can be converted into $\mathrm{J\,K^{-1}\,g^{-1}}$ units with $C_p\,(\mathrm{J\,K^{-1}\,g^{-1}}) = 0.0316\,(C_p/R)$.

below T_{NI} with different effective T_c values, neither of which is equal to the first-order transition temperature T_{NI}. For $T < T_{\mathrm{NI}}$, typical α_{eff} values lie in the range 0.3–0.4 but depend strongly on the fitting range. For $T > T_{\mathrm{NI}}$, α_{eff} values are even less well defined and lie between 0.1 and 0.5 [1, 2]. Furthermore, $(T_c^+ - T_{\mathrm{NI}})$ and $(T_c^- - T_{\mathrm{NI}})$ values obtained from C_p fits are consistently about one-tenth the magnitude of those obtained from several other properties [5, 11]. Thus our detailed understanding of the N–I transition is rather poor. The experimental behavior is well characterized, but more detailed theoretical modeling is still needed.

7.4.2 Blue phase transitions

Chiral liquid crystals (molecules with an asymmetric center that exhibit optical activity) give rise to a special chiral nematic phase (N*), also called cholesteric, in which the director precesses to yield a helical structure with a pitch substantially larger than the molecular dimensions. This helical pitch varies considerably with the molecular structure. When the pitch is

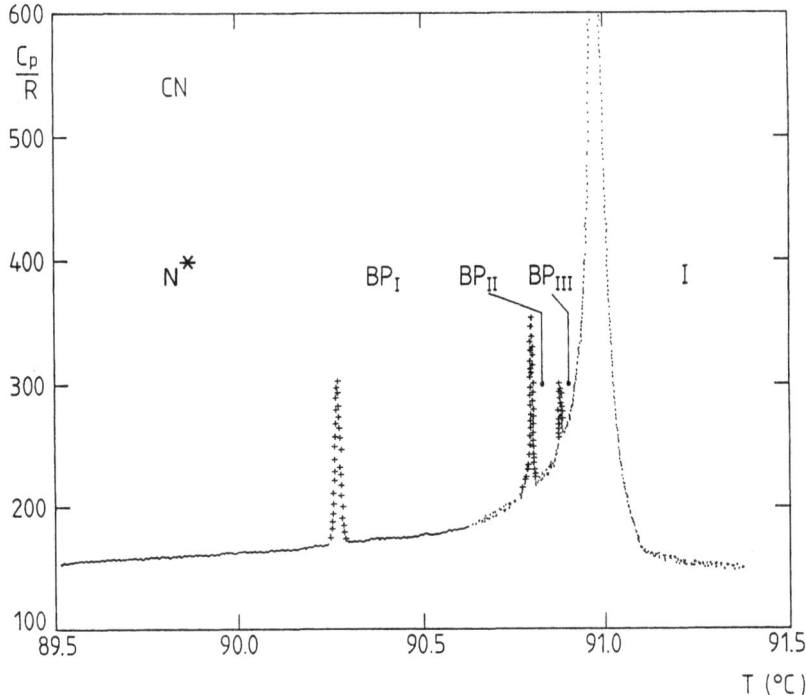

Figure 7.20. Variation of C_p/R for cholesteryl nonanoate [72]. All four transitions are first order, but there is a substantial pretransitional excess heat capacity associated with the BP_{III}–I transition. Points denoted by + are in two-phase coexistence regions and do not correspond to true C_p values. C_p in $J K^{-1} g^{-1}$ units is equal to 0.016 (C_p/R) for all other points.

quite long, there is a direct N*–I transition which is very similar in character to the N–I transition. When the pitch is sufficiently short (typically less than 0.5 μm), a set of intermediate blue phases (BP) are observed between the isotropic and chiral nematic phases. In order of increasing temperature, these phases are denoted BP_I, BP_{II}, and BP_{III}. The first two have complicated three-dimensional cubic structures of defects [6, 71], whereas BP_{III} (the 'fog phase') appears to be amorphous.

Phase transitions involving blue phases are expected theoretically to be first order except for the BP_{III}–I transition in which case a first-order line can terminate at an isolated critical point [60]. The results of a high-resolution study of CN with adiabatic scanning calorimetry [72] are shown in Fig. 7.20. The N*–BP_I, BP_I–BP_{II}, and BP_{II}–BP_{III} transitions are all first order with small latent heats and no pretransitional effects: $\Delta H = 34$ mJ g^{-1} for N*–BP_I, 11 mJ g^{-1} for BP_I–BP_{II}, 3.6 mJ g^{-1} for BP_{II}–BP_{III}. It is clear

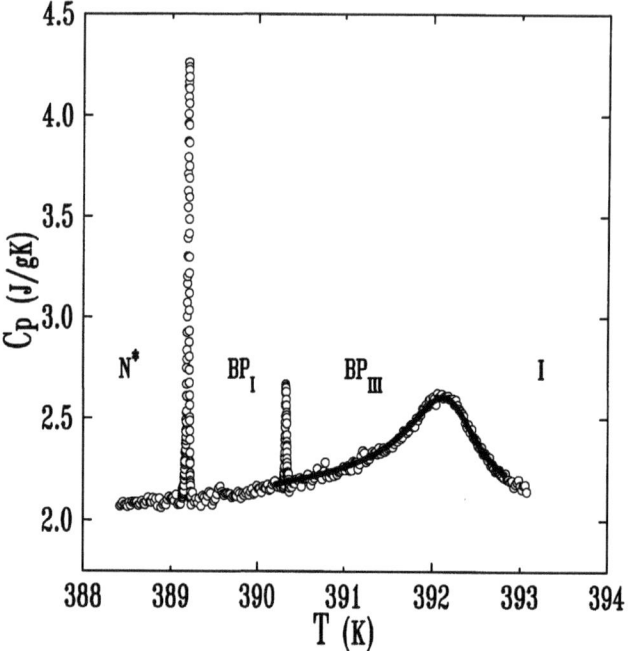

Figure 7.21. Heat capacity variations in the N*–BP$_I$–BP$_{III}$–I region for S,S-MBBPC [74a]. ○ non-adiabatic scanning data, — ac data. Note the sharp C_{eff} peaks from non-adiabatic scanning through the N*–BP$_I$ and BP$_I$–BP$_{III}$ regions due to latent heat effects at these first-order transitions. No such effect occurs at the BP$_{III}$–I 'transition', and this is a continuous supercritical evolution not a true transition in this material.

from Fig. 7.20 that there are substantial pretransitional heat capacity effects associated with the BP$_{III}$–I transition, which means that a large amount of energy is going into changing the local nematic order. Indeed, this BP$_{III}$–I transition in CN is quite similar to the N–I transition in non-chiral materials: there are large pretransitional C_p wings, there is two-phase coexistence over ~50 mK, and there is a latent heat ΔH (BP$_{III}$–I) of 323 ± 50 mJ g^{-1}.

Further calorimetric studies have shown that ΔH(BP$_{III}$–I) decreases as the chirality of the molecule increases and that the magnitude of ΔH(BP$_{III}$–I) varies considerably as a function of composition in binary mixtures of R- and S-enantiomers [73]. A recent study of S,S-MBBPC, which is highly chiral and has a very short pitch, yielded the results given in Fig. 7.21. Note that this compound exhibits BP$_I$ and BP$_{III}$ phase. By combining ac calorimetry and non-adiabatic scanning calorimetry, one can show that there are sharp first-order N*–BP$_I$ ($\Delta H = 56.4$ mJ g^{-1}) and BP$_I$–BP$_{III}$ ($\Delta H = 11.3$ mJ g^{-1}) transitions, but no thermodynamic BP$_{III}$–I transition occurs in the S,S-enantiomer of MBBPC [74]. Instead a super-critical evolution occurs. This implies that the BP$_{III}$ and I phases must have the same macroscopic symmetry so that a first-order transition line can terminate in an isolated critical point (analogous to the situation for the liquid–gas transition in simple fluids). An investigation of mixtures of S,S-MBBPC and its racemate has located the BP$_{III}$–I critical point and determined its critical behavior [74b].

7.4.3 Nematic–smectic-A transitions

The N–SmA transition has been the most extensively studied of all liquid crystal phase transitions. The close analogy between this transition and the normal–superconductor transition in metals was first pointed out in 1972 by de Gennes [6], and there has been great interest in the critical behavior at this deceptively simple kind of one-dimensional freezing. Considerable progress has been made, but this still remains one of the most challenging problems in the statistical mechanics of condensed matter. General discussions of the N–SmA problem are available in Refs. [1, 2, 6, 7]; detailed theoretical treatments are presented in Refs. [61, 62, 75], and a review of experimental results with systematic trends in the theoretical fitting parameters is given in Ref. [56].

The simplest model for the N–SmA transition would be the isotropic three-dimensional XY model, since the smectic order parameter $\psi = |\psi|e^{i\varphi}$, defined by $\rho(z) - \rho_0 = \rho_0 \psi e^{iq_0 z}$, has two components – a magnitude $|\psi|$ and a phase angle φ – just like superfluid helium. However, there are two important sources of deviations from isotropic three-dimensional XY behavior. The first is crossover from second-order to first-order behavior via a tricritical point due to coupling between the smectic order parameter ψ and the nematic order parameter Q. As formulated by de Gennes [6], the smectic free energy is given in the *mean-field approximation* by

$$G = G_N + a_0 t |\psi|^2 + b_0 |\psi|^4 + (\delta Q)^2 / 2\chi - C|\psi|^2 \delta Q, \qquad (7.29)$$

where a_0, b_0, and the coupling constant C are all positive. The quantity G_N is the nematic free energy, $\delta Q = Q - Q_0$ is the extra nematic order induced by the formation of smectic layers, and χ is the temperature-dependent nematic susceptibility whose value at T_{NA} depends on the width of the nematic range. Minimization of G with respect to δQ yields

$$G = G_N + a_0 t |\psi|^2 + b|\psi|^4 + c|\psi|^4, \tag{7.30}$$

with $b \equiv b_0 - C^2\chi/2$. For narrow nematic ranges, $\chi(T_{NA})$ is large and $b < 0$, resulting in a first-order transition. For wide nematic ranges, $\chi(T_{NA})$ is small and $b > 0$, which implies a second-order transition. The tricritical point corresponds to the case $b = 0$. However, even when $b > 0$ one does not expect XY critical behavior except for an asymptotic region very close to T_c. In practice, most experimental data lie in the reduced temperature range $10^{-5} < t < 10^{-2}$ and α_{eff} values are observed, where $\alpha_{XY} < \alpha_{eff} < \alpha_{TC}$. The second source of deviations from isotropic three-dimensional XY behavior arises from the coupling between director fluctuations δn and the smectic order parameter ψ [61, 75]. This coupling is intrinsically anisotropic and influences the behavior of the smectic susceptibility and correlation lengths $\xi_\parallel$ and $\xi_\perp$ much more than the heat capacity. Again there is a broad cross-over – in this case from isotropic XY to a weakly anisotropic regime $(\nu_\parallel > \nu_\perp)$ to the strong coupling limit with highly anisotropic correlation behavior $(\nu_\parallel = 2\nu_\perp)$. Crudely speaking, a narrow nematic range should strengthen this coupling and lead to large deviations from XY behavior, and a wide nematic range implies weaker coupling so that the system should straddle the isotropic and weakly anisotropic regimes.

The calorimetric data confirm the importance of crossover phenomena. Figure 7.18 shows a typical example of three-dimensional XY heat capacity behavior for materials with large nematic ranges $(T_{NA}/T_{NA} \lesssim 0.93)$. Figures 7.6 and 7.16 show $C_p(NA)$ data for nCB compounds with smaller nematic ranges. Analysis of 8CB, where $T_{NA}/T_{NI} = 0.977$, has yielded $\alpha_{eff} = 0.31 \pm 0.03$ [19]. Figure 7.10 provides an example of tricritical behavior in a system with $T_{NA}/T_{NI} = 0.978$. Note that tricritical systems do not exhibit a universal value of T_{NA}/T_{NI}; indeed, tricritical values of this so-called McMillan ratio range from 0.942 to 0.994 for different homologous series [1, 2, 55a]. The evolution from second order to tricritical to first order is illustrated by the behavior of 4O.8 + 6O.8 mixtures [31]. The phase diagram in Fig. 7.22(a) shows that the nematic range is diminished by adding 6O.8 to 4O.8. Pure 4O.8 has a nematic range of 14.72 K $(T_{NA}/T_{NA} = 0.958)$, and $\Delta C_p(NA)$ is well described by $\alpha_{eff} = 0.13$. Pure 6O.8 has a nematic range of only 0.87 K $(T_{NA}/T_{NA} = 0.998)$, and the N–A transition is strongly first order with a latent heat $\Delta H = 3700 \, \text{J mol}^{-1}$. Figure 22(b) provides an overview of the evolution in $\Delta C_p(NA)$ as the mole fraction of 6O.8 is increased from $X_1 = 0$ (pure 4O.8) to $X_7 = 0.50$. A detailed view of the tricritical mixture with $X_5 = 0.35$ is given in Fig. 7.10. Fits to the data in Fig. 7.22(b) yield the following α_{eff} values: 0.13, 0.22, 0.30, 0.45, 0.50

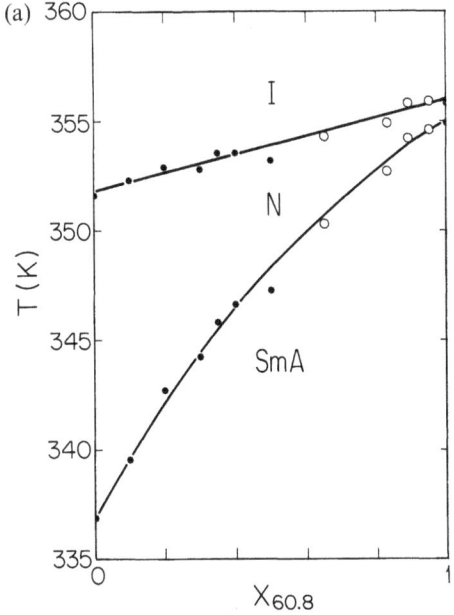

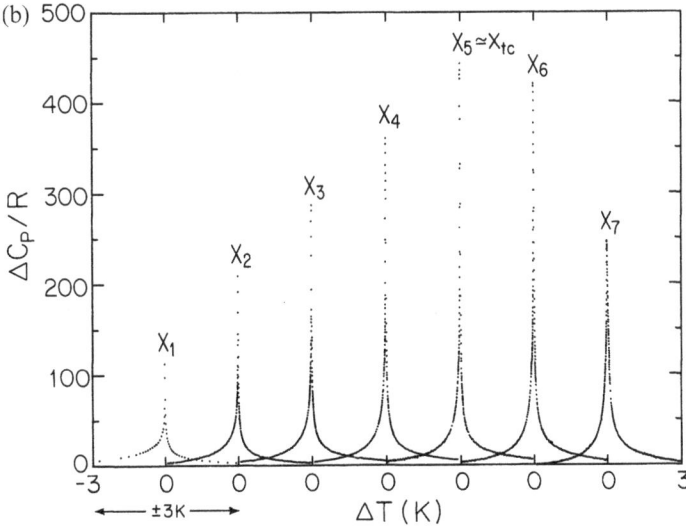

Figure 7.22.   (a) Phase diagram for 4O.8 + 6O.8 mixtures. (b) Excess heat capacity
$\Delta C_p/R$, where ΔC_p corresponds to the molar quantity [31]. ΔC_p values in units of
$J K^{-1} g^{-1}$ are obtained by multiplying $\Delta C_p/R$ by $R/\bar{M} \approx 0.022$, where $\bar{M}$ is the
'average' molecular weight of each mixture in grams. Data are shown over the range
$-3 K < T - T_c < 3 K$. $X_{6O.8}$ values are $X_1 = 0$, $X_2 = 0.10$, $X_3 = 0.20$, $X_4 = 0.30$,
$X_5 = 0.35$, $X_6 = 0.40$, $X_7 = 0.50$.

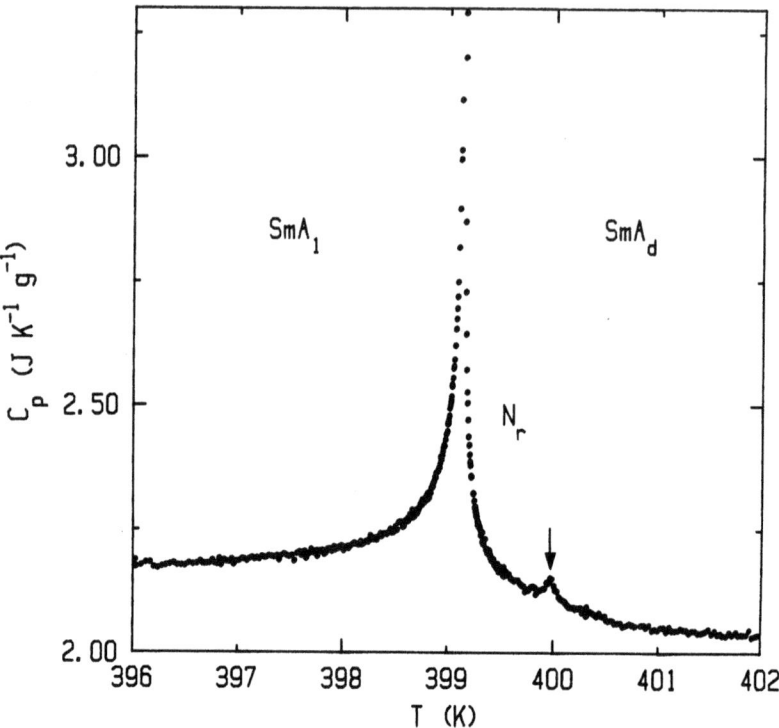

Figure 7.23.   Heat capacity in the SmA_d–N_r–SmA_1 transition region for a $DB_8ONO_2 + DB_{10}ONO_2$ mixture with $X_{10} = 0.5133$ [33]. Note the very small SmA_d–N_r peak marked by the arrow.

for X_1 to X_5 respectively. Mixtures with X_6 and X_7 are weakly first order with latent heats ΔH and pretransitional enthalpies δH given (in $J\,mol^{-1}$) by $\Delta H = 110$ and $\delta H = 1020$ for $X_6 = 0.40$ and by $\Delta H = 420$ and $\delta H = 845$ for $X_7 = 0.50$.

Another, more qualitative, application of calorimetry to the study of N–SmA systems involves heat capacity measurements on a binary mixture of $DB_8ONO_2 + DB_{10}ONO_2$ with 51.33 mole percent of the decyl homolog [33]. This system is a so-called frustrated smectic that can exhibit more than one SmA phase. As shown by Fig. 7.23, the 51.33% mixture undergoes the SmA_d–N_r–SmA_1 sequence in the vicinity of 399–400 K. The N_r–SmA_1 peak is quite large while the N_r–SmA_d peak is very small. The dramatic differ-ence in the integrated area of these two peaks is in excellent agreement with semi-quantitative spin-gas model predictions [76]. Further details about this behavior are given in Refs. [4] and [33].

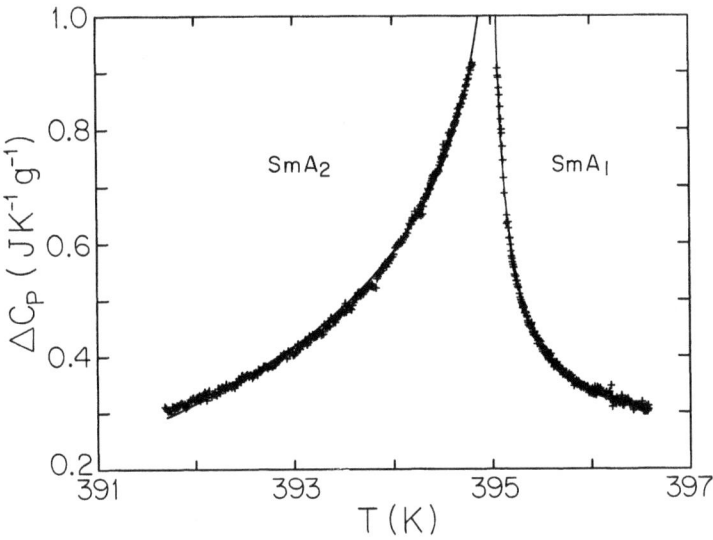

Figure 7.24. Excess heat capacity associated with the SmA_1–SmA_2 transition in a $DB_5CN + TB4A$ mixture with $X_{TB4A} = 0.1214$ [76]. The smooth curve is a theoretical fit with $\alpha = -0.124$, the Fisher renormalized Ising value α_R(Ising).

7.4.4 SmA–SmA transitions

The discussion of 'frustrated smectics', systems that exhibit SmA (and SmC) polymorphism due to the presence of *two* smectic order parameters with incommensurate wavevectors, will be quite brief. A detailed account of critical behavior in such liquid crystals is given in Ref. 4.

The SmA_1–SmA_2 transition is the classic example of the new phase transitions observed in frustrated smectics. This transition was predicted theoretically to be second order belonging to the three-dimensional Ising universality class [66]. Heat capacity and x-ray studies of $DB_nCN + TB4A$ [54, 76] show that Fisher renormalized Ising behavior is actually observed; see Fig. 7.24 for C_p data on a typical mixture. Thus agreement between theory and experiment is excellent since $(dT_c/dX)^2$ is large in these systems and Fisher renormalization is therefore expected.

Instead of a direct SmA_1–SmA_2 transition, frustrated smectics can also exhibit the sequence SmA_1–$Sm\tilde{A}$–SmA_2 involving an intermediate fluid antiphase. An example of this behavior is provided by $DB_5CN + C_5$ stilbene mixtures, and the overall heat capacity variation for such a mixture is given in Fig. 7.25. The N-SmA_1 heat capacity variation shown here is similar to that shown in Fig. 7.18 and is well described in terms of XY critical behavior. Note, however, the great difference between $C_p(A_1$–$A_2)$ in Fig.

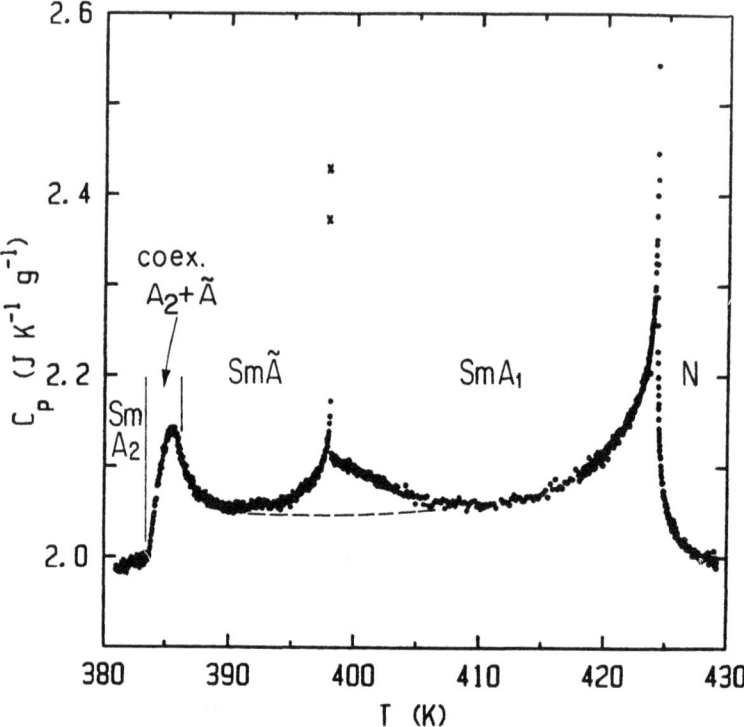

Figure 7.25.   Heat capacity of $DB_5CN + 49.2\%$ C_5 stilbene [78]. The two points marked with crosses indicate coexistence of two phases at a weakly first-order SmA_1–$Sm\tilde{A}$ transition. The dashed curve represents C_p(background) for the SmA_1–$Sm\tilde{A}$ region.

7.24 and complex C_p behavior below 410 K in Fig. 7.25. The $Sm\tilde{A}$–SmA_2 transition is first order but involves an unusually wide coexistence region that was previously ascribed to a new phase, SmA_{cren}. Recent x-ray studies [77] have established the detailed nature of this $Sm\tilde{A} + SmA_2$ coexistence and provided some information about the complex nature of the SmA_1–$Sm\tilde{A}$ transition. A detailed view of the excess heat capacity ΔC_p associated with this SmA_1–$Sm\tilde{A}$ transition is given in Fig. 7.26 for two closely related mixtures. This transition is predicted to be first order due to a fluctuation-induced Brazovskii instability, and indeed two-phase coexistence is observed over a very narrow temperature range [77, 78]. However, there are large and unusual pretransitional fluctuation effects in the SmA_1 phase that are confirmed by the x-ray study.

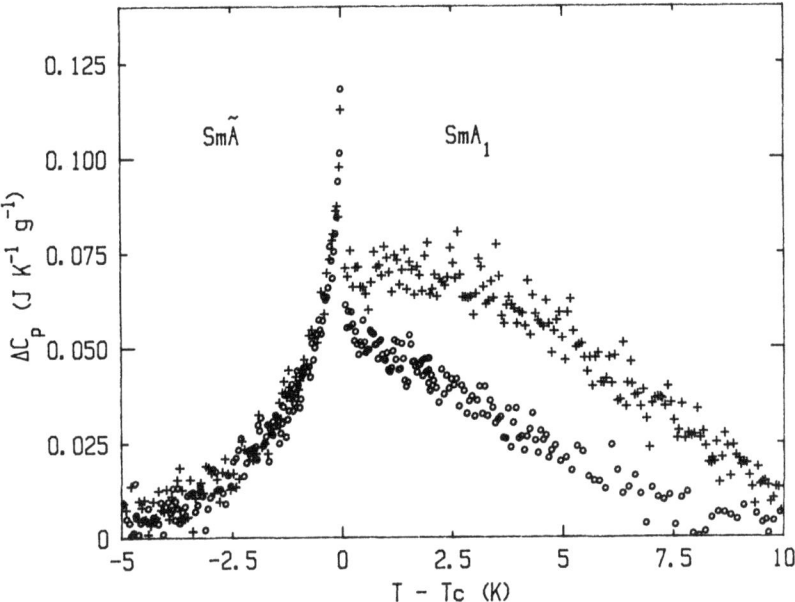

Figure 7.26. $\Delta C_p = C_p - C_p(\text{background})$ associated with the SmA$_1$–Sm$\tilde{\text{A}}$ transition in two liquid-crystal mixtures [78]. Anomalous data points obtained in the coexistence region are not included. Open circles are used for DB$_5$CN + C$_5$ stilbene and plus signs for DB$_6$CN + C$_5$ stilbene.

7.4.5 SmA–SmC transitions

Since the tilt order parameter $\psi = \theta e^{i\phi}$, where θ is the tilt angle and ϕ is the azimuthal director orientation, has two components, one would expect that second-order SmA–SmC transitions should belong to the three-dimensional XY universality class (i.e., exhibit critical behavior like that at the lambda transition in helium). However, SmA$_m$–SmC$_m$ transitions in nonpolar non-chiral materials are second-order transitions that are very well described by Landau theory with a large sixth-order coefficient c (see Section 7.3.1). This is illustrated by the heat capacity data shown in Fig. 7.27 for 2M45OBC [79], which undergoes a typical SmA$_m$–SmC$_m$ transition. The explanation for this mean-field behavior is the fact that the Ginzburg condition indicates that the critical region is extremely small ($t_{\text{crit}} < 10^{-5}$). In contrast to this SmA$_m$–SmC$_m$ behavior, the SmA$_2$–SmC$_2$ heat capacity data are well described by Landau theory with $c \approx 0$ (i.e., a simple step). Figure 7.28 illustrates both behaviors and shows that the magnitude and sharpness of Landau $\Delta C_p(\text{AC})$ peaks varies greatly from one

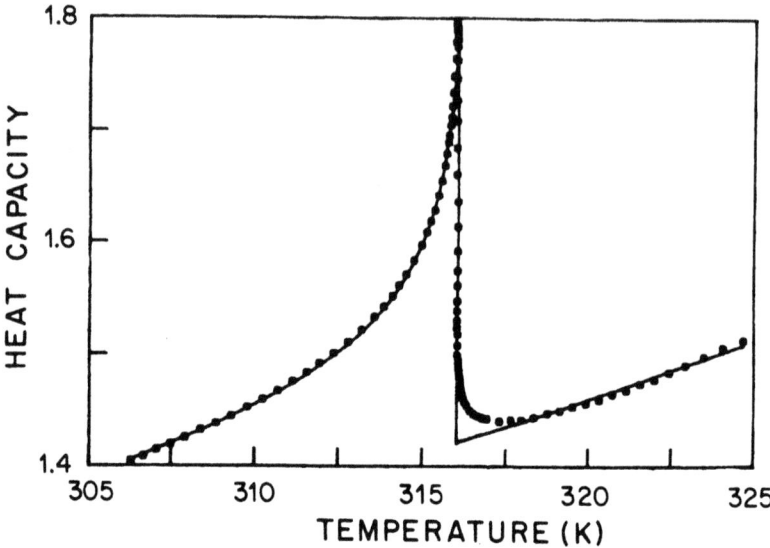

Figure 7.27.   Heat capacity (in units of $J\,K^{-1}\,cm^{-3}$) near the SmA_m–SmC_m transition in racemic 2M45OBC [79]. These data are typical of those for non-polar non-chiral liquid crystals, and the solid line represents a fit with the Landau expression given in Eq. (7.22).

material to another. There have been recent reports of critical energy fluctuations at the SmA–SmC transition in bulk and thick film samples [81]. In a partially perfluorinated compound, the data were fit with a mean-field model including terms characteristic of Gaussian fluctuations [81a]. In contrast to this, calorimetric studies of the SmA–SmC$_\alpha^*$ transition in three antiferroelectric liquid crystals show an XY-like heat capacity near T_c with indications of crossover behavior as the reduced temperature range is increased [81b].

In the case of chiral compounds, there is a chiral smectic–C (SmC*) phase with a helical precession of the direction of the director tilt with respect to the layer normal. Such chiral compounds can exhibit strongly first-order SmA–SmC* transitions as indicated by the behavior of C7 shown in Fig. 7.11. By varying the composition of a binary mixture one can 'tune' the value of the Landau coefficient b from negative to positive. The heat capacity of a tricritical C7 + $\overline{10}\,O\overline{4}$ mixture (for which $b = 0$) is shown in Fig. 7.29. Note the striking qualitative difference between this mean-field tricritical behavior and the Gaussian tricritical fluctuation-dominated behavior shown in Fig. 7.10.

The N–SmA–SmC multicritical point is treated in detail in Refs. 1 and 2

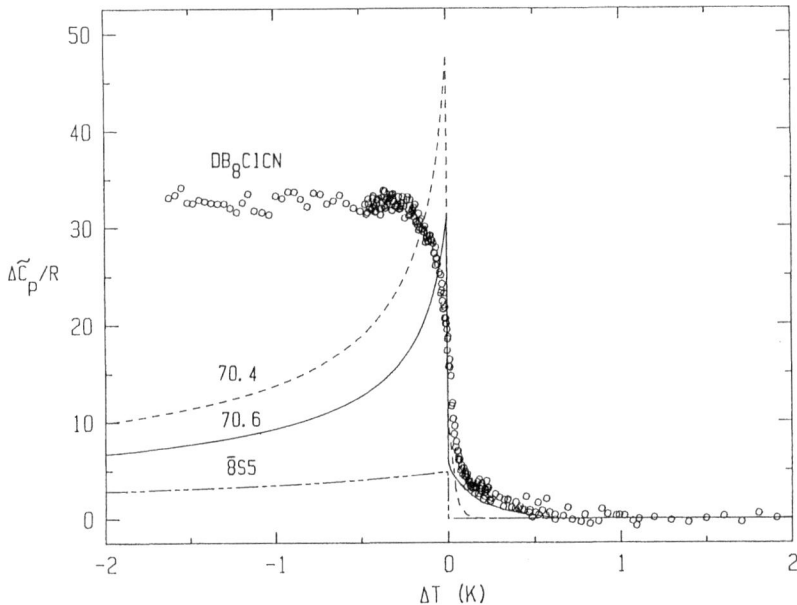

Figure 7.28.   Excess heat capacity in the vicinity of the SmA$_2$–SmC$_2$ transition for the bilayer material DB$_8$ClCN [80] compared with the ΔC_p variation at SmA$_m$–SmC$_m$ transitions in 7O.4, 7O.6, and $\overline{8}$S5.

and will not be discussed here. It should, however, be noted that this special point, which has the simultaneous character of a mean-field tricritical point (as determined calorimetrically) and a Lifshitz point (as determined from x-rays), is not yet well understood theoretically. When chiral compounds are involved, the phase diagram in the vicinity of an expected N*–SmA–SmC* point is significantly more complicated. Theory [82] has predicted the existence of several kinds of twist-grain-boundary (TGB) phases – TBG$_A$ and TGB$_C$ – and these have been confirmed. Recent calorimetric studies with both ac and non-adiabatic scanning techniques [83] provide interesting new results that shed light on phase transitions involving the TGB$_A$ and tilted TGB$_C$ phases and also the chiral line liquid N$_L^*$. TGB$_A$ and tilted TGB$_C$ phases are the liquid crystal analogs of the Abrikosov flux vortex lattice, and N$_L^*$ is the analog of the Abrikosov vortex liquid.

7.4.6 Smectic–hexatic transitions

Hexatic liquid crystal phases, which are smectic phases exhibiting long range bond-orientational (BO) order but only short range in-plane

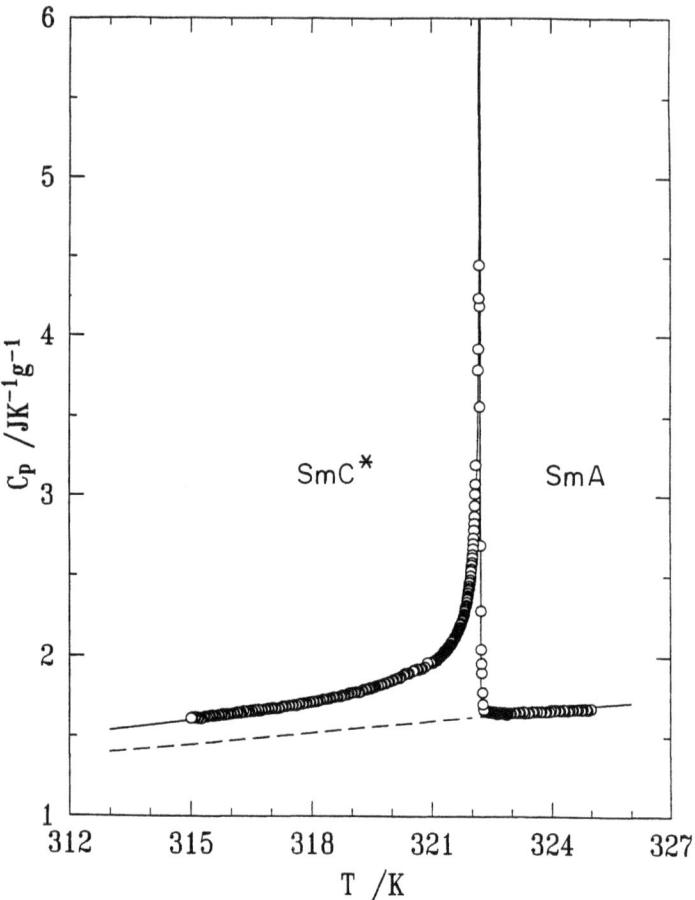

Figure 7.29. Heat capacity in a near tricritical C7 + $\overline{1}0O\overline{4}$ mixture with 11.1 mole percent $\overline{1}0O\overline{4}$ [32]. The solid curve is a Landau mean-field tricritical fit to these data, and the dashed line presents the background variation $C_p^o(T)$.

positional order, are not yet well understood. Since the BO order parameter $\Psi = |\Psi|\exp(i6\Psi)$ has XY symmetry, one would expect $\alpha = \alpha_{XY} = -0.007$ for the critical heat capacity exponent. The results for 65OBC shown in Fig. 7.30 are typical of all SmA–HexB transitions, and these data yield the large value $\alpha = 0.60$ [84]. An extensive series of investigations by Huang and coworkers on nmOBC and other compounds having a HexB phase [84–89] show that all SmA–HexB transitions appear to be second order with α values lying in the range 0.48–0.67, although in some cases the C_p peak is rounded over an interval of 50–200 mK for reasons that are not clear. The idea that this transition is second order but nearly tricritical (recall that

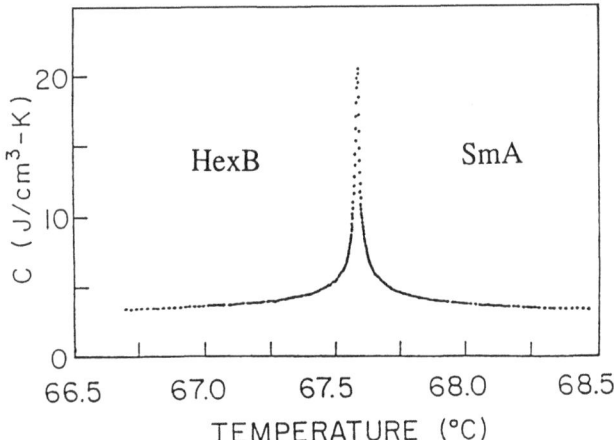

Figure 7.30.   Temperature dependence of the heat capacity per unit volume for
bulk 65OBC near its SmA–HexB transition [84].

$\alpha_{TC} = 0.5$) is not consistent with the existence of a continuous line of
SmA–HexB critical points with $0.5 < \alpha < 0.65$ [5, 6]. A more recent very
detailed high-resolution calorimetric study of 65OBC [90] shows that the
SmA–HexB transition is very weakly first order and suggests that the effective critical exponents in *nm*OBC compounds are consistent with *quasicrit-
ical* behavior (strain-smeared weak first order).

In addition to studies of bulk SmA–HexB transitions, Huang *et al.* have
carried out brilliant measurements on free-standing films of thickness from
~100 layers down to two layers. See the data in Fig. 7.12 for thin films of
3(10)OBC as an example. Theoretical models have proposed the importance of coupling Ψ to herringbone order [91a], crystalline order [91b], or
the layer displacement [91c]. The applicability of these models to bulk
SmA–HexB transitions is unclear [90], but a recent two-dimensional Monte
Carlo simulation of an XY model with coupled hexatic and herringbone
order [92] yields a new type of phase transition that appears to be compat-
ible with the critical behavior observed for two-layer films [93].

Tilted hexatic phases, SmF and SmI, combine BO order Ψ with a long
range tilt of the director with respect to the smectic layer normal where Ψ
is coupled to the tilt angle via a term of the type $h |\theta|^{6}\Psi$. In SmI the tilt is
toward a nearest neighbor, and in SmF the tilt direction is between two
neighbors. The presence of BO order in SmI(SmF) does not, however,
change the point group symmetry from that of SmC. Thus the situation is
qualitatively analogous to a liquid–gas transition in a simple fluid. As
a function of composition in a mixture, there can be a first-order

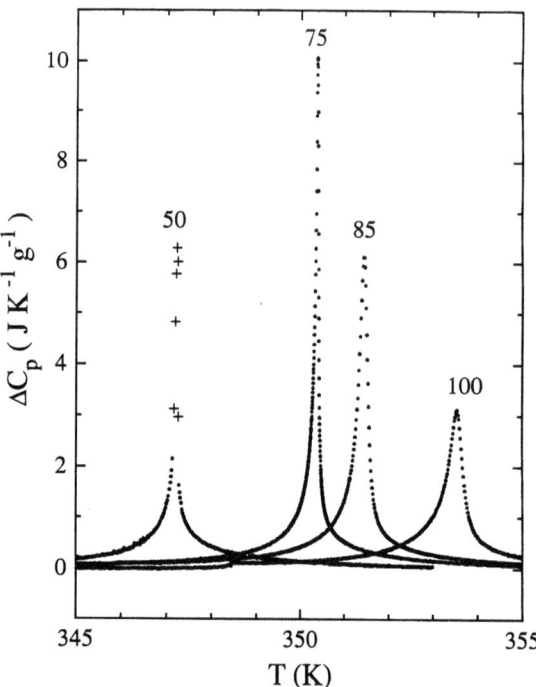

Figure 7.31. Excess heat capacity ΔC_p associated with the SmC–SmI transition in 8SI + 8OSI mixtures with weight percent 8OSI concentrations $X = 50$, 75, 85, 100 [40a]. These data were obtained with the ac technique at $\omega_o = 0.916$. The plus symbols mark two-phase coexistence data points for the first-order $X = 50$ sample. The critical composition is very close to $X = 75$.

SmC–SmI(SmF) transition which terminates at an isolated critical point, beyond which there is no thermodynamic phase transition but only a super-critical evolution of both BO order and tilt in a single phase that is conventionally called 'SmC' at high temperatures and 'SmI(SmF)' at lower temperatures [70]. This critical point is predicted to belong to the same new universality class as SmA_d–SmA_2 critical points [67, 70].

The SmC-SmI critical point has been realized in mixtures of 8SI + 8OSI [40a]. The SmC–SmI transition in 8SI is strongly first order, and that in 8OSI is supercritical. Figure 7.31 shows the ΔC_p(SmC–SmI) variation for three 8SI + 8OSI mixtures and pure 8OSI. Analysis of the critical $X = 75$ data is complicated by issues concerning the path of approach to the critical point and a critical frequency dependence of ΔC_p very close to T_c but the final conclusion is that static $\Delta C_p(0)$ data vary like $t^{-\gamma}$ in this system with $\gamma = \gamma_{MF} = 1$ [40a].

Acknowledgments

This work was supported in part by National Science Foundation grants DMR-9311853 and DMR-9400334 (MRSEC). The author wishes to thank J. Thoen for many helpful discussions and for a critical reading of this manuscript.

References

1. J. Thoen, Thermal Investigations of Phase Transitions in Thermotropic Liquid Crystals, *Int. J. Mod. Phys.* **B9**, 2157 (1995). Among many other things this contains an exceptionally detailed bibliography.
2. J. Thoen, Calorimetric Studies of Liquid Crystal Phase Transitions: Steady State Adiabatic Techniques, in *Phase Transitions in Liquid Crystals*, S. Martellucci and A. N. Chester (Eds.), NATO ASI Ser. B290, Chapter 10, Plenum, New York (1992).
3. C. W. Garland, Calorimetric Studies of Liquid Crystal Phase Transitions: ac Techniques, in *Phase Transitions in Liquid Crystals*, S. Martellucci and A. N. Chester (Eds.), NATO ASI Ser. B290, Chapter 11, Plenum, New York (1992).
4. C. W. Garland, Critical Behavior of Polymorphic Smectic-A Liquid Crystals, in *Geometry and Thermodynamics*, J.-C. Tolédano (Ed.), NATO ASI Ser. B229, pp. 221–254, Plenum, New York, (1990).
5. M. A. Anisimov, *Critical Phenomena in Liquids and Liquid Crystals*, Chapter 10, Gordon and Breach, Philadelphia (1991).
6. P. G. de Gennes and J. Prost, *The Physics of Liquid Crystals*, 2nd Edn., Clarendon Press, Oxford (1993).
7. P. M. Chaikin and T. C. Lubensky, *Principles of Condensed Matter Physics*, Cambridge University Press, Cambridge (1995).
8. G. Vertogen and W. deJeu, *Thermotropic Liquid Crystals, Fundamentals*, Springer-Verlag, Berlin (1988).
9. P. Barois, J. Pommier, and J. Prost, Frustrated Smectics, in *Solitons in Liquid Crystals*, L. Lam and J. Prost (Eds.), Chapter 6, Springer-Verlag, New York (1992).
10. M. A. Anisimov, A. V. Voronel, and T. M. Ovodova, *Sov. Phys. JETP* **35**, 356 (1972); M. Sorai, S. Asahina, C. Destrade, and N. H. Tinh, *Liq. Cryst.* **7**, 163 (1990).
11. M. A. Anisimov, *Mol. Cryst. Liq. Cryst.* **162A**, 1 (1988) and references cited therein.
12. V. N. Raja, R. Shashidhar, B. R. Ratna, G. Heppke, and Ch. Bahr, *Phys. Rev. A* **37**, 303 (1988).
13. G. Nounesis, C. W. Garland, and R. Shashidhar, *Phys. Rev. A* **43**, 1849 (1991).
14. G. Nounesis, S. Kumar, S. Pfeiffer, R. Shashidhar, and C. W. Garland, *Phys. Rev. Lett.* **73**, 565 (1994).
15. T. Chan, Ph.D. thesis in Physics, Massachusetts Institute of Technology (1995).
16. P. F. Sullivan and G. Seidel, *Phys. Rev.* **173**, 679 (1968); J. D. Baloga and C. W. Garland, *Rev. Sci. Instrum.* **48**, 105 (1977); J. M. Viner, D. Lamey, C. C. Huang, R. Pindak, and J. W. Goodby, *Phys. Rev. A* **28**, 2433 (1983).
17. J. A. Lipa, C. Edwards, and M. J. Buckingham, *Phys. Rev. Lett.* **25**, 1086 (1970); *Phys. Rev. A* **15**, 778 (1977).
18. E. Bloemen, J. Thoen, and W. Van Dael, *J. Chem. Phys.* **73**, 4628 (1980).

19. J. Thoen, H. Marijnissen, and W. Van Dael, *Phys. Rev. A* **26**, 2886 (1982).
20. J. Thoen, E. Bloemen, H. Marijnissen, and W. Van Dael, *Proceedings of the 8th Symposium on Thermophysical Properties*, Nat. Bur. Stand. Maryland, 1981 (Am. Soc. Mech. Eng., New York, 1982); see also M. A. Anisimov, V. P. Voronov, A. O. Kulkov, and F. Kholmurodov, *J. Phys.* (Paris) **46**, 2137 (1985).
21. Y. A. Kraftmaker, *Zh. Prikl. Mekh. Tekh. Fiz.* **5**, 176 (1962).
22. P. F. Sullivan and G. Seidel, *Phys. Rev.* **173**, 679 (1968).
23. C. A. Schantz and D. L. Johnson, *Phys. Rev. A* **17**, 1504 (1978).
24. J. M. Viner, D. Lamey, C. C. Huang, R. Pindak, and J. W. Goodby, *Phys. Rev. A* **28**, 2433 (1983); C. C. Huang, J. M. Viner, and J. C. Novak, *Rev. Sci. Instrum.* **56**, 1390 (1985).
25. R. Geer, T. Stoebe, T. Pitchford, and C. C. Huang, *Rev. Sci. Instrum.* **62**, 415 (1991).
26. G. B. Kasting, K. J. Lushington, and C. W. Garland, *Phys. Rev. B* **22**, 321 (1980).
27. C. W. Garland, *Thermochim. Acta* **88**, 127 (1985).
28. I. Hatta and A. J. Ikushima, *Jpn. J. Appl. Phys.* **20**, 1995 (1981); M. Castro and J. A. Puertolas, *J. Therm. Anal.* **41**, 1245 (1994).
29. J. D. Baloga and C. W. Garland, *Rev. Sci. Instrum.* **48**, 105 (1977).
30. K. J. Stine, Ph.D. thesis in Chemistry, Massachusetts Institute of Technology (1988).
31. K. J. Stine and C. W. Garland, *Phys. Rev. A* **39**, 3148 (1989).
32. T. Chan, Ch. Bahr, G. Heppke, and C. W. Garland, *Liq. Cryst.* **13**, 667 (1993).
33. K. Ema, G. Nounesis, C. W. Garland, and R. Shashidhar, *Phys. Rev. A* **39**, 2599 (1989).
34. K. J. Stine and C. W. Garland, *Mol. Cryst. Liq. Cryst.* **188**, 91 (1990).
35. X. Wen, C. W. Garland, and M. D. Wand, *Phys. Rev. A* **42**, 6087 (1990).
36. T. Stoebe, I. M. Jiang, S. N. Huang, A. J. Jin, and C. C. Huang, *Physica A* **205**, 108 (1994).
37. D. Djurek, J. Baturic-Rubcic, and K. Franulovic, *Phys. Rev. Let.* **33**, 1126 (1974).
38. K. Ema, T. Uematsu, A. Sugata, and H. Yao, *Jpn. J. Appl. Phys.* **32**, 1846 (1993) and references cited therein.
39. K. Ema, H. Yao, I. Kawamura, T. Chan, and C. W. Garland, *Phys. Rev. E* **47**, 1203 (1993).
40. (a) H. Yao, T. Chan, and C. W. Garland, *Phys. Rev. E* **51**, 4585 (1995); (b) H. Yao, K. Ema, and C. W. Garland, *Rev. Sci Instrum.* **69**, 172 (1998).
41. S. T. Shin, S. Kumar, D. Finotello, S. S. Keast, and M. E. Neubert, *Phys. Rev. A* **45**, 8683 (1992).
42. A Rosencwaig, *Photoacoustics and Photoacoustic Spectroscopy*, John Wiley & Sons, New York (1980).
43. U. Zammit, M. Marinelli, F. Scudieri, and F. Mercuri, Photothermal Calorimetry: Simultaneous Measurement of Specific Heat and Thermal Conductivity in *Phase Transitions in Liquid Crystals*, S. Martellucci and A. N. Chester (Eds.), NATO ASI Ser. B290, Chapter 12, Plenum, New York (1992).
44. (a) M. Marinelli, U. Zammit, F. Scudieri, S. Martellucci, J. Quartieri, F. Bloisi, and L. Vicari, *Nuovo Cimento D* **9**, 557 (1987); (b) U. Zammit, M. Marinelli, R. Pizzoferrato, F. Scudieri, and S. Martellucci, *Phys. Rev. A* **41**, 1153 (1990).
45. C. Glorieux, E. Schoubs, and J. Thoen, *Mater. Sci. and Eng. A* **122**, 87

(1989); J. Thoen, C. Glorieux, E. Schoubs, and W. Lauriks, *Mol. Cryst. Liq. Cryst.* **191**, 29 (1990).

46. N. O. Birge, *Phys. Rev. B* **34**, 1631 (1986); N. O. Birge and S. R. Nagel, *Rev. Sci. Instrum.* **58**, 1464 (1987); D. H. Jung, T. W. Kwon, D. J. Bae, I. K. Moon, and Y. H. Jeong, *Meas. Sci. Techn.* **3**, 475 (1992).

47. X. Wen, C. W. Garland, R. Shashidhar, and P. Barois, *Phys. Rev. B* **45**, 5131 (1992).

48. C. Bagnuls and C. Bervillier, *Phys. Rev. B* **32**, 7209 (1985); C. Bagnuls, C. Bervillier, D. I. Meiron, and B. G. Nickel, *ibid.* **35**, 3585 (1987).

49. C. W. Garland, G. Nounesis, and K. J. Stine, *Phys. Rev. A* **39**, 4919 (1989).

50. (a) G. Nounesis, K. I. Blum, M. J. Young, C. W. Garland, and R. J. Birgeneau, *Phys. Rev. E* **47**, 1910 (1993); (b) C. W. Garland, G. Nounesis, M. J. Young, and R. J. Birgeneau, *ibid.* **47**, 1918 (1993) and references cited therein.

51. K. J. Lushington, G. B. Kasting, and C. W. Garland, *Phys. Rev. B* **22**, 2569 (1980); A. R. Kortan, H. von Känel, R. J. Birgeneau, and J. D. Litster, *J. Phys. (Paris)* **45**, 529 (1984).

52. G. Sigaud, F. Hardouin, M. F. Achard, and H. Gasparoux, *J. Phys. (Paris)* **40**, C3-356 (1979).

53. M. E. Fisher, *Phys. Rev.* **176**, 257 (1968); M. E. Fisher and P. E. Scesney, *Phys. Rev. A* **2**, 825 (1970).

54. C. W. Garland, C. Chiang, and F. Hardouin, *Liq. Cryst.* **1**, 81 (1986).

55. (a) M. E. Huster, K. J. Stine, and C. W. Garland, *Phys. Rev. A* **36**, 2364 (1987); (b) J. P. Hill, B. Keimer, K. W. Evans-Lutterodt, R. J. Birgeneau, and C. W. Garland, *Phys. Rev. A* **40**, 4625 (1989).

56. C. W. Garland and G. Nounesis, *Phys. Rev. E* **49**, 2964 (1994).

57. V. I. Ginzburg, *Fiz Tverd. Tela* **2**, 2031 (1960) [*Sov. Phys. Solid State* **2**, 1824 (1961)].

58. S. Chandrasekhar, *Liquid Crystals*, 2nd Edn., Chapter 6, Cambridge University Press, Cambridge (1992).

59. R. G. Priest and T. C. Lubensky, *Phys. Rev. B* **13**, 4159 (1976).

60. T. C. Lubensky and H. Stark, *Phys. Rev. E* **53**, 714 (1996).

61. T. C. Lubensky, *J. Chim. Phys.* **80**, 31 (1983) and references cited therein.

62. J. Prost, *Adv. Phys.* **33**, 1 (1984).

63. J. Prost and P. Barois, *J. Chim. Phys.* **80**, 65 (1983).

64. J. Swift, *Phys. Rev. A* **14**, 2274 (1976).

65. G. Grinstein, T. C. Lubensky, and J. Toner, *Phys. Rev. B* **33**, 3306 (1986) and references cited therein.

66. J. Prost, *J. Phys. (Paris)* **40**, 581 (1979); J. Wang and T. C. Lubensky, *Phys. Rev. A* **29**, 2210 (1984).

67. Y. Park, T. C. Lubensky, P. Barois, and J. Prost, *Phys. Rev. A* **37**, 2197 (1988); Y. Park, T. C. Lubensky, and J. Prost, *Liq. Cryst.* **4**, 435 (1989).

68. J. Prost and J. Toner, *Phys. Rev. A* **36**, 5008 (1987).

69. (a) R. Bruinsma and G. Aeppli, *Phys. Rev. Lett.* **48**, 1625 (1982); (b) A. Aharony, R. J. Birgeneau, J. D. Brock, and J. D. Litster, *Phys. Rev. Lett.* **57**, 1012 (1986); (c) J. V. Selinger, *J. Phys. (Paris)* **49**, 1387 (1988).

70. A. D. Défontaines and J. Prost, *Phys. Rev. E* **47**, 1184 (1993).

71. H. Stegemeyer, Th. Blümel, K. Hiltrop, H. Onusseit, and F. Prosch, *Liq. Cryst.* **1**, 3 (1986) and references cited therein.

72. J. Thoen, *Phys. Rev. A* **37**, 1754 (1988).

73. G. Voets and W. Van Dael, *Liq. Cryst.* **14**, 617 (1993); G. Voets, Ph.D. thesis, Katholieke Univ., Leuven, Belgium (1993).

74. (a) Z. Kutnjak, C. W. Garland, J. L. Passmore, and P. J. Collings, *Phys. Rev. Lett.* **74**, 4859 (1995); (b) Z. Kutnjak, C. W. Garland, C. G. Schatz, P. J. Collings, C. J. Booth, and J. W. Goodby, *Phys. Rev. E* **53**, 4955 (1996).
75. B. R. Patton and B. S. Andereck, *Phys. Rev. Lett.* **69**, 1556 (1992); B. S. Andereck and B. R. Patton, *Phys. Rev. E* **49**, 1393 (1994).
76. P. Das, G. Nounesis, C. W. Garland, G. Sigaud, and N. H. Tinh, *Liq. Cryst.* **7**, 882 (1990).
77. M. J. Young, L. Wu, G. Nounesis, C. W. Garland, and R. J. Birgeneau, *Phys. Rev. E* **50**, 368 (1994).
78. K. Ema, C. W. Garland, G. Sigaud, and N. H. Tinh, *Phys. Rev. A* **39**, 1369 (1989).
79. S. C. Lien, C. C. Huang, and J. W. Goodby, *Phys. Rev. A* **29**, 1371 (1984).
80. Y. H. Jeong, K. J. Stine, C. W. Garland, and N. H. Tinh, *Phys. Rev. A* **37**, 3465 (1988).
81. (a) L. Reed, T. Stoebe, and C. C. Huang, *Phys. Rev. E* **52**, R2157 (1995); (b) K. Ema, M. Ogawa, A. Takagi, and H. Yao, *Phys. Rev. E* **54**, R25 (1996) and K. Ema, H. Yao, A. Fukuda, Y. Takanishi, and H. Takezoe, *Phys. Rev. E* **54**, 4450 (1996).
82. S. R. Renn and T. C. Lubensky, *Phys. Rev. A* **38**, 2132 (1988) and *Mol. Cryst. Liq. Cryst.* **209**, 349 (1991).
83. T. Chan, C. W. Garland, and H. T. Nguyen, *Phys. Rev. E* **52**, 5000 (1995); L. Navailles, C. W. Garland, and H. T. Nguyen, *J. Phys. II (France)* **6**, 1243 (1996).
84. G. Nounesis, C. C. Huang, and J. W. Goodby, *Phys. Rev. Lett.* **56**, 1712 (1986).
85. C. C. Huang and T. Stoebe, *Adv. Phys.* **42**, 343 (1993) and references cited therein.
86. T. Pitchford, G. Nounesis, S. Dumrongrattana, J. M. Viner, and C. C. Huang, *Phys. Rev. A* **32**, 1938 (1985).
87. C. C. Huang, G. Nounesis, and D. Guillon, *Phys. Rev. A* **33**, 2602 (1986).
88. C. C. Huang, G. Nounesis, R. Geer, J. W. Goodby, and D. Guillon, *Phys. Rev. A* **39**, 3741 (1989).
89. G. Nounesis, R. Geer, H. Y. Liu, C. C. Huang, and J. W. Goodby, *Phys. Rev. A* **40**, 5468 (1989).
90. H. Haga, Z. Kutnjak, C. S. Iannacchione, S. Qian, D. Finotello, and C. W. Garland, *Phys. Rev. E* **56**, 1808 (1997).
91. (a) R. Bruinsma and G. Aeppli, *Phys. Rev. Lett.* **48**, 1625 (1982); (b) G. Aeppli and R. Bruinsma, *Phys. Rev. Lett.* **53**, 2133 (1984) and A. Aharony, R. J. Birgeneau, J. D. Brock, and J. D. Litster, *Phys. Rev. Lett.* **57**, 1012 (1986); (c) J. V. Selinger, *J. Phys. (Paris)* **49**, 1387 (1988).
92. I. M. Jiang, S. N. Huang, J. Y. Ko, T. Stoebe, A. J. Jin, and C. C. Huang, *Phys. Rev. E* **48**, R3240 (1993).
93. T. Stoebe, C. C. Huang, and J. W. Goodby, *Phys. Rev. Lett.* **68**, 2944 (1992); T. Stoebe and C. C. Huang, *Phys. Rev. E* **50**, R32 (1994).

8

Freely suspended film experiments

JOEL D. BROCK

School of Applied & Engineering Physics, Cornell University, Ithaca, NY, USA

The colour plate referred to in this chapter can be downloaded from
www.cambridge.org/9780521187947

8.1 Introduction

Freely suspended films of liquid crystal provide an excellent system in which to study the in-plane structure of the smectic phases. In the smectic phases with fluid-like layers (e.g., SmA, SmC), films of thickness varying from two to thousands of molecular layers can be drawn across an orifice. These films are stable for days or even weeks. A soap bubble is the proto-typic example of a freely suspended film. The soap film is built up of three layers: two surface layers of surfactant (soap) and an interior water layer. In contrast, a film of thermotropic liquid crystal can have any number of layers. All of these layers, however, are made of liquid crystal. The variable thickness of a freely suspended film has obvious applications when study-ing dimensional crossover. However, there are several other advantageous features of these samples. The first is the high quality of the sample align-ment, i.e., narrow mosaic. Surface tension at the liquid crystal to vacuum interfaces constrains the film to be quite flat. The film is prevented from thinning by the forces which keep the smectic layers from permeating one another. The second is that the film is essentially self-cleaning. Impurities and defects tend to be pushed to the edges of the film and expelled. This effect is extremely useful when studying phase transitions.

For concreteness, I will assume that the experimenter is going to use x-rays to study the liquid crystal film. However, the considerations for implementing light scattering, torsional oscillator, electron diffraction, or heat capacity measurements are all similar.

8.2 Chamber design and principles

The sample chambers used for x-ray scattering measurements of the struc-ture of freely suspended liquid crystal films must meet demanding design specifications. First, they must have the ability both to draw films and to measure accurately the film thickness *in situ*. Second, the background scattering needs to be extremely low. Third, they must be able to control temperature on the millikelvin scale with a short time constant to minimize dead-time while the synchrotron is running. Finally, they must be mechanically stable to prevent rupture of the freely suspended film. The next several sections will deal with these issues in reverse order.

8.2.1 Generic design

The structure of a conventional Dewar is a relatively simple design which is consistent with our requirements and we have based our chambers on this geometry. Specifically, as shown in Fig. 8.1, there are two concentric cylindrical cans. The outer can is made of stainless steel and has a cylindrical beryllium x-ray window, electrical and vacuum feed-thrus, and fused quartz optical windows. The beryllium window slides on over two Viton O-rings which make the vacuum seal. Heat is provided to the outer can by several thermo-foil resistive heaters placed on the stainless steel cylinder both above and below the beryllium window.

The inner can is made of copper and, like the outer can, has a beryllium x-ray window. Optical access is provided by holes in the copper cylinder. Thermo-foil heaters are again placed on the copper cylinder both above and below the beryllium window. Temperature gradients in the chamber are monitored using standard thermistors, e.g., YSI 44015 (1 MΩ at 25 °C), placed above and below the beryllium window on the inner can. The sample temperature is monitored with an ultra-precision thermistor, e.g., Thermometrics SP60BB105K-B4 (1 MΩ at 25 °C, $d/dt \ln R \leq 0.1\%$/year), placed in thermal contact with the sample holder. The precision thermistor must be calibrated against some other absolute thermometer.

The lid of the inner can is made out of copper. The thermistor which is used to provide an error signal to the temperature controller is placed on this lid. A stainless steel tube connects the lid of the inner can to the top of the oven. Thermally insulating gaskets can be installed to provide additional thermal insulation between the lid of the inner can and the stainless steel tube.

To perform the x-ray diffraction measurements, the chamber is mounted onto an Eulerian cradle as shown in Fig. 8.2.

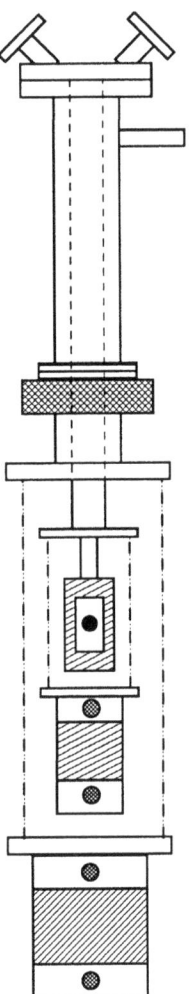

Figure 8.1.   Generic design of x-ray scattering chamber.

8.2.2 Mechanical stability

The basic goals are to ensure that our system does not vibrate at frequencies which are near the resonant modes of the film and that it is strong enough to support itself. We will model the film as a membrane. The vibrations of the supporting tube are the most likely candidates for driving the transverse oscillations of the membrane. A beam can vibrate longitudinally, torsionally, and transversely. Symmetry suggests that the torsional

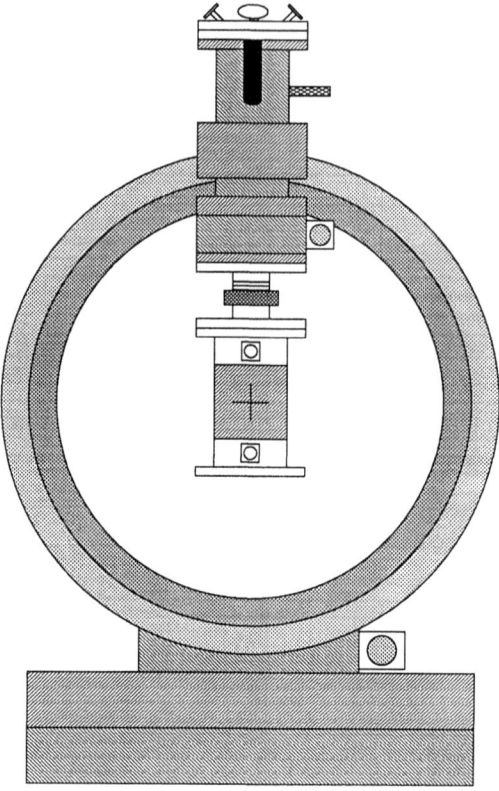

Figure 8.2.   Liquid crystal film chamber mounted on Eulerian cradle.

and transverse modes of the supporting rod will be most effective at driving the membrane oscillations. We will discuss first the membrane resonance of the film and subsequently the normal modes of the chamber.

Membrane

Transverse displacements of a circular membrane $\zeta(r, \theta, t)$ obey the wave equation [1]

$$\frac{\partial^2 \zeta}{\partial r^2} + \frac{1}{r}\frac{\partial \zeta}{\partial r} + \frac{1}{r^2}\frac{\partial^2 \zeta}{\partial \theta^2} = \frac{1}{c^2}\frac{\partial^2 \zeta}{\partial t^2}, \tag{8.1}$$

where $1/c^2 = \sigma_0/\tau_0$ and σ_0 is the mass/unit area and τ_0 is the surface tension of the membrane. The wave equation for a circular membrane has the standing wave solutions

$$\zeta(r,\theta,t) = J_m(kr)\cos(m\theta)\cos(\omega t + \delta), \tag{8.2}$$

where $k = \omega/C$ and $J_m(u)$ is the Bessel function of order m. Requiring as a boundary condition that $\zeta(a,\theta,t) = 0$ creates the constraint

$$k_{m,n}a = \alpha_{m,n} \qquad (8.3)$$

where $\alpha_{m,n}$ is the nth zero of the Bessel function of order m.

Typical values of the surface tension and mass density are on the order of $\gamma = 20\text{--}26$ dyn/cm and $\rho = 1$ g/cm^3. Experimentally, the surface tension of a film is observed to be localized near the surface [2] and there are two surfaces in a film; therefore, $\tau_0 \approx 50$ dyn/cm. The mass/area of a film of n layers, each $d = 30$ Å thick is given by

$$\sigma_0 = \rho(nd) = n\,(30 \times 10^{-8})\ \text{g}^2/\text{cm}. \qquad (8.4)$$

Substituting in these values and solving for the fundamental frequency, we obtain for a film 1 cm in diameter and n layers thick

$$\omega_{0,1} = \frac{\alpha_{0,1}}{a}\sqrt{\frac{\tau_0}{\sigma_0}}$$

$$= \frac{2.405}{0.5}\sqrt{\frac{50}{(1) \times (n) \times 30 \times 10^{-8}}}\ \text{s}^{-1} \qquad (8.5)$$

$$= \frac{6.21 \times 10^4}{\sqrt{n}}\ \text{s}^{-1}.$$

This is the lowest frequency mode likely to be excited by transverse oscillations of the supporting rod. Similarly,

$$\omega_{1,1} = \frac{\alpha_{1,1}}{a}\sqrt{\frac{\tau_0}{\sigma_0}}$$

$$= \frac{3.832}{0.5}\sqrt{\frac{50}{(1) \times (n) \times 30 \times 10^{-8}}}\ \text{s}^{-1} \qquad (8.6)$$

$$= \frac{9.89 \times 10^4}{\sqrt{n}}\ \text{s}^{-1}.$$

This is the lowest frequency mode likely to be excited by torsional oscillations of the supporting rod.

Our next job is to calculate the fundamental frequencies of the oscillations of the support structure.

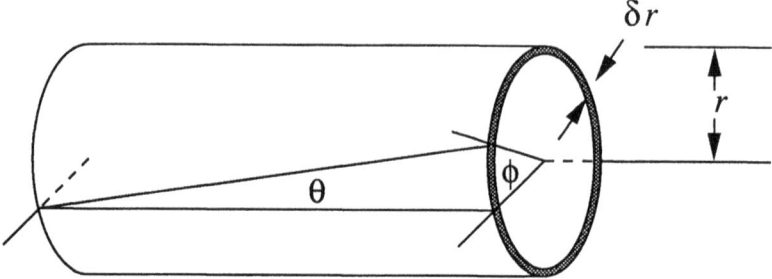

Figure 8.3. A cylindrical shell in torsion.

Torsional oscillations

Torsional displacements $\phi(x,t)$ on a slender tube obey the wave equation [1]

$$\frac{\partial^2\phi}{\partial x^2} = \frac{\rho}{\mu}\frac{\partial^2\phi}{\partial t^2},\tag{8.7}$$

where μ is the shear modulus. The geometry of this problem is illustrated in Fig. 8.3. Note that this equation is independent of the radius of the tube. It is simplest to calculate the normal mode frequencies by assuming *free* boundary conditions: i.e., $\partial\phi/\partial x = 0$. In this case, the wave equation has the standing wave solution

$$\phi(x,t) = A_n \cos(k_n x)\cos(\omega_n t + \delta_n),\tag{8.8}$$

where $k_n\ell = n\pi$ guarantees that $\partial\phi/\partial x = 0$ when $x = \ell$. Thus, the normal mode frequencies of the torsional oscillations are given by

$$\omega_n = n\frac{\pi}{\ell}\sqrt{\frac{\mu}{\rho}}.\tag{8.9}$$

For stainless steel, $\mu = 7.57 \times 10^{11}$ dyn/cm^2, $\rho = 7.91$ g/cm^3 and, in our chamber, $\ell = 30$ cm. Thus, the fundamental torsional frequency is

$$\omega_1 = \frac{\pi}{30}\sqrt{\frac{7.57 \times 10^{11}}{7.91}}\ \text{s}^{-1}$$

$$= 3.24 \times 10^4\ \text{s}^{-1}.\tag{8.10}$$

Bending oscillations

In the limit where the support rod is reasonably slender, transverse oscillations $\eta(x,t)$ on a beam obey the approximate wave equation [1]

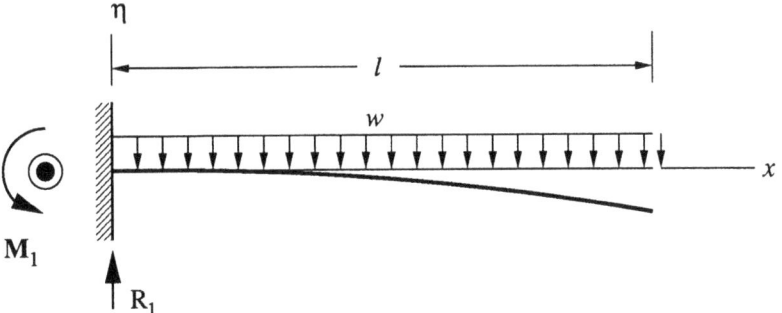

Figure 8.4.   A beam bent by its own weight.

$$\frac{\partial^4 \eta}{\partial x^4} = -\frac{\rho A}{EI}\frac{\partial^2 \eta}{\partial t^2}, \tag{8.11}$$

where I is the geometric moment of inertia. The geometry of this problem is illustrated in Fig. 8.4. A general solution of this wave equation having a sinusoidal time factor is

$$\eta(x,t) = \{A\sinh(kx) + B\cosh(kx) + C\sin(kx) + D\cos(kx)\}\cos(\omega t + \delta), \tag{8.12}$$

where

$$k = \left(\frac{\rho A \omega^2}{EI}\right)^{1/4}. \tag{8.13}$$

The *clamped-free* boundary conditions are appropriate for our chamber.

$$\frac{\partial^3 \eta}{\partial x^3} = 0 \quad \text{at} \quad x = \ell,$$

$$\frac{\partial^2 \eta}{\partial x^2} = 0 \quad \text{at} \quad x = \ell, \tag{8.14}$$

$$\frac{\partial \eta}{\partial x} = 0 \quad \text{at} \quad x = 0,$$

$$\eta = 0 \quad \text{at} \quad x = 0.$$

The boundary conditions on η and η' at $x = 0$ require that $D = -B$ and $C = -A$ respectively. The boundary conditions on η'' and η''' at $x = \ell$ require that

$$A[\sinh(k\ell) + \sin(k\ell)] + B[\cosh(k\ell) + \cos(k\ell)] = 0,$$
$$A[\cosh(k\ell) + \cos(k\ell)] + B[\sinh(k\ell) - \sin(k\ell)] = 0. \tag{8.15}$$

These homogeneous equations give the ratio B/A provided their determinant vanishes, i.e., provided

$$\cosh(k\ell)\cos(k\ell) = -1. \tag{8.16}$$

The first root of this transcendental equation occurs when $k\ell = 1.875$. Thus, the frequency of the fundamental mode is given by

$$\omega_1 = \left(\frac{1.875}{\ell}\right)^2 \sqrt{\frac{EI}{\rho A}}$$

$$= \left(\frac{1.875}{30}\right)^2 \sqrt{\frac{(19.6 \times 10^{11})(1.23)}{(7.91)(0.2)}} \text{ s}^{-1} \tag{8.17}$$

$$= 4.82 \times 10^3 \text{ s}^{-1}.$$

8.2.3 Thermal stability

Historically, liquid crystals have been very useful systems for precise studies of critical phenomena. These measurements near a phase transition require excellent stability. The goal is to get as close as possible to the phase transition temperature. A good figure of merit to aim for is a reduced temperature of $t \equiv |T - T_c|/T_c \sim 10^{-6}$. For room temperature phase transitions, this requires better than millikelvin temperature control. Compared to other kinds of measurements, x-ray measurements on freely suspended films have the additional complication that the sample chamber must be rotated during the measurement. In the absence of rotations, asymmetries in the thermal environment, which can produce temperature gradient variations at the sample position, can be eliminated by adjusting the heater power of various heating elements. Since this is no longer possible when the chamber must be rotated, care needs to be taken to avoid orientation-dependent thermal variations.

General thermal design

The basic thermal design of the chamber consists of concentric, independently controlled stages. The basic design is sketched in Fig. 8.5. The sample is suspended across an orifice in the center of the chamber. The temperature of the sample is controlled indirectly by controlling the temperature of its environment. Using a feedback loop, the outer can is maintained at a nearly constant temperature roughly 10 K below the desired sample temperature. Using the outer can as a constant temperature heat reservoir

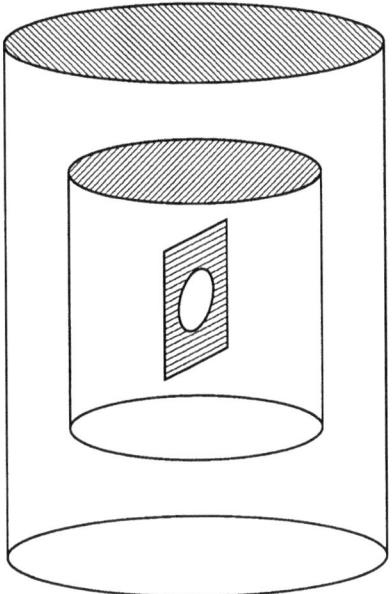

Figure 8.5. Basic thermal design.

and thermally coupling the inner and outer cans via a thermal exchange gas, the temperature of the inner can may be precisely controlled. The sample thus finds itself in a constant temperature environment and we simply measure its temperature.

The following design points need to be kept in mind.

- The heat capacity of the outer can should be much greater than the heat capacity of the inner can so that it can serve as a thermal reservoir.
- For the same reason, the heat capacity of the inner can should be much greater than the heat capacity of the sample mount.
- The heaters on the inner can should be able to produce more heating power than the heat loss rate to the outer can.
- The overall thermal time constant of the sample must be short enough to efficiently use synchrotron beam time (e.g., $\tau \leq 1$ minute).

Symmetry

The obvious way to deal with orientation-dependent temperature variations is to make the sample chamber as symmetric as possible under rotations. The key point to keep in mind is that it is only the thermal properties which must be rotationally invariant. Since the main heat loss mechanism

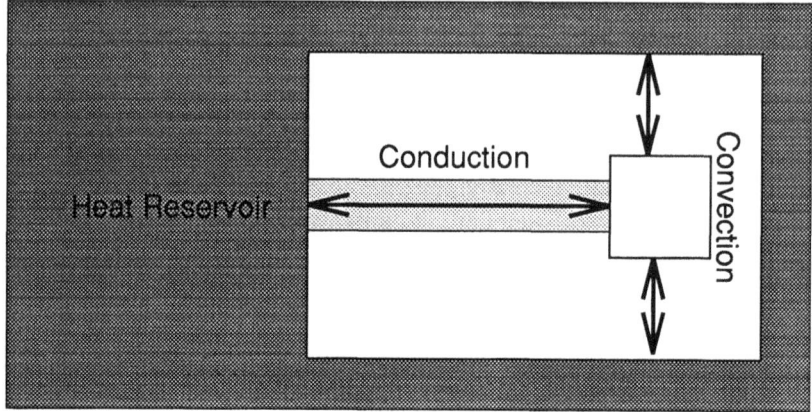

Figure 8.6.   Schematic depiction of heat flow in cylindrical sample chamber.

of the outer can is convection by the room air, the ideal chamber would be spherical. Clearly, a spherical chamber is impractical to build. Cylindrical is the best we can do. However, we can keep the sample in a symmetric thermal environment by using the random motions of an exchange gas to smear out the influence of the asymmetric geometry. To do this, we need the rate at which heat is conducted along the mechanical supports to be negligible compared to the rate at which heat is transported by the exchange gas.

For example, in a cylindrically symmetric chamber, the sample is usually mechanically supported from only one side. Keeping in mind that it is only the thermal loss mechanisms which must be symmetric, we require that the rate of heat transport via thermal conduction up and down the mechanical support is small compared to that by convection through the exchange gas. This situation is depicted schematically in Fig. 8.6 and described analytically by

$$\left(\frac{\delta Q}{\delta t}\right)_{\text{conduction}} \ll \left(\frac{\delta Q}{\delta t}\right)_{\text{convection}}, \tag{8.18}$$

where Q is heat energy.

Materials

We can guarantee that Eq. (8.18) is satisfied by choosing our materials and geometry carefully. The design requirements for the mechanical support of the inner can are that it (i) provide a rigid mechanical support, and (ii) be a poor thermal conductor.

The rate at which heat energy, Q, is added to the sample via conduction along the supporting rod is given by

$$\left(\frac{\delta Q}{\delta t}\right)_{\text{conduction}} = j_{\text{cond}} A_{\text{rod}}, \tag{8.19}$$

where A_{rod} is the cross-sectional area of the mechanical support and j_{cond} is the energy current density. Empirically, j_{cond} is proportional to the thermal gradient. The proportionality constant is the thermal conductivity, κ. For our simple geometry, we have

$$j_{\text{cond}} = -\kappa_{\text{rod}}\left(\frac{T - T_{\text{outer}}}{\ell}\right), \tag{8.20}$$

where κ_{rod} is the thermal conductivity, ℓ is the length of the supporting rod, T is the temperature of the inner can, and T_{outer} is the temperature of the outer can. Thus, using the heat capacity of the inner can, C, we find that the rate of change of the temperature of the inner can is

$$\frac{dT}{dt} = \frac{1}{C}\frac{\delta Q}{\delta t}$$

$$= -\frac{1}{C}\kappa_{\text{rod}} A_{\text{rod}}\left(\frac{T - T_{\text{outer}}}{\ell}\right). \tag{8.21}$$

Defining $T' = T - T_{\text{outer}}$, Eq. (8.21) can be rewritten as

$$\frac{dT'}{T'} = -\frac{\kappa_{\text{rod}} A_{\text{rod}}}{\ell C}dt, \tag{8.22}$$

where C is the heat capacity of the sample. Equation (8.22) clearly exhibits the expected exponential relaxation of the temperature and identifies the time constant as

$$\tau_{\text{rod}} = \frac{\ell C}{\kappa_{\text{rod}} A_{\text{rod}}}. \tag{8.23}$$

Clearly, to make the effects of thermal transport via conduction through the support rod negligible, we want τ_{rod} to be much greater than the corresponding time constant describing the relaxation of the temperature due to heat transport by the exchange gas, τ_{gas}.

To estimate τ_{gas}, we begin by noting that the thermal conductivity of an ideal gas is *independent* of the pressure[1] of the gas as long as the mean free

---

[1] This apparent paradox is resolved by noting that although the number of gas molecules available to transport energy is doubled when the number of molecules is doubled, the mean free path of each molecule is simultaneously halved so that each molecule transports energy half as effectively. The net rate of energy transport is left unchanged.

path is both much greater than the molecular diameter and much smaller than the chamber dimensions [3]. Using the kinetic theory of gases and assuming ideal behavior, the mean free path, ℓ_{MFP}, of a gas atom is given by

$$\ell_{\text{MFP}} \approx \left(\sqrt{2}n\sigma_0\right)^{-1}, \tag{8.24}$$

where n is number density and σ_0 is the cross-section. We can use this relation to estimate the mean free path for a typical gas at room temperature ($\approx 300\,\text{K}$) and atmospheric pressure ($\approx 10^6\,\text{dyn/cm}^2$). The number density n is obtained from the equation of state for an ideal gas.

$$n = \frac{P}{k_{\text{B}}T} = \frac{10^6}{(1.4 \times 10^{-16})(300)} = 2.4 \times 10^{19}\,\text{molecules/cm}^3. \tag{8.25}$$

Approximating the impact parameter by the molecular diameter, d, and using $d \approx 2 \times 10^{-8}\,\text{cm}$ as a typical value, we find that

$$\sigma_0 \approx \pi(2 \times 10^{-8})^2 \approx 12 \times 10^{-16}\,\text{cm}^2. \tag{8.26}$$

Therefore, at atmospheric pressure,

$$\ell_{\text{MFP}} \approx 3 \times 10^{-5}\,\text{cm}. \tag{8.27}$$

We need to be careful as we reduce the pressure that ℓ_{MFP} remains smaller than the radial distance between chamber elements. Thus, if the characteristic chamber dimension is on the order of $1\,\text{cm}$, we must operate at pressures greater than 25×10^{-3} Torr.

Assuming that we are in the appropriate pressure range, the thermal conductivity of an ideal gas is given by

$$\kappa_{\text{gas}} = \frac{2}{3\sqrt{\pi}} \frac{c}{\sigma_0} \sqrt{\frac{k_{\text{B}}T}{m}}, \tag{8.28}$$

where $c = \frac{3}{2}k_{\text{B}}$ is the heat capacity of a single ideal gas molecule and m is its mass. Using $m_{\text{He}} = 4.0026$ amu, we estimate

$$\kappa_{\text{gas}} \approx \frac{2}{3\sqrt{\pi}} \frac{\left(\frac{3}{2}\right)(1.4 \times 10^{-16})}{12 \times 10^{-16}} \sqrt{\frac{(1.4 \times 10^{-16})(300)}{(4.0)/(6.0 \times 10^{23})}} \frac{\text{ergs}}{\text{cm s K}}$$

$$= 5.2 \times 10^{-4} \frac{\text{W}}{\text{cm K}}. \tag{8.29}$$

Using this value for κ and assuming that heat is transported only via the exchange gas, we can calculate the thermal time constant of the system.

$$\frac{dT}{dt} = \frac{1}{C}\frac{\delta Q}{\delta t}$$

$$= -\frac{1}{C}\kappa_{gas} 2\pi a\ell' \frac{(T - T_{outer})}{b - a}. \tag{8.30}$$

Here, a is the radius of the inner can, b is the radius of the outer can, and ℓ' is the length of the inner can. Just as for the case of heat transport via conduction, we have

$$\tau_{gas} = \frac{(b - a)C}{2\pi\kappa_{gas}a\ell'}. \tag{8.31}$$

Thus, our requirement that thermal conduction through the supporting rod be negligible compared to thermal conduction via the exchange ($\tau_{rod} \gg \tau_{gas}$) becomes

$$\frac{\ell}{\kappa_{rod} A_{rod}} \gg \frac{b - a}{\kappa_{gas} 2\pi a\ell'}. \tag{8.32}$$

In a typical chamber used in our laboratory, the supporting rod is made out of stainless steel pipe and the chamber has the following dimensions: $\ell = 30\,\text{cm}$; $A_{rod} = 0.2\,\text{cm}^2$; $b = 4.5\,\text{cm}$; $a = 2.9\,\text{cm}$; and, $\ell' = 10\,\text{cm}$. At room temperature, the thermal conductivity of stainless steel is $\kappa_{SS} \approx 1.5 \times 10^{-1}\,\text{W cm}^{-1}\text{K}^{-1}$. Thus, $\tau_{rod}/\tau_{gas} \approx 57$ and we are in good shape.

8.2.4 Background scattering

The second criterion the chamber must meet is that the background scattering rate must be sufficiently low to observe the weakly scattering liquid crystal sample. At room temperature and pressure ($T = 300\,\text{K}$, $P \sim 10^6\,\text{dyn cm}^{-2}$), a scattering volume of N_2 contains $\sim 10^{18}$ electrons. The geometry is illustrated in Fig. 8.7. A freely suspended film in the same size x-ray beam contains $2n \times 10^{13}$ molecules where n is the number of smectic layers. A typical thermotropic liquid crystal has on the order of 250 electrons/molecule. Thus, for an n-layer film, the ratio of the number of electrons in the scattering volume due to air to the number of electrons in the scattering volume due to the freely suspended film is on the order of $10^{18} : 5n \times 10^{15}$. By dropping the pressure to 4 Torr, the ratio becomes on the order of $1 : n$. Thus, to get a reasonable signal to noise level in thin films, the sample chamber must be evacuated to pressures on the order of 100×10^{-3} Torr or less.

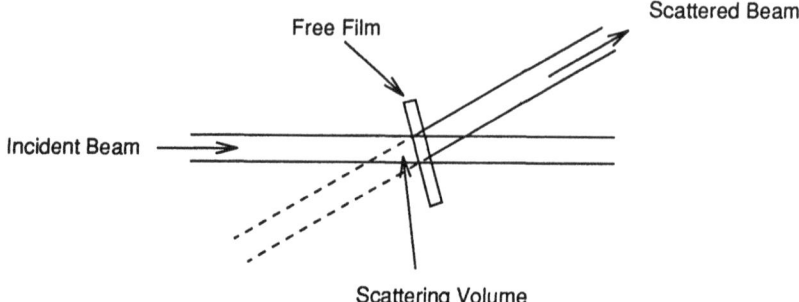

Figure 8.7.   Scattering volume.

8.2.5 Vacuum levels

The choice of operating pressure is thus constrained from below by the need to keep the mean free path of the exchange gas molecules less than a typical chamber dimension and from above by the need to minimize the background scattering rate. Empirically, pressures in the $25 - 100 \times 10^{-3}$ Torr range work well.

8.2.6 Optical and x-ray access

Optical access is provided by fused quartz windows. Fused quartz windows exhibit less strain-induced birefringence and are thus preferable to other types of optical windows. X-ray access to the sample is provided by concentric cylindrical shells of beryllium. When designing the x-ray windows, the design considerations are that the window needs to be thicker than $0.010''$ $(0.025 \, cm)$ for structural strength and to avoid pin holes. The maximum thickness is completely determined by absorption. The absorption in the beryllium window is due almost entirely to iron impurities. Standard purity beryllium is not usually sufficient. Good results have been obtained by machining a cylinder $\approx 0.010''$ thick out of a solid piece of beryllium. The vacuum seal is made either by having the beryllium compress O-rings at either end of the cylinder or by gluing with Torr Seal.

Minimization of solid angle via software

The Eulerian cradles on modern four-circle diffractometers can orient single crystal samples such that any reciprocal space vector is in the diffracting condition. The angles of such a diffractometer are indicated in Fig. 8.8. The basic reference for calculating the angles given the h,k,ℓ for four-circle diffractometers is the paper by Busing and Levy [4]. The notation used here

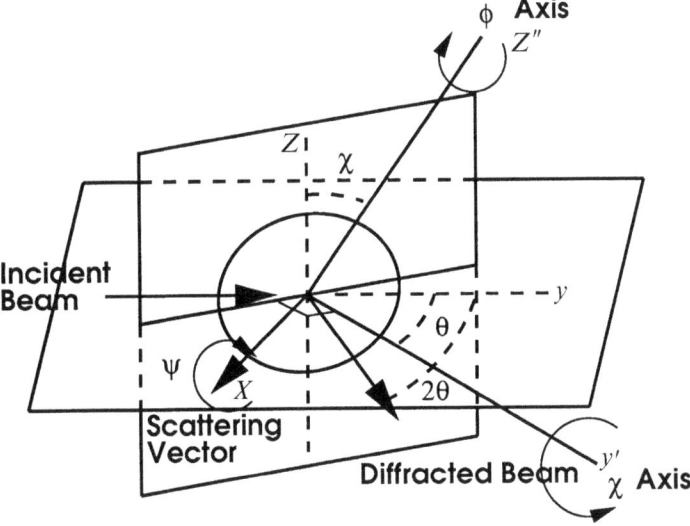

Figure 8.8.   Schematic representation of a four-circle diffractometer.

will conform to their notation as closely as possible. Briefly, let **v** be the column vector describing some physical vector **G** in terms of the right-handed reciprocal lattice vectors $\mathbf{b}_i$ so that

$$\mathbf{G} = \sum_{i=1}^{3} v_i \mathbf{b}_i. \tag{8.33}$$

The first transformation to be made changes to a Cartesian coordinate system. If we choose the x axis to be parallel to $\mathbf{b}_1$, the y axis to be in the plane of $\mathbf{b}_1$ and $\mathbf{b}_2$, and the z axis perpendicular to that plane, then

$$\mathbf{v}_c = \mathbf{B}\mathbf{v}. \tag{8.34}$$

In the next transformation, we rotate to a second Cartesian coordinate system which has the property that when all the spectrometer angles are set to zero, the y axis is along the incident beam and the z axis is out of the scattering plane. We define **U** to be the rotation matrix which relates this coordinate system to the samples Cartesian system

$$\mathbf{v}_0 = \mathbf{U}\mathbf{v}_c = \mathbf{U}\mathbf{B}\mathbf{v}. \tag{8.35}$$

Thus, the orientation matrix (**UB**) describes the sample orientation relative to the zeroed diffractometer. The diffractometer angles θ, χ, and ϕ are simply the Euler angles [5]. When all the angles are set to zero, the x axis is along the scattering vector, the y axis is along the primary beam, and the z

axis is vertical. The first rotation is by ϕ about the z axis. The second rotation is by χ about the y' axis. The final rotation is by θ about the z'' axis. These rotations are illustrated in Fig. 8.8.

Given **UB** it is possible to calculate a set of diffractometer angles (2θ, θ, χ, ϕ) which rotate a particular reciprocal space vector **G** into the diffracting position. There are an infinite number of possible solutions to this problem; thus, in all cases, we must supply an additional constraint. Crystallographers, for example, frequently do this by requiring that $\omega = \theta - 2\theta/2 = 0$.

We have a different special case which stems from our chamber geometry. In contrast to the 360° access in the meridional plane of the chamber, out of the meridional plane of the chamber, the finite height of the cylindrical x-ray window limits the angular access of x-rays to the sample. Thus, we wish to orient the sample chamber in a manner that both meets the requirements for diffraction and maximizes the angle between the diffraction plane and the axis of the chamber. In our case, we will define a unit vector $\hat{n}$ which points along the axis of the cylindrical chamber. Defining the azimuthal angle, ψ, to be zero when $\hat{n}$ lies in the scattering plane, a 90° rotation about the scattering vector will orient the axis of the chamber as far out of the scattering plane as possible.

Computing the diffractometer angles for some specified value of the azimuthal angle, ψ, is straightforward. The rotation matrix **R** is defined by

$$\mathbf{R} = \mathbf{\Omega}\mathbf{X}\mathbf{\Phi}, \qquad (8.36)$$

where $\mathbf{\Omega}$, $\mathbf{X}$, and $\mathbf{\Phi}$ are rotation matrices specifying the three Euler rotations. Physically, **R** transforms the coordinates of a vector from the zeroed spectrometer coordinate system to the final orientation. For the vector to be in the scattering position, we must have

$$\mathbf{h} = \mathbf{R}\mathbf{h}_0 = \begin{pmatrix} q \\ 0 \\ 0 \end{pmatrix}. \qquad (8.37)$$

If we choose a particular diffractometer configuration which satisfies this condition and define $\psi = 0$, we can then evaluate

$$\mathbf{R}_0 = \mathbf{\Omega}_0\mathbf{X}_0\mathbf{\Phi}_0. \qquad (8.38)$$

To rotate the sample about the scattering vector through an angle ψ measured from this zero position, we generate the rotation matrix

$$\mathbf{R} = \mathbf{\Psi}\mathbf{R}_0, \qquad (8.39)$$

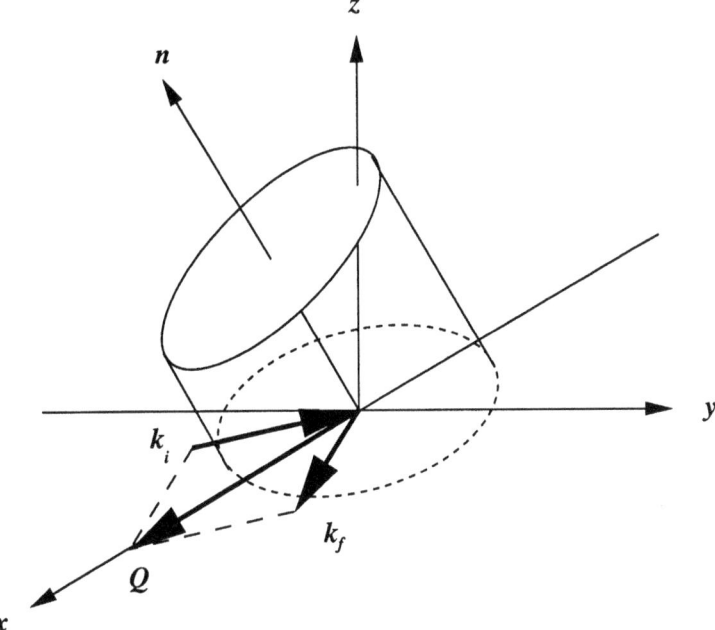

Figure 8.9. Specifying azimuthal angle to maximize x-ray access to sample.

where

$$\Psi = \begin{pmatrix} 1 & 0 & 0 \\ 0 & \cos\psi & \sin\psi \\ 0 & -\sin\psi & \cos\psi \end{pmatrix} \qquad (8.40)$$

We obtain the required instrument angles by explicitly evaluating Eq. (8.39). Performing the multiplication [4], we find that

$$\phi = \tan^{-1}(-R_{32}, -R_{31}),$$
$$\chi = \tan^{-1}[(R_{31}^2 + R_{32}^2)^{1/2}, R_{33}], \qquad (8.41)$$
$$\omega = \tan^{-1}(R_{13}, -R_{23}).$$

Formally, the instrument setting which defines $\psi = 0$ is defined by choosing a reference vector $\mathbf{h}_0$ which is not parallel to $\mathbf{h}$. We then choose the zero of ψ by requiring that $\mathbf{h}$ is in the scattering position and $\mathbf{h}_0$ is in the scattering plane with $y'' > 0$. In our particular problem, we define the reference vector $\mathbf{h}_0$ to be along the axis of the chamber $\hat{\mathbf{n}}$ and set $\psi = 90°$ so that the primary and diffracted beams make the largest possible angle with $\hat{\mathbf{n}}$. This geometry is illustrated in Fig. 8.9.

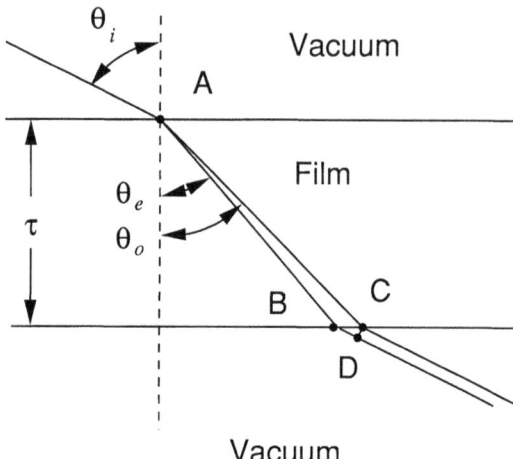

Figure 8.10.   Geometry for ellipsometric measurement of film thickness.

8.3 Film thickness measurements

An absolutely crucial piece of information for analyzing data from a freely suspended film is the number of smectic layers in the film. Unfortunately, there is no single method capable of accurately measuring the number of layers over the entire range of film thicknesses possible. For thick films ($n \geq 100$), an ellipsometry measurement can measure n to an accuracy of ± 5 layers. For intermediate thickness films ($20 < n < 100$), the color of visible light reflected from the film can determine n to ± 3 layers. For thin films ($n \leq 20$), the intensity of laser light reflected from the film can determine n exactly. The value of n can then be correlated with the intensity of x-rays scattered by the SmA phase or with the line width in the CrJ phase as a further consistency check.

8.3.1 *Optical: ellipsometry*

The physics of an ellipsometric measurement is contained in the different optical path lengths seen by the different linear polarizations due to the birefringence of the SmA material. Linearly polarized light with a polarization vector parallel to the nematic director sees a different index of refraction than light with a polarization vector in the plane of the smectic layers. The situation is drawn in Fig. 8.10. The refracted ray splits into an ordinary and an extraordinary ray. The difference in path length seen by the two rays is

$$\Delta\Phi = \Phi_{AB} + \Phi_{BD} - \Phi_{AC}, \tag{8.42}$$

where Φ_{AB} is the path length for the extraordinary ray to get from point A to point B and so on. Using standard optical path length methods,

$$\Delta\Phi = k\tau \left\{ \frac{n_e(\theta_e)}{\cos\theta_e} - \frac{m_o}{\cos\theta_o} + \sin\theta_i(\tan\theta_0 - \tan\theta_e) \right\}, \tag{8.43}$$

where $k = 2\pi/\lambda$ is the wavenumber of the incident light, τ is the thickness of the film, n_0 is the ordinary index of refraction and $n_e(\theta)$ is the extraordinary index of refraction. For the angles appropriate for the free film chamber, $n_0 = 1.5$ and $n_e = 1.7$, we have

$$\Delta\Phi = 2.36 \times 10^{-5}\pi\tau \text{ radians}, \tag{8.44}$$

where τ is measured in Å.

A typical Soleil–Babinet compensator can measure $\Delta\Phi$ to an accuracy of $\pi/1325$ radians. Thus, in principle, we can measure film thicknesses in units of ~32 Å. However, in practice, we rarely know the values of n_0 or n_e. 8OCB is one of the prototypic thermotropic molecules. The refractive indices of 8OCB are typical of thermotropic materials. Therefore, we can use Schaetzing's [6] values: $n_0 = 1.50$, $n_e = 1.68$. The uncertainty in the birefringence is the major source of error in this method, limiting us to uncertainties of the order of ± 5 layers.

8.3.2 Optical: reflected color

The reflectivity of a homogeneous dielectric film with an index of refraction n and thickness τ is a function of wavenumber and is given by [7]

$$R = |r|^2 = \frac{2r^2 + 2r^2 \cos 2\beta}{1 + r^4 - 2r^2 \cos 2\beta}, \tag{8.45}$$

where

$$\beta = \frac{2\pi}{\lambda} n\tau \cos\theta_2 \tag{8.46}$$

and

$$r = \frac{\cos\theta_1 - n\cos\theta_2}{\cos\theta_1 + n\cos\theta_2}. \tag{8.47}$$

Here, $\theta_1 = \theta_i$, $\theta_2 = \theta_{o,e}$.

If we assume the incident light to be unpolarized and to have a blackbody intensity distribution at temperature T, we can calculate the reflected intensity as a function of wavelength (λ), film thickness (r), angle of

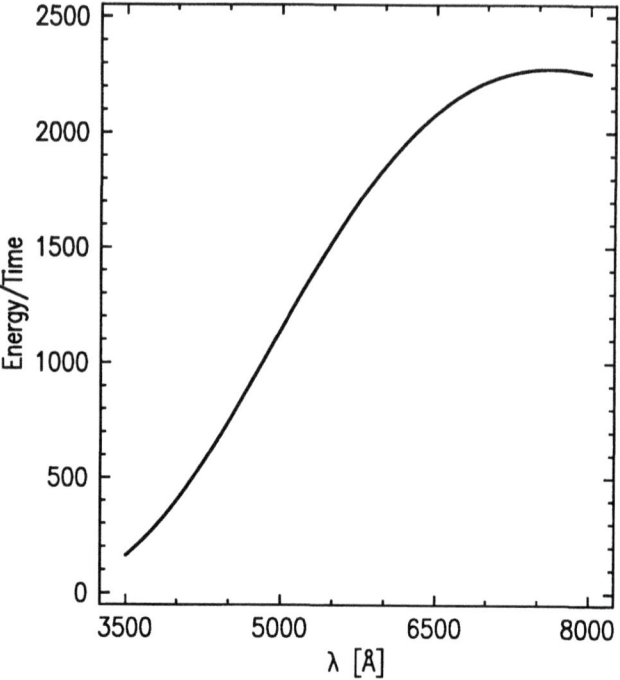

Figure 8.11. For an $n = 20$ layer film illuminated with a 100 W light bulb, the appropriate parameters are: $T = 2800$ K; $\tau = 620$ Å; $\theta_i = 65°$; $n_o = 1.5$; and $n_e = 1.7$.

incidence (θ_i), and the index of refraction. See Fig. 8.11. The question, 'What *color* will the human eye perceive this distribution of wavelengths to be?' is actually quite complex and subtle.

Any perceived color can be built up from a basis of three primary colors. The intensity distributions corresponding to the imaginary colors X, Y, Z used by the CIE [8] are plotted in Fig. 8.12. These *imaginary* colors have the advantage that all the real colors can be made up out of positive amounts of X, Y, and Z. If red, green, and blue had been used, negative amounts would have been required for certain perceived colors. The calculated reflected intensity distribution can be projected onto these three color functions (e.g., $X = \int d\lambda X(\lambda)I(\lambda)$). The human eye is not sensitive to light out of the spectral range defined by $X(\lambda)$, $Y(\lambda)$, $Z(\lambda)$. The three projections X, Y, and Z form a vector in a three-dimensional 'color-space'. This three-dimensional space can be projected onto a plane by normalizing to the total intensity. Thus, the coordinates

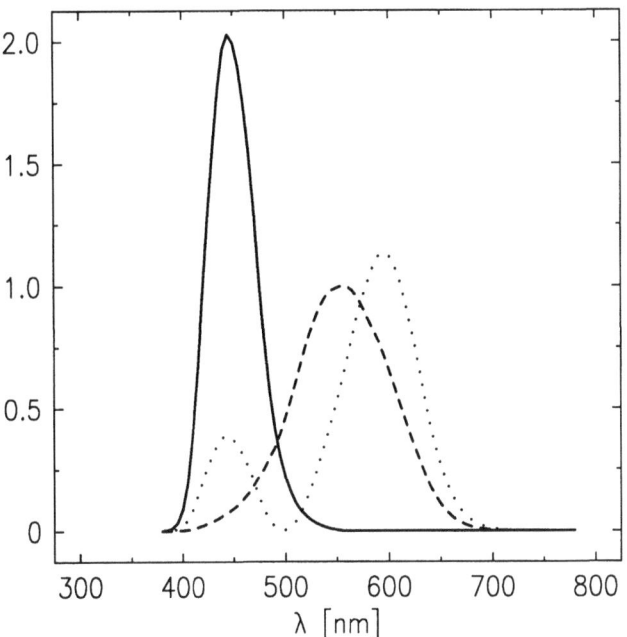

Figure 8.12. CIE color functions: solid line, Z; dashed line, Y; dotted line, X.

$$\bar{x} \equiv \frac{X}{X+Y+Z}, \tag{8.48}$$

$$\bar{y} \equiv \frac{Y}{X+Y+Z}, \tag{8.49}$$

give a complete description of the perceived color in the limit of constant total reflected intensity. The perceived color is determined by plotting ($\bar{x}$, $\bar{y}$) on a CIE color map [9].

Obviously, as $X+Y+Z \to 0$, the eye will perceive the reflected color as white. Therefore, this analysis applies only to a limited range of reflected intensity, corresponding to intermediate thickness films. An example is illustrated in Fig. 8.13 (color) which assumes the same parameters as Fig. 8.11 but plots the color as a function of film thickness for ($20 \le n \le 100$).

8.3.3 Optical: reflectivity

In the limit of the film thickness τ being much smaller than the wavelength of the light, the electrons will radiate coherently. Thus, the scattering sum

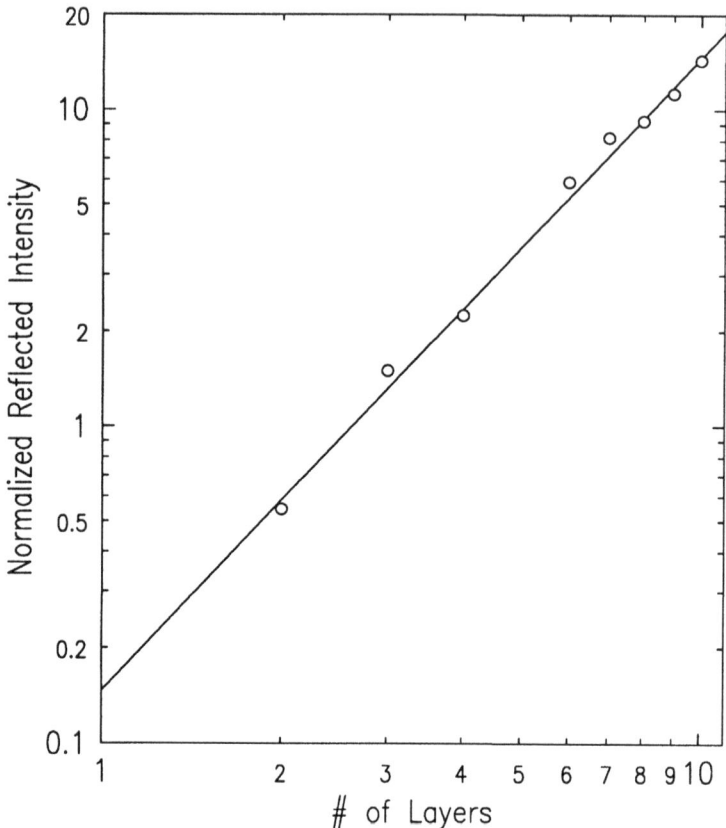

Figure 8.14.   Reflected intensity vs. film thickness.

rule tells us that the reflected intensity must go as the number of scatterers squared. The number of scatterers goes as the number of layers. Thus, the reflected intensity must go as the number of layers squared for very thin films. A key advantage of this method is that it determines the number of layers exactly. Only discrete values of the reflected intensity are allowed and there is a one-to-one correspondence with film thickness in the applicable range.

A plot of reflected intensity vs. layer number is shown in Fig. 8.14. It is extremely important to normalize out any absorption by material condensed on the quartz windows or fluctuations in laser power. A two layer film reflects about 5 μW of a 1.5 mW laser beam. Any stray light or unaccounted losses can be disastrous to the measurement.

8.3.4 X-ray techniques for measuring film thickness

Since the extinction length is so enormous, one can, in principle, use the finite width of any Bragg peak to measure the sample thickness. The simplest case would be the width of the lowest order smectic density wave peak. Using kinematic scattering theory, the width of a diffraction peak is determined by the mean square Fourier amplitude of the charge-density. Thus,

$$
\begin{aligned}
\rho_{\mathbf{q}} &= \int d\mathbf{r}\, e^{-i\mathbf{q}\cdot\mathbf{r}}\, \rho(\mathbf{r}) \\
&= \left\{ \int dx\, dy\, e^{-i(q_x x + q_y y)} \bar{\rho}(x,y) \right\} \sum_{n=0}^{N-1} e^{-iq_z nd} \\
&= F(q_\parallel) \left\{ \frac{1 - e^{-iq_z Nd}}{1 - e^{-iq_z d}} \right\},
\end{aligned}
\tag{8.50}
$$

where $\bar{\rho}$ is the charge per unit area of a smectic layer and F is its Fourier transform. Thus, we obtain

$$
\mathscr{S}(\mathbf{q}) = \langle |\rho_{\mathbf{q}}|^2 \rangle = \langle |F(\mathbf{q})|^2 \rangle \left\{ \frac{\sin^2\left(\dfrac{q_z Nd}{2} \right)}{\sin^2\left(\dfrac{q_z d}{2} \right)} \right\}.
\tag{8.51}
$$

Equation (8.51) is plotted in Fig. 8.15 for a film 10 layers thick. We estimate that the full width at half maximum of each of these peaks is roughly equal to the distance from the peak center to the first zero, δq. The zeros occur when the argument of the numerator is an integer multiple of π. Therefore, the first zero occurs when

$$
\frac{\delta q Nd}{2} = \pi.
\tag{8.52}
$$

Therefore,

$$
\tau = Nd = \frac{2\pi}{\delta q} = \frac{\pi}{\text{HWHM}}.
\tag{8.53}
$$

In many chambers, the smectic density wave Bragg peak is inaccessible. However, this technique may still be used if a true three-dimensional crystal phase exists. A scan in the $\hat{n}$ direction through any Bragg peak has the same behavior.

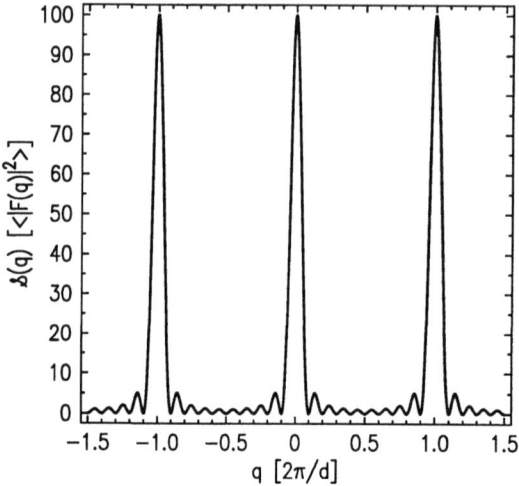

Figure 8.15. Reflected x-ray intensity for a 10-layer thick film.

Another type of x-ray thickness measurement utilizing the intensity rather than the line width can be performed on the fluid smectic phases. Since the layers are decoupled, the total scattered intensity at the peak of the fluid structure factor is proportional to the number of smectic layers. By measuring the scattered intensity at the peak of the fluid structure factor for a film of known thickness, one can then obtain the thickness of other films by simple scaling arguments.

8.4 Preparing films

8.4.1 Drawing thin films

The technique for drawing films was developed by the Harvard group [10] who used circular holes painstakingly drilled by hand in glass cover slips. We have found that a circular hole drilled and reamed in a $0.025''$ $(0.006\,\mathrm{cm})$ thick stainless steel plate also works well. Both sides of the steel are polished to a mirror finish and the edges of the hole should be checked for nicks or scratches that might pin an alignment defect in the sample. A small amount of sample material is melted around the edges of the sample holder before sealing the chamber. The chamber is then evacuated and heated into the SmA temperature range. A wiper blade pulls the liquid crystal material across the hole in the stainless steel, drawing the freely suspended film. The thickness of the film is determined by the rate of pulling, the amount of material on the edges and the temperature. High quality thin films are pulled rather quickly, high in the SmA temperature range, with only small amounts of material present. High quality thick films are pulled slowly, in the lower portion of the SmA temperature range, with ample amounts of material present.

After drawing a film, it is necessary to wait for the film to anneal. A thin film ($n \leq 10$) will anneal in 30 minutes to a few hours. A thick film ($n \geq 10$) may require several hours to more than a day to anneal. Repeatedly taking the sample through a phase transition can help to speed this process.

8.4.2 *Preparing thick films*

For thick films, a closely related technique has been pioneered by the group at Warsaw Technical University [11]. The liquid crystal sample in powder form is placed at room temperature inside a circular hole, usually less than a millimeter in diameter, in a metal plate. The sample is pressed into place by hand using a piston. Monitoring the sample optically to make sure that the sample does not flow out of the sample holder, the sample is heated into either the nematic or the isotropic phase and then cooled slowly down to the smectic phase. Typical cooling speeds are on the order of 1 °C in 6 hours. The thickness of the film is determined by the thickness of the metal plate. The success of this technique is very sensitive to any surface contamination of the metal plate. Contact with other surfaces will cause the liquid crystal to flow out of the hole.

For samples thinner than about 100 μm, this technique is capable of producing monodomain samples. For thicker samples (e.g., up to 1 mm), either a magnetic or an electric field can be used as for the thin film technique.

8.5 Sample orientation

8.5.1 *Surface*

Single domain samples are a great asset in many experiments. Freely suspended films always have excellent alignment of the smectic layers. And, thinning forces tend to drive incomplete layers out of the film, making the film a uniform thickness. For many experiments which have high spatial resolution, e.g., electron diffraction studies of the HexB phases [12], this surface alignment is sufficient.

In other cases, the in-plane structure also needs to be aligned. There are two other fields conveniently available: electric and magnetic.

8.5.2 Electric field

A ferroelectric sample can be aligned by the application of a dc electric field. Two simple electrodes and a high voltage supply are all that is required. The disadvantage is that any ions in the sample will electro-migrate, disrupting the order.

8.5.3 Magnetic field

Another alignment technique is magnetic alignment. Although it is logically possible to use electromagnets (as is done for bulk SmA phase experiments), it is usually much more convenient to use small permanent rare-earth magnets such as $SmCo_5$. The fields created by two permanent magnets can be calculated simply using magnetic circuit techniques if one constructs a magnetic return yoke to trap all of the magnetic flux lines. An example of such a circuit which also utilizes pole pieces to concentrate the field in the gap is illustrated in Fig. 8.16.

To calculate the field strength in the gap, we apply Ampère's circuit law

$$\oint \mathbf{H} \cdot d\boldsymbol{\ell} = I_{enc}^{free} \tag{8.54}$$

to the closed loop shown. Since there are no free currents I^{free} enclosed in the loop, the line integral of $\mathbf{H} \cdot d\boldsymbol{\ell}$ around the magnetic circuit must be zero. Thus

$$H_i \ell_i + 2 h_m \ell_m + H_g \ell_g = 0, \tag{8.55}$$

where the subscripts i, m, and g refer, respectively, to the soft iron yoke, the permanent magnets, and the gap.

Applying another of Maxwell's equations,

$$\oint_S \mathbf{B} \cdot \hat{\mathbf{n}} \, dS = 0 \tag{8.56}$$

we find that if we neglect leakage flux, the flux of $\mathbf{B}$ is constant, and

$$B_1 A_1 = B_2 A_2, \tag{8.57}$$

where $A_{1,2}$ is the cross-sectional area of the circuit. Combining Eq. (8.55) and (8.57) and noting that in the gap $\mathbf{H}_g = \mathbf{B}_g / \mu_0$ and in the yoke $\mathbf{H}_i = \mathbf{B}_i / \mu_i$ where $\mu_i \approx 1000 \times \mu_0$, we obtain

$$\frac{B \ell_i}{\mu_i} + 2\left(\frac{B}{\mu_0} - M\right) \ell_m + \frac{B \ell_g}{\mu_0}\left(\frac{A_g}{A_m}\right) = 0, \tag{8.58}$$

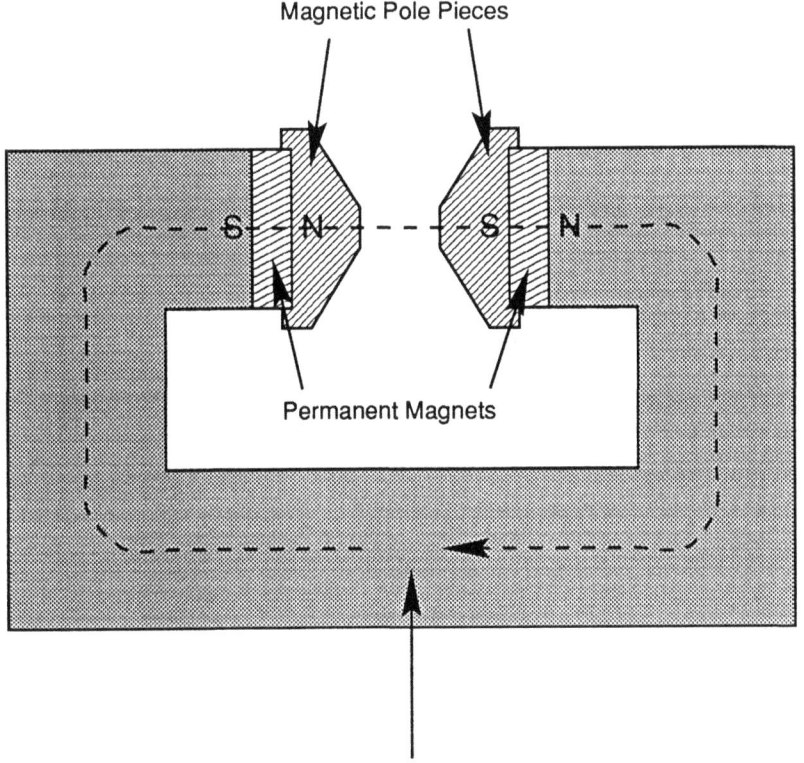

Figure 8.16.   Magnetic circuit energized by a pair of permanent magnets.

where we have set $B_i = B_m = B_g(A_g/A_m)$ and used the relationship

$$\mathbf{H} = \frac{\mathbf{B}}{\mu_0} - \mathbf{M} \tag{8.59}$$

inside the permanent magnets.

Solving Eq. (8.58) for M we obtain

$$M = B\left\{\frac{\dfrac{\ell_i}{\mu_i} + \dfrac{2\ell_m}{\mu_0} + \dfrac{\ell_g}{\mu_0}\left(\dfrac{A_g}{A_m}\right)}{2\ell_m}\right\}. \tag{8.60}$$

As noted above, in a ferromagnet the relationship between $\mathbf{M}$ and $\mathbf{B}$ is not linear and must be measured. To complete the solution of the magnetic circuit problem, we plot Eq. (8.60) on the measured magnetization curve.

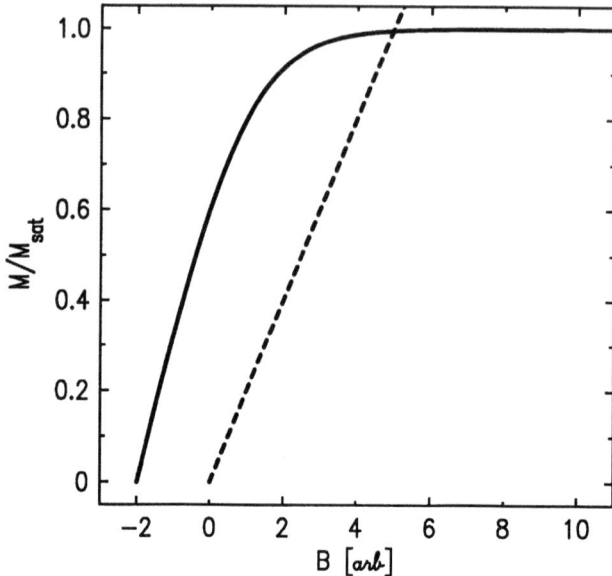

Figure 8.17.   Graphical solution to magnetic circuit problem: —— hysteresis curve; – – – – magnetic circuit.

As shown in Fig. 8.17, Eq. (8.60) is a straight line and the solution is the intersection of the two curves.

8.6 Example experiment: hexatic order in freely suspended films

In order to illustrate the application of many of the previously discussed ideas and techniques, in this section I will review some of my work on x-ray diffraction studies of the HexI phase. This section is largely an adaptation of portions of Ref. 21.

In 1979, D. E. Moncton and R. Pindak [13–15] proposed that, in the thin film limit, freely suspended films of thermotropic liquid crystal should provide an ideal substrate-free system for studying two-dimensional melting. They began a series of studies on the melting of very thin films at the SmA → HexB phase transition, using a synchrotron x-ray source. The expected in-plane diffraction patterns for the fluid SmA and hexatic HexB structures are illustrated in Fig. 8.18. Consider first the normal fluid iso-tropic scattering function. There are three features of note. First, in a close packed two-dimensional system, one expects a peak at a momentum transfer $Q \sim 4\pi/(3^{1/2}a)$ where a is the average molecular separation. Second, if the fluid is isotropic, the intensity and shape of this peak must be

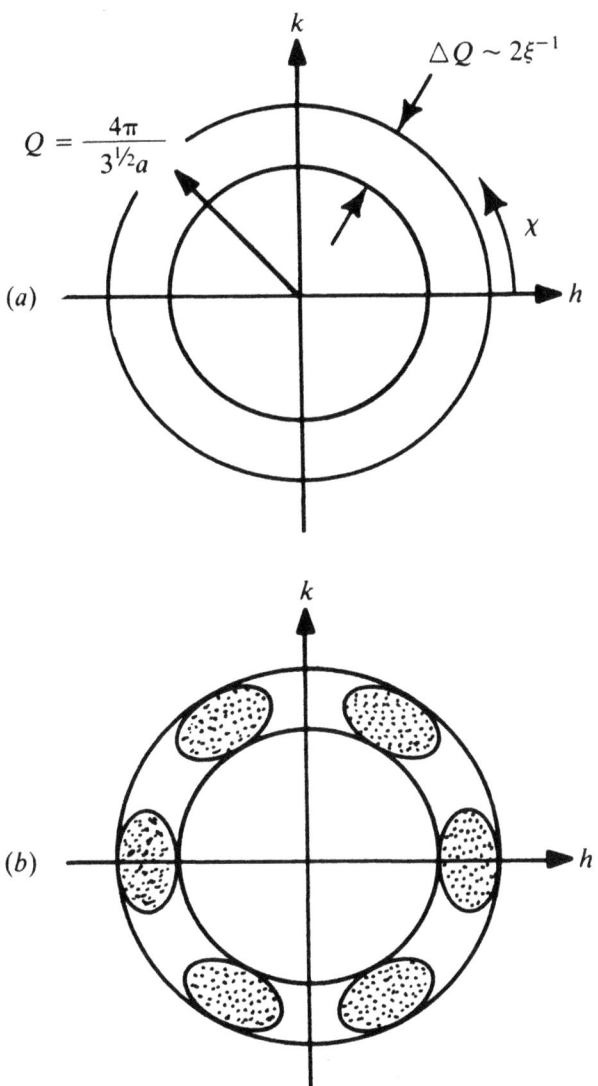

Figure 8.18. (a) Schematic illustration of the in-plane x-ray scattering functions for a fluid smectic liquid crystal. The width of the diffraction peak, ΔQ, measures twice the inverse of the distance over which the molecules are positionally ordered. (b) As above, for a bond-orientationally ordered hexatic.

independent of direction. Therefore, if one scans the angular variable χ, one should not see any variation in the strength of the scattering. Third, the width of the ring, ΔQ, should be $\sim 2/\zeta$ where ζ is the length scale on which positional correlations between the molecules decay. Such diffraction patterns are seen ubiquitously in fluids. The hexatic diffraction pattern is illustrated in the lower panel of Fig. 8.18. The position and widths ΔQ are as for the fluid phase described above. However, in the bond-orientationally ordered hexatic phase, the fluid develops a six-fold modulation in the angular variable χ, that is, the angular anisotropy is broken in the hexatic phase. In spite of this angular anisotropy, the system is still a fluid, due to the short range positional order – albeit, a very viscous fluid. At a lower temperature the material may freeze, condensing the modulated ring of liquid scattering into sharp spots. Sharp diffraction (Bragg) peaks are the signature of a crystal.

Moncton and Pindak's first set of experiments showed that, contrary to the hexatic hypothesis, the smectic-B (previously denoted as S_B) phase found in the liquid crystal compound N-(4-n-butyloxybenzylidene)-4-n-octylaniline (commonly called 4O.8) has three-dimensional long range positional order and is therefore a true crystalline phase [13]. However, Pindak *et al.* [16] later used x-ray scattering to observe a hexatic S_B phase in a freely suspended film of the liquid crystal n-hexyl-4'-n-pentyloxybiphenyl-4-carboxylate which is usually given the simpler label 65OBC. Thus S_B refers to two distinct phases which are now called the CrB (crystal) and the HexB (hexatic) phases, respectively. Although these latter experiments identified a bond-orientationally ordered phase, Pindak and coworkers were not able to obtain quantitative information on the bond-orientational order. Even though the bond-orientational order was well established in their sample, a variable number of domains was present within the area probed, making quantitative measurements of the bond-orientational order impossible. Clearly, *single domain* hexatic samples are necessary to obtain quantitative information on bond-orientational order.

Two major technical advances occurred which allowed major progress in this field. The first clear demonstration that the HexI phase in very thin films has hexatic bond-orientational order was an elegant experiment by Dierker, Pindak and Meyer [17]. A $+2\pi$ point disclination in the director field is a common defect in the SmC phase; in it, the orientation of the director follows a circumferential path around the defect core. In the HexI phase, the tight distortion near the defect core produces a large bond-orientational strain due to coupling between the tilt and the bond-orientational order. This strain can be relaxed, at the expense of director disclination line energy, by creating radial lines of 60° disclinations which separate regions of relatively uniform director orientation. Using depolarized laser reflection microscopy to study a freely suspended film, Dierker *et al.* observed this defect in a very thin film of the liquid crystal compound racemic 4-(2′methylbutyl)phenyl 4-*n*-octylbiphenyl-4-carboxylate (commonly labelled 8SI), thus demonstrating that the HexI phase has hexatic bond-orientational order.

The second advance was the discovery that by applying a small magnetic field, the tilt direction of the SmC phase in films could be oriented at the SmA→SmC phase transition, and that as the liquid crystal cools from the SmC phase to the HexI phase, the hexatic axes develop at fixed angles relative to the molecular tilt and therefore the magnetic field. The net result is a *single domain* HexI-phase sample. The first quantitative measurement of a bond-orientational order parameter was a systematic set of synchrotron x-ray diffraction experiments using freely suspended films of the liquid crystal 8OSI in a small (~1 kG) magnetic field [18]. 8OSI has a rich phase sequence [19] with the following transition temperatures (in °C):

$$I \xrightarrow{174.5} N \xrightarrow{170.0} SmA \xrightarrow{133.4} SmC \xrightarrow{79.9} HexI \xrightarrow{75.1} CrJ \xrightarrow{61.9} CrK$$

which includes the essential phase sequence SmA → SmC → HexI needed to obtain single domain hexatic samples.

There are two limitations when using the *tilted hexatic* phases to study bond-orientational order. First, if the phase transition in the idealized field-free case were a continuous transition, the tilt field would introduce a linear term in the free energy which would formally destroy the phase transition. Second, the two-fold symmetry of the tilt field will destroy the six-fold symmetry of the hexatic axes. However, as long as the coupling between the bond orientational order and the molecular tilt is weak there is no serious difficulty; the critical behavior may be extracted by considering the full equation of state. The loss of six-fold symmetry may be a more serious problem. One must observe the scattering on a case-by-case basis to determine if the loss of six-fold symmetry is significant.

8.6.1 *Experiments on thick films*

As discussed earlier, the in-plane x-ray scattering function of the SmA phase is a diffuse ring characteristic of the fluid order in the smectic layers. In the case of the SmC phase, the shape of the diffuse ring is more complex. Due to the tilt of the molecules, the molecular form factor is tilted with respect to the plane of the smectic layers; effectively tilting the diffuse ring.

Accordingly, it is necessary that scans probing the bond-orientational order be done in the *tilted* plane in reciprocal space which contains the maximum of the molecular form factor. For simplicity, we still refer to these more complicated scans probing the bond orientational order as χ scans. The finite size of the beryllium window may limit how far out of the smectic plane we can obtain data. We can maximize this distance by using the orientation matrix techniques described above. An important point to bear in mind is that, in an x-ray scattering experiment, any modulation in a χ scan indicates long range bond-orientational order. The orientation of the axes must persist over lengths comparable to the dimensions of the illuminated area; otherwise, different regions of the illuminated area will average out the orientational structure.

Figure 8.19 shows χ scans at several temperatures near the SmC $\rightarrow$ HexI phase transition in a thick film of 8OSI [18]. At high temperatures, the ring is uniform to within counting statistics. As the sample is cooled, a measurable sinusoidal modulation of the ring develops. At $T \sim 77.5\,°C$ the χ scan shows definite peaks every $60°$ indicating substantial amounts of hexatic bond-orientational ordering. At high temperatures, longitudinal scans through the same peak show a broad diffuse scattering profile indicative of short range positional order. As the sample is cooled, the width of the peak narrows simultaneously with the development of the bond-orientational order, suggesting that the enhanced positional correlations are due to a coupling to the bond-orientational order. The peak width never approaches the resolution of the diffractometer ($\Delta Q = 2 \times 10^{-4}\,\text{Å}^{-1}$); the positional order is always short-range. Thus the material is indeed a *bulk hexatic* in the HexI phase; that is, it has the positional order of a fluid and the bond orientational order of a crystalline solid. When the material is cooled below $\sim 73\,°C$ into the CrJ phase, the broad peaks of Fig. 8.19 and of the corresponding longitudinal scan become sharper than the resolution of the diffractometer. Thus the *bulk* CrJ phase is actually *crystalline* rather than a liquid crystal.

8.6.2 Experiments on thin films

As discussed above, one of the attractive features of the free standing film technique is that one can produce stable films whose thickness varies from two molecules to macroscopic values. Since the notion of a hexatic phase originates in the Halperin–Nelson model for two-dimensional freezing [20], one naturally asks how the bulk smectic states evolve as the film becomes

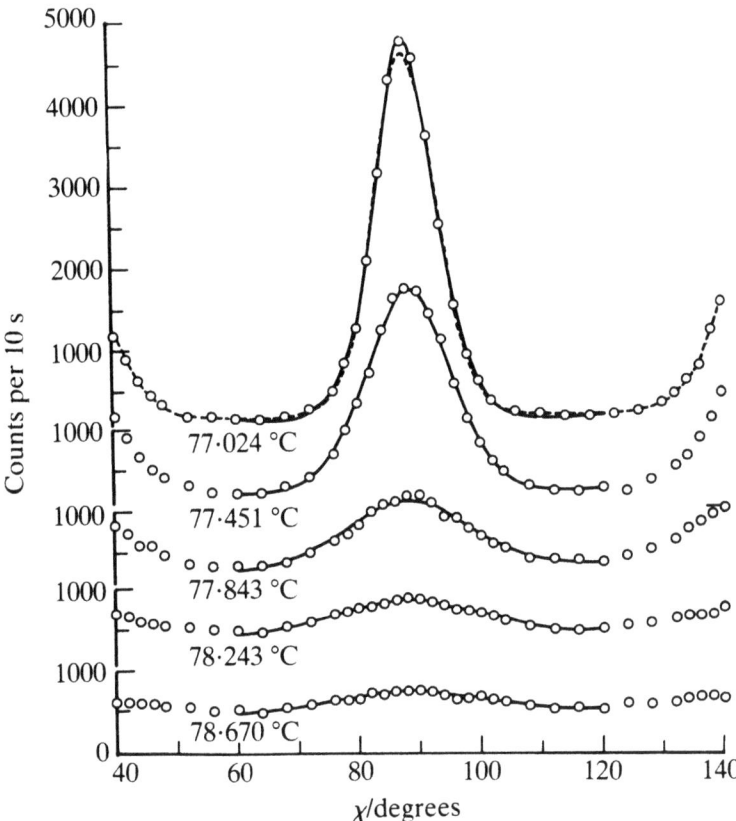

Figure 8.19. Angular scans in a thick film of 8OSI in the immediate vicinity of the SmC → HexI phase transition for given temperatures. These data are from Ref. 18.

thinner and thinner. Such experiments were performed on single domain films of 8OSI with thicknesses of 22 and 4 molecules [21]. Results for the 22-layer film are shown in Fig. 8.20. At 80.16 °C the material is in the SmC phase and the χ scan is indeed isotropic as expected. As the film is cooled, angular structure develops, demonstrating that the material is going into a bond orientationally ordered phase. This is exactly similar to the three-dimensional HexI phase data shown in Fig. 8.19 except that the bond-orientational order is much less well developed. Measurements of ΔQ (Fig. 8.18) show that the material is positionally disordered, that is, it is a fluid. Identical results are obtained in the four-layer film. Thus 22 layers already put on the two-dimensional limit for this material. Again, when the film is cooled below 72 °C, it goes into the crystalline CrJ phase.

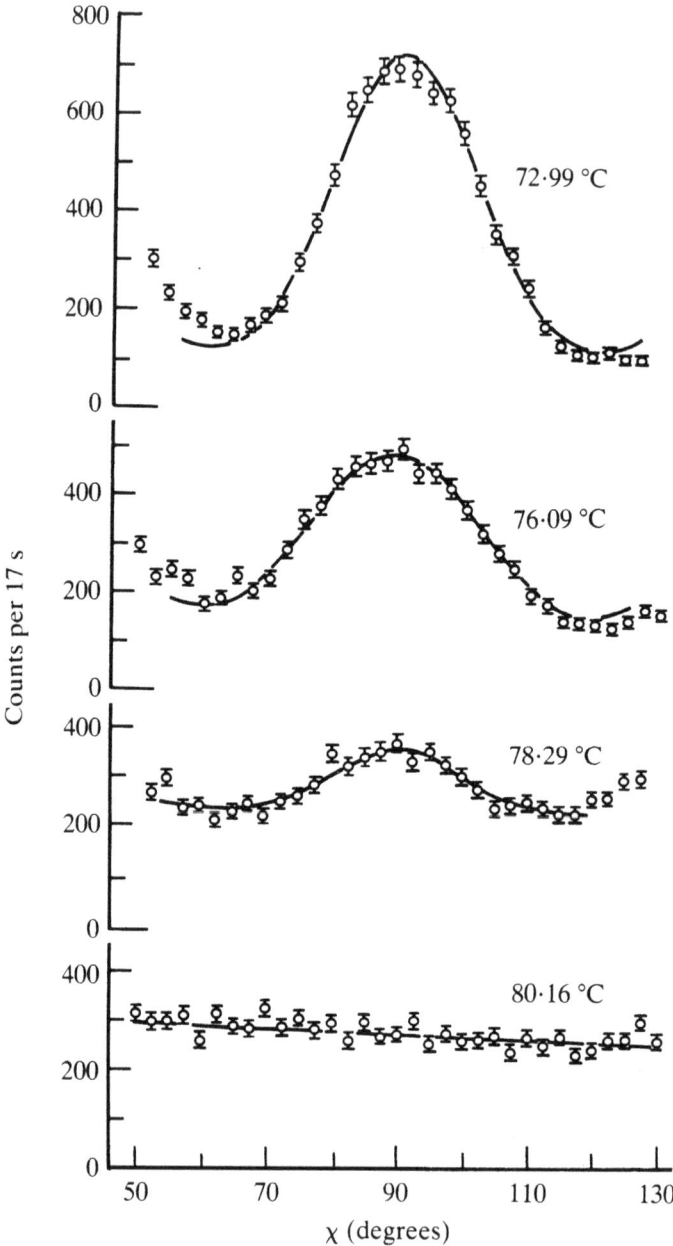

Figure 8.20. Angular scans in a film of 8OSI of thickness 22 molecules. These show the evolution from a two-dimensional isotropic fluid (80.16 °C) to a two-dimensional hexatic phase (78.29–72.99 °C). At lower temperatures the material forms a two-dimensional solid. Data from Ref. 21.

8.6.3 Measuring the bond-orientational order parameter

The availability of single-domain HexI phase samples makes it possible to characterize the bond-orientational order quantitatively. This can be done quite generally by performing a non-linear least-squares fit of the χ scan data between 60° and 120° to the Fourier cosine series [18]:

$$S(\chi) = I_0 \left\{ \frac{1}{2} + \sum_{n=1}^{\infty} C_{6n} \cos 6n(90° - \chi) \right\} + I_{\text{BG}}, \tag{8.61}$$

where χ is the angle between the in-plane component of the scattering vector **q** and the magnetic field **H**. The coefficients C_{6n} measure the amount of $6n$-fold ordering in the sample. I_{BG} is a small background correction. With the constant term chosen as $\frac{1}{2}$ in Eq. (8.61), each C_{6n} approaches 1 for perfect bond-orientational order. Each of the C_{6n} is an independent bond-orientational order parameter. The temperature dependence of the first seven members of the set of bond-orientational order parameters, $\{C_{6n}\}$, is shown in Fig. 8.21; as noted above, all of the C_{6n} evolve continuously with temperature. The fits show explicitly that as the temperature falls the system smoothly develops first six-fold order, then twelve-fold order, then eighteen-fold order and so on. There is no abrupt transition; rather, the evolution is smooth and continuous. This behavior confirms the prediction that coupling between the tilt and the hexatic fields induces hexatic ordering in the SmC phase, destroying the sharp SmC → HexI phase boundary [23, 24].

The most astonishing result produced by this analysis was the empirical discovery of the simple scaling relation: $C_{6n} = C_6^{\sigma_n}$ [18]. The average values of σ_n and their standard deviations are given in Table 8.1 and plotted in the inset of Fig. 8.21. With this scaling, all C_{6n}^{1/σ_n} out to $n = 7$ fall on the same curve to within the fitting error over the complete temperature range with no adjustment in the amplitude.

It turns out that a complete theoretical understanding of the new set of exponents, $\{\sigma_n\}$, can be obtained using standard approaches to critical phenomena. A three-dimensional renormalization group calculation [25] of the critical behavior of the three-dimensional XY model in a weak ordering field, h, predicts that, in the critical region,

$$\sigma_n = n + x_n n(n-1), \tag{8.62}$$

with $x_n \approx 0.3 - 0.008n$. Of course, the system is generally not asymptotically close to the phase transition. However, even far from the critical region, the correction to the mean-field result ($\sigma_n = n$) scales like a temperature dependent constant raised to the power $n(n-1)$. Thus one may write

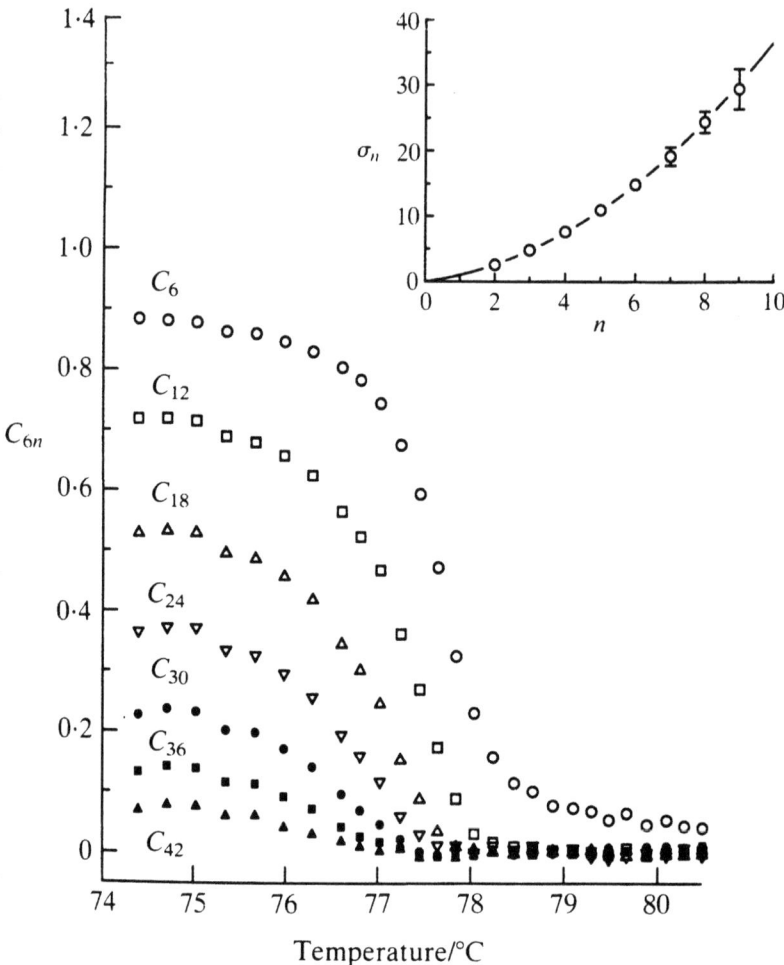

Figure 8.21.   First seven Fourier coefficients (Eq. (8.75)) describing the hexatic ordering in a thick film of 8OSI. The inset shows the scaling exponents σ_n versus harmonic number n together with the best fit to the theoretical functional form $\sigma_n = n + \lambda n(n-1)$. Data from Ref. 22.

$$\sigma_n = n + \lambda(T)n(n-1). \tag{8.63}$$

As an illustration, the measured values of σ_n are plotted in the inset of Fig. 8.21, and the best fit to this form (with constant λ) is plotted as a solid line. Obviously, the agreement is quite good. Alternatively, the raw data may be re-fitted to Eq. (8.61) using the scaling form,

$$C_{6n} = C_6^{n + \lambda(T)n(n-1)}. \tag{8.64}$$

Table 8.1 *Harmonic scaling exponents.*

n	σ	$\delta\sigma$	n	σ	$\delta\sigma$
1	1.00	0.00	5	10.96	0.44
2	2.57	0.05	6	14.85	0.70
3	4.80	0.14	7	19.2	1.4
4	7.59	0.27			

The solid lines in Fig. 8.19 and 8.20 are the results of such a fit. Again, $\lambda(T)$ turns out to be practically temperature independent and close to the theoretical asymptotic value, $\lambda \simeq 0.3$. Thus, the behavior of the bond-orientational order in the HexI phase of thick films is well described by a three-dimensional XY model in an orienting field. The beauty of the analysis is that the finite ordering field in the tilted hexatic has not hidden interesting behavior but rather has made possible a direct measurement of the C_{6n}; the right-hand side of Eq. (8.61) follows even when $h \neq 0$, and C_6 is a function of both T and h.

Acknowledgments

I gratefully acknowledge the advice and instruction of Ron Pindak, Larry Sorenson, Jeff Collett, John Budai, Cirus Safinya, Sam Sprunt, Satyendra Kumar, David Litster, Ben Ocko, Robert Birgeneau and Eric Sirota on liquid crystal, x-ray, and optical experiments. The design discussed in this chapter is a linear descendent of their earlier sample chambers.

References

1. W. C. Elmore and M. A. Heald, *Physics of Waves*, Dover, Mineola, NY (1985).
2. C. C. Huang, private communication.
3. F. Reif, *Fundamentals of Statistical and Thermal Physics*, McGraw-Hill Book Company, New York (1965).
4. W. R. Busing and H. A. Levy, *Acta Cryst.* **22**, 457 (1967).
5. Jerry B. Marion, *Classical Dynamics of Particles and Systems*, 2nd Edn. Academic Press, New York (1970) Section 12.7.
6. R. Schaetzing, Ph.D. thesis, M.I.T., 1980, p. 151.
7. M. Born and E. Wolf, *Principles of Optics*, 6th Edn., Pergamon Press, New York (1980) Section 1.6.4.
8. Commission Internationale de L'Eclairage: CIE Publication No. 15, *Colorimetry* by committee 1.3.1, Bureau Central de la CIE, 4 Ave du Recteur Poincaré 75, Paris 16me.
9. K. L. Kelly, Color Designations for Lights, *J. Research NBS* **31**, 271 (1943) RP1565.

10. C. Y. Young, R. Pindak, N. A. Clark, and R. B. Meyer, *Phys. Rev. Lett.* **40**, 773 (1978).
11. J. Przedmojksi and S. Gierlotka, *Liq. Cryst.* **3**, 409 (1988); S. Gierlotka, J. Przedmojski, and B. Pura, *Liq. Cryst.* **3**, 1535 (1988).
12. See, for example, M. Cheng, J. T. Ho, S. W. Hui, and R. Pindak, *Phys. Rev. Lett.* **59**, 1112 (1987).
13. D. E. Moncton and R. Pindak, *Phys. Rev. Lett.* **43**, 701 (1979).
14. D. E. Moncton and R. Pindak, in *Ordering in Two Dimensions: Proceedings of an International Conference Held at Lake Geneva, U.S.A.*, S. K. Sinha (Ed.), North Holland, New York (1980).
15. D. E. Moncton, R. Pindak, S. C. Davey, and G. S. Brown, *Phys. Rev. Lett.* **49**, 1865 (1982).
16. R. Pindak, D. E. Moncton, S. C. Davey, and J. W. Goodby, *Phys. Rev. Lett.* **46**, 1135 (1981).
17. S. B. Dierker, R. Pindak, and R. B. Meyer, *Phys. Rev. Lett.* **56**, 1819 (1986).
18. J. D. Brock, A. Aharony, R. J. Birgeneau, K. W. Evans-Lutterodt, J. D. Litster, P. M. Horn, G. B. Stephenson, and A. R. Tajgbakhsh, *Phys. Rev. Lett.* **57**, 98 (1986).
19. A. J. Leadbetter, J. P. Goughan, B. Kelly, G. W. Gray, and J. Goodby, *J. Phys. (Paris)* **40**, C3-178 (1979).
20. B. I. Halperin and D. R. Nelson, *Phys. Rev. Lett.* **41**, 121 (1978).
21. J. D. Brock, D. Y. Noh, B. R. McClain, J. D. Litster, R. J. Birgeneau, A. Aharony, P. M. Horn, and J. C. Liang, *Z. Phys. B* **74**, 197 (1989).
22. J. D. Brock, R. J. Birgeneau, J. D. Litster, and A. Aharony, *Contemp. Phys.* **30**, 321 (1989).
23. D. R. Nelson and B. I. Halperin, *Phys. Rev. B* **21**, 5312 (1980).
24. R. Bruinsma and D. R. Nelson, *Phys. Rev. B* **23**, 402 (1981).
25. A. Aharony, R. J. Birgeneau, J. D. Brock, and J. D. Litster, *Phys. Rev. Lett.* **57**, 1012 (1986).

9

X-ray surface scattering studies of
liquid crystals

SUNIL K. SINHA

Advanced Photon Source, Argonne National Laboratory, Argonne, IL 60439, USA

9.1 Introduction

Most studies of liquid crystals have dealt with their fascinating bulk properties, particularly their exotic types of structural order and the associated phase transitions between them. The manner in which such structural properties are modified at the surface or interface of thin films has received less attention, but is perhaps equally important from both the scientific and technological points of view. Meanwhile, the use of x-ray and neutron scattering to study surfaces, interfaces, and thin films of both solids and liquids has developed rapidly over the last several years. In this field, there are three broad areas one may consider: namely, x-ray reflectivity (XR) or neutron reflectivity (NR) (restricted to specular reflection of the incident radiation), grazing incidence x-ray and neutron diffraction (GIXD, GIND), and off-specular or diffuse scattering (DS). It is the purpose of this article to acquaint the reader with the basic elements of the theory and practice of these methods, with some illustrations of their application to the study of surface and thin film structures of liquid crystals and polymeric systems.

Specular reflectivity measurements are capable of yielding (down to molecular length scales) the density profile of a solid or liquid in the direction normal to the surface or interface. It can be used to probe phenomena such as surface crystallization, surface-induced ordering, or surface roughness. Grazing incidence diffraction measurements are used to study in-plane periodic structures at or near a surface or interface, and may be used to study the structural details of surface crystallization, as well as its depth dependence. Finally, off-specular or diffuse scattering may be used to study the structural details of in-plane fluctuations, such as capillary waves on liquid surfaces, the appearance of random surface inhomogeneities, the morphology of surface roughness and the like.

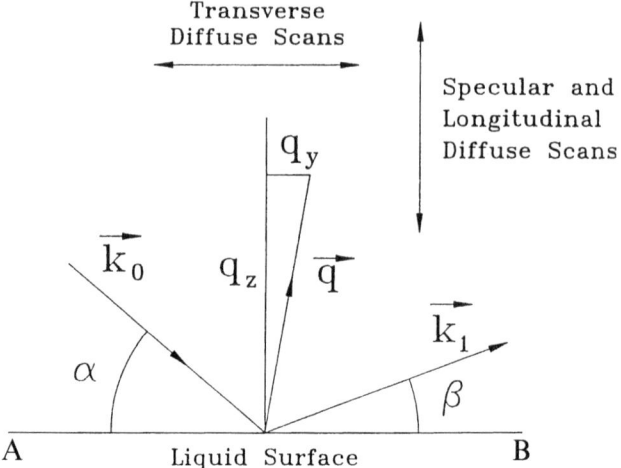

Figure 9.1.   Schematic diagram of a surface scattering experiment, with incident wavevector denoted by $\mathbf{k}_0$, scattered wavevector denoted by $\mathbf{k}_1$, and scattering wavevector denoted by $\mathbf{q}$. The grazing angle of incidence with the surface is α, and the grazing angle of scattering is β. Also shown are the 'specular ridge' and trajectories in $\mathbf{q}$-space for longitudinal and transverse diffuse scans.

Figure 9.1 illustrates schematically a surface scattering experiment in the geometry conventionally used for studies of specular reflectivity and diffuse scattering. The figure is a hybrid representation of both real and reciprocal (wavevector) space. The real-space part is the physical surface (or interface) represented by the line AB ($z = 0$) to denote its disposition. (We consider the z coordinate in what follows to *increase* from $z = 0$ into the medium below.) The rest of the figure is to be interpreted in reciprocal space. The wavevector of the incident radiation is identified by $\mathbf{k}_0$ and that of the scattered radiation by $\mathbf{k}_1$. The angles these vectors make with the surface are identified by α (grazing angle of incidence) and β (grazing angle of scattering), respectively. The plane formed by the normal to the surface $\mathbf{n}$ and by $\mathbf{k}_0$ is called the *plane of incidence*. The scattering wavevector $\mathbf{q}$ is defined by

$$\mathbf{q} = \mathbf{k}_1 - \mathbf{k}_0. \tag{9.1}$$

The plane formed by $\mathbf{k}_0$ and $\mathbf{k}_1$ is called the plane of scattering. If $\mathbf{k}_1$ lies in the plane of incidence (as shown in Fig. 9.1) and $\alpha = \beta$, we have the case of specular reflection, and $\mathbf{q}$ is along the direction of $\mathbf{n}$, sometimes also known as the specular ridge. Note that for elastic scattering processes (to which we restrict ourselves in the article) $\mathbf{k}_1 = \mathbf{k}_0$. The diffractometer on which the experiment is performed may be used to control the direction of $\mathbf{k}_0$, $\mathbf{k}_1$

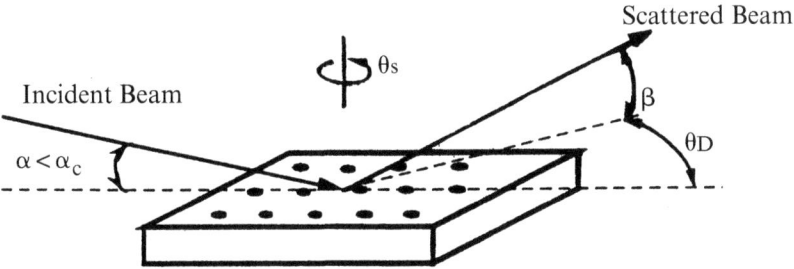

Figure 9.2. Schematic diagram for grazing incidence diffraction. $\mathbf{k}_1$ is offset from the plane of incidence by an angle of θ_D projected into the plane of the surface.

relative to the surface so that scans may be made of the scattered intensity as a function of $\mathbf{q}$, either along the specular ridge (keeping $\alpha = \beta$) as for reflectivity experiments, or, in the case of diffuse scattering studies, along tracks offset from the specular ridge (longitudinal diffuse scans) or perpendicular to it (transverse diffuse scans), both of which are schematically illustrated in Fig. 9.1. Note that so-called 'rocking-curve' scans (obtained by rotating the direction of $\mathbf{n}$ in-plane while keeping $\mathbf{k}_0$ and $\mathbf{k}_1$ fixed) amount to scanning $\mathbf{q}$ in the arc of a circle about the origin (since $\mathbf{q}$ is always measured relative to the sample axes) which, for small enough angular ranges, approximates transverse diffuse scans. If $\mathbf{k}_1$ is not restricted to the plane of incidence (as illustrated in Fig. 9.2), we achieve the conditions of grazing incidence diffraction (GID). In such experiments, α is restricted to be slightly below the critical angle α_c for the reflection (so that the beam is evanescent below the surface and is most sensitive to the surface structure), and $2\theta_D$ is varied so that $\mathbf{q}_{\|}$, the in-plane component of $\mathbf{q}$, matches one of the two-dimensional (2D) reciprocal lattice vectors of the ordered structure on the surface. For a purely 2D structure, the 'Bragg reflections' extend along rods in reciprocal space emanating from those reciprocal lattice points and oriented normal to the surface. (For a three-dimensional (3D) crystal structure truncated at a surface, there would be streaks of intensity along similar rods normal to the surface and passing through the 3D reciprocal lattice points. These are known as 'truncation rods' [1, 2].) In GID experiments, the intensity along a rod can also be measured and the dependence of the intensity on q_z is related to the modulus squared of the Fourier transform of the z dependence of the scattering density of the 2D (or quasi-2D) structure. For a single ordered layer of molecules, this Fourier transform is the projection of the individual molecular form factor along the z axis.

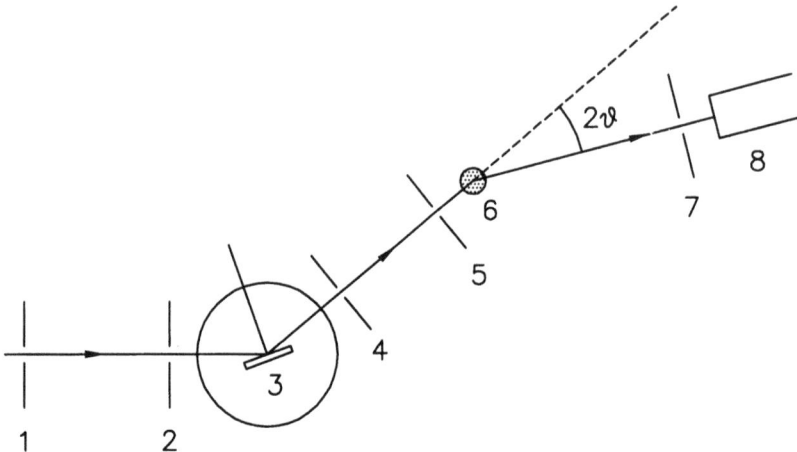

Figure 9.3. Schematic diagram for a surface scattering experiment carried out with a two-circle diffractometer. 1, 2, 4, 5, 7 denote collimating slits; 3 is the mono-chromating crystal; 6 denotes the sample and 8 the detector at a scattering angle denoted by 2θ.

In Section 9.2 we shall outline some of the experimental details regarding such types of experiments. In Section 9.3 we shall develop the formalism for describing and interpreting specular reflectivity measurements and discuss examples of such studies. In Section 9.4 we shall do the same for off-specular (GID and diffuse scattering) measurements. Finally, in Section 9.5, we shall conclude with a brief summary of this chapter.

9.2 Experimental details

X-ray experiments are generally carried out with tube sources, rotating-anode sources, or synchrotron radiation. Neutron experiments are carried out at reactor sources or pulsed neutron sources. Figure 9.3 shows a schematic of a two-circle diffractometer used to study specular reflectivity or in-plane diffuse scattering at conventional x-ray sources or reactor-based neutron sources. The incident beam is monochromated with a single-crystal monochromator and collimated with slits to define $\mathbf{k}_0$ on the sample. The scattered beam $\mathbf{k}_1$ is defined by slits also and the detector can rotate about the sample center. Independent rotation of the sample and detector allows one to set α and β independently and thus to carry out any of the scans in the reciprocal space of the sample discussed in the previous section. Sometimes a position-sensitive linear or 2D detector is used on the scattered side, so that the scattered intensity can be simultaneously measured over a range of βs (including out-of-plane directions). Detector counts are usually normalized to a given number of counts in a monitor detector placed in the incident beam to allow for fluctuations in beam intensity. Often pyrolytic graphite (002) crystal or perfect silicon (Si) or germanium (Ge) crystals are used as monochromators. For attaining higher resolution, sometimes an analyzer crystal is used before the detector. The scattering plane is, in general, the horizontal plane. For synchrotron sources on the other hand, the scattering plane is usually vertical to take advantage of the naturally small beam divergence in the vertical direction characteristic of synchrotron radiation, and in general a double-crystal monochromator consisting of two parallel or channel-cut perfect single crystals of Si or Ge is used. Since the specularly reflected beam can often be of the same intensity as the incident beam (for total reflection) or close to it, care must often be taken (particularly at synchrotron sources) to avoid damage to the detector by inserting suitable attenuators before the detector.

Higher-order contamination of the nominally monochromatic beam is sometimes a problem. This is radiation of wavelength $\lambda/2$ (where λ is the wavelength of the primary reflection) in the case of graphite (002) monochromators or of wavelength $\lambda/3$ in the case of Si(111) or Ge(111) monochromators, and can be a problem particularly in the case of synchrotron sources or reactor neutron sources. For neutrons, $\lambda/2$ radiation is usually suppressed by working at a neutron wavelength of $\lambda = 2.35$ Å and using a pyrolytic graphite filter in the incident beam to remove second-order contamination [3]. For synchrotron radiation, the energy cut-off of the focusing mirror often considerably reduces higher-energy contamination, but other methods may be used such as pulse-height discrimination or the use of energy-dispersive solid state detectors. Since, as we shall see, the reflectivity from most samples drops precipitously with the magnitude of the scattering vector q, second-order or third-order contamination (at $2q$ and $3q$, respectively) will not usually seriously affect the shape of the measured reflectivity curve, but may present a problem in *normalizing* the reflected intensity by the incident beam intensity to obtain the absolute reflectivity. (This is particularly true when substantial beam attenuation is introduced to protect the detector, since the attenuator will transmit far greater amounts of the higher energy contamination than of the primary radiation).

The angles α, β (see Fig. 9.1) are set by fixing the detector scattering angle (often referred to as $2\theta_s$) and the sample table rotation angle (often referred to as θ_s). In our notation, $\theta_s = \alpha$ and $2\theta_s = \alpha + \beta$, so that specular reflectivity measurements are carried out using conventional 'θ–2θ' scans on the two-circle diffractometer, where $2\theta_s$ is always set equal to twice the value of θ_s.

Other types of scans may be set by using the relations derived from Eq. (9.1) with respect to the sample axes:

$$k_0 (\sin \alpha + \sin \beta) = q_z,$$
$$k_0 (\cos \alpha - \cos \beta) = q_x, \qquad (9.2)$$

where the x and z axes are as defined in Fig. 9.1. ($k_0 = 2\pi/\lambda$.)

For specular reflectivity measurements, certain corrections have to be made before converting raw data to true reflectivity. The first is connected with the fact that for small enough values of α, the reflecting surface will not intercept all of the incident beam, thus reducing the actual reflectivity by a factor of ($\ell \sin \alpha/W$), where ℓ is the length of the reflecting surface in the plane of incidence and W is the total width of the incident beam. This correction (referred to as the 'beam-footprint correction') ceases to apply once $\sin \alpha > W/\ell$. The second correction is due to the fact that even at the specular condition, the measured intensity is the sum of specular reflection and diffuse scattering from the surface (and possibly also from the bulk of the sample). The non-specular part can be estimated by measuring a corresponding point slightly off the specular condition (i.e., with the same q_z) and assuming that the diffuse scattering at that point is not significantly different from its value on the specular ridge. (This assumption sometimes is invalid, in which case more elaborate methods must be used.) This estimate of the diffuse scattering can be achieved by a slight mis-set of the sample orientation α on either side of the specular reflection, while keeping $2\theta_s(= \alpha + \beta)$ constant, or by an actual reciprocal space scan along q_z, using Eqs. (9.2) with q_x set equal to $\pm\delta$ (δ being a small quantity). The mis-set in α must be just large enough so that any contamination by the true specular reflection is negligible, i.e., beyond the instrumental resolution width of the specular reflection. This diffuse background may usually be substracted off point by point from the measured specularly reflected intensities, and after making the 'beam-footprint correction' discussed above, may be divided by the incident beam intensity to get the *true* specular reflectivity. There is no correction for the instrumental resolution in this procedure, as long as one sits always at the actual specular condition. This is because the detector, when measuring the incident beam intensity, receives the same fraction of the rays that it does from the specularly reflected beam when set for specular reflection at $2\theta_s$, since the latter case simply corresponds to an angular displacement of the source in the plane of incidence by $-2\theta_s$. Absolute reflectivities measured in this way should be very close to unity below the critical angle for total reflection (except for the effects of absorption). Usually for well-aligned experiments, values greater than 90% are measured. Lower values can sometimes be obtained due to significant amounts of $(\lambda/2)$ or $(\lambda/3)$ contamination in the beam if attenuators are being used. Sometimes, particularly in the case of high-resolution experiments, small misalignments cause a nominal specular scan to veer off its actual specular ridge in reciprocal space as a function of q_z. Rocking curves at various points on the scan can usually be used to check whether this is occurring. If an analyzer crystal is used in this configuration, it must be borne in mind that the image of the monochromator crystal is inverted in the specular reflection process so that the 'non-dispersive' (focusing) configuration now becomes a 'bounce' in the *same* sense as in the monochromator crystal, instead of in the opposite sense as normally used.

For careful studies of the specular or diffuse scattering, one must bear in mind that the observed intensity $I(\mathbf{q})$ for any scattering vector $\mathbf{q}$ is a convolution or 'smearing' of the actual scattering function $S(\mathbf{q})$ with the instrumental resolution function $R(\mathbf{q})$, for example,

$$I(\mathbf{q}) = \int d\mathbf{q}' \, R(\mathbf{q} - \mathbf{q}') S(\mathbf{q}'). \tag{9.3}$$

This resolution function is often approximated by a Gaussian function whose half-height contour is represented by an ellipsoid in reciprocal space. The width and orientation of this resolution ellipsoid may be obtained by using Eq. (9.2) to calculate the spread in q_x, q_z due to small (independent) spreads in α, β due to finite collimation of the beam. (We neglect here effects due to a spread in the energy of the incident beam.) Thus, we have

$$\left. \begin{aligned} \langle \Delta q_z^2 \rangle &= k_0^2 \cos^2 \alpha \langle \Delta \alpha^2 \rangle + k_0^2 \cos^2 \beta \langle \Delta \beta^2 \rangle \approx k_0^2 \left[\langle \Delta \alpha^2 \rangle + \langle \Delta \beta^2 \rangle \right] \\ \langle \Delta q_x^2 \rangle &= k_0^2 \sin^2 \alpha \langle \Delta \alpha^2 \rangle + k_0^2 \sin^2 \beta \langle \Delta \beta^2 \rangle \approx k_0^2 \left[\alpha^2 \langle \Delta \alpha^2 \rangle + \beta^2 \langle \Delta \beta^2 \rangle \right] \\ \langle \Delta q_y^2 \rangle &= k_0^2 \langle \Delta \gamma^2 \rangle \end{aligned} \right\} \tag{9.4}$$

$$\begin{aligned} \langle \Delta q_x \Delta q_z \rangle &= k_0^2 \sin(2\beta) \langle \Delta \beta^2 \rangle - k_0^2 \sin(2\alpha) \langle \Delta \alpha^2 \rangle \\ &\approx 2k_0^2 \left[\beta \langle \Delta \beta^2 \rangle - \alpha \langle \Delta \alpha^2 \rangle \right], \end{aligned} \tag{9.5}$$

where $\langle \Delta \gamma^2 \rangle$ is the sum of the mean-square angular divergence of the incident and scattered beams out of the plane of incidence. (We have used the approximation that α, β are small angles.) Equations (9.4) and (9.5) may be used to obtain the parameters of the Gaussian resolution function. We shall not go into details here.

Ideally, the collimations should be set so that the spreads of α and β are matched. The resolution function is then symmetric about the specular ridge. However, if for instance the detector opening dominates the resolution width, the resolution function broadens considerably on going from one side of the specular ridge (small β) to the other (large β) in a rocking curve, or transverse scan, making the diffuse scattering spectra look asymmetric. For many purposes, the out-of-plane collimation is kept wide open so that q_y is in effect integrated over. However, it may also be collimated by means of slits in the incident and scattered beams. From Eq. (9.4) it may be seen that the *in-plane transverse* resolution width $\langle \Delta q_z^2 \rangle$ is very small because of the factors α^2 or β^2. Typically, this means that one may resolve features in the diffuse scattering in the transverse direction as low as $\sim 10^{-4}$ Å^{-1} away from the specular ridge, for reasonably tight collimation. (With synchrotron sources it is feasible to get resolutions of $\sim 10^{-5}$ Å^{-1} in this direction.) Thus, using this technique it is possible to resolve surface inhomogeneities, fluctuations, or ordering on length scales up to tens of micrometers, which is remarkable considering the wavelength of the radiation used to probe the structure ($\sim 1 - 4$ Å)!

Geometrical considerations prevent one from probing large values of $q_\parallel$ for small values of q_z, as can be seen from an examination of Eqs. (9.2) (since α, β must both be positive). To reach these, one must employ the grazing incidence geometry illustrated in Fig. 9.2, where one can move the detector out of the plane of incidence, as well as up and down above the plane of the surface. Liquid surfaces are often studied with diffractometers of this type [4]. Obviously a liquid surface must be horizontal and thus cannot be 'rocked' in the conventional sense. Instead, the incident beam direction (or α) is tilted from the horizontal by means of a second monochromator or mirror, and the scattered beam direction (or β) similarly set by moving the detector and slit system up and down. The liquid surface height is simultaneously adjusted to maintain the 'footprint' of the beam centered on the liquid surface. Figure 9.4 illustrates a liquid surface diffractometer of this type at the National Synchrotron Light Source at Brookhaven National Laboratory, constructed and operated by a team from Harvard University and Brookhaven National Laboratory. Using such a diffractometer, one can freely probe both q_z and $q_\parallel$ space, although the relevant in-plane resolution is now in the q_y direction (out of the plane of incidence) and by Eq. (9.4), this is much worse than what is achievable in the plane of incidence.

Liquid surface neutron diffractometers also exist on reactor-based and pulsed neutron sources. For pulsed neutron sources, rather than reactors, the method used to measure surface scattering is rather different than described above. A *white* pulsed beam of neutrons (suitably filtered to remove very fast neutron contamination) is collimated and incident on the surface at a *fixed* grazing angle of incidence α. Usually, a fixed linear or 2D position-sensitive detractor is used to detect the scattered neutrons as a function of β and time of arrival (relative to the source pulse). From the

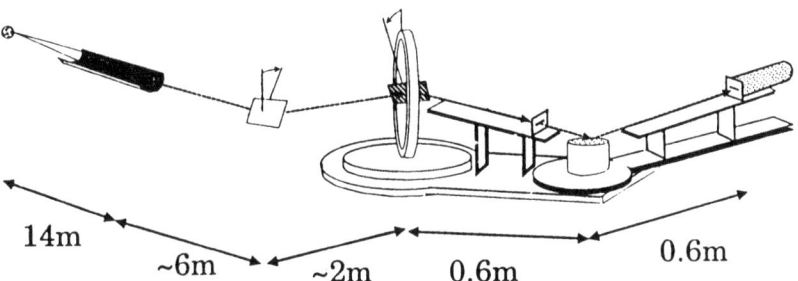

14m   ~6m   ~2m   0.6m   0.6m

Figure 9.4. Schematic of a liquid surface diffractometer similar to the Harvard–BNL Diffractometer at the National Synchrotron Light Source, Brookhaven. The incident beam is tipped down onto the liquid surface by a second monochromator, and the sample and detector heights can be changed automatically to achieve desired values of α and β. The detector can also be moved in the horizontal plane for grazing incidence diffraction experiments. (After M. L. Schlossman and P. S. Pershan, in *Light Scattering by Liquid Surfaces and Complementary Techniques*, Ed. D. Langevin, Marcel Dekker, New York (1992), p. 365.)

total path length and time of flight, the velocity and the wavelength of the neutron involved in the scattering process is obtained. Thus by Eq. (9.2), the value of $\mathbf{q}$ and the intensity as a function of scattering wavevector is obtained. This method is sometimes also used at a reactor source in conjunction with a neutron chopper to pulse the beam. The advantage of this method is that neither the incident beam nor the sample orientation is changed during the scans, so the 'footprint' of the beam on the sample stays constant. An alternative way of carrying out grazing incidence diffraction which does not require the detector to move simultaneously in two planes is to use a four-circle diffractometer on which the sample is mounted and to rotate the plane of the surface being studied about an axis in the scattering plane, i.e., the plane defined by $\mathbf{k}_0$, $\mathbf{k}_1$. This serves to rotate the plane of incidence out of the scattering plane.

In studying surface scattering, one sometimes has to distinguish it from bulk scattering, since even bulk diffuse scattering can be comparable to or larger than most types of surface scattering (unless α is below the critical angle, in which case the beam never penetrates into the bulk). This may be done in terms of the symmetries of different types of scattering. Thus, for instance, for liquid or disordered bulk samples, the bulk scattering intensity is a function of $|\mathbf{q}|$ only, and thus depends only on $(\alpha + \beta)$, since

$$|\mathbf{q}| = (4\pi/\lambda)\sin\left(\frac{\alpha + \beta}{2}\right). \tag{9.6}$$

Thus, for a transverse scan across the specular ridge or a rocking curve, the bulk scattering will be constant, while surface-related scattering will in general show a peak of symmetric structure about the specular ridge.

9.3 Specular reflectivity

We now discuss the theory of specular reflectivity and the information one may learn from such experiments. We assume for this purpose a surface or interface with a density or composition which is uniform in the plane of the surface (the x-y plane) but may vary in the z direction. In actual fact, the spatial density even for a nominally smooth surface cannot be uniform because of the molecular composition of matter, but since $q_{\parallel} \approx 0$, it can be shown that the scattering around specular is not sensitive to density fluctuations on molecular length scales. Similarly, for the regime $q_z \lesssim \frac{1}{a}$ (where a is a molecular spacing) we can replace the density by a continuously varying function of z. These restrictions will be dropped for large q_z or $q_{\parallel}$ (GID experiments). For small q_z a rigorous calculation of the reflectivity can be performed in terms of a z-dependent refractive index, which is related to the density by the relation,

$$n(z) = 1 - \frac{b\lambda^2}{2\pi}\rho(z). \tag{9.7}$$

For x-rays, b is the Compton scattering length of an electron ($= e^2/mc^2$) and $\rho(z)$ is the (x,y averaged) z-dependent electron density of the material. For neutrons, b is the nuclear scattering length, and $\rho(z)$ is the density of nuclei. (For non-elemental materials, $b\rho(z)$ is replaced by the appropriate compositional-weighted average.) We shall consider x-rays incident with s-polarization (E-field parallel to the surface) and denote the electric field by $\varphi(\mathbf{r})$. For small grazing angles of incidence, it may be shown that the same results apply for p-polarized radiation. For neutrons, we denote the neutron wave function by $\varphi(\mathbf{r})$. (In this article, we ignore all effects due to magnetic scattering of neutrons or x-rays or due to anisotropic refractive indices. Thus, we do not explicitly need to consider the neutron spin here.) In both cases, the function $\varphi(\mathbf{r})$ obeys the stationary wave equation:

$$\nabla^2 \varphi(\mathbf{r}) + n^2(\mathbf{r})k_0^2\varphi(\mathbf{r}) = 0, \tag{9.8}$$

where k_0 is the wavevector *in vacuo*.

In what follows, we shall explicitly treat the case of x-rays and refer to $\rho(z)$ as the electron density. (For neutron scattering, it is understood that we must use the weighted nuclear density.) Since n is only a function of z, the

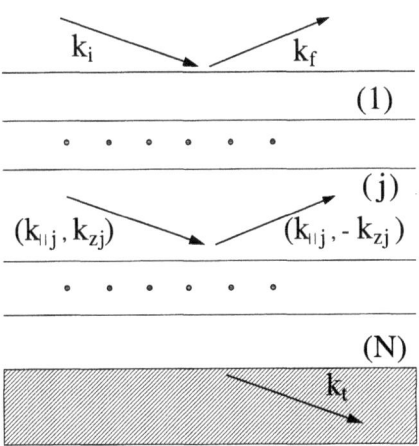

Figure 9.5. Schematic of an *N*-layer film (stratified medium) with different refractive indices for each layer.

solution separates into an (x,y)-dependent function and a z-dependent function:

$$\varphi(\mathbf{r}) = \chi(x,y)\psi(z). \tag{9.9}$$

Substituting, in Eq. (9.8), we see that we may write $\chi(x,y)$ as the plane-wave solution:

$$\chi(x,y) = e^{\pm i k_{\parallel} \cdot \boldsymbol{\rho}}, \tag{9.10}$$

where $\mathbf{k}_{\parallel}$, $\boldsymbol{\rho}$ are the in-plane components of $\mathbf{k}$ and $\mathbf{r}$ respectively, and $\psi(z)$ is a solution of

$$\frac{d^2\psi(z)}{dz^2} + \left[n^2(z)k_0^2 - k_{\parallel}^2 \right]\psi(z) = 0. \tag{9.11}$$

It may not always be possible to obtain an analytical solution of Eq. (9.11) for arbitrary $n(z)$.

Let us consider a set of layers of uniform electron density ρ_i normal to the z axis (see Fig. 9.5) and consider solutions of Eq. (9.11) in each such layer (i). These are given by

$$\psi_i(z) = A_i\, e^{ik_{i,z}z} + B_i\, e^{-ik_{i,z}z}. \tag{9.12}$$

Here

$$k_{i,z}^2 = k_{0,z}^2 - q_{i,c}^2/4 \tag{9.13}$$

$$k_{0,z}^2 = k_0^2 - k_{\parallel}^2 \tag{9.14}$$

and

$$q_{i,c}^2 = 16\pi b\rho_i. \tag{9.15}$$

$k_{0,z}$ is the normal component of the incident wavevector in free space. These solutions are matched using the appropriate boundary conditions of each interface, namely that $\psi(z)$ and $\dfrac{d\psi}{dz}(z)$ are continuous across each interface. (By our assumptions, at this stage, the interfaces are all smooth.) We also use the condition that in the last medium there is only a transmitted wave, i.e., no $e^{-ik_{i,z}z}$ solution. For the simple case of a single interface between two uniform media, situated at $z = 0$, this yields

$$r \equiv B_1/A_1 = (k_{1,z} - k_{2,z})/(k_{1,z} + k_{2,z}) \tag{9.16}$$

and

$$t = A_2/A_1 = 2k_{1,z}/(k_{1,z} + k_{2,z}). \tag{9.17}$$

r and t are the (complex) reflectance and transmittance coefficients respectively. The reflectivity (defined by the ratio of the reflected intensity to the incident intensity) is given by

$$R = |r|^2, \tag{9.18}$$

while the transmission is given by

$$T = |t|^2. \tag{9.19}$$

Equations (9.16)–(9.19) constitute the well-known Fresnel theory of reflectivity for an ideal smooth interface [5].

For a free surface between a medium of electron density ρ and vacuum,

$$k_{1,z} = k_{0,z} = k_0 \sin\alpha \tag{9.20}$$

so that if

$$\sin^2\alpha < \frac{q_c^2}{4k_0^2} = b\lambda^2\rho/\pi = \sin^2\alpha_c. \tag{9.21}$$

$k_{2,z}$ is by Eq. (9.13) purely imaginary, corresponding to an evanescent wave in the medium. In this case, $|r|^2 = 1$, and we have the phenomenon of total *external* reflection. α_c is called the *critical angle* (and is wavelength dependent) while the corresponding value of the scattering vector on the vacuum side is

$$q_c = 2k_0 \sin\alpha_c = 4(\pi b\rho)^{1/2}, \tag{9.22}$$

and is called the *critical wavevector* and depends only on the electron density of the medium.

The Fresnel coefficients r and t satisfy the identity

$$t = 1 + r \qquad (9.23)$$

arising from the continuity of $\psi(z)$ of the interface. For $\alpha \gg \alpha_c$, i.e., $k_{0z} \gg q_c/2$, we may expand Eq. (9.16), to obtain

$$r \approx \frac{q_c^2}{16k_{0z}^2} = \frac{q_c^2}{4q_z^2}, \qquad (9.24)$$

since for specular reflectivity

$$q_z = 2k_{0z}. \qquad (9.25)$$

Thus

$$R \approx \frac{q_c^4}{16q_z^4}. \qquad (9.26)$$

The q_z^{-4} dependence of the reflectivity of a smooth interface is related to the q^{-4} law known as Porod's law for small angle scattering from randomly oriented interfaces [6].

We now return to the problem of calculating the reflectivity of N layers of uniform media, as illustrated in Fig. 9.5. This is by now a standard problem in optics which has been treated in the classic works of Abeles [7], Born and Wolf [8], and Parratt [9] nearly 40 years ago. (See also Ref. 10.) For each layer, the propagation vector in the z direction is given by Eq. (9.13). (The in-plane component $k_{\parallel}$ is constant for all interfaces.) One then applies the boundary condition at successive interfaces (starting with the one most 'downstream' from the incident radiation) to solve for the coefficients A_{i-1}, B_{i-1} (see Eq. (9.12)) in the $(i-1)$th layer in terms of those for the ith layer. This yields

$$\begin{pmatrix} A_{i-1} \\ B_{i-1} \end{pmatrix} = M(i) \begin{pmatrix} A_i \\ B_i \end{pmatrix}, \qquad (9.27)$$

where the matrix $M(i)$ for the ith interface (located at $z = z_i$) is defined by

$$M(i) \equiv \begin{pmatrix} \dfrac{k_{i-1}+k_i}{2k_{i-1}} \exp[-i(k_{i-1}-k_i)z_i] & \dfrac{k_{i-1}-k_i}{2k_{i-1}} \exp[-i(k_{i-1}+k_i)z_i] \\[2ex] \dfrac{k_{i-1}-k_i}{2k_{i-1}} \exp[i(k_{i-1}+k_i)z_i] & \dfrac{k_{i-1}+k_i}{2k_{i-1}} \exp[i(k_{i-1}-k_i)z_i] \end{pmatrix}. \qquad (9.28)$$

(We have dropped the subscript z on the wavevectors, so that in the above expression all wavevectors refer only to components normal to the layers.) Thus, iterating Eq. (9.26) to the Nth layer, we have

$$\begin{pmatrix} A_{i-1} \\ B_{i-1} \end{pmatrix} = X(i) \begin{pmatrix} A_N \\ O \end{pmatrix} \equiv \left[\sum_{j=1}^{N} M(j) \right] \begin{pmatrix} A_N \\ O \end{pmatrix}, \tag{9.29}$$

remembering that for the last layer there is no reflected wave. The reflectance coefficient $r(i)$ for the ith interface is (B_{i-1}/A_{i-1}). Thus, by Eq. (9.28),

$$r(i) = (X_{21}(i)/X_{11}(i)). \tag{9.30}$$

The reflectivity of the whole stack of N layers is thus given by

$$R = |r(1)|^2 = |X_{21}(1)/X_{11}(1)|^2. \tag{9.31}$$

Thus the calculation is reduced to evaluating the product of N (2×2) matrices. Equation (9.28) may also be used to calculate the fields inside each layer. If one is only interested in the reflectivity, one may use Eqs. (9.28) and (9.29) to derive an iterative relation between $r(i)$ and $r(i-1)$:

$$\begin{aligned} r(i) &= \frac{M_{21}(i) X_{11}(i+1) + M_{22}(i) X_{21}(i+1)}{M_{11}(i) X_{11}(i+1) + M_{12}(i) X_{21}(i+1)} \\ &= \frac{M_{21}(i) + M_{22}(i) r(i+1)}{M_{11}(i) + M_{12}(i) r(i+1)} \\ &= \exp(2ik_{i-1} z_i) \frac{r_0(i) + \exp(-2ik_i z_i) r(i+1)}{1 + r_0(i) \exp(-2ik_i z_i) r(i+1)}, \end{aligned} \tag{9.32}$$

where

$$r_0(i) \equiv (k_{i-1} - k_i)/(k_{i+1} + k_i). \tag{9.33}$$

Thus, an alternative scheme to calculate the reflectivity is to start from the Nth interface, set $r(N+1) = 0$ and iterate Eq. (9.32) back to $r(1)$ [11].

These solutions for the reflectivity of a stack of layers are implemented by fast numerical routines and used to evaluate the reflectivity of systems ranging from thin films and bilayers to multilayers. They also provide a way of numerically approximating the reflectivity for an *arbitrary* electron density profile $\rho(z)$. We may divide $\rho(z)$ into slices along the z axis and approximate the actual profile by a sequence of layers of constant density as shown in Fig. 9.6. By making the slices as thin as we wish (and accepting greater numerical computation) we may approximate fairly accurately the actual density profile and obtain the overall reflectivity using the above methods.

These methods, while cumbersome, give accurate results for the reflectivity even when it is unity or close to it, since no assumption has been made about the weakness of the scattering. For large q_z, on the other hand, the

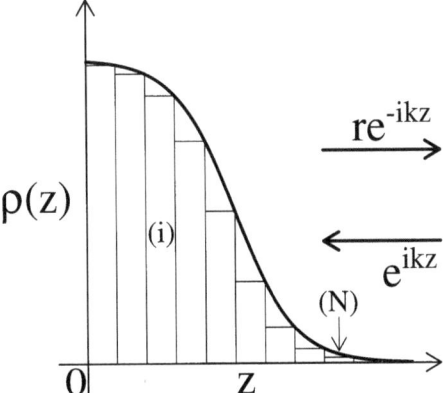

Figure 9.6. The 'slicing method', whereby an interface with a continuously varying electron or scattering density is represented by a histogram of N layers of uniform density.

assumption of a continuous density profile breaks down, as discussed above. Fortunately, in this regime, the reflectivity is weak and we may use the Born approximation [12], which provides a simple analytic expression for the reflectivity and is typically valid for $q_z \gtrsim 2q_c$. In this approximation, the scattering is taken to be a perturbation of the free plane wave traveling through the medium. The perturbing potential by Eq. (9.8) is given by

$$V(\mathbf{r}) = k_0^2(1 - n^2(\mathbf{r})) \simeq 4\pi b\rho(z).$$ (9.34)

Standard scattering theory [12] then yields in lowest order the following result for the differential cross-section (or number of particles scattered into unit solid angle per unit time per unit incident flux),

$$\frac{d\sigma}{d\Omega} = b^2 |\langle 2|\rho(z)|1\rangle|^2,$$ (9.35)

where the states $1\rangle$ and $|2\rangle$ are respectively the incident and scattered plane waves. Evaluation of the matrix element yields,

$$\frac{d\sigma}{d\Omega} = 4\pi^2 Ab^2 \left| \int_{-\infty}^{\infty} dz\rho(z)\, e^{iq_z z} \right|^2 \delta(q_x)\delta(q_y),$$ (9.36)

where A is the total surface area. The delta functions in q_x and q_y show that only specular reflection is allowed as a consequence of the assumption that the electron density is a function of z only. We can convert Eq. (9.35) into an expression for the reflectivity, if we integrate over the scattered solid angle $d\Omega$, use the fact that the incident beam intensity hitting the sample is the incident beam flux times $(A \sin \alpha)$, and that the element of solid angle

$$d\Omega = (k_0^2 \sin \beta)^{-1} dq_x dq_y. \tag{9.37}$$

Thus we obtain,

$$R = \frac{4\pi^2 b^2}{k_0^2 \sin \alpha \sin \beta} \left| \int_{-\infty}^{\infty} dz \, \rho(z) e^{iq_z z} \right|^2 = \frac{16\pi^2 b^2}{q_z^2} \left| \int_{-\infty}^{\infty} dz \, \rho(z) e^{iq_z z} \right|^2, \tag{9.38}$$

where we have used the fact that at the specular conditions $k_0 \sin \alpha = k_0 \sin \beta = q_z/2$. By partial integration Eq. (9.37) may also be rewritten

$$R = \frac{16\pi^2 b^2 \rho_0^2}{q_z^4} \left| \frac{1}{\rho_0} \int_{-\infty}^{\infty} dz \, \frac{d\rho(z)}{dz} e^{iq_z z} \right|^2, \tag{9.39}$$

where ρ_0 is the asymptotic (constant) value of $\rho(z)$ deep inside the medium. Remembering that the first factor on the right is just the limiting value (for $q_z \gg q_c$) of the Fresnel reflectivity R_F (see Eq. 9.25), we may then extrapolate to small q_z to write generally [13],

$$R = R_F \left| \frac{1}{\rho_0} \int_{-\infty}^{\infty} dz \, \frac{d\rho(z)}{dz} e^{iq_z z} \right|^2. \tag{9.40}$$

For a single smooth surface, $d\rho/dz$ is simply $\rho_0 \, \delta(z)$ and we recover Eq. (9.24). It is important to note that Eq. (9.38) is the actual Born approximation expression for R in the limit when the reflectivity is weak ($q_z \gtrsim 2q_c$) while Eq. (9.39) is a kind of hybrid expression which is expected to be better when $q_z \simeq q_c$. It may be justified in terms of the distorted wave Born approximation (DWBA) [12, 14, 15]. In this approximation, we may regard $\rho(z)$ in terms of a uniform solid with a sharp interface, i.e., a step function at $z = 0$ plus a perturbation. The perturbation theory for the scattering is then carried out relative to the zero-order process (which in this case is the Fresnel reflection from the sharp interface) rather than relative to the free plane wave states of the Born approximation.

In the preceding, we have discussed both accurate, numerical methods as well as approximate but analytical expressions for the reflectivity of an interface with an arbitrary density profile $\rho(z)$. It is instructive to look at several simple cases as illustrated in Fig. 9.7, which shows the reflectivity for several types of interface, as calculated in the Born approximation Eq. (9.38). For the case of a simple uniform film of thickness t, and electron density ρ_1 on top of a uniform bulk substrate of density ρ_0, the Born approximation yields,

$$R = \frac{16\pi^2 b^2}{q_z^4} |\rho_1 + (\rho_0 - \rho_1)e^{iq_z t}|^2$$

$$= \frac{16\pi^2 b^2 \rho_0^2}{q_z^4} \left[1 - \frac{4\rho_1(\rho_0 - \rho_1)}{\rho_0^2} \sin^2 \frac{q_z t}{2} \right], \qquad (9.41)$$

which puts modulations in the reflectivity curve of period $2\pi/t$ in q_z as shown in Fig. 9.7. These are the well-known Kiessig fringes characteristic of reflectivity from a uniform film and the origin of these fringes is the interference between the waves reflected from the two interfaces. The exact expression (from Eq. (9.31) is

$$R = \left| \frac{(k_0 - k_1)(k_1 - k_2) + (k_0 + k_1)(k_1 - k_2)e^{2ik_1 t_2}}{(k_0 + k_1)(k_1 + k_2) + (k_0 - k_1)(k_1 - k_2)e^{2ik_1 t_2}} \right|^2, \qquad (9.42)$$

where $k_0, k_1\, k_2$ are respectively the z components of the propagation vectors in air, the film and the substrate respectively, as given by Eq. (9.13). Multiple films with different thickness produce multiple modulations of the reflectivity as a function of q_z (e.g., in Fig. 9.7(c)). Periodically repeated layers, or bilayers (e.g., as found in Langmuir–Blodgett films or multilayers) produce maxima or peaks which begin to resemble Bragg reflections at $q_z = n2\pi/\Lambda$ (n=integer, Λ=period of multilayer), with Kiessig fringes for the total film thickness still showing up in between. The Born approximation expression for the reflectivity of a simple bilayer of thickness t_1, t_2 and density ρ_1, ρ_2 on a substrate ρ_0 is

$$R = \frac{16\pi^2 b^2}{q_z^4} |\rho_1 + (\rho_2 - \rho_1)e^{iq_z t_1} + (\rho_0 - \rho_2)e^{iq_z(t_1 + t_2)}|^2, \qquad (9.43)$$

and for N repeated bilayers of this type with period $\Lambda = t_1 + t_2$ on a substrate of density ρ_0 it is given by

$$R = \frac{16\pi^2 b^2}{q_z^4} \left| \rho_0 e^{iNq_z\Lambda} + (1 - e^{iNq_z\Lambda}) \left(\rho_2 + [\rho_1 - \rho_2] \frac{[1 - e^{iq_z t_1}]}{[1 - e^{iq_z\Lambda}]} \right) \right|^2. \qquad (9.44)$$

We shall not give the iterative or DWBA expressions for R for this case explicitly here as they are not particularly transparent. However, one difference with the Born approximation expression which should be commented on is that as q_z approaches q_c from above, the period of the Kiessig

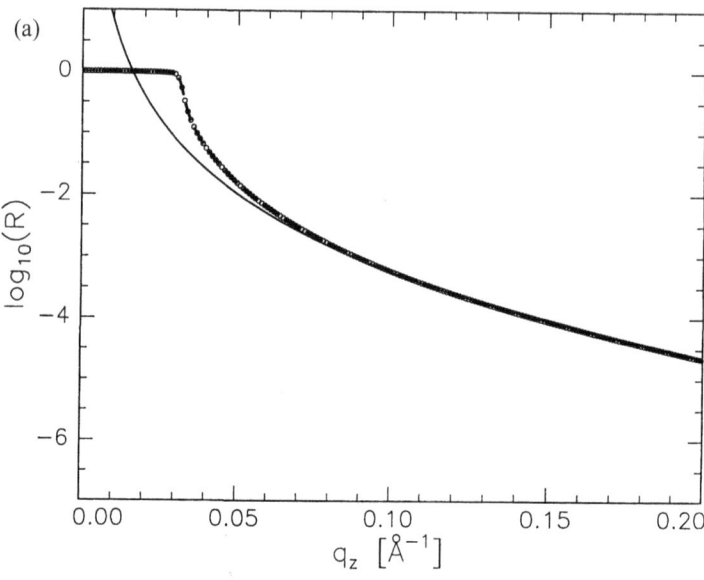

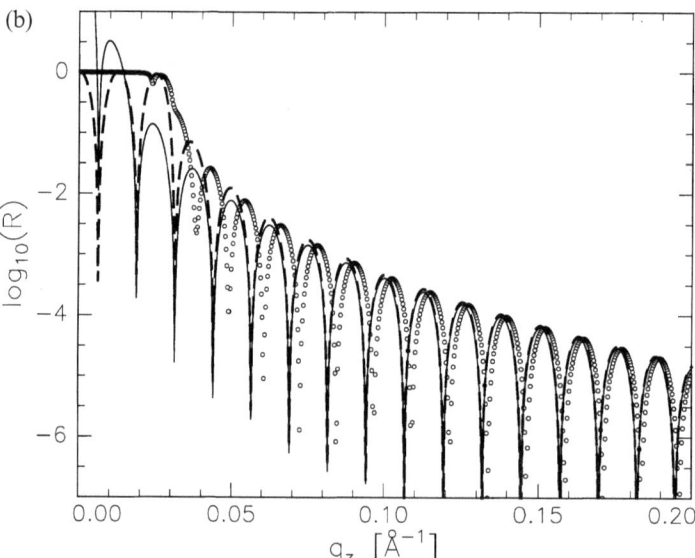

Figure 9.7. Calculations of the reflectivity in the Born approximation (dashed curve), the DWBA (dotted curve) and using the exact expression (solid curve) for (a) a single interface, (b) a thin film on a substrate of different density.

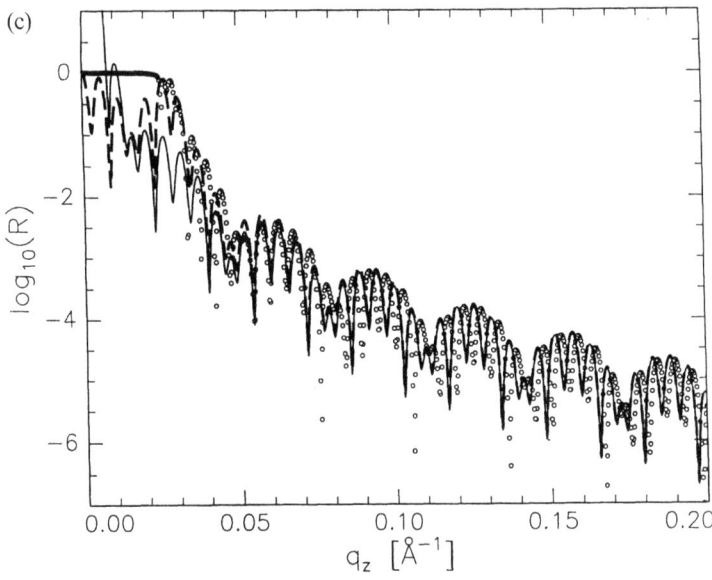

Figure 9.7. (*cont.*) (c) two films of different thicknesses and densities on a substrate of different density to either. A roughness of 4 Å for each interface has been assumed.

fringes and the positions of the Bragg peaks are slightly shifted, the shift being approximately that obtained by replacing the free space q_z by $\bar{q}_z$, the *average* q_z in the film or the multilayer. At large q_z, $\bar{q}_z \approx q_z$ so there is no shift. For small q_z, the shift maybe regarded as being due to *refraction* in the medium.

An important special case relates to a surface with roughness. A unified treatment of specular and off-specular scattering from a rough surface is given in the next section. For the moment we *assume* that, for the purposes of calculating the *specular* reflectivity, we may average the electron density of the rough surface over the (x,y) coordinates, leading to a *graded* density profile $\rho(z)$ in the normal (z) direction. For many rough surfaces, the distribution of heights above (and below) the average $z = 0$ plane is Gaussian, leading to the relation

$$\frac{1}{\rho_0}\frac{d\rho(z)}{dz} = \frac{1}{\sqrt{2\pi}\sigma}e^{-z^2/2\sigma^2}, \tag{9.45}$$

where σ denotes the root mean square (rms) roughness. (ρ_0 is the asymptotic electron density inside the medium with the rough surface.) A direct Fourier transform then yields, using Eq. (9.39), that

$$R = R_F e^{-q_z^2 \sigma^2}, \tag{9.46}$$

i.e., the Fresnel reflectivity is modified by a factor which is very similar to the Debye–Waller factor encountered in calculating the intensities of Bragg reflections from a crystal in which the atoms are vibrating about their mean positions. A more accurate (self-consistent) application of the DWBA leads to the modified expression [15, 16]

$$R = R_F \left| e^{-q_z \bar{q}_z \sigma^2/2} \right|^2, \tag{9.47}$$

where $\bar{q}_z = 2k_{1z}$, i.e., the normal component of the wavevector transfer in the medium. Equation (9.47) was first derived for a rough surface (using a completely different method) in a classic series of papers by Nevot and Croce [17]. An interesting property of Eq. (9.47), as opposed to Eq. (9.46), is that R is still unity below the critical angle in the absence of absorption (where $\bar{q}_z$ is purely imaginary). This cannot be completely correct, since even below the critical angle there will be some loss due to off-specular scattering by the roughness fluctuations. The conditions under which the Nevot–Croce expression (9.47) is more appropriate, and when the Rayleigh expression (9.46) is more appropriate, have been discussed by De Boer [18].

For either a single interface or multiple interfaces, it should be noted that a calculation to arbitrary accuracy of the specular reflectivity may be carried out by sufficiently fine slicing of the graded interface density profile at each interface, as given by the integral of an expression such as Eq. (9.45), which yields an error function. In practice, for profiles containing several rough interfaces, it is much more convenient and probably as accurate to use modified reflectivities at each interface (as given by Eq. (9.47)) in expressions such as Eq. (9.31).

We may also note that Eq. (9.35) is valid even at very large q_z, even when we are probing densities on molecular length scales in the z direction. In such a case it may be more convenient to recast it in terms of the explicit positions of the centers of the *layers* of molecules parallel to the surface, as

$$\frac{d\sigma}{d\Omega} = 4\pi A b^2\, \delta(q_x)\delta(q_y) \left| \sum_{n=1}^{\infty} f_n(\mathbf{q}) e^{iq_z z_n} \right|^2, \tag{9.48}$$

where the layers run from $n=1$ at the surface to $n=\infty$ at $z \to \infty$, z_n being the average position of the nth layer, with the molecular form factor of that layer denoted by $f_n(\mathbf{q})$. (The dependence on n of $f_n(q)$ allows us to take into account the possible change of molecular *tilt* as we move from the surface layers into the bulk.) Equation (9.48) is a special example of a *truncation rod* going through the origin of reciprocal space and, for instance, gives rise to the smectic peaks, along the (00ℓ) direction. For a smectic phase confined to a single layer or to a few layers at the surface, this gives rise to a considerable broadening of the smectic peak along the q_z direction, while for a bulk ordered smectic phase, this would yield a sharp peak with tails along q_z owing to the restriction of the sum in Eq. (9.48) to the half space $z_n \geq 0$. However, fluctuations in the positions of the molecules will affect the detailed line shapes of the smectic peak.

Now that we have discussed how the reflectivity can be calculated from a given density profile and what the general characteristics of the function $R(q_z)$ are for different density profiles, we turn to the interpretation of reflectivity experiments. Here, one is in general attempting to solve the *inverse* problem, namely, knowing the function $R(q_z)$ trying to obtain the density profile $\rho(z)$ which gives rise to it. There is by now a fairly extensive literature on these types of inverse problems [19–25]. At first sight, at least within the Born approximation (Eq. (9.38)), it appears that the solution $\rho(z)$ cannot be unique, since what is measured is the *modulus squared* of its Fourier transform. Thus we have in principle no *phase* information, as is also the case in the usual 'phase problem' in three-dimensional crystallography. Putting various restrictions on the analytic behavior of $\rho(z)$ and using the full reflectivity curve rather than just the region of validity for the Born approximation is claimed to render the solution to the inverse problem unique in the *mathematical sense*. In practice, such uniqueness is compromised by (a) the accuracy of measurement of $R(q_z)$ and (b) by the finite range of q_z over which $R(q_z)$ is measured. Thus for instance, if $R(q_z)$ is measured out to a maximum value of $q_z(\text{max})$, the spatial resolution over which one may resolve structure in $\rho(z)$ is $\sim \pi/q_z(\text{max})$. It should be remarked here that we mean structure 'resolved' in an unbiased sense in the absence of any information regarding $\rho(z)$. Obviously, if the form of $\rho(z)$ were known, then the actual $\rho(z)$ could be determined with far more limited information in terms of $R(q_z)$. In fact, the most common method of obtaining $\rho(z)$ from $R(q_z)$ is to assume or guess a functional form for $\rho(z)$ and using some kind of least-squares fitting to the data or similar minimization process to determine the numerical parameters in the appropriate functions [19, 23]. Other methods attempt to avoid biasing the choice of the final $\rho(z)$ by procedures such as maximum entropy [20], or using sets of known functions with arbitrary coefficients (such as sine and cosine functions [21], or splines [22]) to represent $\rho(z)$.

Other methods of inverting reflectivity data are directed towards compensating for the absence of phase information by using independent but modified measurements of the reflectivity [24, 25]. Thus, for x-rays one can tune the x-ray energy through an absorption edge of one of the constituent atoms at the interface, thus changing both the real and imaginary parts of their contribution to the refractive index [24]. For neutrons, a commonly used technique is deuteration of either the whole or selective parts of a hydrocarbon molecule, again changing their scattering length densities [25]. In fact one can often (for a single surface) go from completely matching the refractive indices of the adjacent media (in which case the reflectivity identically goes to zero everywhere) to having maximum contrast. For example, one can choose a suitable mixture of H_2O and D_2O such that its refractive index is unity, so that there is no reflectivity at the bare air/water interface, thus making the measurement more sensitive to a thin film at the interface. Another possibility is to have magnetized substrates under the film one is studying and measure reflectivities for spin polarized neutron beams with the spins respectively parallel or antiparallel to the substrate magnetization, thus changing the reflectance of the substrate (but not the film) [26]. Let us consider all these methods together within the framework of the Born approximation (Eq. (9.38)). Let us write

$$b\rho(z) = \overline{\rho}_s(z) + \overline{\rho}_f(z), \tag{9.49}$$

where $\overline{\rho}_s(z)$ and $\overline{\rho}_f(z)$ are the effective scattering length densities of the substrate and film respectively (including dispersion corrections in the case of anomalous x-ray reflectivity or including deuteration effects or neutron spin-dependent effects in the case of neutrons). Then Eq. (9.38) reads

$$R(q_z) = \frac{16\pi^2}{q_z^2} |\overline{r}_s(q_z) + \overline{r}_f(q_z)|^2, \tag{9.50}$$

where

$$\overline{r}_s(q_z) = \int_{-\infty}^{\infty} dz \overline{\rho}_s(z) e^{iq_z z} \tag{9.51}$$

and

$$\overline{r}_f(q_z) = \int_{-\infty}^{\infty} dz \overline{\rho}_f(z) e^{iq_z z}. \tag{9.52}$$

If $\overline{r}_f(q_z)$ is unknown, but $\overline{r}_s(q_z)$ can be calculated or is known from varying the substrate scattering length density in a known manner, and $R(q_z)$ measured for two (or more) different (but known) forms of the function $\overline{r}_s(q_z)$, then for each q_z, the real and imaginary parts of $\overline{r}_f(q_z)$ may be obtained by solving Eq. (9.50) in terms of the two (or more) measured values of $R(q_z)$

and the corresponding $\bar{r}_s(q_z)$. Then the full $\bar{\rho}_s(z)$ may be simply obtained by a direct inverse Fourier transform of $\bar{r}_f(q_z)$. There are several difficulties encountered in the implementation of such a scheme:

1. The Born approximation does not work at all well for $q_z \lesssim q_c$, thus yielding a distortion in the obtained density profile on length scales $\geq q_c^{-1}$. This may be improved by use of the DWBA (for details see Refs. [24, 25]), but even the latter does not work well unless the film thickness is $\leq 100\,\text{Å}$ typically.
2. The act of putting the film on the substrate may sometimes change the substrate surface itself so that $\bar{\rho}_s(z)$ (and thus $\bar{r}_s(q_z)$) may not be exactly what is assumed or measured for the bare substrate.
3. For large q_z the substrate reflectance and thus its interference with $\bar{r}_f(q_z)$ becomes very small and the determination of the *phase* of $\bar{r}_f(q_z)$ from Eq. (9.50) becomes fairly inaccurate. To some extent, this may be overcome by use of a periodic or multilayer substrate which maintains significant reflectivity (at least at the multilayer Bragg reflections) out to fairly large q_z. Thus, use of a Si/Ge multilayer substrate and the anomalous scattering at the Ge edge enables use of the method to $q_z \simeq 0.5\text{Å}^{0-1}$.

Recently Majkrzak and Berk [27] and also Dehaan *et al.* [28] have made a more rigorous advance on this method, without using the Born or DWBA approximations for the reflectivity. They have shown that, if one may reflect absorption (which makes their method more valid for neutrons than x-rays), that one may rigorously derive the full complex reflectance as a function of q_z from reflectivity profile measurements of the unknown film against three known substrates (in practice, this can be achieved by using ferromagnetic substrates and neutron reflectivities measured with the neutron spin polarized parallel, and antiparallel and normal to the substrate magnetization). Knowing the full phase of the reflectance, one may then employ known inversion methods to obtain the scattering density profile [29].

Note that selective deuteration of the molecules of the sample film or substrate also poses problems in the sense that it involves carrying out measurements on different samples which might not be identical (apart from labeling), particularly since deuteration can sometimes change the structure or phase behavior of the system itself. Nevertheless, the method has often been successfully employed, particularly in the case of surfactants on liquid surfaces.

We conclude this section with a discussion of some experimental results which illustrate the principles developed above. Many of these relate to surface ordering phenomena in liquid crystals. In the isotropic phase of a liquid crystal, the free surface can produce an alignment to a nematic phase which penetrates further into the bulk as the temperature is lowered towards the bulk isotropic–nematic transition temperature (T_{IN}) [30]. We may say that the nematic phase *wets* the surface slightly above T_{IN}. Similarly, the smectic phase can begin to wet the surface in the nematic phase as T tends to the nematic–smectic transition temperature and this effect has been studied by x-ray specular reflectivity techniques by Pershan, Als-Nielsen and coworkers [31], using a liquid surface reflectometer. Figure 9.8 shows the measured specular reflectivity from the surface of the liquid crystal octyloxycyanobiphenyl (8OCB) as the temperature is decreased towards the nematic–smectic A transition temperature T_{NA}. Plotted are both $R(q_z)$ and $R(q_z)/R_F(q_z)$ where R_F is the theoretical Fresnel reflectivity. For $q_z \ll Q_q$ (where Q_0 is 2π divided by the periodicity of the smectic layers), $R(q_z)$ follows the Fresnel reflectivity $R_F(q_z)$ fairly well. For $q_z \sim Q_0$, the smectic peak appears with a width proportional to $\xi_\|$, the correlation length in the direction normal to the smectic layers. This peak sharpens as T_{NA} is approached. For $q_z \gtrsim Q_0$, $R(q_z)$ dips sharply *below* $R_F(q_z)$. This is due to destructive interference between the surface reflectivity and that from the smectic layers. To fit the data, the authors used the formula given in Eq. (9.40) with

$$\rho(z) = \rho_0(z) + \rho_1(z), \tag{9.53}$$

where the first term was taken to be given by

$$\frac{1}{\rho_\infty} \frac{d\rho_0}{dz} = \frac{d}{dz} \Theta(z - z_0) B_s \exp\left[-\frac{z - z_0}{\xi_\|(T)} \right] \sin[Q_0(z - z_0)], \tag{9.54}$$

($\Theta(z - z_0)$ being the step function which is 1 if $z > z_0$, and 0 otherwise) and the second term was taken to be

$$\frac{1}{\rho_\infty} \frac{d\rho_1}{dz} = C_1(\pi z)^{-1} \exp\left[\frac{-(\sigma z)^2}{2} \right] \sin(Q_1 z), \tag{9.55}$$

where C_1 is defined to ensure $\int \rho_1^{-1} d\rho_1/dz = 1$.

The Fourier transform then yields

$$R(q_z) = R_F(q_z) |\Phi(q_z)|^2, \tag{9.56}$$

where

$$\Phi(q_z) = \Phi_0(q_z) + \Phi_1(q_z) \tag{9.57}$$

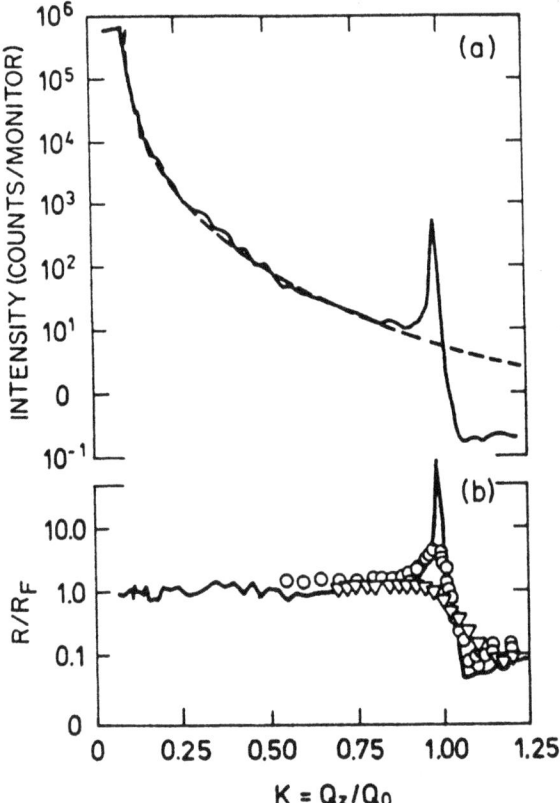

Figure 9.8. (a) Specular x-ray reflectivity from the surface of 8OCB at $T - T_{NA} = 0.05\,°C$. The Q_z scale is normalized to $Q_0 = 0.199\,°Å^{-1}$. The dashed line is the calculated Fresnel reflectivity (b) $R(Q_z)/R_F(Q_z)$ vs. Q_z/Q_0. The solid line is for $T - T_{NA} = 0.05\,°C$, open circles for $T - T_{NA} = 2.8\,°C$ and the triangles for $T - T_{NA} = 11.06\,°C$. (After Pershan *et al.*, Ref. 31.)

and

$$\Phi_0(q_z) = i\frac{B_s}{2}\exp(-iq_z z_0)\left[\frac{\xi_{\parallel} Q_0 - 1}{(q_z - Q_0)\xi_{\parallel} - i} - \frac{\xi_{\parallel} Q_0 + 1}{(q_z + Q_0)\xi_{\parallel} + i}\right], \quad (9.58)$$

$$\Phi_1(q_z) = \left(\frac{C_1}{2}\right)\left[\text{erf}\left[\frac{q_z + Q_1}{\sqrt{2}\,\sigma}\right] - \text{erf}\left[\frac{q_z - Q_1}{\sqrt{2}\,\sigma}\right]\right]. \quad (9.59)$$

The term ρ_0 represents a sinusoidal (smectic) density decaying into the bulk with correlation length $\xi_{\parallel}$, with a phase which is controlled by z_0 while ρ_1 represents a surface smeared with a roughness σ, but also with a sinusoidally varying component which is damped rapidly into the bulk. It is basically an empirical form chosen to represent the experimental data, and is presumed to correct for the fact that the form of Eq. (9.54) for ρ_0 does not accurately represent the smectic oscillations near the surface. Fitting of the data using Eq. (9.56) yields a value for $\xi_\|$ essentially identical to the bulk correlation length measured in x-ray scattering experiments on bulk samples in the region near the phase transition. It also yields a value for z_0 which is $0.25d$, d being the layer spacing of the smectic-A phase ($= 2\pi/Q_0$), yielding a maximum in ρ_0 exactly $d/2$ away from the surface. Q_1 turned out to be equal to Q_0 within a small percentage. An interesting result of this experiment was the fact that the specular peak at $q_z = Q_0$ was extremely sharp in the transverse direction, while the bulk critical scattering showed a finite transverse correlation length $\xi_\perp$. Thus the smectic order at the surface had a much larger in-plane correlation than in the bulk.

In another experiment on the system dodecylcyanobiphenyl (12CB) which does not have a smectic phase, but a first-order transition from the bulk isotropic phase to the smectic-A phase at $T_{IA} = 57.7\,°C$, Ocko *et al.* [32] found a specular reflectivity that clearly showed interference effects developing due to smectic layers forming at the surface above T_{IA} as the transition temperature was approached (Fig. 9.9). The data showed quantization effects as a function of temperature (Fig. 9.10) due to discrete smectic layers forming. Up to five smectic layers were observed as the temperature was lowered. The data was again fitted with expression (9.40) with the form taken for $\rho(z)$ being given by

$$\rho(z) = \Theta(z - z_0) + H_n(z)B_s \sin(2\pi z/d) \qquad (9.60)$$

where $\Theta(z)$ is the step function defined in Eq. (9.54), d the smectic layer spacing and $H_n(z) = 1$ for $0 < z < nd$ and zero elsewhere. The above expression was convoluted with a Gaussian profile to represent surface or interlayer roughness and the resulting reflectivity fitted to the experimental data by adjusting z_0, B_s and the roughness parameter. These fits are shown in Fig. 9.9, in a sequence from $n = 0$ (no layer) at the highest temperature, to $n = 5$ for the lowest temperature. The results thus showed an (incomplete) wetting of the free surface of the isotropic phase of 12CB by the smectic-A phase, but in discrete layers, rather than with an exponential decay into the bulk. Ocko has also used x-ray specular reflectivity experiments to demonstrate such layering transitions at smectic-A–solid interfaces [33].

In view of the fact that the free surface has a disordering effect on a solid, and that surface melting appears to be common in ordinary solids [34], the ordering of liquid crystals at the free surface is somewhat unexpected. Holyst [35] has proposed a theory which may account for the layer-by-layer

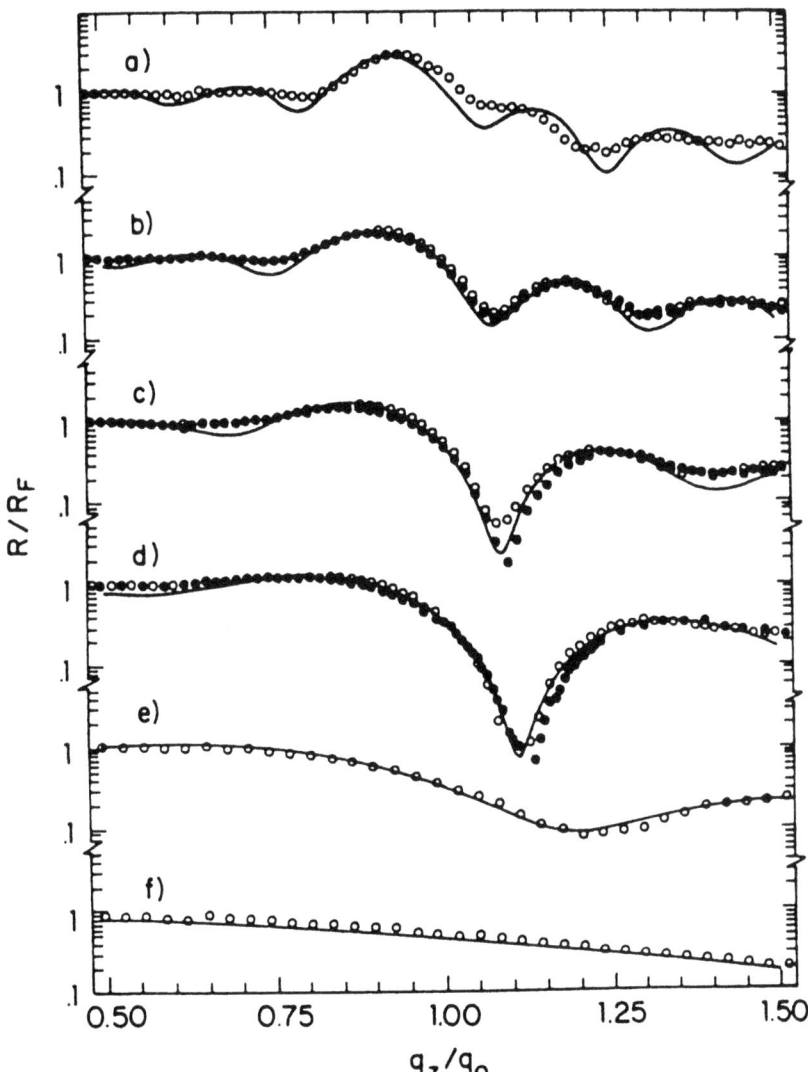

Figure 9.9.   The specular x-ray reflectivity divided by the Fresnel reflectivity from the surface of 12CB vs. q_z/q_0 ($q_0 = 0.1605$ Å^{-1}) for different values of $t = (T - T_{IA})/T_{IA}$, ranging from $t = 3 \times 10^{-5}$ (curve (a)) to $t = 6.1 \times 10^{-2}$ (curve (f)). The solid lines are for a model with sinusoidal density modulation terminated after an integral number of periods, ranging from five for curve (a) through zero for curve (f). (After Ocko *et al.*, Ref. 32.)

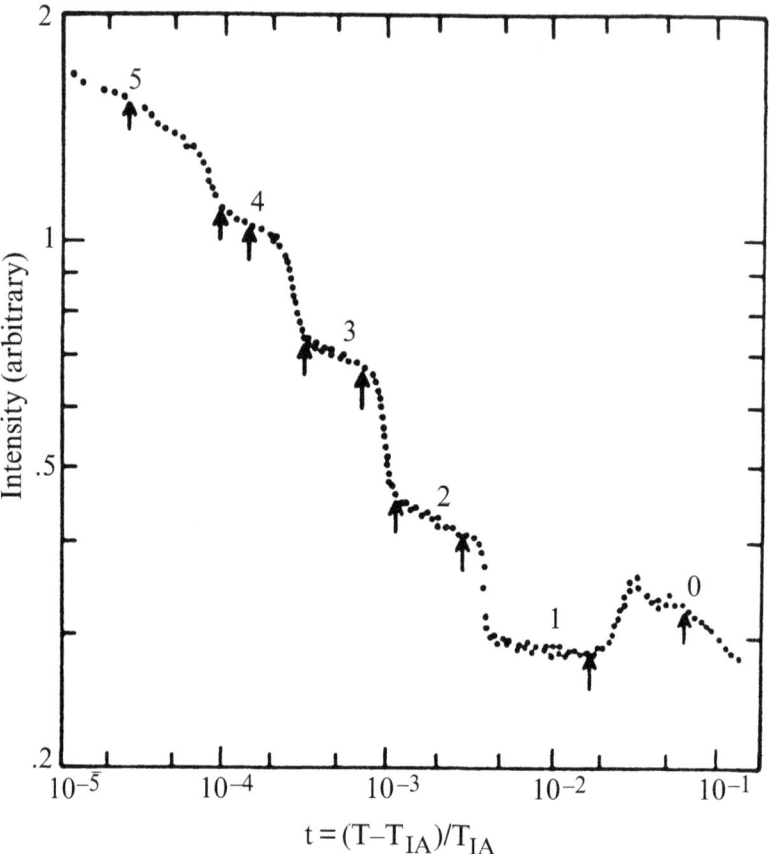

Figure 9.10.   The reflected intensity from the surface of 12CB at $q_z = 0.15$ Å^{-1} for various values of reduced temperatures above the smectic phase transition, indicating the quantized nature of the layering consistent with Fig. 9.9. (After Ocko *et al.*, Ref. 32.)

freezing in terms of the quenching of the out-of-plane surface fluctuations by surface tension effects.

Liquid crystals are not the only systems to show surface ordering effects. Even the simplest chain-like hydrocarbon molecules, namely the normal alkanes (which possess no nematic or smectic phases) show such behavior. Figure 9.11 shows the specular x-ray reflectivity from the surface of three liquid alkanes (carbon numbers = 18, 20, 24) at roughly 4 °C above their respective bulk crystallization temperatures T_b [36]. At the higher temperature, the reflectivity is well fitted by the simple Nevot–Croce form as expected for a simple liquid surface with an effective roughness due to

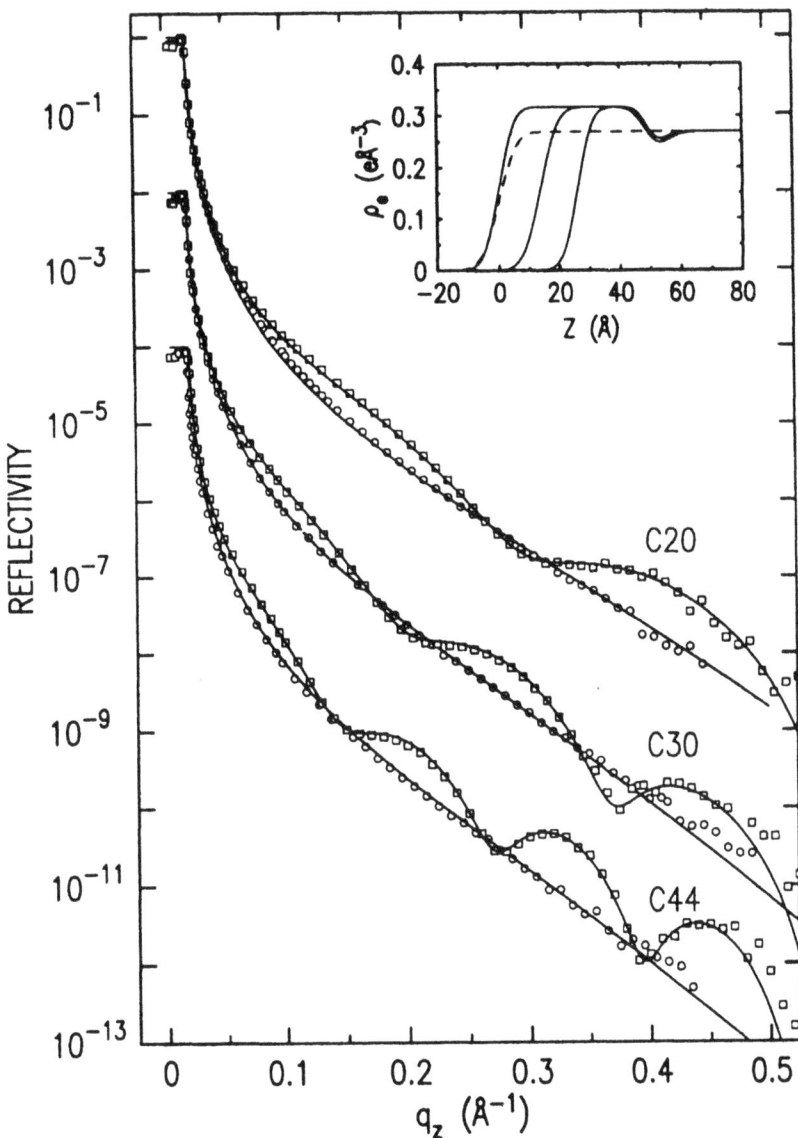

Figure 9.11. Specular x-ray reflectivity data for liquid alkanes with carbon numbers $n = 20$, 30, and 44 at $T_m + 4\,°C$ (open circles) and $T_m + 3\,°C$ (open squares), where T_m is the bulk freezing temperature for each. The inset shows model fits in the crystalline (solid) and liquid (dashed) surface phases of the electron densities at the interfaces.

capillary wave fluctuations. However, at about 3 °C above T_b, oscillations in the reflectivity indicate the presence of a slightly denser layer on the liquid surface, with a thickness approximately equal to the chain length of the molecule. (The density profile fitted to the reflectivity for the C24 chain in this regime is shown in the inset to Fig. 9.11.) The electron density of this layer is close to that of the bulk 'rotator' crystalline phase, where the chains are oriented in a hexagonal structure normal to the layers, but the bonds along the chain are disordered. These reflectivity experiments thus suggest strongly that a single layer of this crystalline phase has formed on the surface of the bulk liquid slightly above T_b. This was confirmed by in-plane grazing incidence x-ray diffraction (GIXD) experiments, discussed in a later section, and also by surface tension measurements [37]. Unlike the 12CB liquid crystal experiment discussed above, only *one* layer of the crystalline phase was seen to form before the bulk phase was reached via a strongly first-order transition. Subsequent reflectivity and GIXD experiments showed that for long chain alcohols, a similar surface crystalline phase also formed slightly above T_b. This surface phase, however, consisted of a single *bilayer* of the hexagonally close-packed and tilted alcohol chains [38]. Reflectivity experiments have also shown surface crystallization in mixtures of alkane chains of different lengths [39]. For mixtures of chains of fairly comparable lengths, the thickness of the layer yields an *average* of the chain length of the two species. For fairly disparate chain length mixtures on the other hand, one observes surface crystallization of either the majority component only or no surface crystallization at all when the concentrations are nearly equal. This has been explained [39] as being due to the extra energy cost of cocrystallizing chains of very different length adjacent to each other. GIXD experiments, to be discussed later, show that in these monolayer surface crystalline phases, there also exist tilted phases, i.e., where the alkane or alcohol molecules, instead of being oriented normal to the surface, develop a tilt angle towards nearest-neighbor or next-nearest-neighbor positions.

Along with x-ray reflectivity experiments to probe ordering at bulk liquid surfaces, there have also been a number of experiments to probe the development of ordering in *thin films* of liquid crystals (supported on solid substrates, e.g., glass or silicon). Here the relative strengths of the interactions of the liquid crystal molecules with each other and with the substrate may also determine the sequence of observed phases. Thus, for example, Shi, Cull and Kumar [40] have studied x-ray specular reflectivity from films of diheptylazoxybenzene (D7-AOB) absorbed on flat glass substrates. D7-AOB goes through crystalline (CrK), smectic-A (SmA), nematic (N) and

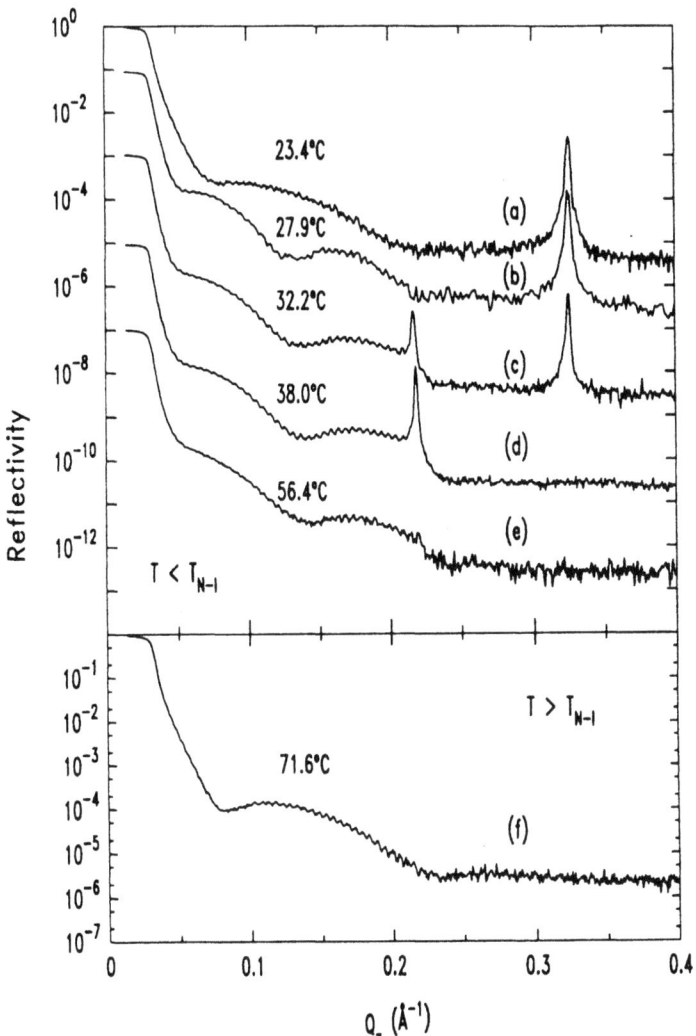

Figure 9.12. Specular x-ray reflectivity profiles from a film of D7-AOB liquid crystal on a glass substrate at various temperatures as discussed in the text. (After Y. Shi *et al.*, Ref. 40.)

isotropic (I) phases as a function of increasing temperature, with $T_{CA} = 34\ °C$, $T_{AN} = 54.5\ °C$, and $T_{NI} = 71\ °C$. Figure 9.12 shows reflectivity profiles of one such film as a function of temperature. It is seen that the reflectivity consists of (a) a broad oscillation in q_z characteristic of a thin surface absorbed layer (the initial thickness was found to be 37.3 ± 0.6 Å from fitting, but then doubled to 75 Å as the temperature was raised to $27.9\ °C$

and then went back to its original value above T_{NI}); (b) superimposed rapidly oscillating Kiessig fringes corresponding to a layer of thickness 1200 Å; (c) a crystalline peak at $q_z = 0.325$ Å^{-1}, coexisting with and then giving way to a smectic-A peak at $q_z = 0.216$ Å^{-1} above T_{CA}. The authors attributed the underlying thin layer to a crystalline bilayer (of thickness 37.3 Å) on the substrate, slightly compressed from its bulk value, together with a superimposed layer of islands (as in the Stranski–Krastanov growth mode [41]) which go from crystalline to smectic to nematic to isotropic (with some regions of coexistence) and which coalesce to form the thicker film on top of the absorbed bilayer. The latter 'melts' far above T_{NI} and the authors ascribed to this layer the 'surface memory effect' known for liquid crystal devices [42].

There have also been several measurements of specular reflectivity from *free standing* liquid crystal films in the smectic-A and smectic-C phases. (In-plane ordering in such films was studied several years ago by Pindak, Moncton and others [43–45] using transmission geometry, where the **q**-vector lies in the plane of the film. But the density profile normal to the smectic layers was studied, using the reflectivity geometry, only relatively recently.) Thus, Gierlotka, Lambooy and De Jeu [46] studied thin films of the liquid crystals 12CB and 8CB in the smectic-Ad phase. They used the formula (9.40) for the reflectivity, where the electron density was taken to be uniform with a harmonic modulation $A \cos(2\pi z/d)$ normal to the smectic layers superimposed (where $z = 0$ is here defined as the mid-plane of the film and d is the period of the modulation, related to the Bragg peak wavevector q_0, by $q_0 = 2\pi/d$). This form for the electron density yields for the reflectivity,

$$R(q) = R_F(q)|F(q)|^2, \tag{9.61}$$

where $R_F(q)$ is the Fresnel reflectivity, and

$$F(q) = 2\sin(qD/2) + A[q/(q+q_0)]\sin[(q+q_0(D'/2]$$
$$+ A[q/(q_0-q)]\sin[(q_0-q)D'/2], \tag{9.62}$$

where $D' = Nd$, N being the number of smectic layers. In the simplest case, the modulation period is commensurate with the slab thickness and $D' = D$. The final reflectivity pattern is extremely sensitive to the positioning of the density modulation relative to the film surfaces. Of interest is the case when the film thickness D is incommensurate with the number of layers, which means that the outer layers are altered with respect to the bulk structure. Thus Fig. 9.13(a) shows the reflectivity profile for a film of 12CB which

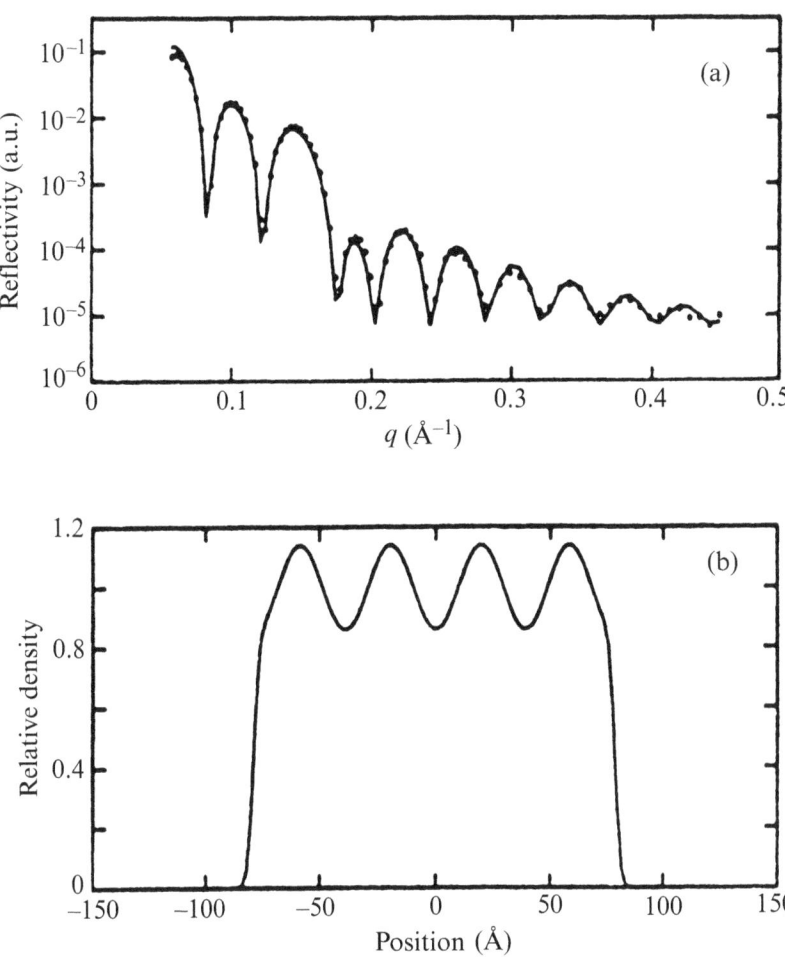

Figure 9.13. (a) X-ray reflectivity of a commensurate four-layer film of 12CB at 55.0 °C, showing measured points and a model fit as discussed in the text; (b) associated electron density profile. (After S. Gierlotka *et al.*, Ref. 46.)

turns out to be commensurate in thickness with exactly four layers, and the corresponding electron density profile from fitting this reflectivity curve is shown in Fig. 9.13(b). On the other hand, Fig. 9.14 shows the reflectivity profile and corresponding electron density profile from an incommensurate four layer film of 8CB which shows additional material in the outer layers.

For a *thick* film of smectic liquid crystal, which corresponds to the bulk phase, the one-dimensional order between smectic layers cannot be infinite range by the Landau–Peierls theorem regarding the lack of long-range

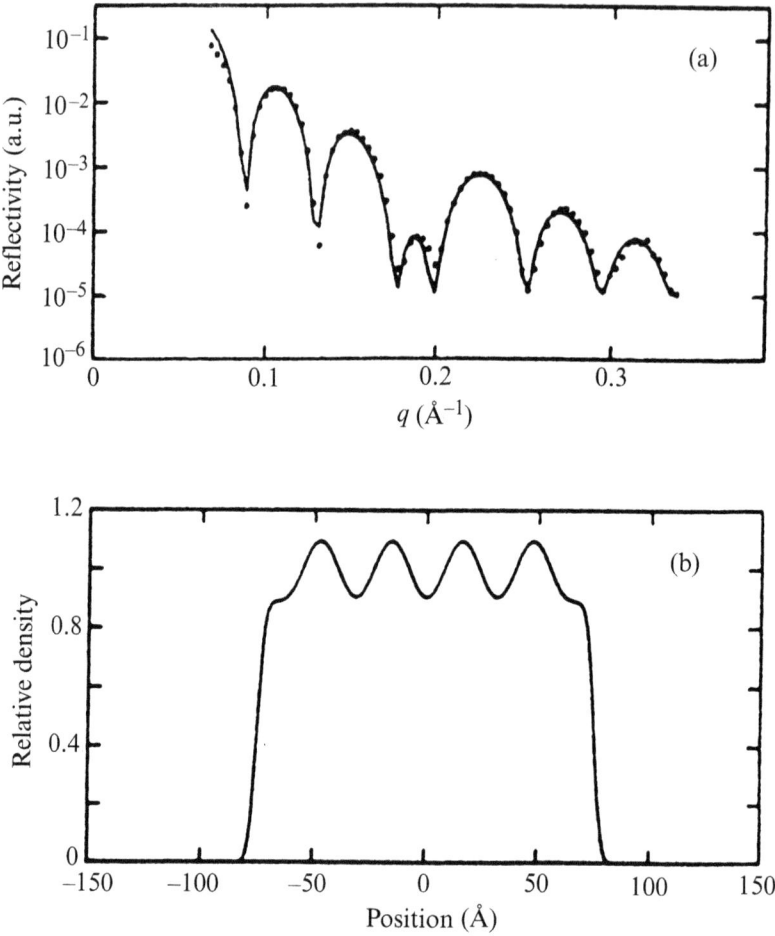

Figure 9.14.   (a) X-ray reflectivity of an incommensurate four-layer film of 8CB at 23.0 °C showing measured points and a model fit; (b) associated electron density profile. (After S. Gierlotka *et al.*, Ref. 46.)

positional order in low-dimensional systems [47]. In fact, the mean-square fluctuation in the layer positions diverges logarithmically with the film thickness. However, at the film surface, surface tension effects reduce the mean-square displacements and this effect has been calculated by Holyst, Tweet and Sorenson [48]. This group also carried out specular reflectivity for free-standing liquid crystal films of (70.7) which were in the bulk smectic-C phase in the interior, with the surface layers being in the more tilted smectic-I phase [49]. To analyze the data, the authors used a laterally averaged electron density profile which was of the form:

$$\rho(z) = \zeta \left(2\pi\sigma_i^2\right)^{-\frac{1}{2}} \exp\left[-\frac{(z-z_1)^2}{2\sigma_i^2}\right] \otimes M_i(z), \qquad (9.63)$$

where $\otimes$ defines convolution, σ_i^2 defines the mean-square displacement of the ith layer, and $M_i(z)$ is the molecular electron density projected along the z-axis, which depends also on the tilt of the molecules. (This expression is equivalent to the form introduced earlier in Eq. (9.48).) Tweet *et al.* were able to fit their reflectivity curves for several film thicknesses, with the σ_i calculated from their theoretical formulation [48] assuming parametrized values for B and K, the compressive and bending smectic elastic constants respectively, the surface tension γ (the product BK was made to depend on the number of layers), and with the molecular tilt as a function of z given by the expression

$$\phi(z) = \frac{(\phi_s - \phi_B)\cosh(z/\xi)}{\cosh(D/2\xi)} + \phi_B \qquad (9.64)$$

where ζ is the tilt decay length and ϕ_s and ϕ_B are the tilts at the surface and bulk respectively and represent fixed parameters. Using this model, the authors were able to get a remarkably good fit for the complete reflectivity profiles (over several orders of magnitudes) for several films ranging in thickness from 3 to 35 layers, illustrating the detailed structural information available from experiments of this type.

It should be mentioned that surface-induced phase transitions in freestanding liquid crystalline films have also been investigated in detail by the University of Minnesota group [50]. However, their experiments involve delicate heat capacity signatures of these phase transitions rather than scattering studies.

Finally we shall discuss some neutron specular reflectivity results on block copolymer films which show a lamellar structure, as studied by Russell and coworkers [51, 52]. The films are made up of symmetric copolymers of polystyrene (PS) and polymethyl-methacrylate (PMMA) which, when properly annealed, tend to micro-phase separate. The thin film geometry then forces a lamellar structure parallel to the substrate, with a PMMA layer segregating to the silicon oxide layer on the substrate and a PS layer residing at the free surface. If L is the thickness of the bilayer (i.e., the bulk copolymer lamellar periodicity) the thickness of the overall film is thus quantified to be $(n + 1/2)L$, where n is an integer. If the thickness of the film is not exactly equal to such a value, holes or islands of height L form on the top of the film to accommodate the excess or deficient material. X-rays see very little contrast between the PS and PMMA layers in such a

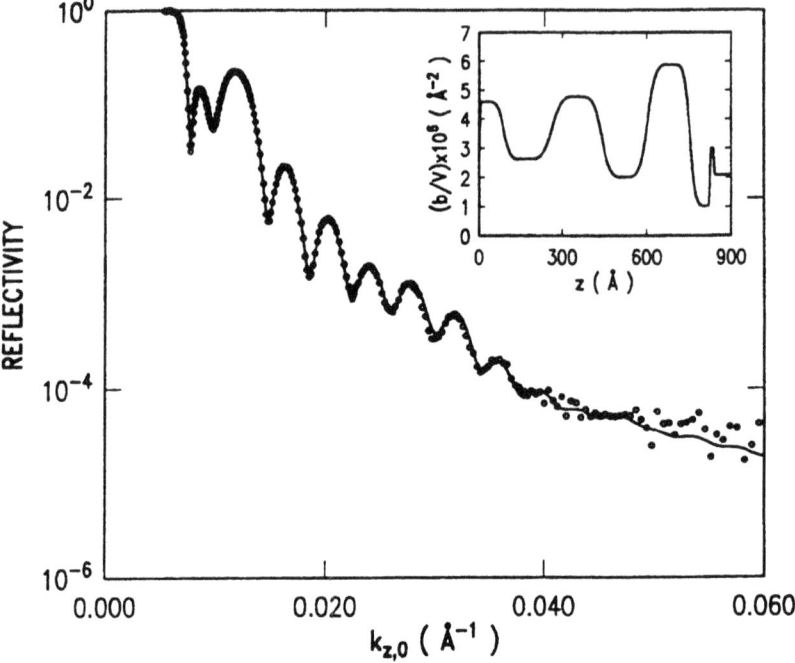

Figure 9.15.   Neutron reflectivity from an 800 Å film of P(d-S-b-MMA) annealed for 0.1 hour and model fit (solid curve). The inset shows the corresponding neutron scattering length density fitted to the data, where $z = 0$ represents the free surface. (After A. M. Mayes *et al.*, Ref. 52.)

multilayer lamellar structure, and are thus used to study the surface morphology of these islands or holes, which will be discussed later in this chapter. On the other hand, by perdeuterating the PS block, one can arrange for a rather large neutron scattering length density contrast between the layers, resulting in the appearance of Bragg-like peaks (and associated Kiessig fringes) from these systems, an example of which is shown in Fig. 9.15 for an 800 Å film after an annealing time of 0.1 hour. The inset shows the scattering length density profile which was least squares fitted to the reflectivity curve using the methods discussed above. As annealing proceeds, the lamellar ordering into d-PS and PMMA layers builds up gradually from the substrate towards the free surface.

Neutron reflectivity methods have also been used to investigate surface-induced layering in microemulsions [53]. In these systems, which consist of oil, water (deuterated for scattering contrast) and surfactant, a phase called a disordered bicontinuous phase exists for certain concentrations and temperatures, consisting of domains of water and oil (separated by surfactant) and repeated in a quasiperiodic fashion with periodicity D and coherence length ζ. Neutron specular reflectivity experiments from the surface of a bulk microemulsion [53] in the single-phase region show the one-dimensional analogue of this effect, namely a layering of D_2O-rich and oil-rich regions parallel to the surface and extending deep into the bulk of the liquid (see Fig. 9.16). The density profiles deduced from the reflectivity curves were produced by 'inversion', using a highly efficient least squares minimization algorithm using the 'slicing' formalism for the reflectivity discussed earlier in this section [25].

An interesting example of using several different specular reflectivity curves to tie down uniquely the density profile of the system is the method used by Vaknin and coworkers [54] to study the layering of DPPC (phosphatidylcholine) monolayers on water. In this case the neutron reflectivity of deuterated DPPC on both H_2O and D_2O was measured, as well as the x-ray reflectivity of the same monolayer under similar conditions. The fits to all three curves provided the parameters for a model which was unique and consisted of a liquid condensed (LC) phase of the monolayer with the head groups of the molecules interpenetrated with the subphase water and a tilt angle of the hydrophobic chains from the surface normal of 33 ± 3 degrees.

9.4 Off-specular scattering

While specular reflectivity is a powerful and popular technique for studying interfaces and thin films, it yields no information about the lateral structure at the interface. Thus, the detailed morphology of surface roughness, surface fluctuations, molecular ordering in the plane of the surface, etc. can only be explored by looking at the off-specular (q_x or $q_y \neq 0$) components of the scattering.

We first consider off-specular scattering from a single surface in the Born approximation. This yields for the scattering cross section

$$\frac{d\sigma}{d\Omega} \equiv S(\mathbf{q}) = b^2 |\int_V d\mathbf{r} \; \rho(\mathbf{r}) e^{-i\mathbf{q}\cdot\mathbf{r}}|^2, \tag{9.65}$$

where $\rho(\mathbf{r})$ is the electron density and the integration is over the sample volume. If the wavevector transfer is small compared to the inverse of the molecular size or separation, we may replace $\rho(\mathbf{r})$ by a constant ρ_0 *inside* the medium and zero outside. Equation (9.65) may then be rewritten as

$$S(\mathbf{q}) = b^2 \; \rho_0^2 \int_V \int_V d\mathbf{r} \; d\mathbf{r}' \; e^{-i\mathbf{q}\cdot(\mathbf{r}-\mathbf{r}')}, \tag{9.66}$$

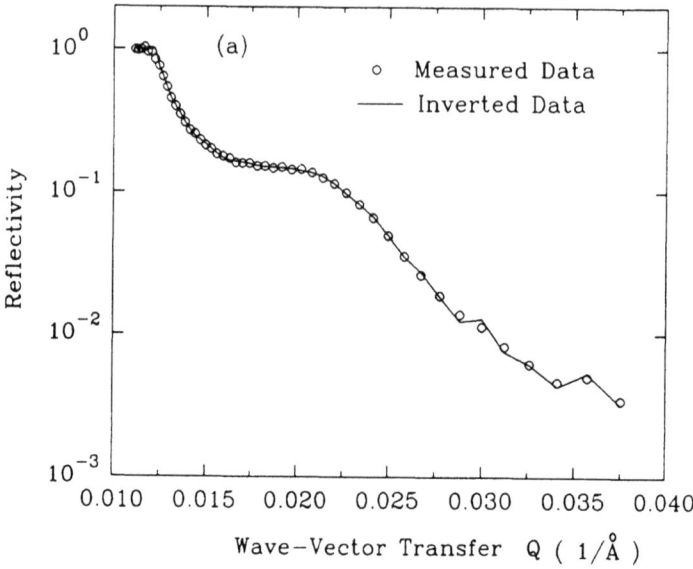

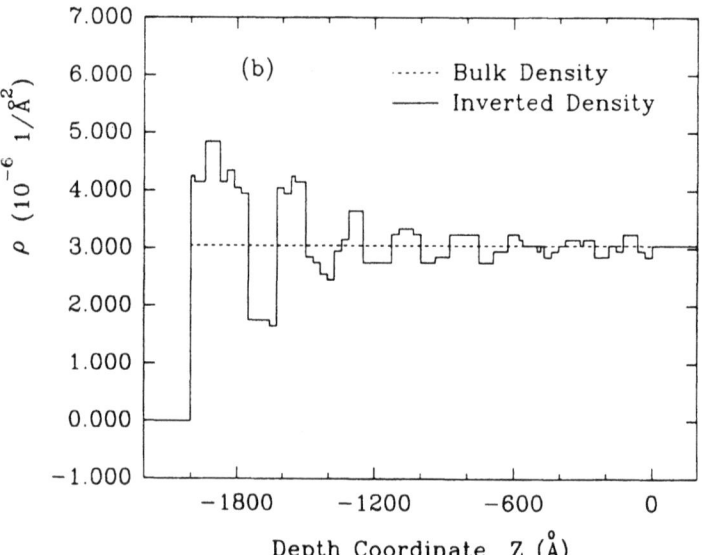

Figure 9.16. (a) The neutron reflectivity from an *n*-octane/D_2O/$C_{10}E_4$ micro-emulsion at $T = 23.2\,°C$ at a weight fraction of $C_{10}E_4$ of 10.55%. Measured points are open circles, and solid curve is obtained from the model fitted to the data. (b) The scattering length density obtained from the model fit. The dashed line indicates the scattering length density in the bulk. (After X.-L. Zhou *et al.*, Ref. 53.)

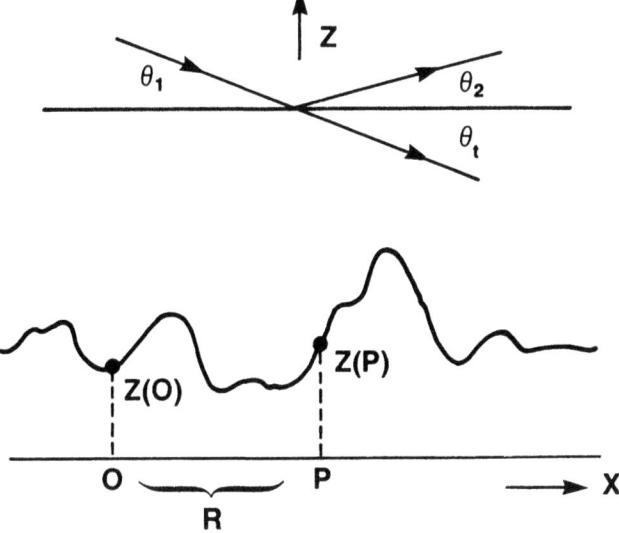

Figure 9.17. Schematic of a rough surface, with the incident, scattered, and transmitted beams indicated.

where the integration is over the volume enclosing the medium. This may be transformed to a surface integral using Stokes theorem to yield

$$S(\mathbf{q}) = b^2 \rho_0^2 \frac{1}{q_z^2} \int_S (d\mathbf{S} \cdot \hat{\mathbf{z}}) \int_S (d\mathbf{S'} \cdot \hat{\mathbf{z}}) e^{-i\mathbf{q} \cdot (\mathbf{r} - \mathbf{r'})}, \qquad (9.67)$$

where $\hat{\mathbf{z}}$ is the vector normal to the average surface and the integrals are over the surface. (The integrals over the bottom surface of the medium at $z \to \infty$ may be ignored by imagining a small absorption coefficient in the medium.) Since $d\mathbf{S} \cdot \hat{\mathbf{z}}$ is simply the element of area projected onto the reference (x,y) plane, Eq. (9.67) may be written as [15]

$$S(\mathbf{q}) = b^2 \rho_0^2 \frac{1}{q_z^2} \int\int\int\int dx\, dy\, dx'\, dy'\; e^{-iq_z[z(x,y)-z(x',y')]} e^{-i[q_x[(x-x')+q_y(y-y')]}, \qquad (9.68)$$

where we define $z(x,y)$ as the height function of the surface above the *average* surface at point (x,y) on the plane (see Fig. 9.17). The morphology of the surface determines the statistical properties of the integrand in Eq. (9.68). For instance, as is often the case, if $z(x,y)$ is a *Gaussian* random variable, and the statistical average $\langle e^{-iq_z[z(x,y)-z(x',y')]} \rangle$ depends only on the relative separation of (x,y) and (x',y'), we may write

$$\langle e^{-iq_z[z(x,y)-z(x',y')]} \rangle = e^{-\frac{1}{2}q_z^2\, g(\mathbf{R})}, \qquad (9.69)$$

where

$$\mathbf{R} \equiv (x - x', y - y') \qquad (9.70)$$

and

$$g(\mathbf{R}) \equiv \left\langle [z(\mathbf{R}) - z(\mathbf{O})]^2 \right\rangle \qquad (9.71)$$

is the mean-square height deviation function. Thus, by Eq. (9.68)

$$S(\mathbf{q}) = b^2 \, \rho_0^2 \frac{A}{q_z^2} \int_A d\mathbf{R} \, e^{-\frac{1}{2}q_z^2 \, g(\mathbf{R})} \, e^{-i\mathbf{q}_\rho \cdot \mathbf{R}}, \qquad (9.72)$$

where A is the total area of the average surface, and $\mathbf{q}_\rho$ is the in-plane component of $\mathbf{q}$. Equation (9.72) expresses the scattering in terms of the two-dimensional Fourier transform of $\exp(-1/2 \, q_z^2 \, g(\mathbf{R}))$. $g(\mathbf{R})$ in turn can be related to the height–height correlation function of the surface, since by Eq. (9.71),

$$g(\mathbf{R}) = \left\langle [z(\mathbf{R})]^2 \right\rangle + \left\langle [z(\mathbf{O})]^2 \right\rangle - 2\left\langle z(\mathbf{O})z(\mathbf{R}) \right\rangle = 2\sigma^2 - 2\left\langle z(\mathbf{O})z(\mathbf{R}) \right\rangle, \quad (9.73)$$

where σ^2 is the mean-square height function or mean-square roughness. Equation (9.72) then yields

$$S(\mathbf{q}) = b^2 \, \rho_0^2 \frac{A}{q_z^2} e^{-q_z^2 \sigma^2} \int d\mathbf{r} \, e^{q_z^2 C(\mathbf{R})} \, e^{-i\mathbf{q}_\rho \cdot \mathbf{R}}, \qquad (9.74)$$

where $C(\mathbf{R})$ is the height–height correlation function $\left\langle z(\mathbf{O})z(\mathbf{R}) \right\rangle$. If, as is usual, $C(\mathbf{R}) \to 0$ as $R \to \infty$, the integral in Eq. (9.74) contains a delta function corresponding to an integrand of unity which can be explicitly separated out to yield the *specular* component

$$S(\mathbf{q})_{\text{spec}} = \frac{4\pi^2 A b^2 \rho_0^2}{q_z^2} e^{-q_z^2 \sigma^2} \, \delta(q_x)\delta(q_y), \qquad (9.75)$$

which can be easily shown to yield the previously derived (Born approximation) expression (Eq. 9.46)) for the specular reflectivity of a rough surface (see Eqs. (9.35)–(9.39)). The remaining (diffuse) part is given by

$$S(\mathbf{q})_{\text{diff}} = \frac{A b^2 \rho_0^2}{q_z^2} e^{-q_z^2 \sigma^2} \int d\mathbf{R} \left[e^{q_z^2 C(\mathbf{R})} - 1 \right] e^{-i\mathbf{q}_\rho \cdot \mathbf{R}}. \qquad (9.76)$$

We now discuss specific forms for $C(\mathbf{R})$. A form which is commonly used is [15]

$$C(\mathbf{R}) = \sigma^2 \, e^{-(R/\xi)^{2h}}, \qquad (9.77)$$

where σ is the rms roughness defined in Eq. (9.73), h is called the roughness exponent and generally h is between 0 and 1, and ξ is the so-called roughness cut-off length over which the height fluctuations are correlated. By Eq. (9.73), in the regime $R \ll \xi$, we have

$$g(\mathbf{R}) = \frac{2\sigma^2}{\xi^{2h}} R^{2h}, \tag{9.78}$$

which is the characteristic of a statistically *self-affine* surface, for which the rms height deviation between two points scales as a power of the lateral separation of the points. Application of the DWBA (as in the case of specular reflectivity in the previous section) leads to a modification of Eq. (9.76) of the form [15, 16]

$$S(\mathbf{q})_{\text{diff}} = \frac{Ab^2\rho_0^2}{q_z^2}|t(\alpha)|^2|t(\beta)|^2 e^{\text{Re}\,(\bar{q}_z)^2\,\sigma^2} \int^A d\mathbf{R}[e^{|q_z|^2\,C(\mathbf{R})} - 1]e^{-i\mathbf{q}_\rho \cdot \mathbf{R}}, \tag{9.79}$$

where $\bar{q}_z$ is the normal component of the wavevector transfer *in* the medium, (Re) denotes the real part, and $t(\alpha), t(\beta)$ are the transmission coefficients for the interface for grazing angles α, β respectively. In the original application of the DWBA, $t(\alpha), t(\beta)$ were obtained as the Fresnel transmission coefficients (Eq. (9.17)) for the *ideal* (smooth) interface, although alternative expressions have been suggested which represent $t(\alpha)$ self-consistently as the transmission coefficient of the rough interface, even though a rigorous expression for the latter is unknown. Frequently, experiments are done with the instrumental resolution relaxed in the direction normal to the plane of scattering (in the reflectivity configuration), so that effectively all components of $\mathbf{q}_\rho$ in this direction (say, the q_y direction) are integrated over. In such a case, the integral over q_y in Eq. (9.79) yields a delta function in y, so that the integral over $\mathbf{R}$ can be replaced by an integral over x yielding [15]

$$S(\mathbf{q})_{\text{diff}}^{\text{int}} = \frac{2\pi Ab^2\rho_0^2}{q_z^2}|t(\alpha)|^2|t(\beta)|^2\,e^{\text{Re}\,(\bar{q}_z)^2\,\sigma^2} \int dx[e^{|\bar{q}_z|^2\,C(x)} - 1]e^{-iq_x x}. \tag{9.80}$$

Figure 9.18 illustrates the form of the diffuse scattering cross-section for a typical rough surface for both longitudinal and transverse scans, as calculated from Eq. (9.80). Often, there are two side peaks in the transverse scans. These do not occur along truncation rods normal to the surface and are not manifestations of any lateral surface ordering. Rather, they occur when either α or β are equal to the critical angle for the reflection from the surface. They are known as 'Yoneda wings' after the researcher who first noted them [55]. They occur because of maxima in the functions

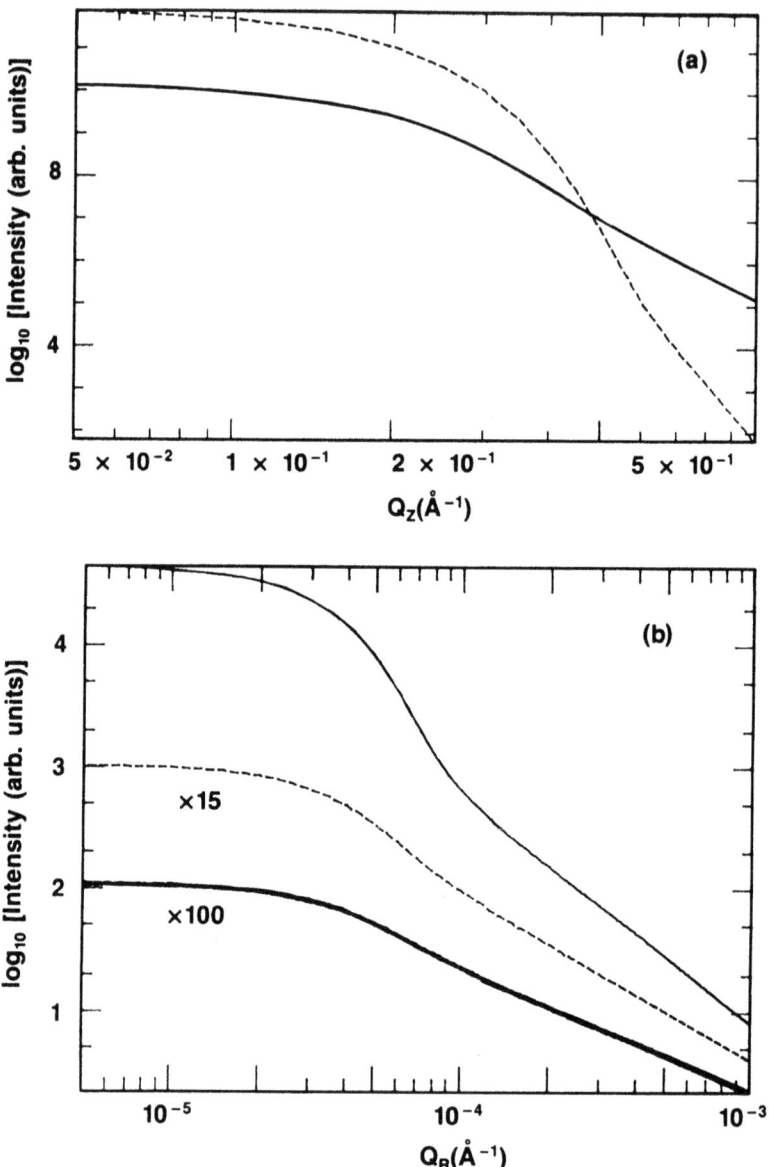

Figure 9.18.   Calculated form of diffuse scattering for a surface with typical roughness parameters for (a) longitudinal diffuse scattering (q_p constant) vs. q_z (solid
curve is for the case where $h = 0.5$, $\sigma = 7\,\text{Å}$, $\xi = 7000\,\text{Å}$; dashed curve represents the
case where $h = 0.2$, $\sigma = 7\,\text{Å}$, $\xi = 7000\,\text{Å}$) and (b) transverse diffuse scattering (q_z
constant) vs. q_p (solid line represents the case where $h = 0.5$, $\sigma = 7\,\text{Å}$, $\xi = 7000\,\text{Å}$; dashed
line represents the case with no cut-off; squares represent the case where $h = 0.2$,
$\sigma = 7\,\text{Å}$, $\xi = 7000\,\text{Å}$). (After S. K. Sinha *et al.*, Ref. 15.)

$|t(\alpha)|^2$ or $|t(\beta)|^2$ in Eq. (9.80) and this can only be explained in the context of the DWBA. Physically, these peaks arise because, at the critical angle, the incident and reflected beams are equal and in phase, leading to a doubling of the electric field at the surface and consequently increased scattering from surface irregularities. It is important to realize that the specular and diffuse components in the above model arise as sharp (i.e., resolution limited) and broader peaks both centered on the specular rod ($q_x = q_y = 0$). The diffuse scattering must be subtracted from under the specular peak to obtain the 'true' specular (by the methods discussed in Section 9.2), or else the reflectivity will appear to drop off much more slowly with q_z yielding a smaller roughness than the 'true' roughness, i.e., that averaged over a coherence area of the radiation across the surface. The effective roughness will depend on how much of the diffuse scattering is accepted by the diffractometer when it is set for specular, i.e., on the instrumental resolution.

A case where it is more or less impossible to separate the diffuse scattering from the specular in the above manner is the case of a fluid surface possessing capillary wave fluctuations. A phenomenological expression for the free energy of a free fluid surface is [56]

$$F = F_0 + \int d^2\mathbf{r}\left[\tfrac{1}{2}\rho g z^2(\mathbf{r}) + \tfrac{1}{2}\gamma(\nabla z(\mathbf{r}))^2\right], \tag{9.81}$$

where F_0 is the free energy of the unperturbed surface, $z(\mathbf{r})$ is the height fluctuation at lateral position $\mathbf{r}$ on the surface, γ is the surface tension, ρ the fluid density and g the acceleration due to gravity. Writing the second term in Eq. (9.81) in terms of its Fourier transform, we get F as a sum of normal mode contributions:

$$F = F_0 + \int d^2\mathbf{q}\left[\kappa^2 + q^2\right]z(\mathbf{q})z(-\mathbf{q}), \tag{9.81}$$

where $\kappa = \sqrt{\dfrac{\rho g}{\gamma}}$ is the inverse of the *capillary length*. Equation (9.82) leads to the evaluation of the thermodynamic average $\langle z(q)z(-q)\rangle$ as [56]

$$\langle z(q)z(-q)\rangle = \frac{\pi B}{\kappa^2 + q^2}, \tag{9.83}$$

where

$$B = (kT/\pi\gamma). \tag{9.84}$$

This leads to an evaluation of the height–height correlation function [57, 58]

$$C(\mathbf{R}) \equiv \langle z(O)\, z(R) \rangle = \pi B \int \frac{d^2\mathbf{q}}{(2\pi)^2} \frac{e^{i\mathbf{q}\cdot\mathbf{R}}}{\kappa^2 + q^2} = \tfrac{1}{2} B K_0\,(\kappa R), \qquad (9.85)$$

where $K_0(x)$ is the modified Bessel function of the first kind.

If one attempts to write down the total mean-square height fluctuation or surface roughness due to capillary waves by integrating Eq. (9.83) over all $\mathbf{q}$, one must introduce an upper wavevector cut-off q_{max} in order to prevent divergences (see also Ref. 59) and one then obtains for the total mean-square roughness of the liquid surface

$$\sigma^2 = \tfrac{1}{2} B \ln(q_{max}/\kappa). \qquad (9.86)$$

At length scales $\ll \kappa^{-1}$ (which are in practice those relevant for scattering experiments) the Bessel function in Eq. (9.85) may be replaced by [57] $-0.577 - \ln(\kappa R/2)$. To prevent short (molecular) length scale problems, we also introduce a lower length scale cut-off [58]. Thus we finally write

$$C(\mathbf{R}) = -\tfrac{1}{2} B \ln\!\left[\tfrac{1}{2}\,\kappa [R^2 + r_0^2]^{\frac{1}{2}}\right] - \tfrac{1}{2}(0.577), \qquad (9.87)$$

where r_0 is defined to give the correct surface roughness, as given in Eq. (9.86) as $R \to 0$ which implies that $r_0 \sim q_{max}^{-1}$.

Substituting this in Eq. (9.79), we may calculate the scattering in the DWBA, after folding with the resolution function. As shown by Sanyal *et al.* [57], the logarithmic form yields asymptotic power laws for the scattering function as a function of q_ρ for fixed q_z, as in the well-known case of a 2D crystal. These are smeared out at small q_ρ due to instrumental resolution effects [57]. The exponent of the power law (without resolution effects) is $-(2 - \eta)$, where η is $\left(\tfrac{1}{2} B q_z^2\right)$ which can be calculated knowing the surface tension. Good agreement was obtained with such a calculated scattering function (Fig. 9.19): McClain *et al.* [58] have measured both the specular and off-specular scattering from an oil/water/microemulsion interface with very low surface tension and consequently large amplitude capillary wave fluctuations and also find good agreement with the above theory. Braslau *et al.* [60] had earlier studied the specular and diffuse scattering from water and shown that the mean square roughness σ_{eff}^2 obtained from specular reflectivity experiments (by fitting with expressions of the form of Eq. (9.46)) is actually given by

$$\sigma_{eff}^2 = \tfrac{1}{2} B \ln(q_{max}/q_{min}), \qquad (9.88)$$

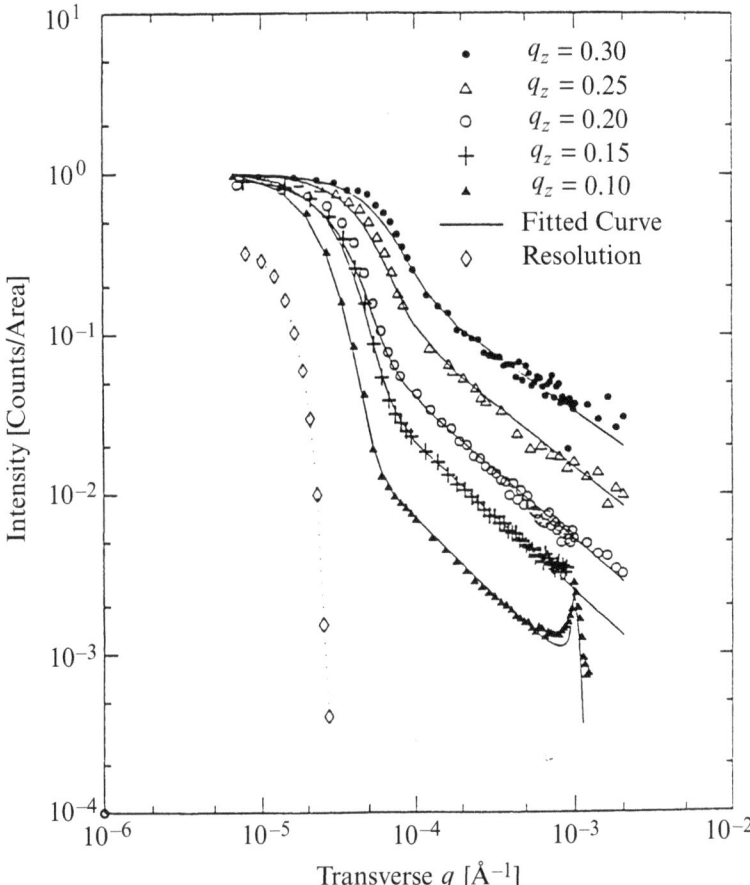

Figure 9.19. Measured (symbols) and calculated (solid curves) transverse x-ray diffuse scattering from the surface of liquid ethanol at room temperature, shown on a log–log plot vs. in-plane q_y for several values of q_z. The data has been normalized to unity at $q_y = 0$, and background has been subtracted. One in-plane direction, q_x has been integrated over by wide slits. The model represents the capillary wave theory folded with instrumental resolutions. (After M. K. Sanyal *et al.*, Ref. 57.)

where q_{min} is the *instrumental resolution* cut-off, which pre-empts the capillary length cut-off κ, since it is in general much larger. Thus, we have in principle to correct for the instrumental resolution when we measure the roughness of a liquid surface using *specular* reflectivity. The dependence of the cut-off q_{max} on molecular dimensions has been checked by Ocko *et al.* for liquid alkanes [61].

Finally, we recognize that in many instances, the roughness which exists on a surface may not be characterized by Gaussian randomness, e.g., one may have a surface characterized by regular or irregular *islands* of *constant* height or randomly varying heights. The same may be true of a surface with irregular *pits* or holes embedded on it. In such a case, Eq. (9.69) is no longer valid, and we must go back to the basic Eq. (9.68).

Let us define $\gamma_0(x,y)$ as the two-dimensional analogue of the Debye correlation function, related to the probability of crossing over from an island to 'no island' within a relative lateral separation of (x,y) and ϕ as the fractional coverage of islands. If the islands are all of the same height (Δ) we obtain by methods which we cannot reproduce in detail here [62, 63, 69]

$$S(\mathbf{q}) = S(\mathbf{q})_{\mathrm{spec}} + S(\mathbf{q})_{\mathrm{diff}}, \tag{9.89}$$

where

$$S(\mathbf{q})_{\mathrm{spec}} = \frac{A4\pi^2\rho_0^2}{q_z^2}\delta(q_x)\delta(q_y)\left[1 - 4\phi(1-\phi)\sin^2(q_z\Delta/2)\right] \tag{9.90}$$

and

$$S(\mathbf{q})_{\mathrm{diff}} = \frac{A\rho_0^2}{q_z^2}(1-\phi)\sin^2(q_z\Delta/2)\iint dx\,dy\,\gamma_0(x,y)e^{-i(q_x x + q_y y)}. \tag{9.91}$$

Note that the longitudinal diffuse scattering ($q_x, q_y \approx 0$) has a modulation along q_z with period $(2\pi/\Delta)$ which is exactly out-of-phase with a similar modulation in the specular reflectivity.

The above theory can be easily generalized to the case of a film with islands deposited on a substrate, and to include roughness fluctuations as well. The expression for the specular reflectivity may be written as [64]

$$R(q_z) = \frac{16\pi^2}{q_z^4}\left\{\rho_1^2 e^{-q_z^2\sigma_1^2}\left[1 - 4\phi(1-\phi)\sin^2(q_z\Delta/2)\right] + (\rho_2 - \rho_1)^2 e^{-q_z^2\sigma_2^2}\right.$$
$$\left. + 2\rho_1(\rho_1 - \rho_2)e^{-\frac{1}{2}q_z^2(\sigma_1^2+\sigma_2^2)}\left[\phi\cos(q_z t + \Delta) + (1-\phi)\cos(q_z t)\right]\right\}, \tag{9.92}$$

where t is the total film thickness, ρ_1 is the film electron density, ρ_2 that of the substrate and σ_1, σ_2 are respectively the roughness values at the film/air and film/substrate interfaces. This reflectivity expression yields both the rapid Kiessig fringes, as well as modulations due to the islands on the surfaces. Figure 9.20 shows the specular reflectivity and longitudinal diffuse scattering from a polymer film decorated with such islands, where both the Kiessig fringes (which appear in phase in the specular and the diffuse scattering due to conformal roughness of the film and substrate) and the 'island

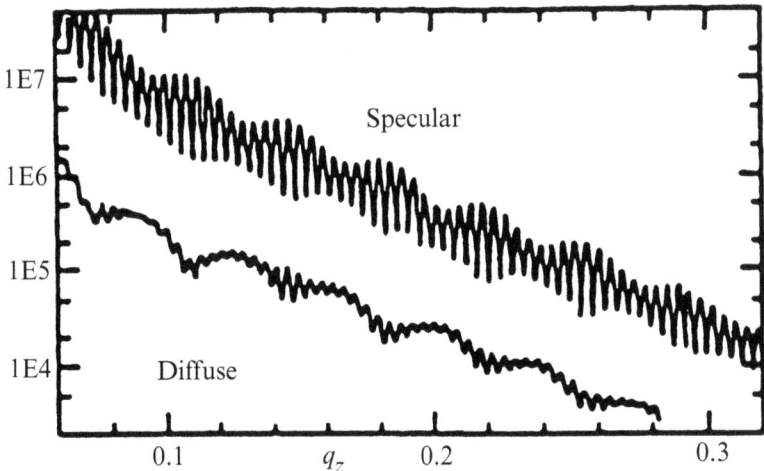

Figure 9.20.   Specular and longitudinal diffuse x-ray scattering as a function of q_z for a PS/PMMA (50:50), MW = 30 000, film on a silicon substrate. The film is decorated with islands. The fitted curves are not shown for clarity since they are almost indistinguishable from the experimental curves. (After S. K. Sinha *et al.*, Ref. 62.)

modulations' which are out of phase in the specular and in the diffuse) are observed [62].

One may generalize Eqs. (9.90) and (9.91) to the case where the islands have a *distribution* of heights rather than a constant height [62], but we shall not go into such details here.

We now discuss the off-specular scattering from thin films and multi-layers.

The diffuse scattering from free-standing thin films was first discussed explicitly by Daillent and Belorgey [65], who showed how the off-specular scattering can yield definite information about the correlations between the fluctuations of the free surfaces (which yield Kiessig-like fringes in the off-specular scattering as well). These authors also experimentally studied the diffuse scattering of x-rays by 'black-soap' films, and showed that the results could be fitted by a theoretical model for the fluctuations of liquid-like membranes.

In Section 9.3, we have shown how the specular reflectivity from a medium uniform in the x, y directions but stratified in the z direction may be calculated exactly and in the Born approximation. A uniform and smooth thin film on a uniform smooth substrate is a particular example of such a system and the specular reflectivity (Eqs. (9.41) and (9.42)) will show the characteristic Kiessig fringes or modulations of spacing $\Delta q_z = 2\pi/d$,

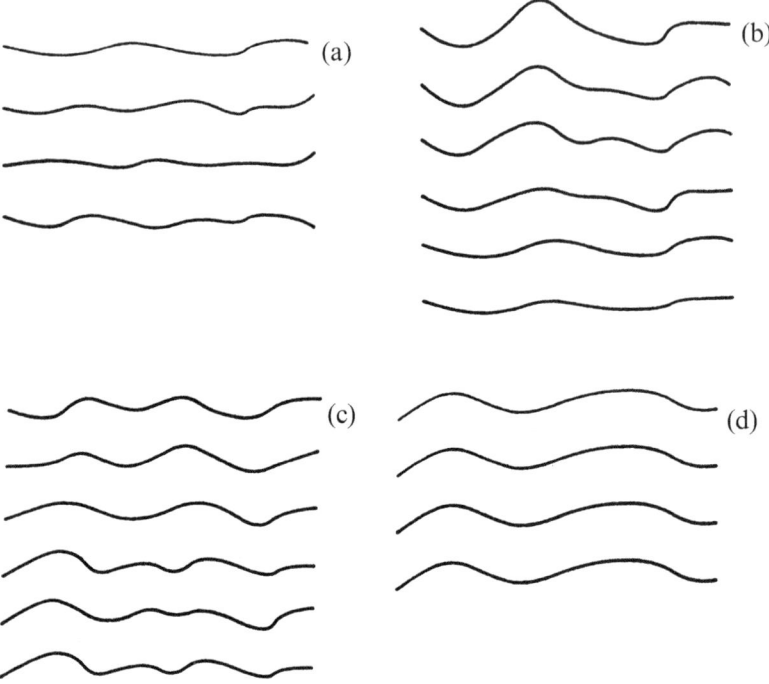

Figure 9.21. Sketch of multiple interfaces indicating degrees of conformal roughness from (a) uncorrelated to (d) completely conformal.

where d is the film thickness. If we now introduce roughness at the film/substrate and film/air interfaces, we may ask how such a film would scatter in the off-specular directions. (We have already seen in Section 9.3 that an approximate way of treating the modification to the specular reflectivity is to multiply the reflectance of each interface in the expression for the total reflectivity by an appropriate Debye–Waller factor of the form given by Eq. (9.47)). The answer depends on to what extent the roughness height fluctuations of the top interface are correlated with those of the lower interface, i.e., to what extent the roughness fluctuations are *conformal* (see Fig. 9.21).

For Gaussian roughness and in the Born approximation, the generalization of Eq. (9.76) for multiple (N) interfaces is [66–68]

$$\left(\frac{d\sigma}{d\Omega}\right)_{\text{diff}} = \frac{Ab^2}{q_z^2} \sum_{i,j=1}^{N} (\Delta\rho_i)(\Delta\rho_j^*) e^{-\frac{1}{2}q_z^2\left(\sigma_i^2+\sigma_j^2\right)} e^{-iq_z(z_i-z_j)} \int d\mathbf{R} \left[e^{q_z^2 C_{ij}(\mathbf{R})}-1\right] e^{-i\mathbf{q}_p\cdot\mathbf{R}},$$

(9.93)

where i, j sum over the various interfaces, σ_i^2 is the mean-squared roughness for interface i, $(\Delta\rho_i)$ is the effective electron difference (which may include an imaginary part due to absorption) across the ith interface (always measured in the same sense), $\bar{z}_i$ is the average height of the ith interface, and C_{ij} (**R**) is now the height–height correlation function generalized to a *pair* of interfaces (i,j) at lateral separation **R**, i.e.,

$$C_{ij}(\mathbf{R}) \equiv \langle \delta z_i(\mathbf{o}) \delta z_j(\mathbf{R}) \rangle, \qquad (9.94)$$

$\delta z_i(\mathbf{R})$ being the vertical height fluctuation of interface i about the mean at position **R**. Now, if the surfaces are completely uncorrelated $C_{ij}(\mathbf{R})$ vanishes for $i \neq j$, and the diffuse scattering from Eq. (9.92) reduces to the independent sum of diffuse scattering contributions from each interface. On the other extreme, where all interfaces are completely correlated, so that $C_{ij}(\mathbf{R})$ $= C(R)$, independent of i and j, the expression in Eq. (9.93) factors into the diffuse scattering from each interface modulated in the q_z direction by the structure factor for the specular reflectivity so that all Kiessig fringes, Bragg reflections etc., which are present in the specular reflectivity are faithfully reproduced also in the off-specular scattering, as illustrated in Fig. 9.22. In the first case, the Kiessig fringes in the specular reflectivity from a thin wetting film of water on glass are reproduced in the off-specular longitudinal diffuse scattering scans. In the second, the Bragg peaks in the reflectivity from a multilayer film are also reproduced in the off-specular scattering. In general, the existence of fringes or peaks in the q_z dependence of the diffuse scattering is an indication of a degree of conformality between the interfaces. In the vicinity of the critical angle (for thin films) or Bragg reflections (in the case of multilayers), the simple Born approximation given by Eq. (9.93) must be replaced by a DWBA based expression, which may be found discussed in Refs. 16, 65, 69–72, but is beyond the scope of this article. We will only note that, as for the Yoneda wings in the diffuse scattering from a single rough surface, such dynamical effects can introduce extra peaks or oscillations into the structure of the diffuse scattering, including modulations in the intensities as a function of q_ρ for transverse diffuse scans in the case of thin films.

There have been relatively few experiments on x-ray diffuse scattering from liquid crystalline films. A recent experiment by Shindler *et al.* [73] on specular *and* diffuse scattering from free-standing smectic-A films has provided a remarkably complete fit of the displacement–displacement correlation function using recent theoretical models [74, 75] of such systems, which express the displacement–displacement correlation functions in terms of the surface tension γ and the elasticity parameters K and B introduced in Section 9.3. Using these models and a simple model for the density profile of the liquid crystal molecule, the authors have achieved, with a

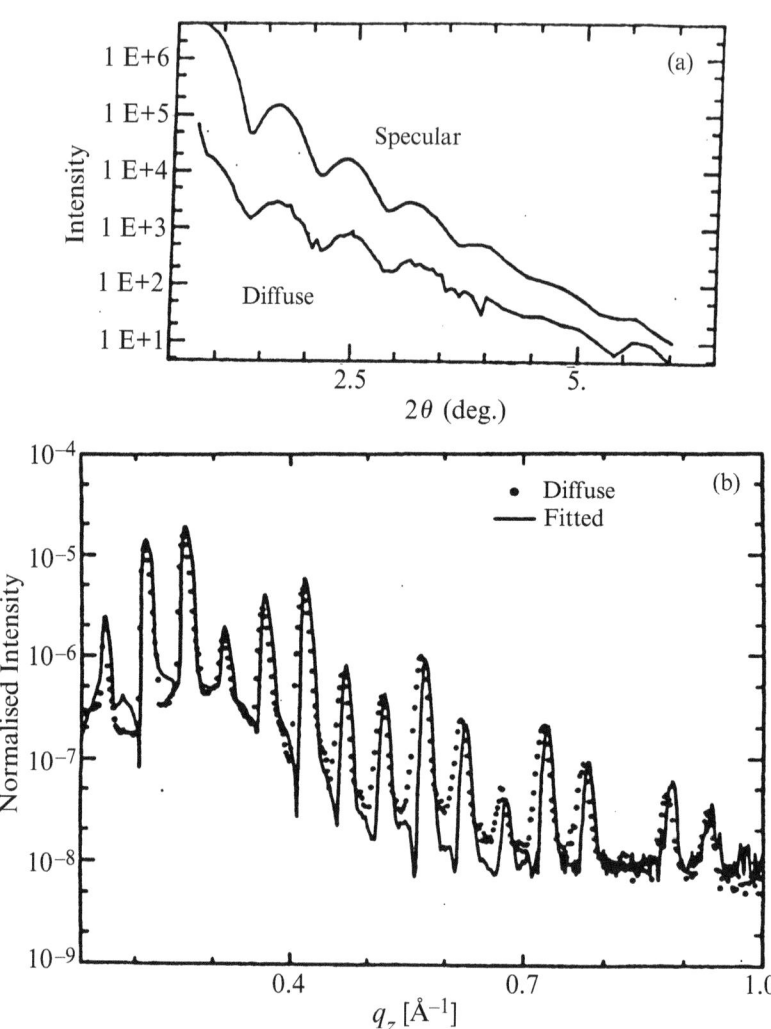

Figure 9.22. (a) Specular and (b) longitudinal diffuse scattering vs. q_z, from a GaAs/AlAs multilayer which possessed completely conformal interface roughness, showing how the diffuse scattering displays the Bragg-like peaks of the specular reflectivity. (After S. K. Sinha *et al.*, Ref. 67.)

single set of parameters, an impressive fit to both the specular reflectivities and longitudinal-diffuse and transverse-diffuse off-specular scattering for films of several different thicknesses. It appears that theory and experiment for the fluctuations of free-standing smectic-A films can be now regarded as being in satisfactory agreement.

Recently, Geer *et al.* [76] carried out a detailed study of the static undulations in a liquid crystal polymer film on a substrate by studying the non-specular diffuse scattering. Unlike the case of free-standing liquid crystal films, they found that the layer fluctuations for the liquid crystal film corresponded to completely conformal layer undulations driven by the substrate roughness, i.e., they found that the off-specular scattering could be well fitted by the expression (9.93) where the height–height correlation function $C_{ij}(\mathbf{R})$ between layers (i,j) was independent of i and j and equal to that for the substrate. (These authors used an expansion of the exponential in the integral of Eq. (9.93) for ease of calculation of $S(q)$.) The values which they obtained for the parameters characterizing the roughness were $\sigma = 3.6 \pm 0.12$ Å, $\xi = 1327 \pm 18$ Å and $h = 0.25 \pm 0.05$. Thus, for actual liquid crystal films used in applications (which are always on a substrate) the dynamical fluctuations discussed for free-standing films may be a relatively minor contribution to the actual layer undulations.

We now turn to the case of scattering at sufficiently large (in-plane) q_ρ that we cannot approximate the material by a medium of uniform electron density. This is the situation for grazing incidence x-ray diffraction (GIXD) experiments. The Born approximation (Eq. (9.65) now has the form

$$S(\mathbf{q}) = b^2 f^2(\mathbf{q}) \left\langle \sum_{i,j} e^{-i\mathbf{q}\cdot(\mathbf{r}_i - \mathbf{r}_j)} \right\rangle, \tag{9.95}$$

where the bracket denotes a statistical average, $f(\mathbf{q})$ denotes the molecular form factor (for simplicity, we assume a uniform orientation of all molecules contributing to the diffraction peak), and $\mathbf{r}_i, \mathbf{r}_j$ denote the positions of the molecules i, j. Let us assume that the molecules form layered structures parallel to the surface and that only the first N layers possess in-plane correlations or, alternatively, that the film is a free-standing film consisting of only N layers. Thus, for the large $\mathbf{q}_\rho$ considered here, only the contributions of these molecules will be considered in the sum in Eq. (9.95), which can now be written

$$S(\mathbf{q}) = b^2 f^2(\mathbf{q}) \sum_{m,n=1}^{N} e^{-iq_z(z_m - z_n)} \left\langle \sum_{i,j} e^{-iq_z(\delta z_i(m) - \delta z_j(n))} e^{-i\mathbf{q}_\rho\cdot(\mathbf{R}_i(m) - \mathbf{R}_j(n))} \right\rangle, \tag{9.96}$$

where z_m is the average z-coordinate of the center of the mth layer, $\mathbf{R}_i(m)$ is the *lateral* coordinate of the ith molecule in the layer m, and $\delta z_i(m)$ is its fluctuation in the z-direction about z_m. For many situations, the z-direction fluctuations may be decoupled from in-plane fluctuations, yielding

$$S(\mathbf{q}) = b^2 f^2(\mathbf{q}) \sum_{m,n=1}^{N} e^{-iq_z(z_m - z_n)} \sum_{i,j} e^{-\frac{1}{2}q_z^2 C_{ij}(m,n)} \left\langle e^{-i\mathbf{q}_\rho\cdot(\mathbf{R}_i(m) - \mathbf{R}_j(n))} \right\rangle, \tag{9.97}$$

where

$$C_{ij}(m,n) = \langle [\delta z_i(m) - \delta z_j(n)]^2 \rangle. \tag{9.98}$$

In general, the sum over (i,j) in Eq. (9.97) will yield a function of q_ρ which will be the *convolution* of the in-plane structure factor $\sum_{ij} \langle e^{-iq_\rho \cdot (R_i(m) - R_j(n))} \rangle$

with the in-plane Fourier transform of $e^{-\frac{1}{2}q_z^2 C_{ij}(m,n)}$. The latter will have a delta function part due to the asymptotic long range part of C_{ij}, proportional to $e^{-\frac{1}{2}q_z^2(\sigma_m^2 + \sigma_n^2)}$ where σ_m^2 is the mean-square fluctuation in the mth layer. Let us now suppose that only the top layer of the sample ($m = 1$) possesses in-plane long range order. Equation (9.97) then simplifies to

$$S(\mathbf{q}) = b^2 f^2(\mathbf{q}) e^{-q_z^2 \sigma_1^2} \sum_{G_\parallel} S(\mathbf{q}_\rho - \mathbf{G}_\parallel), \tag{9.99}$$

where $\mathbf{G}_\parallel$ is an in-plane reciprocal lattice vector of the two-dimensionally ordered monolayer. (We shall here ignore subtleties such as the lack of true long range order in monolayer systems which would cause the delta functions in Eq. (9.99) to be replaced by power-law singularities around $\mathbf{G}_\parallel$, etc. [77].)

Equation (9.99) shows that $S(\mathbf{q})$ is sharply peaked along *rods* in reciprocal space normal to the surface (i.e., parallel to q_z) and passing through the two-dimensional reciprocal lattice of the monolayer. It is the scattering along these rods which one picks up in a GIXD experiment. The q_z-dependence of the intensity along the rod will depend on $e^{-q_z^2 \sigma_1^2}$ but more strongly on the q_z-dependence of the form factor of the molecule, via the factor $f^2(\mathbf{q})$. Thus, if the molecule is an elongated cylindrical object, $f(\mathbf{q})$ is a function that will have equal amplitude contours on a flat disk normal to the cylinder axis (being the Fourier transform of its electron density). If the molecules are oriented normal to the plane of the surface, the intensity along the rods will be a maximum at $q_z = 0$ and will drop off without a q_z value inversely proportioned to the length of the molecule. (If the in-plane order extends to several layers in from the top layer, then $S(q)$ will also develop modulations along q_z, eventually beginning to peak at q_z values of $n(2\pi/d)$ where d is the layer spacing, and with a width inversely proportional to the thickness of the surface-ordered region.) If, on the other hand, the molecules are *tilted* from the surface normal, the corresponding contours of $f^2(\mathbf{q})$ will be *tilted* disks normal to the molecular axis, and the intensity along the rods will peak at *finite* q_z for $\mathbf{q}_\rho$ parallel to the direction of tilt, but still at $q_z = 0$ for $\mathbf{q}_\rho$ normal to the tilt direction.

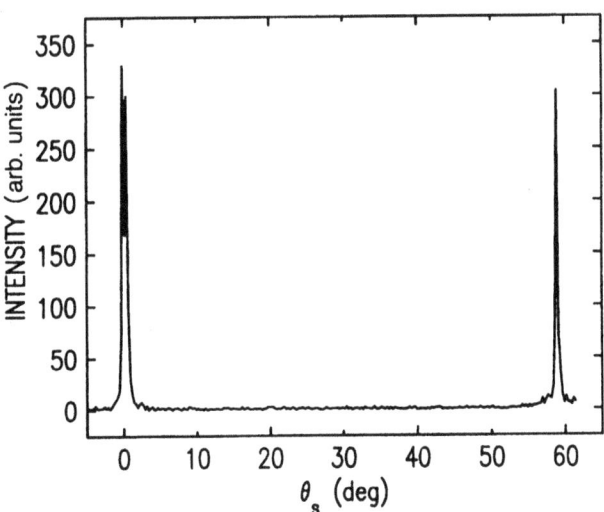

Figure 9.23. Grazing incidence diffraction peaks corresponding to the (10) rods of the hexagonal surface crystalline structure of a C_{24} alkane as a function of sample reflections, showing the six-fold symmetry. (After X. Z. Wu *et al.*, Ref. 36.)

As an example of a GIXD experiment we consider again the experiments of Wu *et al.* on liquid alkanes [36, 37]. In Section 9.3, we discussed the results of specular reflectivity measurements on liquid normal alkanes with carbon numbers > 14, which indicated a denser layer forming on the surface of the bulk liquid a few degrees above the bulk freezing temperature, as evidenced by the appearance of Kiessig-like oscillations in the reflectivity. The proof that this layer exhibits in-plane crystalline order was only confirmed when GIXD experiments were carried out, using the liquid surface diffractometer. Rods of scattering normal to the surface with six-fold symmetry were soon found (see Fig. 9.23) indicating a hexagonal in-plane structure for the ordered surface monolayer, with a lattice constant identical to the structure of a layer of the crystalline 'rotator phase' for the corresponding alkane. Intensity scans along the rods showed a single maximum around $q_z = 0$ for $C_n < 30$ indicative of molecules oriented normal to the surface, while for $C_n > 30$, the appearance of an additional maximum at finite q_z indicated that the molecules were tilted towards nearest neighbors, as discussed above. (Both peaks occur, since the experiment averages over domains where the tilt is parallel to the in-plane **q** and also where it is normal to it.) Similar experiments on GIXD from liquid alcohols have recorded tilted phases in these surface-ordered crystalline bilayers [38].

Factor *et al.* [78] have used GIXD to examine crystalline order in films of aromatic polyamides.

One of the advantages of the GIXD geometry is that by keeping the incident beam at or below the critical angle of total external reflection, so that the beam is evanescent in the bulk of the sample, one can greatly enhance surface scattering effects relative to bulk scattering effects, as pointed out in some of the earliest experiments of this type by Eisenberger and Marra [79]. This has been exploited in surface scattering experiments by several groups [80, 81, 82], and more recently grazing incidence neutron diffraction experiments have also been carried out [83]. Under the conditions of grazing incidence, it is more accurate to use the distorted wave Born approximation (DWBA) than the Born approximations, and this results in a slightly modified expression for the scattering [14, 15, 72, 82–84]:

$$S(\mathbf{q}) = |t(\alpha)|^2 |t(\beta)|^2 S_{Born}(\mathbf{q}) \qquad (9.100)$$

where $S_{Born}(\mathbf{q})$ is the corresponding expression as given by the Born approximation (Eq. (9.95)). $t(\alpha)$, $t(\beta)$ are, as before (see Eq. (9.79)), the Fresnel transmission coefficients for the ideal surface at grazing angles of incidence α, β respectively. These cause $S(q)$ to be at its maximum value when α or β is equal to the critical angle of incidence for total reflection.

There have been few GIXD studies of liquid crystals, because transmission x-ray scattering measurements on thick free-standing films have already revealed much about the in-plane ordering of the outer (surface) layers [43–45].

An interesting variation of x-ray studies of film roughness is the application of the above methods to study anisotropic roughness. (In everything discussed above, the assumption has been made that the average surface morphology is isotropic.) Anisotropy of surface morphology can occur when, for instance, polymer or liquid crystal films are preferentially oriented by buffing [85, 86] or in the case of semiconductor films by the formation of aligned steps on a film grown on a slightly miscut single crystal substrate [87]. This can also occur when photopolymerization is induced using linearly polarized light [88]. In a recent study of x-ray reflectivity from linearly photopolymerized poly (vinyl 4-methoxy-cinnamate) (PVMC) films and 12-8 (poly) diacetylene (PDA) films deposited on glass substrates, Cull *et al.* [88] found specular reflectivities which were considerably different when the plane of scattering was parallel to or perpendicular to the direction of linear polarization of the light. The Kiessig fringes were much more visible in one direction compared to the other, indicating very different effective roughnesses in the two directions. (An example of similar

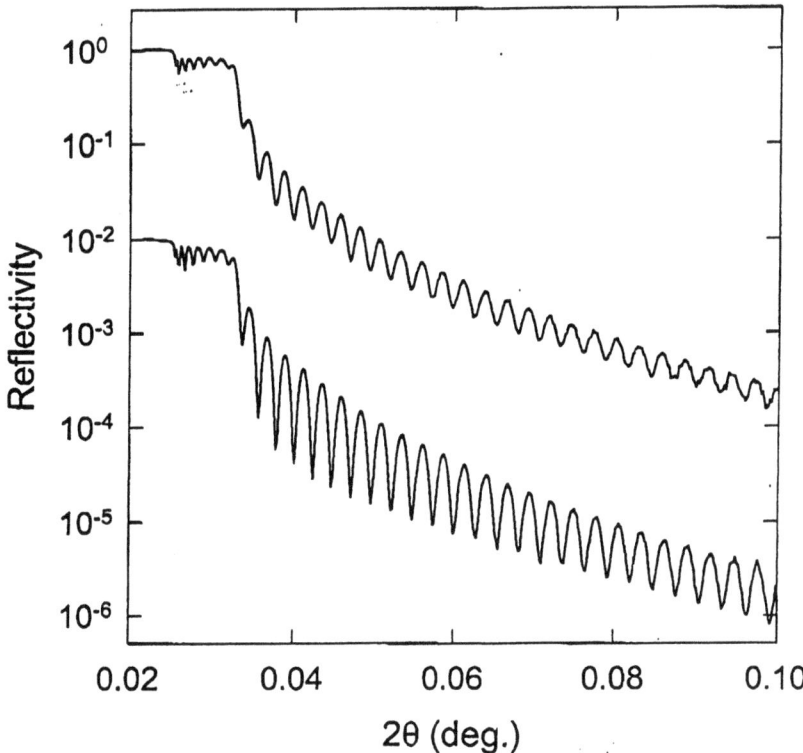

Figure 9.24.   X-ray reflectivity for rubbed polyimide films showing anisotropic roughness. (After Cull *et al.*, Ref. 88.)

experiments on rubbed polyimide films by the same group is shown in Fig. 9.24). By analyzing the reflectivity data, these authors deduced a value of 7.7 ± 0.5 Å and 25.7 ± 0.5 Å in the y- and x-directions respectively at the film–air interface in the PVMC films (where the x-axis is in the plane of the film and parallel to the direction of the light polarization). In the case of the PDA films, the anisotropy was reversed, the effective roughness being 5.1 ± 0.5 Å in the x-direction, but 19.6 ± 0.5 Å in the y-direction. Since the specular reflectivity actually measures the mean-square height fluctuations averaged over the coherence area of the x-rays as projected onto the plane of the film (roughly inversely proportional to the projection of the instrumental resolution in q-space in this plane), it is possible for the measured values of σ to be different in the two orientations, since the coherence area is elongated and narrow along the projection of the incident beam direction (see Eqs. (9.4) and (9.5) and the discussion following). Cull *et al.* accounted for the anisotropy of the surface morphology in terms of the microscopic details of the polymerization processes in the two different cases. Very similar results have been obtained on stepped GaAs/AlAs multilayer films by Sinha *et al.* [87].

9.5 Conclusions

We have shown how the scattering of x-rays or neutrons from surfaces and thin films can be described in terms of the density profiles normal to the surface (which determine the specular reflectivity) and in terms of lateral surface or interface fluctuations (which determine the off-specular or diffuse scattering). Such experiments can reveal details or surface ordering, surface phase transitions, surface roughness and capillary wave fluctuations and the morphology of thin films. While not capable of yielding direct images as in the case of AFM and other imaging microscopies (to which x-ray and neutron scattering may be regarded as complementary techniques), x-ray and neutron scattering have the advantage of being relatively non-invasive probes, particularly of buried interfaces, and also of yielding *global* statistical information about the surfaces being studied.

We have described the various experimental techniques employed in such scattering experiments, particularly at synchrotron radiation and neutron sources. The application of these techniques has already revealed the existence and nature of surface-nucleated nematic/smectic phase transitions at liquid crystal/air and liquid crystal/solid interfaces, of static undulations in liquid crystalline polymer films, of the morphology of ordering and fluctuations in thin liquid crystal films and capillary wave fluctuations at liquid/air and liquid/liquid interfaces. They have also revealed unexpected surface phase transitions at liquid hydrocarbon/air interfaces. With the advent of new higher brilliance synchrotron and neutron sources, there is no doubt that one may anticipate many more detailed studies of the surfaces and thin films of liquid crystals, polymers, complex fluids, membranes and related systems in the years to come.

Acknowledgment

This work has been supported by the Office of Basic Energy Sciences, U.S. Dept. of Energy under contract number W-31-109-ENG-38. I thank O. H. Seeck for help with some of the calculations and figures.

References

1. I. K. Robinson, *Phys. Rev. B* **33**, 3830 (1986); I. K. Robinson and D. J. Tweet, *Rep. Prog. Phys.* **55**, 599 (1992); E. Vlieg, J. F. van der Veen, S. J. Gurman, C. Norris, and J. E. Macdonald, *Surf. Sci.* **210**, 301 (1989).
2. S. R. Andrews and R. A. Cowley, *J. Phys.* C **18**, 6427 (1985); H. You, in *Surface X-Ray and Neutron Scattering*, H. Zabel and I. K. Robinson (Eds.), Springer Proceedings in Physics, Vol. **61**, Springer, Berlin (1992), p. 47; S. Dietrich and A. Haase, *Phys. Rep.* **260**, 102 (1995).
3. S. M. Shapiro and N. J. Chesser, *Nucl. Instrum. Methods* **101**, 183 (1972); B. O. Loopstra, *Nucl. Instrum. Methods* **44**, 181 (1966); G. Shirane and V. J. Minckiewicz, *Nucl. Instrum. Methods* **89**, 109 (1970).
4. J. Als-Nielsen and P. S. Pershan, *Nucl. Instrum. Methods* **208**, 545 (1983); P. S. Pershan, *J. Phys (Paris) Colloq.* **50**, C7-1 (1989).
5. M. Born and E. Wolf, *Principles of Optics*, 5th Edn., Pergamon, Oxford (1975), p. 38.
6. G. Porod, in *Small-Angle X-ray Scattering*, O. Glitter and O. Kratky (Eds.) Academic Press, New York (1982), p. 44.
7. F. Abeles, *Ann. Phys.* **3**, 504 (1958); **5**, 596 (1960).
8. M. Born and E. Wolf, *Principles of Optics*, 5th Edn., Pergamon, Oxford (1975), p. 51.
9. L. G. Parratt, *Phys. Rev.* **95**, 359 (1954).
10. P. Croce and B. Pardo, *Nouv. Rev. Opt. Appl.* **1**, 229 (1970).
11. J. H. Underwood and T. W. Barbee, *AIP Conference Proceedings* **75**, 170 (1981).
12. L. I. Schiff, *Quantum Mechanics*, McGraw-Hill, New York (1968); A. Messiah, *Quantum Mechanics*, Wiley, New York (1962).
13. J. Als-Nielsen, F. Christensen, and P. S. Pershan, *Phys. Rev. Lett.* **48**, 1107 (1984); E. S. Wu and W. W. Webb, *Phys. Rev. A* **8**, 2077 (1973); J. Als-Nielsen, *Z. Phys. B* **61**, 411 (1985).
14. G. H. Vineyard, *Phys. Rev. B* **26**, 4146 (1982).
15. S. K. Sinha, E. B. Sirota, S. Garoff, and H. B. Stanley, *Phys. Rev. B* **38**, 2297 (1988).
16. R. Pynn, *Phys. Rev. B* **45**, 602 (1992).
17. L. Nevot and P. Croce, *Rev. Phys. Appl.* **15**, 761 (1980).
18. D. K. G. de Boer, *Phys. Rev. B* **49**, 5817 (1994); D. K. G. de Boer, A. J. G. Leenaers, and W. W. Van den Hoogenhof, *J. Phys. (Paris) III* **4**, 1559 (1994).
19. G. P. Felcher and T. P. Russell, Methods of Analysis and Interpretation of Neutron Reflectivity Data, *Physica B* **173**, 1 (1991).
20. D. S. Sivia, W. A. Hamilton, and G. S. Smith, *Physica B* **173**, 121 (1991).
21. J. S. Pedersen and I. W. Hamley, *Physica B* **198**, 16 (1994).
22. C. F. Majkrzak, N. F. Berk, J. F. Ankner, S. K. Satija, and T. P. Russell, *SPIE Proc.* **1738**, 282 (1992).
23. W. L. Clinton, *Phys. Rev. B* **48**, 1 (1993).
24. M. K. Sanyal, S. K. Sinha, A. Gibaud, K. G. Huang, B. L. Carvalho, M. Rafailovich, S. Sokolov, X. Zhao, and W. Zhao, *Europhys. Letts.* **21**, 691 (1993); S. K. Sinha, M. K. Sanyal, B. L. Carvalho, M. Rafailovich, J. Sokolov, X. Zhao, and W. Zhao, in *Resonant Anomalous Scattering*, G. Materlik, C. J. Sparks and K. Fischer (Eds.), Elsevier Science B. V. (1994), p. 421.
25. J. Chupa, J. P. McCauley, L. M. Strongin, A. B. Smith, J. K. Blasie, L. J. Peticolas, and J. C. Bean, *Biophys. J.* **67**, 336 (1994).

26. G. P. Felcher, W. D. Dozier, Y. Y. Huang, and X. L. Zhou, in *Surface X-Ray and Neutron Scattering*, Springer Proceedings in Physics, **61**, H. Zabel and I. K. Robinson (Eds.), Springer-Verlag, Berlin (1992), p. 99.
27. C. F. Majkrzak and N. F. Berk, *Phys. Rev. B* **52**, 10827 (1995).
28. V. Dehaan, A. Van Welk, S. Adenwallah, and G. P. Felcher, *Phys. Rev. B* **52**, 10831 (1995).
29. T. M. Roberts, *Physica B* **173**, 157 (1991).
30. P. Sheng, *Phys. Rev. Lett.* **37**, 1059 (1976); K. Miyano, *Phys Rev. Lett.* **43**, 51 (1979); D. W. Allender, G. L. Henderson and D. L. Johnson, *Phys. Rev. A.* **24**, 1086 (1981).
31. P. S. Pershan, A. Braslau, A. H. Weiss, and J. Als-Nielsen, *Phys. Rev. A* **35**, 4800 (1987).
32. B. M. Ocko, A. Braslau, P. S. Pershan, J. Als-Nielsen and M. Deutsch, *Phys. Rev. Lett.* **57**, 94 (1986).
33. B. M. Ocko, *Phys. Rev. Lett.* **64**, 2160 (1990).
34. J. F. van der Veen and J. W. M. Frenken, *Surf. Sci.* **178**, 382 (1986); R. Ohnesorge, H. Lowen, and H. Wagner, *Phys. Rev. A* **43**, 2870 (1991); D. W. Oxtoby, *Nature* **347**, 725 (1990).
35. R. Holyst, *Phys. Rev. B* **46**, 15542 (1992).
36. X. Z. Wu, E. B. Sirota, S. K. Sinha, B. M. Ocko, and M. Deutsch, *Phys. Rev. Lett.* **70**, 958 (1993); B. M. Ocko, X. Z. Wu, E. B. Sirota, S. K. Sinha, O. Gong and M. Deutsch, *Phys. Rev. E* **55**, 3164 (1997).
37. X. Z. Wu, B. M. Ocko, E. B. Sirota, S. K. Sinha, M. Deutsch, B. H. Cao, and M. W. Kim, *Science* **261**, 1018 (1993).
38. M. Deutsch, X. Z. Wu, E. B. Sirota, S. K. Sinha, B. M. Ocko, and O. M. Magnussen, *Europhys. Lett.* **30**, 283 (1995).
39. X. Z. Wu, B. M. Ocko, N. Tang, E. B. Sirota, S. K. Sinha, and M. Deutsch, *Phys. Rev. Lett.* **75**, 1332 (1995).
40. Y. Shi, B. Cull, and S. Kumar, *Phys. Rev. Lett.* **71**, 1773 (1993).
41. E. G. Bauer, *Z. Krist.* **110**, 372 (1958); J. A. Venables, G. D. T. Spiller, and M. Hanbucken, *Rep. Prog. Phys.* **47**, 399 (1984).
42. N. A. Clark, *Phys. Rev. Lett.* **55**, 292 (1985).
43. D. E. Moncton, R. Pindak, S. C. Davey, and G. S. Brown, *Phys. Rev. Lett.* **49**, 1865 (1982); R. Pindak and D. E. Moncton, *Phys. Today*, May (1982) p. 3.
44. D. E. Moncton and R. Pindak, *Phys. Rev. Lett.* **43**, 701 (1979); R. Pindak, D. E. Moncton, S. C. Davey, and J. W. Goodby, *Phys. Rev. Lett.* **46**, 1135 (1981); G. Aeppli, J. D. Litster, R. J. Birgeneau, and P. S. Pershan, *Mol. Cryst. Liq. Cryst.* **67**, 205 (1981).
45. P. S. Pershan, *Structure of Liquid Crystal Phases*, World Scientific Lecture Notes in Physics Vol. **23**, World Scientific, Singapore (1988).
46. S. Gierlotka, P. Lambooy, and W. H. deJeu, *Europhys. Lett.* **12**, 341 (1990); P. Lambooy, S. Gierlotka, I. W. Hamley and W. H. deJeu, in *Phase Transitions in Liquid Crystals*, S. M. Martellucci and A. N. Chester (Eds.), Plenum, New York (1992), Ch. 17.
47. L. D. Landau and E. M. Lifshitz, *Statistical Physics*, Course of Theoretical Physics Vol. 5, Pergamon, Oxford (1980), p. 434.
48. R. Holyst, D. J. Tweet, and L. B. Sorenson, *Phys. Rev. Lett.* **65**, 2153 (1990).
49. D. J. Tweet, R. Holyst, B. D. Swanson, H. Stragier, and L. B. Sorenson, *Phys. Rev. Lett.* **65** 2157 (1990).
50. T. Stoebe, R. Geer, C. C. Huang, and J. W. Goodby, *Phys. Rev. Lett.* **69**, 2090 (1992); T. Stoebe, C. C. Huang, and J. W. Goodby, *Phys. Rev. Lett.* **68** (1992).

51. S. H. Anastasiadis, T. P. Russell, S. K. Satija, and C. Majkrzak, *Phys Rev. Lett.* **62**, 1852 (1989).
52. A. M. Mayes, T. P. Russell, P. Bassereau, S. M. Baker, and G. S. Smith, *Macromolecules* **27**, 749 (1994).
53. X. L. Zhou, L. T. Lee, S. H. Chen and R. Strey, *Phys. Rev. A* **46**, 6479 (1992).
54. D. Vaknin, K. Kjaer, H. Tingsdorf, R. Blankenburg, M. Pipenstock, A. Dietrich, and M. Lösche, *Langmuir* **9**, 1171 (1993).
55. Y. Yoneda, *Phys Rev.* **131**, 2010 (1963); A. N. Nigam, *Phys Rev. A* **4**, 1189 (1965).
56. C. A. Croxton, *Statistical Mechanics of the Liquid Surface*, Wiley, New York (1980); H. T. Davis, in *Waves on Fluid Interfaces*, R. E. Meyer (Ed.), pp. 123–150 Academic Press, New York (1983).
57. M. K. Sanyal, S. K. Sinha, K. G. Huang, and B. M. Ocko, *Phys Rev. Lett.* **66**, 6281 (1991).
58. B. R. McClain, D. D. Lee, B. L. Carvalho, S. G. J. Mochrie, S. H. Chen, and J. D. Litster, *Phys. Rev. Lett.* **72**, 246 (1994).
59. M. Napierkowski and S. Dietrich, *Phys. Rev. E* **47**, 1836 (1993).
60. A. Braslau, P. S. Pershan, G. Swizlow, B. M. Ocko, and J. Als-Nielsen, *Phys. Rev. A* **38**, 2457 (1988).
61. B. M. Ocko, X. Z. Wu, E. B. Sirota, S. K. Sinha, and M. Deutsch, *Phys. Rev. Lett.* **72**, 242 (1994).
62. S. K. Sinha, Y. P. Feng, C. A. Melendres, D. D. Lee, T. P. Russell, S. K. Satija, E. B. Sirota, and M. K. Sanyal, *Proc. Workshop on Colloid and Interface Science: Trends and Applications*, L. Blum (Ed.) *Physica* (1996), to be published.
63. Z. Cai, K. Huang, P. A. Montano, T. P. Russell, J. M. Bai, and G. W. Zajac, *J. Chem. Phys* **98**, 2376 (1993).
64. Y. Liu, W. Zhao, X. Zhang, A. King, A. Singh, M. H. Rafailovich, J. Sokolov, K. H. Dai, E. J. Kramer, S. A. Schwarz, O. Gebizlioglu, and S. K. Sinha, *Macromolecules* **27**, 4000 (1994).
65. J. Daillent and O. Belorgey, *J. Chem. Phys.* **97**, 5824 (1992); J. Daillent and O. Belorgey, *J. Chem. Phys.* **97**, 5837 (1992).
66. S. K. Sinha, *Physica B* **173**, 35 (1993); S. Garoff, E. B. Sirota, S. K. Sinha, and H. B. Stanley, *J. Chem. Phys.* **90**, 7507 (1989).
67. S. K. Sinha, M. K. Sanyal, A. Gibaud, S. K. Satija, C. F. Majkrzak, and H. Homma, in *Surface X-Ray and Neutron Scattering*, H. Zabel and I. K. Robinson (Eds.), Springer Proceedings in Physics **61**, Springer, Berlin (1992) p. 91.
68. Y. H. Phang, R. Kariotis, D. E. Savage, and M. G. Lagally, *J. Appl. Phys.* **72**, 4627 (1992).
69. V. Holy and G. T. Baumbach, *Phys. Rev. B* **49**, 10668 (1994).
70. S. K. Sinha, *J. Phys. III France* **4**, 1543 (1994).
71. G. T. Baumbach, V. Holy, U. Pietsch, and M. Gailhanou, *Physica B* **198**, 249 (1994).
72. S. Dietrich and A. Hasse, *Phys. Rep.* **260**, 162 (1995).
73. E. A. L. Mol, J. D. Shindler, A. N. Shalaginov, and W. H. deJeu, *Phys. Rev. E* **54**, 536 (1996).
74. V. P. Romanov and A. N. Shalaginov, *Sov. Phys. JETP* **75**, 483 (1992).
75. A. N. Shalaginov and V. P. Romanov, *Phys. Rev. E* **48**, 1073 (1993).
76. R. E. Geer, R. Shashidhar, A. F. Thibodeaux, and R. S. Duran, *Phys. Rev. Lett.* **71**, 1391 (1993).
77. L. Gunther, Y. Imry, and J. Lajzerowicz, *Phys. Rev. A* **22**, 1733 (1980).

78. B. J. Factor, T. P. Russell, and M. F. Toney, *Macromolecules* **26**, 2847 (1993).
79. W. C. Marra, P. Eisenberger, and A. Y. Cho, *J. Appl. Phys.* **50**, 6927 (1979); P. Eisenberger and W. C. Marra, *Phys. Rev. Lett.* **16**, 1081 (1981).
80. K. G. Huang, D. Gibbs, D. M. Zehner, A. R. Sandy, and S. G. J. Mochrie, *Phys. Rev. Lett.* **65**, 3313 (1990).
81. B. N. Thomas, S. W. Barton, F. Novak, and S. A. Rice, *J. Chem. Phys.* **86**, 1036 (1987).
82. H. Dosch, *Critical Phenomena at Surfaces and Interfaces: Evanescent X-Rays and Neutron Scattering*, Springer Tracts Mod. Phys. **126**, Springer, Berlin Heidelberg (1992).
83. H. Dosch, *Appl. Phys. A* **61**, 475 (1995); L. Mailänder, H. Dosch, J. Peisl, and R. L. Johnson, *Phys. Rev. Lett.* **64**, 2527 (1991).
84. S. Dietrich and H. Wagner, *Z. Phys. B* **56**, 207 (1994).
85. T. Uchida, M. Hirano, and H. Sakai, *Liq. Cryst.* **5**, 1127 (1998).
86. M. G. Samant, J. Stohr, H. R. Brown, T. P. Russell, J. M. Sands, and S. K. Kumar, *Macromolecules* **29**, 8334 (1996).
87. S. K. Sinha, M. K. Sanyal, S. K. Satija, C. F. Majkrzak, D. A. Neumann, H. Homma, S. Szpala, A. Gibaud, and H. Morkoc, *Physica B* **198**, 72 (1994).
88. B. Cull, Y. Shi, S. K. Kumar, and M. Schadt, *Phys. Rev. E* **53**, 3777 (1996).

10

Chemical structure–property relationships

MARY E. NEUBERT

Glenn H. Brown Liquid Crystal Institute, Kent State University, Kent, OH 44242, USA

10.1 Introduction

Chemists have been involved in the liquid crystals (mesogens) effort from the beginning, since they are the ones who synthesized the organic compounds in which the liquid crystalline phases (mesophases) were discovered. Interest in the synthesis of mesogens increased dramatically once it was discovered that MBBA had a room-temperature nematic phase which could be useful in displays [1]. In order to determine what types of molecular structures formed mesophases, numerous compounds were synthesized and their mesomorphic properties determined to establish structure–property relationships. These could then be useful in selecting which molecular modifications would be likely to yield new (and hopefully better) mesogens. The availability of materials made it possible to learn more about both the properties and structures of mesophases. These studies then generated interest in synthesizing new additional mesogens. Thus, the synthesis of new materials was stimulated by the discovery of numerous new phases: ferroelectric C, biaxial nematic, reentrant, incommensurate, induced, hexatic-B, discotic and blue. It was also stimulated by the interest in the development of new uses for mesogens in FLC (ferroelectric liquid crystal), PDLC (polymer dispersed liquid crystal) displays, and NLO (non-linear optics). Today, the fastest growing areas are polymeric, organometallic and self-assembled mesogens.

During the evaluation of this synthesis endeavor, numerous new mesogens were prepared and much was learned about the relationship between molecular structure and mesomorphic properties. Yet, the accurate prediction of properties for a new compound is still not possible. The best that can be done is to use known trends in the design of a molecule with the desired properties, and hope that this will be one of the many that do follow

393

these known trends, but not be surprised if it does not. We still do not know all of the parameters needed to determine if a molecule will prefer stepwise melting when proceeding through at least one mesophase, or if it will simply melt directly into the isotropic liquid.

Today, so much information is available on the chemistry of liquid crystals that a comprehensive review would require an entire book. Thus, this chapter attempts to cover only a basic foundation of structure–property relationships, using primarily rod-like (calamitic) structures of smaller molecules. Undoubtedly this 'selection' will reflect the author's own interests and experience. There are numerous excellent review articles [2–11, 16a, 17], book chapters [18a, 19, 20, 21, 22a, 23, 24a], and books [25, 26] that discuss the structure–property relationships in liquid crystals. For the physicists who prefer to prepare their own liquid crystals, there is even a collection of procedures for some of the more common syntheses [27].

Collections of transition temperatures for the various molecules synthesized have been compiled since 1960 [28–31]. These should be consulted to fill in the gaps of information not available in this review. As the number of compounds grew and computers became available, it soon became obvious that only computer databases could provide an adequate means of keeping such collections current, and to provide the ease of searching for various structural and/or mesogenic property parameters. Consequently, a few databases have been developed with the most comprehensive being that by Vill [32], although there are more limited ones such as the one for alkyl/alkoxy phenylbenzoates [33]. Since the complete Vill database [32a] was not available at the time this chapter was started, many comparisons were made using the Demus tables [29, 30]. When no references are provided for the data presented, it can be assumed that it has come from either this source or our own data. Much of the biphenyl data comes from the hard copy of one section of the Vill database [32b].

In this chapter, the term SmB is used for both the HexB and CrB phases, to be consistent with early literature. The term more highly ordered smectic phase includes all phases with an order higher than that of smectic-B phases but less than that of three-dimensional crystals.

10.2 Nomenclature

An understanding of the organic chemistry nomenclature for liquid crystals is necessary so that researchers in this area can accurately discuss a specific structure. Two types of nomenclature exist in organic chemistry: common and systematic. Systematic nomenclature was developed, approved and is continually updated by the International Union of Pure and Applied Chemistry (IUPAC). Theoretically, this means that all chemists worldwide will know the specific structure by this systematic name, as it is a method designed to transcend language barriers. However, with this type of nomenclature, often more than one name can be generated. With the large molecules that form liquid crystals, these names can become quite long and difficult to use in everyday discussions. For this reason, various abbreviations developed as new mesogens were synthesized and, consequently, a variety of systems for creating an abbreviation. These are some of the more commonly used ones:

1. An abbreviation consisting of the first letter in each part of the chemical name for the mesogen:

MeO —◯— CH＝N —◯— Bu

4-methoxybenzylidene-4'-butylaniline   abbreviation = MBBA

2. A numbering system indicating the chain length of the two terminal substituents with some way of differentiating an alkoxy from an alkyl chain: MBBA = 1O.4 when the structure is drawn with the aldehyde on the left. These abbreviations apply only to anils. For the esters:

C_4O —◯— C(=O) Z —◯— C_8

there are three systems when Z = O: 4O-O-5, 4O-5 and 4O5.
There are two when Z = S: 4O-S-5 and $4\bar{S}5$

3. An abbreviation is given to the core part of the molecule with a number added for the chain length.

X —◯— C(=O) O —◯— Y   PB = phenylbenzoate

X —◯—◯— CN   CB = cyanobiphenyl
                 X = C_8 is 8CB
                 X = C_8O is 8OCB

C_n —⌃◯— CN   PCH-n

C_n —⌃⌄— CN   CCH-n

None of these systems is versatile enough to accommodate all possible structures, even within a specific core system. Branched chains containing a variety of substituents, more than one terminal chain containing substituents, and lateral substituents add complications to such systems. A better system for the branched chain phenylbenzoates was proposed [34, 35] but never widely used. The latter system can accommodate more variations since it is based on the chemical name. Still, different abbreviations can be generated from the same chemical name, and different chemical names can be used for the same compounds, all of which can result in several abbreviations for the same compound. The following are some examples:

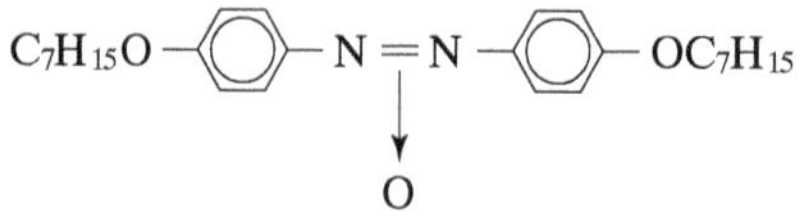

$C_7H_{15}O$ —◯— $N=N$ —◯— OC_7H_{15} diheptyloxyazoxybenzene
      | DHAB, HAB, HOAB, HOAOB 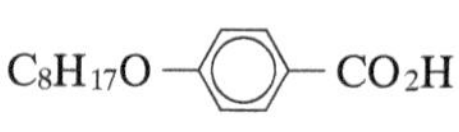
      O

$C_8H_{17}O$ —◯— CO_2H p-octyloxybenzoic acid
            OOBA, OBA

Abbreviations are sometimes randomly assigned to a structure before first checking to see whether one has already been assigned, or if the proposed abbreviation has been assigned to another compound:

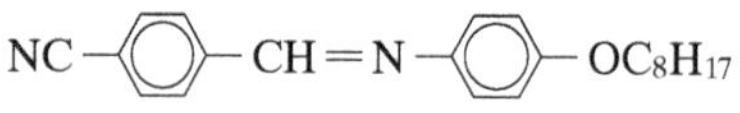

NC —◯— $CH=N$ —◯— OC_8H_{17} 4-cyanobenzylidene-4′-octyloxyaniline
              CBOOA, CBAOB and NBOA

Sometimes an abbreviation is created without using any of the known systems:

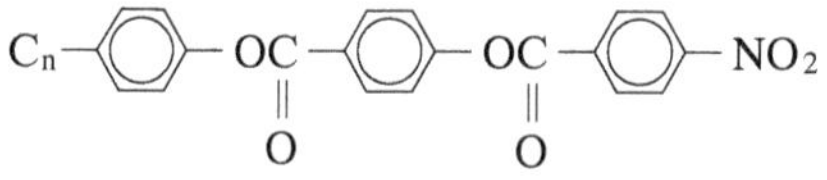

C_n —◯— OC —◯— OC —◯— NO_2 DBnNO$_2$
    || ||
    O   O

and lettering systems based on chemical names can create the same abbreviation for two different compounds, for example:

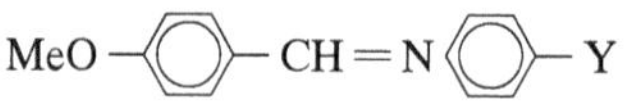

MeO —◯— $CH=N$ ◯— Y MBPA
           could be for Y = propyl or pentyl

Some industrial producers use a combination of numbers and letters to develop their own system known only to them, for example:

8CB=K24, 8OCB=M24

CE8 for C_8H_{17} —◯—◯— CO_2 —◯— $CH_2\overset{*}{C}HEt$
                  |
                 Me

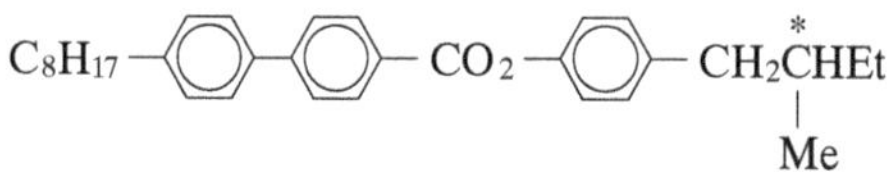

Confusion can arise when a numbering system is used and care is not taken to place the decimal properly. For example:

4O.8 is C_4H_9O —◯— $CH=N$ —◯— C_8H_{17}

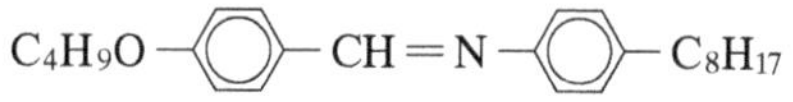

4.O8 is C_4H_9 —◯— $CH=N$ —◯— OC_8H_{17}

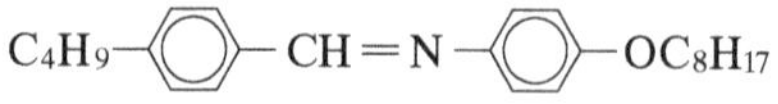

4O-8 is C_4H_9O —[ring]—C(=O)—O—[ring]—C_8H_{17}

A list of the more commonly used abbreviations can be found in the literature [36]. Although these abbreviations may be useful in discussions, the only universally accurate system is the correct chemical name and structure. All authors of papers involving liquid crystals should be aware of this block to universal communication, and include both the structure and the correct systematic chemical name at least once in their papers to avoid confusion.

10.3 Structure–property relationships

What types of compounds form mesophases? Once liquid crystalline phases were discovered, answering this question was the major thrust of research in liquid crystals. Chemists set out to determine the molecular structural features needed in a molecule for it to pack in such a way as to form liquid crystalline phases. Theory suggested that mesophases resulted when long, rigid, rod-like (calamitic) molecules packed in a parallel manner. Numerous compounds consisting of rod-like molecules of the type

$$X-(R_1)-Z-(R_2)-Y$$
$$A \qquad\qquad B$$

were synthesized (and continue to be) to determine just what structural features are needed to observe mesophases. Some general structure–property relationships (trends) were established. These vary in reliability, and none is absolute in its predictability, making it impossible to accurately predict the mesomorphic properties for a new compound. Still, these trends do help to narrow the number of possibilities synthesized in order to obtain a specific set of properties.

The collections of transition temperatures for the various mesogens mentioned earlier are very useful in studying structure–property relationships. Most such relationships, however, use smaller and more restricted collections of molecular variations, since it is much easier to see seemingly reliable trends in small collections. These trends usually became unreliable when applied to larger collections of data. The comprehensive Vill computer database [32] consists of so many structural variations that it is mind-boggling. Attempts have not yet been made to use this for studying structure–property relationships. It will be interesting to see how reliable current trends are when this is done.

These rod-like structures consist of a rigid rod formed by two ring *systems* (R_1 and R_2) that are connected together by a *central group* Z, with two tails (X and Y) called *terminal groups* or *chains* placed in such a position (usually para to the central group) as to give a linear molecule. Usually, at least one of the terminal groups must be a flexible chain. These structures are called the 'classic' liquid crystalline structure and they will be the ones discussed here in detail.

Connecting group (Z)

The Z group was initially called the central group, since many of the compounds had only one such group. Today, many structures are known to have more than one central group, often making a ring system the center of the molecule. Thus, three additional terms have been used: spacer, connecting and bridging groups. Connecting will be used in this chapter to mean the group or groups connecting major ring systems, but not those considered part of the terminal chains, for example:

1. 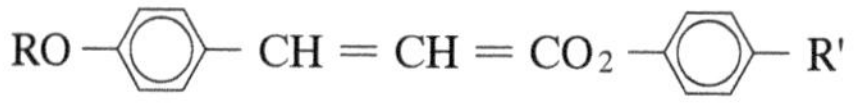 RO—⬡—CH = CH = CO_2—⬡—R'

2. RO—⬡—CH = N—⬡—CH = $CHCO_2R'$

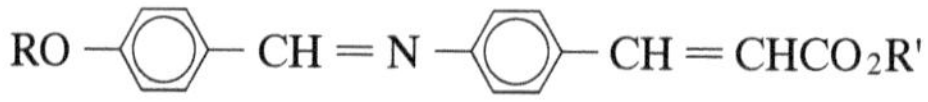

3. 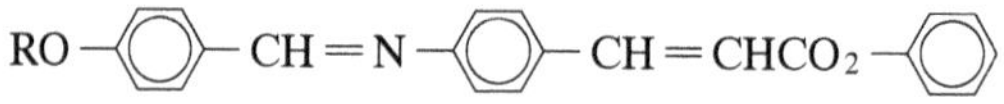 RO—⬡—CH = N—⬡—CH = $CHCO_2$—⬡

In examples 1 and 3, CH=$CHCO_2$ is a connecting group, but in example 2, it is considered part of the terminal chain.

Numerous functional groups have been used as connecting groups; for an extensive list, see Ref.14 and 23. Some of the more common examples are listed in Table 10.1. To obtain a rod-like structure, the connecting group must produce a linear molecule. Thus, with an alkene (Z=C=C) where two isomers (cis and trans) can exist, only the trans isomer is mesogenic. Even if the trans isomer is prepared, it can convert to the cis isomer under certain conditions (such as heat):

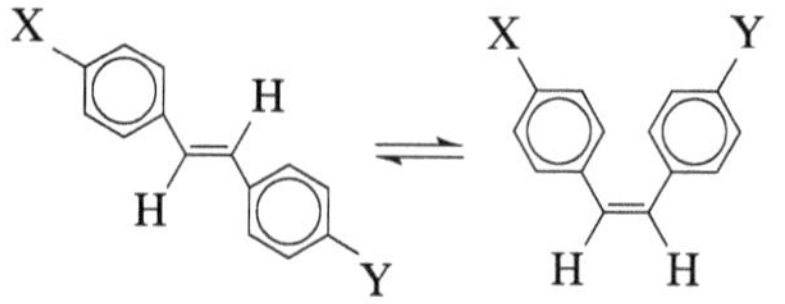

This is why stilbenes (Z=CH=CH) and cinnamates (Z=CH=$CHCO_2$) are not useful in displays. For the same reason, central groups with an odd

Table 10.1 *Connecting groups.*

Group	Common name	Group	Common name
$-(CH_2)_n-$	alkane	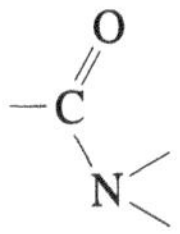 $-C{\overset{O}{\parallel}}{\underset{N<}{}}$	amide
$>C=C<$	olefin (trans isomer) or alkene	$>C=N-N=C<$	azine
$-C\equiv C-$	acetylene, alkyne, tolane	$>CH=CH-CH=N<$	
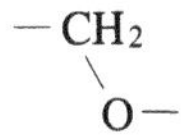 $\overset{-CH_2}{\underset{O-}{\diagdown}}$	ether	$-O_2C(CH_2)_nCO_2-$	
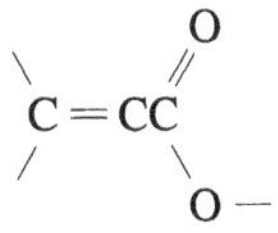 $\overset{\diagdown}{\underset{\diagup}{C}}=C{\overset{O}{\overset{\parallel}{\underset{O-}{C}}}}$	cinnamate (trans isomer)	$-C\equiv C-C\equiv C-$	diacetylene
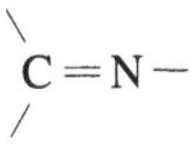 $\overset{\diagdown}{\underset{\diagup}{C}}=N-$	anil, Schiff's base, imine	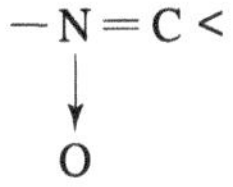 $-N=C<{\underset{\downarrow}{}}{\atop O}$	nitrone
$-N=N-$	azo	$-CONH-NH-$	hydrazide
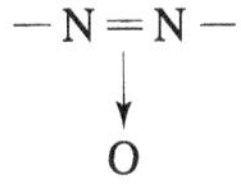 $-N=N-{\atop \underset{O}{\downarrow}}$	azoxy	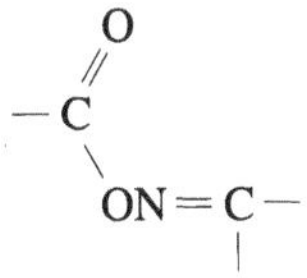 $-C{\overset{O}{\parallel}}{\underset{ON=C-}{\diagdown}}$	oximester
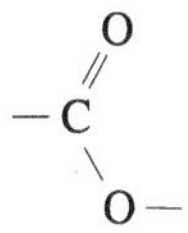 $-C{\overset{O}{\parallel}}{\underset{O-}{\diagdown}}$	ester	$CO-CH=CH-NH-$	enaminoketone
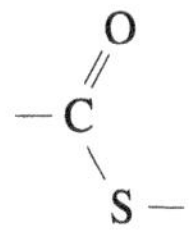 $-C{\overset{O}{\parallel}}{\underset{S-}{\diagdown}}$	thioester	$COCH_2CO$	β-diketone

number of atoms such as $-O-$ and $-CH_2-$ do not produce mesophases. Generally, a fairly rigid connecting group gives the best mesogens, but the more flexible groups such as $-OCH_2-$ and $-CH_2CH_2-$ do show mesomorphic properties. Even compounds with no connecting group at all, as in the biphenyls, yield good mesogens.

$$ X -\!\!\bigcirc\!\!-\!\!\bigcirc\!\!- Y $$

Some comparisons on the effects of the connecting groups in compounds with R_1, R_2 = benzene rings (A, B = H) on mesomorphic properties have been made [37]. These indicate that properties are better in terms of the number of mesophases and their width for $Z=C=N$, than for either $Z=C=C$ or CO_2. This suggests that a connecting group, which causes the benzene rings to be slightly out of plane with each other, produces the best mesomorphic properties. However, x-ray crystallographic studies of the crystalline phase of a variety of mesogens (Table 10.2) suggest little correlation between mesomorphic properties and the angle between the benzene rings (dihedral angle, i.e., the twist or non-coplanarity) in the crystalline phase. The anils ($Z=CH=N$) tend to have the smallest angle, although this varies considerably with different chain lengths. Their homologous series show one of the most extensive polymesomorphisms of all the known mesogens. Yet the azo ($Z=N=N$) compounds with similar dihedral angles show only nematic and smectic-C phases. The phenylbenzoates with a much larger dihedral angle still show good mesomorphic properties, though only rarely do the more highly ordered smectic phases occur. Replacement of the aromatic ring with a cyclohexane ring in the esters gives a dihedral angle similar to that seen in the thioesters, but these show poorer mesomorphic properties. The cyanobiphenyls have dihedral angles smaller than those observed for the phenylbenzoates, but also show fewer mesophases. However, this is probably due to the presence of a terminal CN group, since the biphenyls with $Y=CO_2R$ do show extensive mesomorphism (unfortunately, no x-ray data are available). Clearly, there is no obvious correlation between the planarity of the molecule in the crystalline phase and its mesomorphic properties. Additionally, increasing chain length has been shown to increase the dihedral angle in some azo mesogens [39] indicating that the dihedral angle does not remain a constant value even for a particular connecting group. NMR studies of the dihedral angles in the nematic phases of some alkyl cyanobiphenyls indicate that this angle is essentially retained in the nematic phase [57]. Solid-state ^{13}C NMR studies gave the same dihedral angle in a nitrophenylbenzoate as found by x-ray studies [58].

Table 10.2 *Dihedral angles for mesomorphic compounds in the crystalline phase.*

Structure	X	Y	Angle (°)	Ref.
X–⟨⟩–N=N–⟨⟩–Y	OEt	OEt	2	38
	OEt	O_2CC_4	4.84	39
	OEt	O_2CC_5	10.16	39
	OEt	O_2CC_6	23.2	39
C_7O–⟨⟩–N=N(→O)–⟨⟩–OC_7			12.8	40
RO–⟨⟩–CH=N–⟨⟩–R'	C_4	C_2	61.2	41
	C_4	C_8	13.3, 29.2[a]	42
	C_7	C_6	4.6	41
	C_8	C_4	2.2	43
R–⟨⟩–C(=O)O–⟨⟩–R'	H	H	55.7	44
	C_5	CN	60.0, 59.4[a]	44
	C_7	CN	53.4	46
	MeO	OC_6	66.1	44
	C_4O	CN	69.6	47
	C_8O	NO_2	57.1	44
R–⟨⟩–C(=O)S–⟨⟩–R'	C_5O	C_5	68.7	48
	NC	C_5	69.0	48
	C_5	CN	46.8–50.8	49
C_5–⟨⟩–C(=O)O–⟨⟩–CN			67.3	50
R–⟨⟩–⟨⟩–CN	C_3		42.8	51
	C_4		40.5	52
	C_5		26.3	53
	C_{11}		30.19, 35.83[a]	54
	C_3O		47.5	55
	C_5O		0.82	56

Note:
[a]Two structural conformers were found in the crystal structure.

If a connecting group is symmetrical, then only one structural variant is possible:

$$ArC\equiv CAr' = Ar'C\equiv CAr$$

but if it is not, then two variants can occur:

$$ArCH=NAr' \neq Ar'CH=NAr$$

This is important since an incorrectly drawn structure can mean a different structure than what was originally intended. For example:

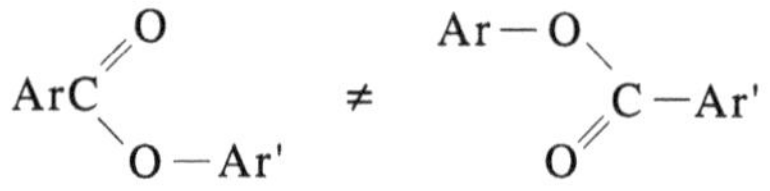

$$ArC\overset{\displaystyle O}{\underset{\displaystyle O-Ar'}{\big<}} \quad \neq \quad \overset{\displaystyle Ar-O}{\underset{\displaystyle O}{\big>}}C-Ar'$$

The connecting group can consist of one functional group such as CH=N, C≡C, N=N or two combined such as: $>C=CO_2-$ (cinnamate) and $>C=N-N=C<$ (azine). More than one functional group can exist separately in a mesogen. These can be the same, as in:

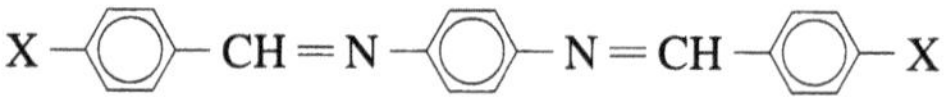

$$X-\bigcirc-CH=N-\bigcirc-N=CH-\bigcirc-X \quad \text{dianil}$$

or different as in

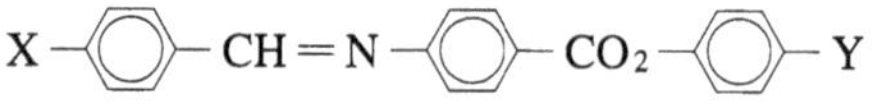

$$X-\bigcirc-CH=N-\bigcirc-CO_2-\bigcirc-Y \quad \text{anil-ester}$$

Two connecting groups make possible more structural variants. For example:

$$X-\bigcirc-Z-\bigcirc-Z'-\bigcirc-Y$$

X = Y, Z = Z' symmetrical such as C≡C (1 variant)
X = Y, Z = Z' but unsymmetrical such as CH=N (3 variants)
Z = CH=N Z' = N=CH
Z = N=CH Z' = CH=N
Z = CH=N Z' = N=CH
X ≠ Y, Z = Z' symmetrical (1 variant)
X ≠ Y, Z ≠ Z' is symmetrical (4 variants):
X, Z = N=N, Z' = CH=N, Y   X, Z = N=N, Z' = N=CH,Y
X, Z = CH=N, Z' = N=N, Y   Z = N=CH, Z' = N=N, Y
X ≠ Y, Z ≠ Z', both are non-symmetrical (8 variants)

The effect of the direction of the ester connecting group on mesomorphic properties in some diesters has been studied [59]:

$$R-\bigcirc-X-\bigcirc-Y-\bigcirc-Z \quad R = C_8O, Z = H$$

X	Y	S_A	N	I
CO_2	CO_2	118	129	140
O_2C	O_2C		118	135
CO_2	O_2C		117	142
O_2C	CO_2		(138)	153

When Z is a strong polar substituent along the molecular axis such as CN or NO_2, reentrant nematics and smectic-A phases, as well as other polar and incommensurate phases can occur [60, 61].

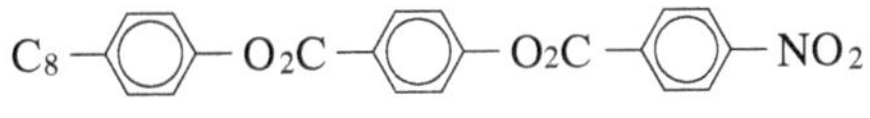

$$C_8 - \bigcirc - O_2C - \bigcirc - O_2C - \bigcirc - NO_2 \quad DB_8NO_2 \quad (Ref. 62)$$

Whether reentrants or incommensurates are preferred depends on the alignment of the dipoles of the connecting groups with the terminal dipole. These phases can also occur when the two connecting groups are different such as X is an ester and Y is an anil, for example,

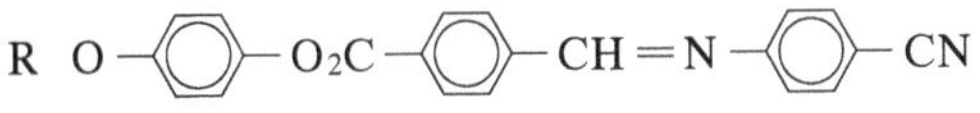

$$R \; O - \bigcirc - O_2C - \bigcirc - CH = N - \bigcirc - CN$$

Often when the number of aromatic rings increases, so do the transition temperatures, viscosity and difficulty of synthesis. Symmetrical variants are generally easier to prepare than asymmetrical ones, but they also have less desirable, higher transition temperatures. Mesogens containing two connecting groups can also have different ring systems than the benzene one, and many mesogens with even more connecting groups are also known [29, 30].

Ring systems (R_1 and R_2)

These range all the way from the early aromatic single benzene rings, to alicyclic, heterocyclic, and a wide variety of combinations. Table 10.3 lists a few examples. Many more can be found in the literature [23, 32a].

Individual benzene rings do not yield good mesogens. At least two rings are needed to give a rod-like molecule. One exception to this is the 4-alkyl-benzoic acids which hydrogen bond to give linear dimers:

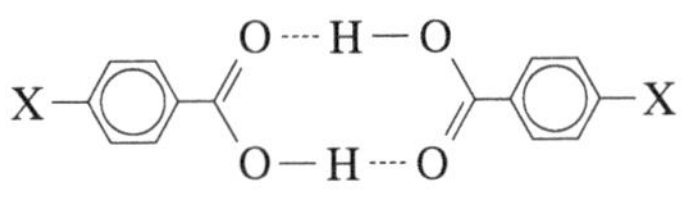

$$X - \bigcirc \overset{O \cdots H - O}{\underset{O - H \cdots O}{}} \bigcirc - X$$

These benzene rings can be linked together either directly as in the biphenyls:

Table 10.3 *Typical ring systems.*

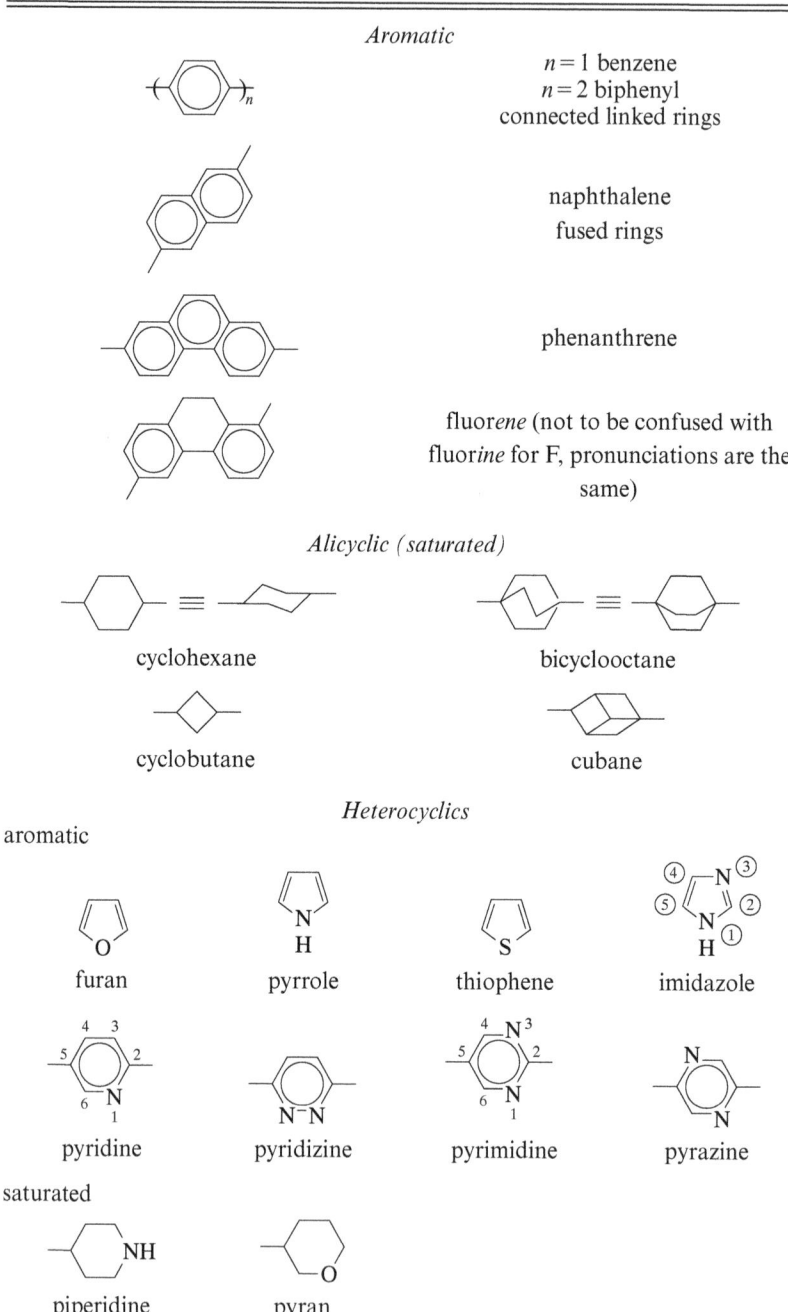

Aromatic

$n = 1$ benzene
$n = 2$ biphenyl
connected linked rings

naphthalene
fused rings

phenanthrene

fluor*ene* (not to be confused with
fluor*ine* for F, pronunciations are the
same)

Alicyclic (saturated)

cyclohexane

bicyclooctane

cyclobutane

cubane

Heterocyclics

aromatic

furan

pyrrole

thiophene

imidazole

pyridine

pyridizine

pyrimidine

pyrazine

saturated

piperidine

pyran

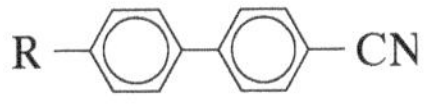

$$R-\langle\bigcirc\rangle-\langle\bigcirc\rangle-CN$$

through connecting groups:

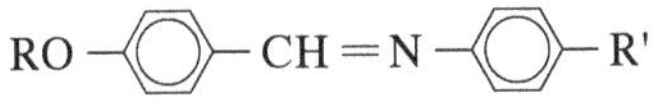

$$RO-\langle\bigcirc\rangle-CH=N-\langle\bigcirc\rangle-R'$$

or by ring fusion:

$$RO-\langle\bigcirc\bigcirc\rangle-CN$$

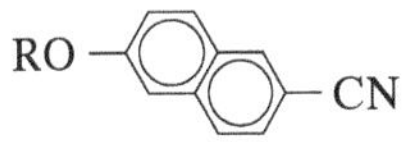

More than two benzene rings can be linked by using more than one connecting group:

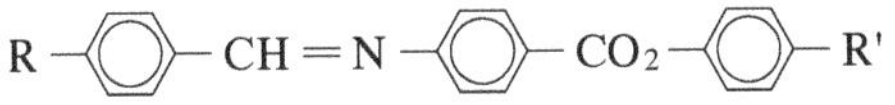

$$R-\langle\bigcirc\rangle-CH=N-\langle\bigcirc\rangle-CO_2-\langle\bigcirc\rangle-R'$$

as already discussed or by combining linked or fused rings with another ring system via connecting groups:

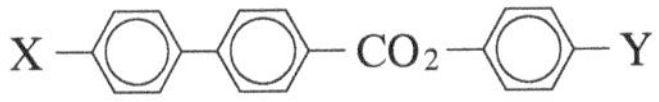

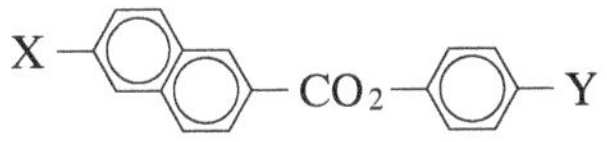

$$X-\langle\bigcirc\rangle-\langle\bigcirc\rangle-CO_2-\langle\bigcirc\rangle-Y$$

$$X-\langle\bigcirc\bigcirc\rangle-CO_2-\langle\bigcirc\rangle-Y$$

The number of possibilities is endless.

Replacing the benzene ring with a saturated, six-membered cyclohexane (alicylic) ring creates a new structure variant with the possibility of cis and trans isomers. The cis isomer, being non-linear, is not mesogenic. For example:

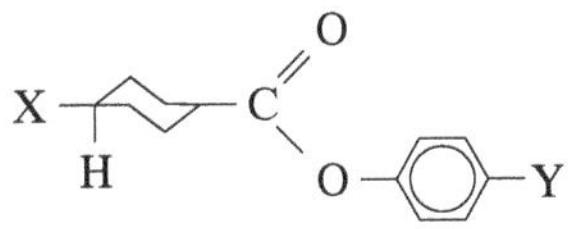

Trans –mesogenic

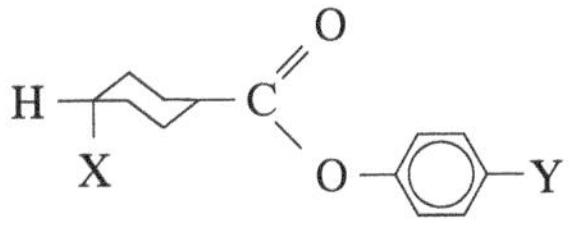

Cis – non-mesogenic

These isomers can be illustrated in a two-dimensional drawing as well:

$$X-\langle\bigtriangleup\rangle-Y \quad \equiv \quad X-\langle\bigcirc\rangle-Y$$

The cyclohexane is also sometimes indicated by $-\langle H\rangle-$

Replacing the benzene ring with a heterocyclic ring dramatically increases the number of possible structure variants. Table 10.3 lists only a few of the simpler possibilities. As an example, the addition of one nitrogen atom to the benzene ring adds two new variants to the *o*-, *m*-, *p*- possibilities (only *p*- is generally mesogenic), namely the position of the nitrogen atom in relationship to the substituents. The substituents can still remain in the para position. For example, there are two possibilities with the pyridine esters derived from pyridine acids:

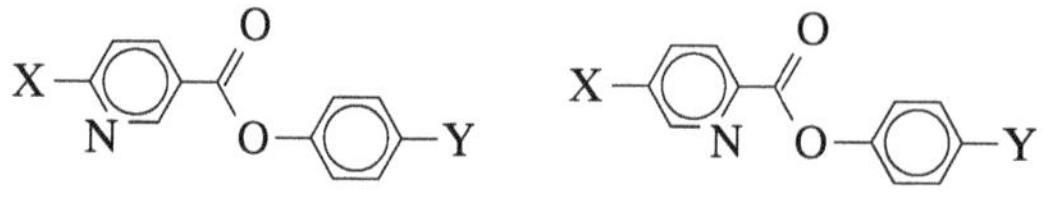

Adding two nitrogen atoms in the para substituted benzene ring yields four possibilities:

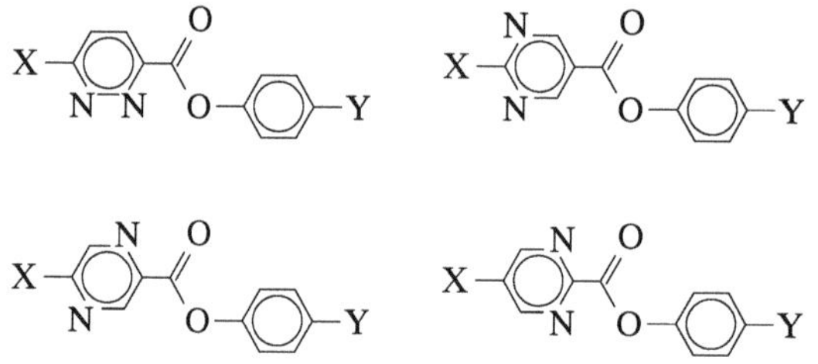

More variants are possible if the two hetero atoms are not identical. With all these ring systems, the size of the ring can vary and the rings can be linked by a simple carbon–carbon bond, connecting groups, or by ring fusion. These three types of ring systems can be combined in an endless number of variations. For example:

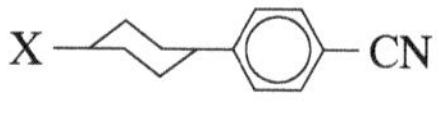

X——◇——◯—CN   Alicyclic-aromatic

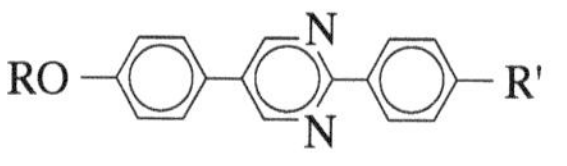

RO—◯—◯—◯—R' Heterocyclic-aromatic

Numerous such variations have been made, and some are easier to synthesize than others.

Initially, it was thought that replacing aromatic rings with saturated alicyclic rings would produce poorer mesogens, since it was felt that the π electron system of the aromatic ring systems was necessary to hold the molecules in a parallel packing arrangement. Nevertheless, many compounds containing these saturated rings show good mesomorphic properties, sometimes better than those of their aromatic counterparts. In fact, this type of structure modification can yield either better or poorer mesogens depending on the location of the alicyclic ring and the type of molecule in which this substitution occurs. It can affect the rigid core by

Table 10.4 *A comparison of mesomorphic properties*
[63] for

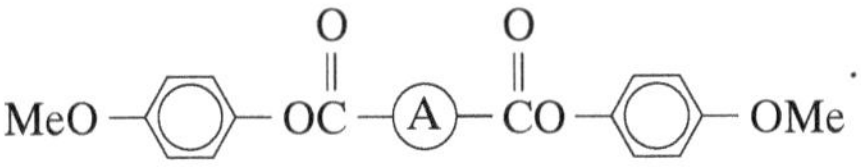

$$\text{MeO} - \bigcirc - \overset{\overset{O}{\|}}{OC} - \textcircled{A} - \overset{\overset{O}{\|}}{CO} - \bigcirc - \text{OMe} .$$

A	Δmp (°C)	Δclp (°C)	N range (°C)
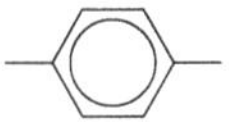	Standard		77
	− 50	− 15.8	66.2
	− 66.9	− 37.7	106.2
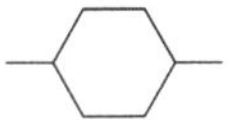	− 70.4	− 44.3	103.1
	− 58.6	− 18.8	116.8
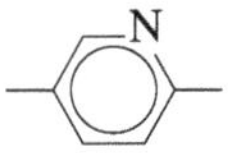	− 82.1	− 32.3	82.1
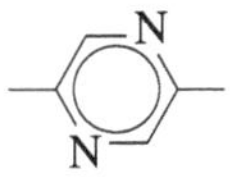	− 8.0	− 38.1	46.9

breaking a long, conjugated system or by increasing the flexibility. This idea was supported by the early studies on the diesters shown in Table 10.4 [63]. Substitution of the central benzene ring with either saturated or hetero-cyclic rings decreased both the melting and clearing temperatures in all the compounds studied. Still, the nematic phase range usually increased. Partially saturated rings gave smaller decreases than the cyclohexane ring. This was also true of the bulky bicyclooctane ring.

In the biphenyl type mesogens (Table 10.5), replacing the benzene ring with a cyclohexane type ring can either raise or lower transition temperatures depending on the terminal substituent. When X is kept constant and $Y = CN$, both melting and clearing temperatures always increase, whereas when $Y = C_5$, these temperatures vary considerably. The increase is unusually (and unexpectedly) large for the dibicyclooctyl series with $Y = C_5$. Unfortunately, data are not yet available for $Y = CN$. Mesophases are poor

Table 10.5 *The effect of replacing benzene with alicyclic rings in 4-pentyl cyanobiphenyl on mesomorphic properties.*

Compound	Δtransition temperatures (°C)		Total mesophase ranges (°C)		Ref.
	Melting	Clearing	S	N	
C5—⟨benzene⟩—⟨benzene⟩—C5			25.7	0	64
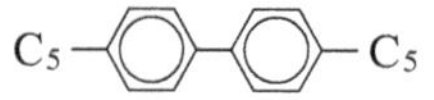 K 26.5 S$_E$ 47.6 S$_B$ 52.2 I standard					
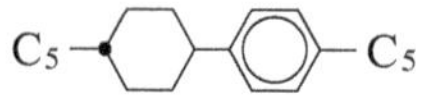 C5—⟨cyclohexane⟩—⟨benzene⟩—C5	− 26.8	− 57.2	m	m	64
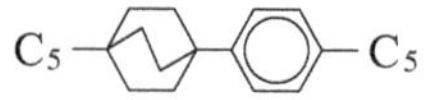 C5—⟨bicyclo⟩—⟨benzene⟩—C5	25.5	− 19.2	0	m	65
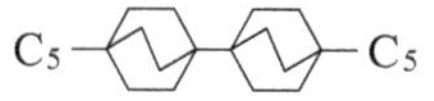 C5—⟨bicyclo⟩—⟨bicyclo⟩—C5	20.0	193.8	200	0	66
C5—⟨cyclohexane⟩—⟨bicyclo⟩—C5	no data				
C5—⟨benzene⟩—⟨benzene⟩—CN	− 3.5	− 17.2	0	12.5	67
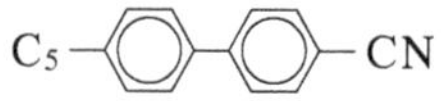 K 22.5 N 35 I 5CB new standard					
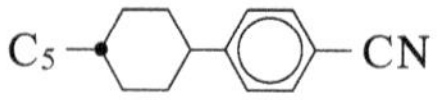 C5—⟨cyclohexane⟩—⟨benzene⟩—CN PCH	8.5	20	0	24	67
C5—⟨benzene⟩—⟨cyclohexane⟩—CN	no data				
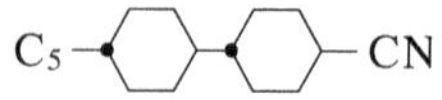 C5—⟨cyclohexane⟩—⟨cyclohexane⟩—CN CCH5	33.5	25	m	22	15
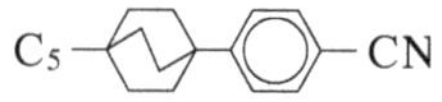 C5—⟨bicyclo⟩—⟨benzene⟩—CN	39.5	65	0	38	67
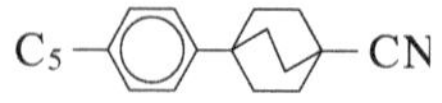 C5—⟨benzene⟩—⟨bicyclo⟩—CN	90.5	15		m	68
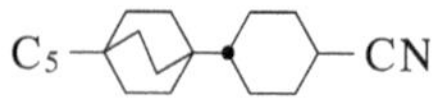 C5—⟨bicyclo⟩—⟨cyclohexane⟩—CN	81.5	94	0	25	69
C5—⟨cyclohexane⟩—⟨bicyclo⟩—CN	no data				
C5—⟨bicyclo⟩—⟨bicyclo⟩—CN	no data				

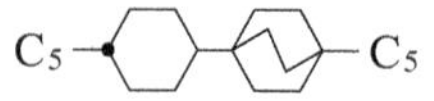

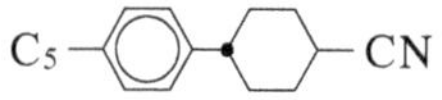

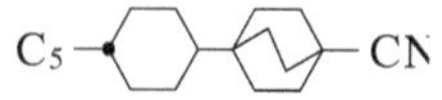

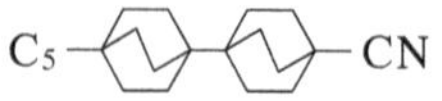

in the cyclohexyl and mono-bicyclooctyl compounds when $Y = C_5$. The aromatic compound shows only smectic phases. However, when $Y = CN$, the nematic phase observed in the aromatic compound usually increases in its range in the saturated analogs. The combination of increased clearing temperatures and wider nematic phase range make these compounds useful additives to the cyanobiphenyls in display mixtures. Saturated rings usually lower the viscosity of a mixture and show good solubility. The bicyclo-octane ring offers improved chemical stability, exhibits a well-defined stereochemistry, and has light transmission in the near UV.

The varying effect of the substitution of a cyclohexane ring for a benzene ring on mesomorphic properties can also be seen in the esters shown in Table 10.6. Decreases in both clearing and melting temperatures occur for the biphenyl type series with $Y = C_3$ (numbers 1–7), but increase in the binaphthyl series when $Y = CN$ (numbers 9–12). The comparisons in Tables 10.5 and 10.6 suggest that the terminal substituent is the determining factor in whether the transition temperatures increase or decrease. Smectic phases are preferred in three of the biphenyl type compounds (numbers 2, 4 and 6), whereas nematics are preferred in the other three (numbers 3, 5 and 7), indicating that this depends on which benzene ring is replaced with a saturated ring. Again, the nematic phase is preferred by the series containing a cyano group (numbers 9–12). Although the trend of decreasing transition temperatures in the order

reported earlier [67] applies in some of these comparisons, it is obvious that it does not occur in all of them.

A comparison of mesomorphic properties for some partially hydrogenated cyclohexane ring esters (Table 10.7) indicates that the position of a single double bond within a cyclohexane ring can have a large effect on both the transition temperatures and the mesophase range. However, which olefin gives the best result also depends on the remaining molecule. In the first series, (compounds 1–4), the 2,3 double bond (number 3) yields the widest mesophase range; whereas in the second series (numbers 5–8), the widest range occurs in the 1,2-olefin (number 6). Yet the widest range nematic phase always occurs in the cyclohexane ester.

Types of bulky saturated ring systems other than bicyclooctane, such as cubane rings, have also produced good mesogens:

C 107.5 N 170.9 I

Table 10.6 *A comparison of mesomorphic properties for some esters.*

Mesogen	Δmp (°C)	Δclp (°C)	S range (°C)	N range (°C)	Ref.
1. C_5—◯—◯—CO_2—◯—C_3	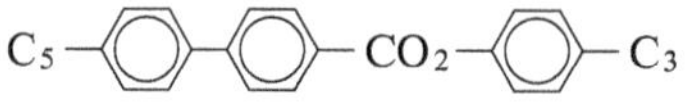 Standard		34	36	
2. C_5—⬡—◯—CO_2—◯—C_3	 −3.4	−9.0	58	37	
3. C_5—◯—⬡—CO_2—◯—C_3	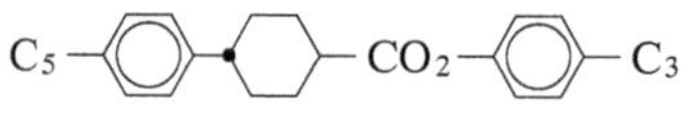 −23.8	−0.7	4.4	88.7	
4. C_3—⬡—⬡—CO_2—◯—C_5	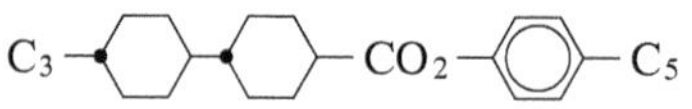 −74	14	125	33	
5. C_5—◯—◯—CO_2—⬡—C_3	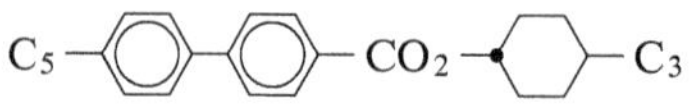 −30	−25	0	75	70
6. C_5—⬡—◯—CO_2—⬡—C_3	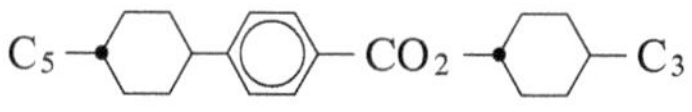 −41	−24	87	0	
7. C_5—◯—◯—CO_2—⬡—C_3	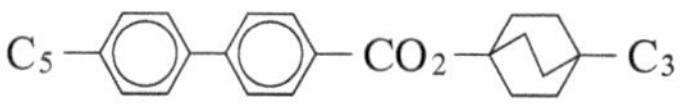 20	10	0	60	
8. C_5—⬡—◯—CO_2—◯—C_3 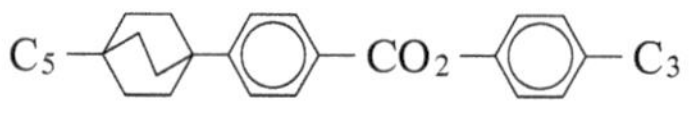	no data				
C_4—Ⓐ—CO_2—◯◯—CN					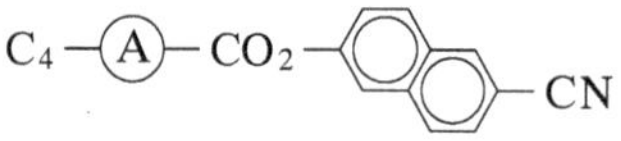 71
9. A = —◯—	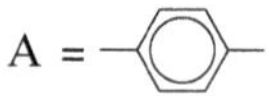 standard		0	62	
10. —⬡—	 4.5	24.5	0	82	
11. —⬡—	 27.5	59.5	0	94	
12. —⬡—⬡—	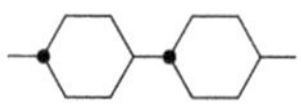 21.5	83.5	0	124	

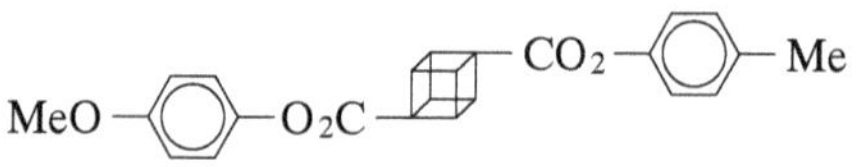

MeO—◯—O_2C—▱—CO_2—◯—Me   C 175.1 N 179 I

Strong intermolecular attractive forces between π electron-rich ring systems are often favorable for producing good mesomorphic properties. This explains why so many mesogens having them have been studied. However, some mesogens are known which have only saturated rings

Table 10.7 *A comparison of mesomorphic properties for cyclohexene esters.*

Mesogen	Δmp (°C)	Δclp (°C)	S range (°C)	N range (°C)
1. C_3—⟨ ⟩—CO_2—⟨ ⟩—OC_4	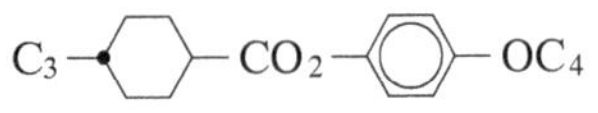 Standard			31
2. C_3—⟨ ⟩—CO_2—⟨ ⟩—OC_4	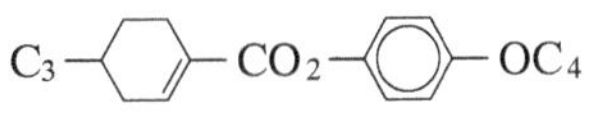 2.5	—		0
3. C_3—⟨ ⟩—CO_2—⟨ ⟩—OC_4	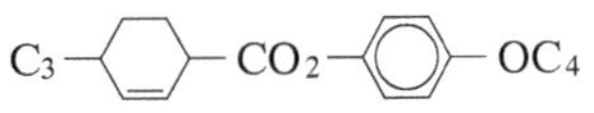 − 20.5	− 26.5		25
4. C_3—⟨ ⟩—CO_2—⟨ ⟩—OC_4	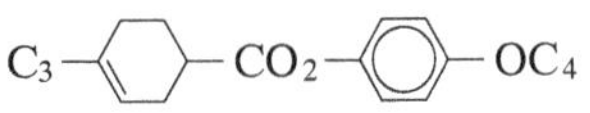 − 33.5	—		0
5. C_5—⟨ ⟩—CO_2—⟨ ⟩—⟨ ⟩—CN	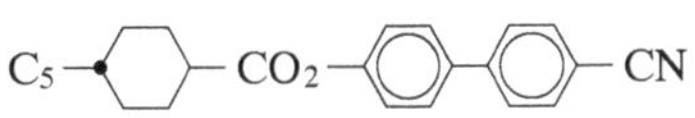 Standard		2.0	155
6. C_5—⟨ ⟩—CO_2—⟨ ⟩—⟨ ⟩—CN	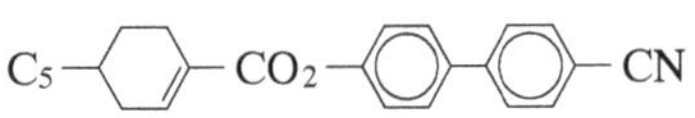 15	28		114
7. C_5—⟨ ⟩—CO_2—⟨ ⟩—⟨ ⟩—CN	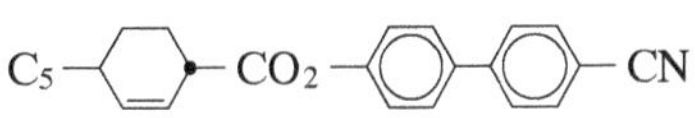 − 40.5	− 57		85
8. C_3—⟨ ⟩—CO_2—⟨ ⟩—⟨ ⟩—CN	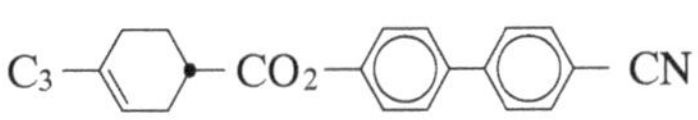 − 25	− 81		101

(Table 10.8) and in some instances no π electrons or polar substituents in the entire molecule (number I with Y = R′ and 3, 5, 7 with X, Y = R, R′), indicating that π electron-rich ring systems are not needed to observe mesomorphic properties. This is best illustrated by the cholesterol compounds (compound 10), some of which were among the first mesogens discovered as they are found in biological systems. These chiral structures often show what are now known to be chiral nematic phases, but were first known as cholesteric phases, after the type of compound in which they were first observed. Needless to say, numerous variations of this structure are now known and mixtures of this type are widely used in thermometers. Ring sizes other than four and six also occur in mesogens, although they are less common. Five-membered rings are common in heterocyclic series, but are usually fused to six-membered rings. This is due to the non-linearity that occurs in using a single five-member ring:

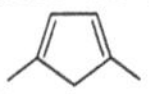

Several studies have been done on the effect on mesomorphic properties of replacing a connected ring system with a fused ring. Two examples of the effect of replacing a biphenyl with a naphthalene ring in esters

Table 10.8 *Mesogens having only saturated ring systems.*

	Mesogen	Ref.
1.	 R$-$(ring)$_n$-Y Y = R′ or CN, n = 2, 3	15, 72
2.	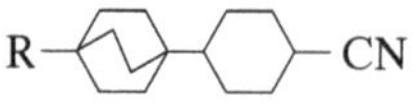 R$-$(ring)$-$(ring)$-$CN	69
3.	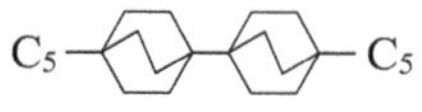 C_5-(ring)$-$(ring)$-C_5$	66
4.	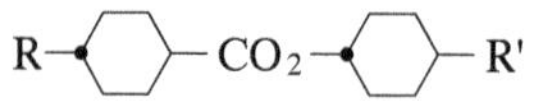 R$-$(ring)$-CO_2-$(ring)$-$R′	
5.	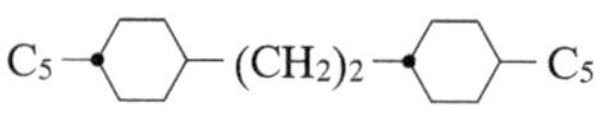 C_5-(ring)$-$(CH$_2$)$_2-$(ring)$-C_5$	15
6.	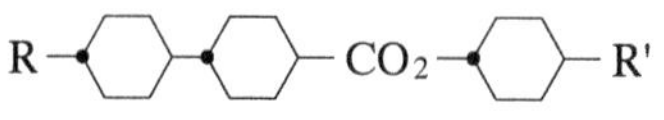 R$-$(ring)$-$(ring)$-CO_2-$(ring)$-$R′	15
7.	 X$-$(fused ring)$-$Y	73
8.	 X$-$(ring)$-$A$-$(ring)$-$B$-$(ring)$-$Y	74
9.	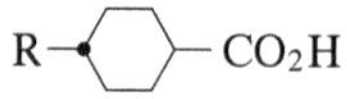 R$-$(ring)$-CO_2H$	75
10.	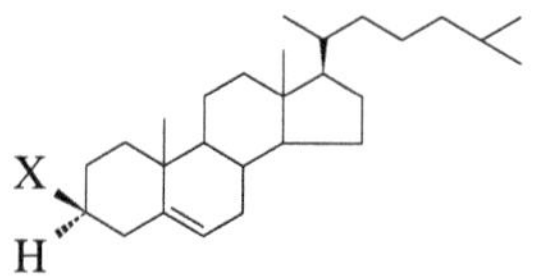 X,H$-$(steroid ring system)	29, 30

are presented in Table 10.9. These studies suggest that the effect of ring substitution on transition temperatures gives the order of

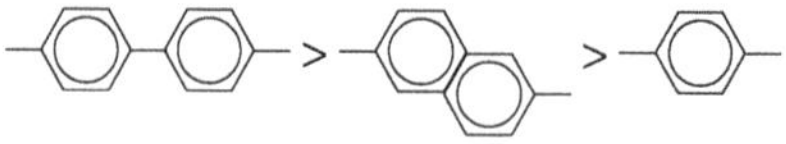

(biphenyl) > (fused ring) > (single ring)

The nematic phase range usually increases while the smectic phase range decreases. Yet, data for the cyanobiphenyl analogs in Table 10.10 indicate that these trends in properties do not apply for these compounds. Melting temperatures increase in the fused ring systems instead of decreasing. This is true of a variety of fused rings. Additionally, these compounds yield

Table 10.9 *Comparison of mesomorphic properties for fused versus connected aromatic rings in esters* X–(A)–CO_2–(◯)–Y .

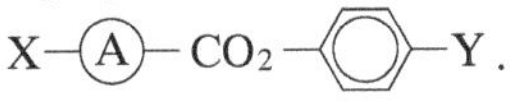

Differences for A = binap-biph[a]

X	Y	mp (°C)	clp (°C)	S range (°C)	N range (°C)
C_4	C_5	− 56.8	− 86.1	− 40.5	11.2
C_5	C_5	− 40.1	− 81.8	− 52	10.3
C_6	C_5	− 46.7	− 84.6	− 60.5	22.6
C_6	OC_6	− 31.9	− 68.8	− 60.5	23.6
C_8O	OC_8	− 24.5	− 34.0	− 24.3	11.2
C_5	CN	− 26.3	− 89.6	− 55	− 63.3

X–(◯)–CO_2–(A)–CN   (Ref. 75)

Differences for A = binap-be[a]   A = biph-binap

X	mp (°C)	clp (°C)	N range (°C)	mp (°C)	clp (°C)	N range[b] (°C)
C_4	14	99.7	61.3	29	99.7	70.7
C_5	32	96.5	57.8	12.5	83.2	70.7
C_6	28.5	91.4	65.4	18	91.2	73.2
C_7	25.5	78.3	65.8	22.5	89.1	66.6

Notes:

[a]be = –(◯)– , biph = –(◯)–(◯)– , binap = (◯◯)– .

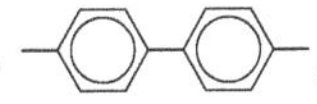

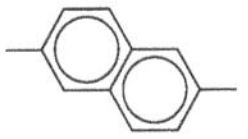

[b]monotropic phases are considered to have zero mesophase range in these calculations.

poorer mesomorphic properties with most having no mesophases at all. Different isomers are sometimes possible with fused rings. Since these can affect the linearity of the molecule, they can also influence the mesomorphic properties. An example occurs in the naphthalene diesters [77].

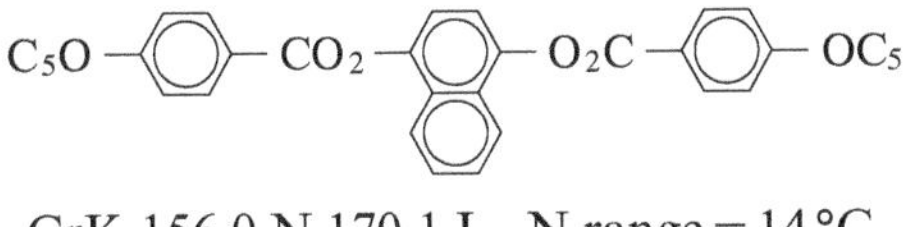

C_5O –(◯)– CO_2 –(◯)– O_2C –(◯)– OC_5

CrK 156.0 N 170.1 I   N range = 14 °C

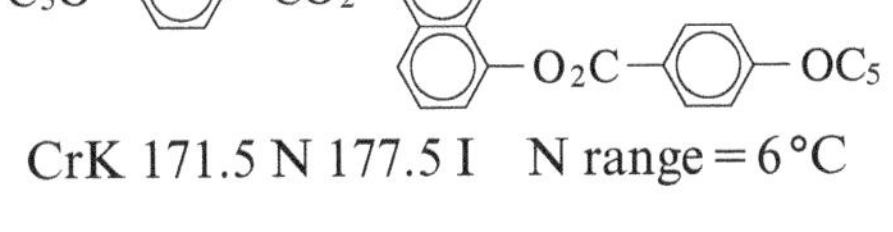

C_5O –(◯)– CO_2 –(◯◯)– O_2C –(◯)– OC_5

CrK 171.5 N 177.5 I   N range = 6 °C

Table 10.10 *The effect of replacing connected benzene rings in 4-substituted cyanobiphenyls with a fused ring system.*

Compound	X	Δmp (°C)	Δclp (°C)	ΔN (°C)
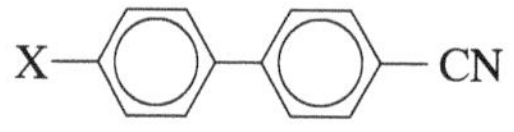 X—⬡—⬡—CN	C_5 C_8O	Standard Standard	K22.5 N35 I K54.5 A67 N80 I	
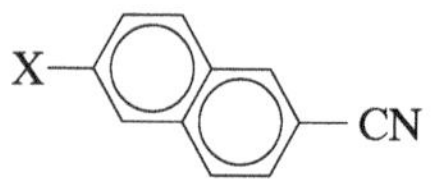 X—⬡⬡—CN	C_5 C_8O	19.5 12.3		
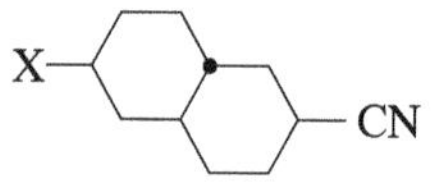 X—⬡⬡—CN (fused)	C_5	19.4		
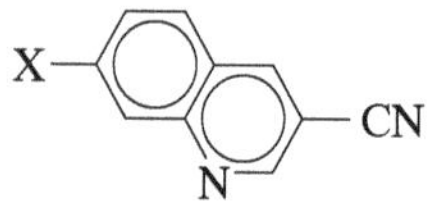 X—⬡⬡—CN (N)	C_8O	16.5		
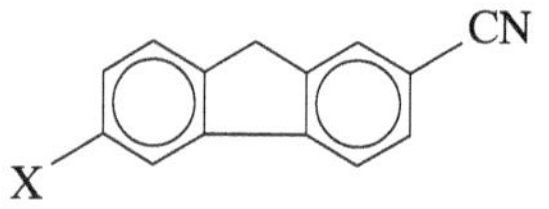 X—(fluorene)—CN	C_5	43.5	− 10.5	12.5

Transition temperature data from Ref. 32b.

Both of these linear isomers are mesogens. The 1,5 isomer has a higher transition temperature suggesting that it is more stable, but the 1,4 isomer has the wider nematic phase range. When compared to the analogous biphenyl diester,

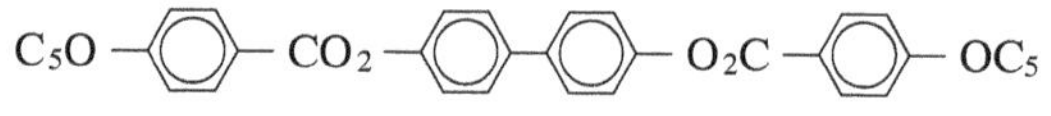

$$C_5O—⬡—CO_2—⬡—⬡—O_2C—⬡—OC_5$$

CrK 189.5 S 194 N 262 I

Sm range $= 4.5$ °C, N range $= 68$ °C, Total $= 72.5$ °C

the effect of the broadening of the molecule by the naphthalene ring on mesomorphic properties is obvious. Although the remaining non-linear isomers in this series were not studied, non-linear isomers of the alkoxy-naphthoic acids were shown earlier to be non-mesogenic [25].

Although the arrangement of the rings in a linear manner generally yields better mesogens, some non-linearity can be tolerated if it is balanced by a longer rod-like section. For example, the structure

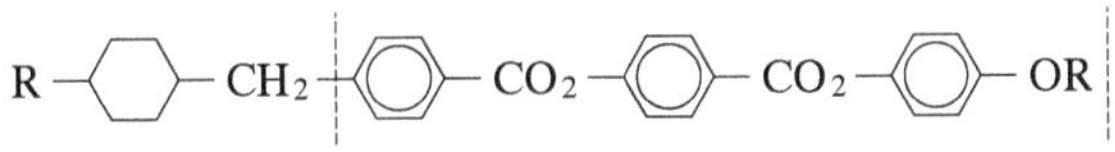

R—⬡—CH₂—|⬡—CO₂—⬡—CO₂—⬡—OR|

Non-linear   Linear

shows mesomorphic properties, even though the CH_2 linking group creates a kink in the molecule [78]. Actually, the occurrence of free rotation around this group creates a shape more like a ball on the end of a long rod:

If these molecules can find a way to pack in a parallel manner, then they can still form mesophases. A little non-linearity might actually create some desirable mesomorphic properties.

Replacing a carbon atom in an aromatic ring with a hetero atom (nitrogen, oxygen and sulfur are the most common) can affect the π electron cloud and introduce new dipoles to the core of the aromatic mesogen. Dipole direction is determined by the size of the ring and the location of the hetero atom. Like the aromatic cores, saturated heterocyclic rings are possible, as are combinations of saturated and unsaturated rings. All these structural features can affect the mesomorphic properties. Both the synthesis of the heterocyclic ring system and its stability can be affected by the location of the hetero atom(s), the type of hetero atom, the number of such atoms, and the size of the ring. Some systems are less stable or more difficult to synthesize. Unlike their carbon counterparts, many heterocyclic rings must be synthesized from non-ring starting materials, and the presence of hetero atoms often requires totally different synthesis schemes. Nevertheless, numerous heterocyclic mesogens have been prepared, some of which are useful in displays.

With the huge number of structural variants possible in heterocyclic ring systems, and the fact that this is a newer area of study than the aromatic ring system, structure–property studies are limited to only a few within a specific type of ring system. Although a comparison of the cyanobiphenyl type analogs is possible at this time, the analogous biphenyl ester comparison is not. In the cyanobiphenyl analogs (Table 10.11), the introduction of a single nitrogen atom into the biphenyl ring gives four possible isomers with each one affecting the mesomorphic properties differently. In all four isomers (and in fact in all the compounds listed) both melting and clearing temperatures are increased by the introduction of nitrogen atoms into the biphenyl ring. This modification also seems to be unfavorable for the nematic phase, and more favorable for smectic phases. The nematic phase is the most favored in compound 2 and the least in compound 5 of those

Table 10.11 *Effect of the addition of ring nitrogen atoms to 4-alkyl cyanobiphenyls[a] on mesomorphic properties[a].*

Compound	S^b	N^b	I^b	S range	N range
1. C_6—(ring)—(ring)—CN		14.3	30.1	0	15.8
2. C_6—(ring)—(ring, N)—CN	42.2	51.7	62.3	9.5	10.6
3. C_6—(ring, N)—(ring)—CN		29	32.5	0	3.5
4. C_6—(ring)—(ring, N)—CN		(44)	58	0	m
5. C_6—(ring, N)—(ring)—CN	(26)		52	m	0
6. C_6—(ring)—(ring, N,N)—CN	86.5 (A)	101.5	103	15	1.5
7. C_6—(ring, N,N)—(ring)—CN		(38.5)	54.5	0	m
8. C_6—(ring)—(ring, N,N)—CN			104	0	0
9. C_6—(ring, N,N)—(ring)—CN			77	0	0
10. C_6—(ring)—(ring, N,N)—CN	no data				
11. C_6—(ring)—(ring, N,N)—CN	no data				
12. C_7—(ring)—(ring, N,N,N,N)—CN^c	55.6 (A)		102.3	46.7	0
13. C_5—(ring, N,N,N,N)—(ring)—CN^c			78.0	0	0

Notes:
[a] Data are from Ref. 32 unless otherwise noted.
[b] S = unspecified smectic phase. Letter identifications are in parentheses when known. N = nematic phase. Values in °C.
[c] Data are from Ref. 78.

having only one nitrogen atom. Additional nitrogen atoms are even less favorable for the nematic phase with either no mesophases occurring or only a smectic phase. Interestingly, the isomer with the widest nematic phase range (compound 2) also has the widest smectic phase range of all the single nitrogen isomers. No data were found for the ortho and para dinitrogen compounds (numbers 10 and 11), although these rings can be found in other types of mesogens. With four nitrogen atoms in the ring (tetrazines), a wide range smectic-A phase occurs, but only when the cyano group is on the tetrazine ring.

The trends of higher clearing temperatures, a preference for smectic phases and the effect of nitrogen position on mesomorphic properties were observed in other nitrogen-containing heterocyclic systems [16b, 24 80]. Still, in the few cases when nematic phases do occur, these compounds can be useful in displays because of their large $\Delta\epsilon$ values, along with lower k_{33}/k_{11} and viscosity values. On the other hand, they are often colored (yellow to red), unstable to light, have a negative dichroic ratio and poor extinction coefficients. Some of these problems could conceivably be eliminated by structure modification.

Terminal substituents (X, Y)

The core of a mesogen, by establishing the primary shape of the molecule and its rigidity, determines the approximate temperature range where mesophases will occur and what types of mesophases are possible. However, a rod-like rigid core rarely produces mesophases. Thus, terminal substituents are needed to balance this rigidity with flexibility. In the standard, two ring systems, mesogens are rarely observed even when one substituent is an aliphatic chain and the other one is simply a hydrogen atom. The addition of another benzene ring, such as in the esters,

can, however, produce mesophases [81]. Replacing the hydrogen atom with a polar substituent can also lead to mesogenic properties. Still, the most studied mesogens are those which have two terminal substituents containing aliphatic chains.

Terminal substituents tend to be used to fine-tune mesomorphic properties. If a core is generally not favorable for forming certain types of mesophases, it rarely is possible to find terminal substituents which will produce these phases. For example, the well known anils (Schiff bases)

show many of the more highly ordered

smectic phases, yet the phenylbenzoates X—⟨ ⟩—C(=O)—O—⟨ ⟩—Y rarely

show smectics of an order higher than a smectic-B phase, regardless of what X and Y are. Terminal substituents are used to raise or lower transition temperatures (e.g., alkoxy, branched chains), create dipoles along the molecular axis or across it (e.g., CN, F), produce chiral mesogens (ex chiral branched chains) or enhance the preference for a specific mesophase (e.g., short alkyl favors nematic). A huge number of variations have been studied. The following discussion covers the more common ones.

Straight alkyl/alkoxy chains

Extensive studies have been done on the effect of alkyl and alkoxy chains, along with their length, on mesomorphic properties. This has been accomplished by determining the mesomorphic properties for a wide variety of homologous series of mesogens. Homologs differ only in the number of methylene (CH_2) groups in an aliphatic chain. These chains can contain other groups, such as an oxygen atom, ketone and ester groups.

To study the effect of chain length on mesomorphic properties, one chain is kept constant while the other one is varied; then a homologous series plot is constructed of the transition temperatures observed for each homolog. A typical example of such a plot is presented for the $C_{10}O/OR'$ phenylbenzoates in Fig. 10.1(a) [82]. The plot for the analogous series with $C_{10}O/OCOR'$ (Fig. 10.1(b)) indicates how much difference a small structural modification in one terminal chain can have on mesomorphic properties. Both of these curves show trends in mesomorphic properties that are typical of nearly all homologous series of rod-like molecules:

1. An odd-even alternating, clearing temperature (M–I) curve which can fall (high temperature mesogens Sm, N–I > 70 °C), remain constant, or rise (low temperature materials Sm, N–I < 70 °C with increasing chain length), the last being rare [22b, 83, 84].
2. A melting temperature curve which generally falls (but often with major fluctuations from a smooth curve) to a minimum and then rises with increasing chain length intersecting the clearing curve with the loss of all enantiotropic mesophases and eventually any monotropic phases.
3. A gradual evolution of mesophases of N, SmA, SmB/SmC phases with increasing chain length, although not all of these mesophases occur in a specific series.

Nematic phases generally occur at short to mid-chain length first increasing in phase length with increasing chain length and then decreasing at longer chain length. As the nematic phase range decreases, smectic phases may appear first increasing in phase length as the alkyl chain length increases, and then decreasing usually with increasing transition temperatures. As the smectic curve rises (Sm–N) and the clearing (N–I) curve either falls or remains constant, they eventually intersect with the disappearance of any enantiotropic nematic phase and the clearing curve now becomes a smectic to isotropic liquid (Sm–I) transition. Smectic-A or -B phases usually occur first, increasing in phase length to a maximum and then decreasing often with a smectic-C phase entering and increasing in phase length. At some point in mid-chain length, the melting temperature begins to rise and continues to do so with an increasing rate as the chain length increases. This rising, melting curve eventually intersects a falling or constant clearing curve with the loss of all enantiotropic mesophases, although monotropic phases can occur (monotropic phases occur *below* the melting temperature).

Various combinations of mesophases are possible in an homologous series. Some series will show only one mesophase, whereas others will show many or only a few. A series with a single smectic-B phase is rare, whereas those with a smectic-A or smectic-C phase are quite common. Nematic and smectic-A phases are observed in more mesogens than any of the other phases. More highly ordered smectic phases can enter at mid-chain length as well. These and the smectic-B phase generally only occur at the mid-chain length, whereas the smectic-C phase often will continue to exist at the longer chain lengths.

Like all trends in mesomorphic properties, there are exceptions to these homologous series trends. A puzzling one occurs in the series

$$RCO_2 - \bigcirc - \underset{N}{\overset{N}{\bigcirc}} - C_8H_{17}$$

where the order of appearance of mesophases in the homologous series is smectics-A, -C, N rather than N, smectics-A, -C [16b].

More data are available for all possible combinations of R/RO (R—R', R—OR', RO—R' and RO—OR') in the phenylbenzoates than in any other known series of mesogens. These make it possible to construct three-dimensional homologous series plots using a computer data base. Figure 10.2(a) shows such a plot for the clearing temperatures for the RO—OR' series and Fig. 10.2(b) presents the melting temperatures. These plots show both the real data and estimated points for any missing homologs. The regularity of the clearing temperature odd-even variation compared with the more haphazard trends in the melting temperatures is obvious from these two curves. A two-dimensional block diagram (Fig. 10.3) showing the homologs (real or estimated) at which the clearing and melting curves

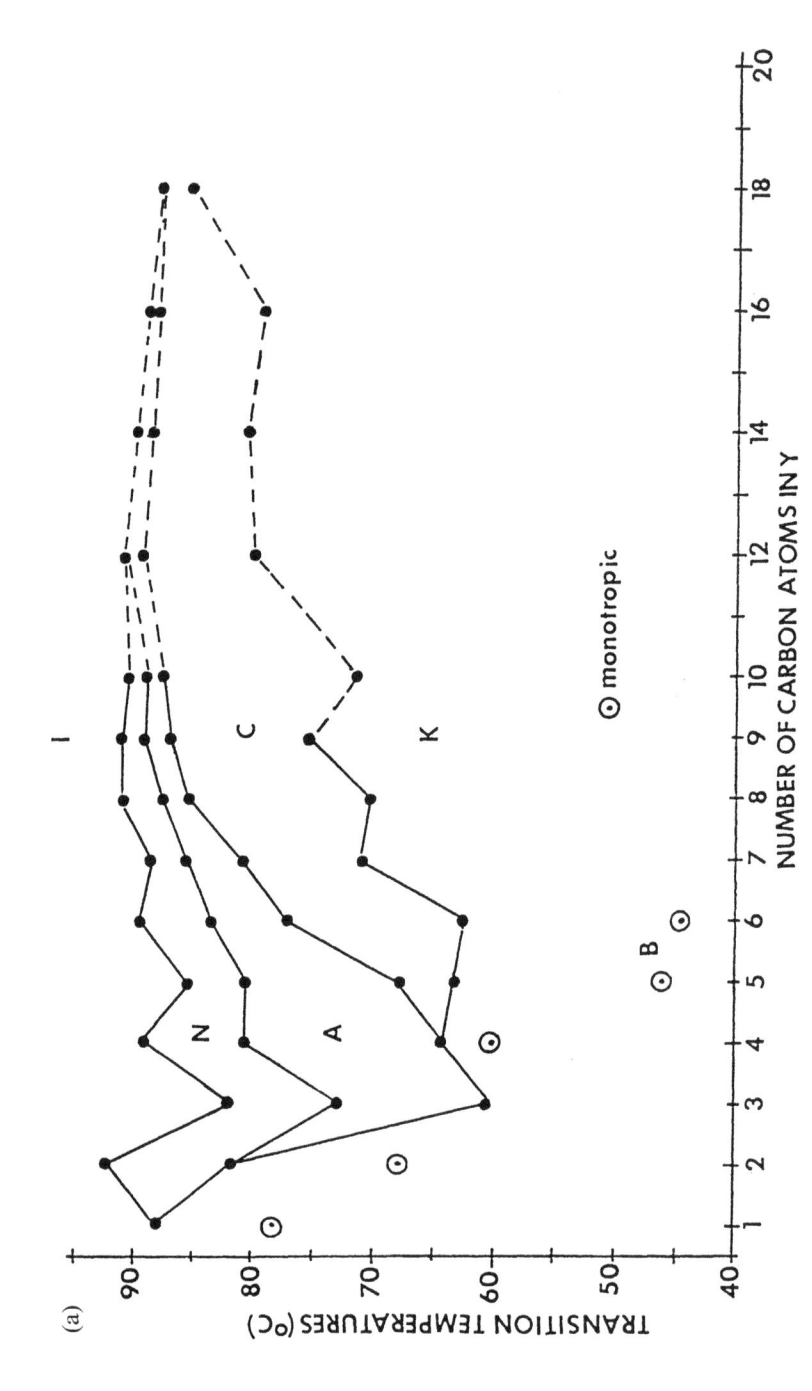

(a)

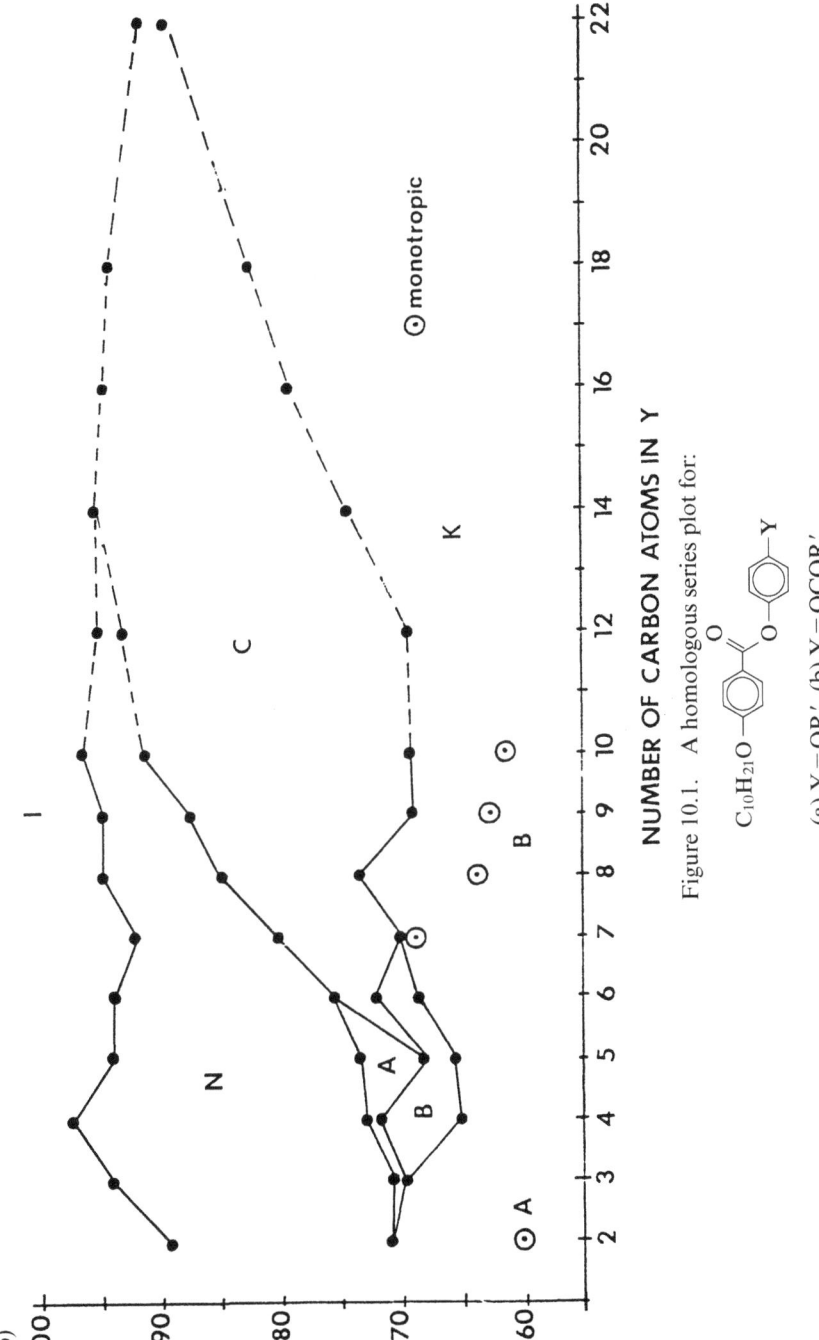

(b)

NUMBER OF CARBON ATOMS IN Y

Figure 10.1. A homologous series plot for:

$C_{10}H_{21}O$ — ⬡ — C(=O)O — ⬡ — Y

(a) Y = OR'. (b) Y = OCOR'.

(First published in Ref. 82; revised with new data, see Refs. 82, 85, 86.)

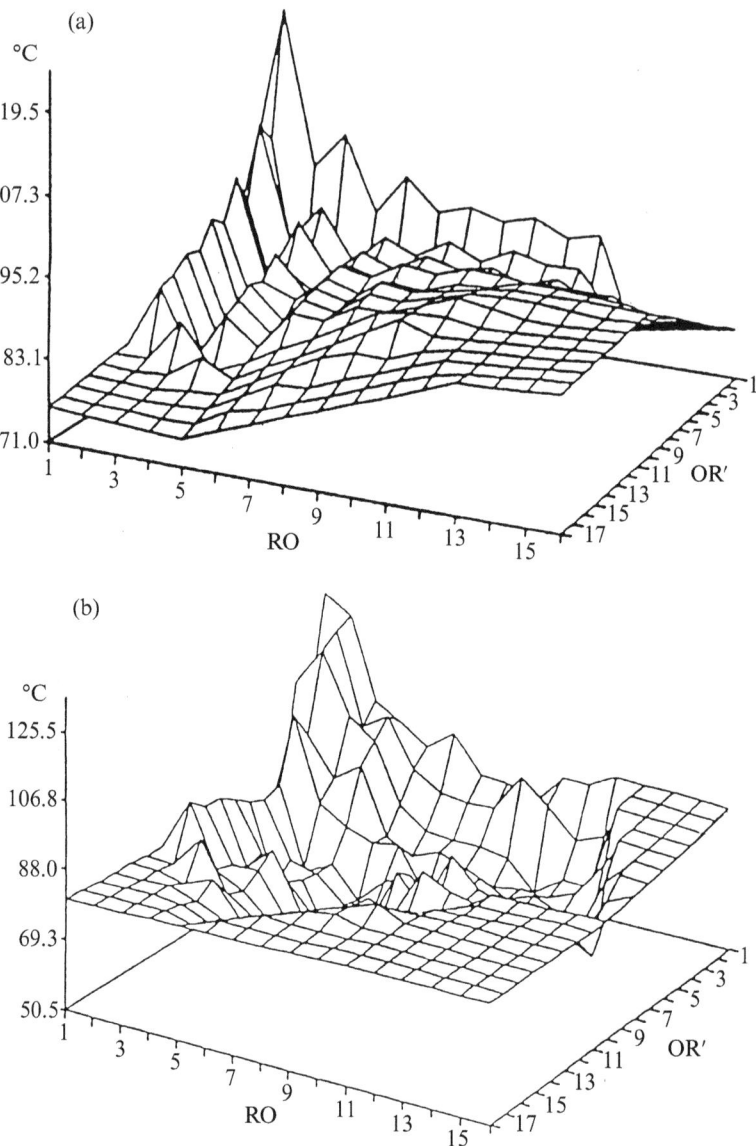

Figure 10.2. Three-dimensional homologous series plots of transition
temperatures for:

(a) Clearing temperatures,
(b) melting temperatures (originally published in Ref. 87).

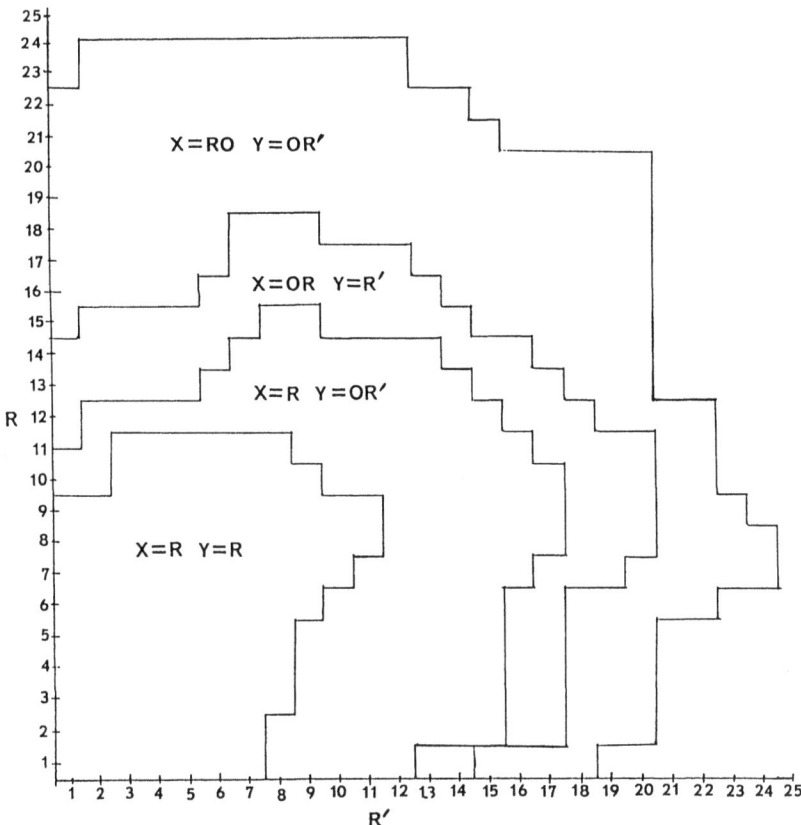

Figure 10.3. Block diagrams showing the predicted maximum chain length at which mesophases will be observed for all four phenylbenzoate series (an updated version of the one published in Ref. 87).

intersect for all four series compares the ability of these compounds to form mesophases at various chain lengths. The role a terminal alkoxy chain plays is obvious: mesophases continue to occur at longer chain lengths than when only alkyl groups are present, and two groups are better than one. Analysis of the ester data also indicates that monotropic phases can occur at up to four carbon atoms beyond the intersection of the melting and clearing curves.

Homologous series can also be plotted in two-dimensional bar graphs to obtain a more detailed picture of the effect of chain lengths on the types of mesophases that occur in a series. Several types of bar graphs have appeared in the literature, but the type shown in Fig. 10.4 presents the most complete data. Both the crystallization and melting temperatures are

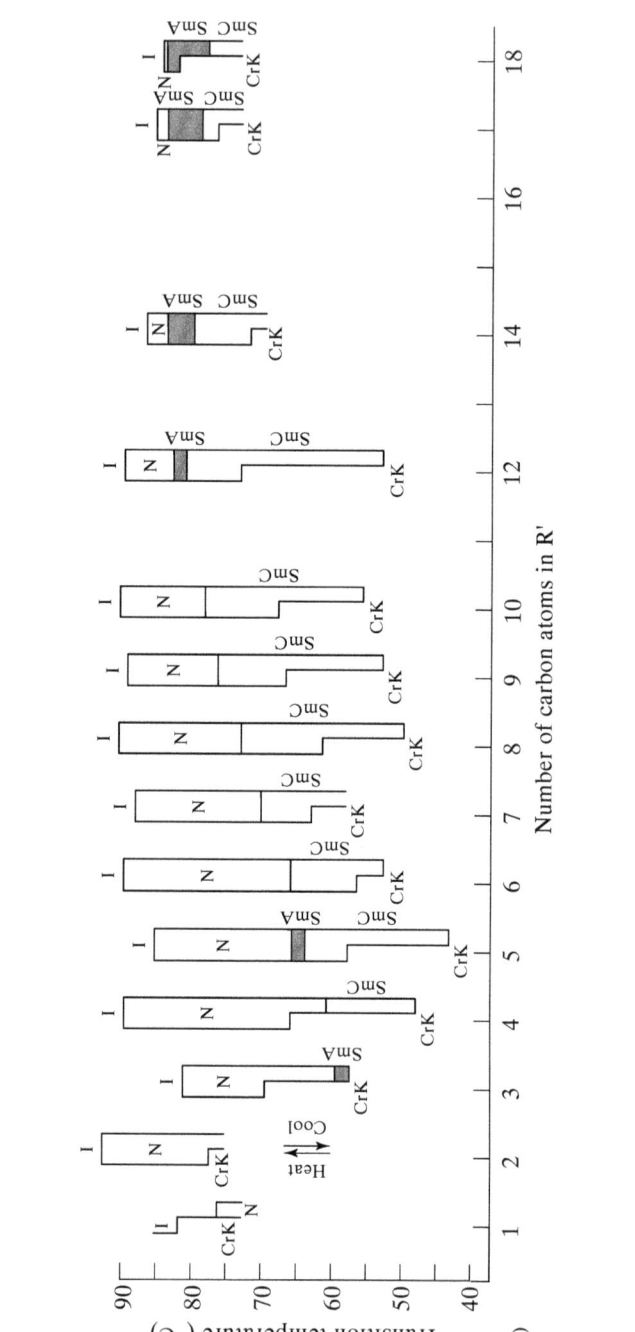

(a)

Transition temperature (°C)

Number of carbon atoms in R'

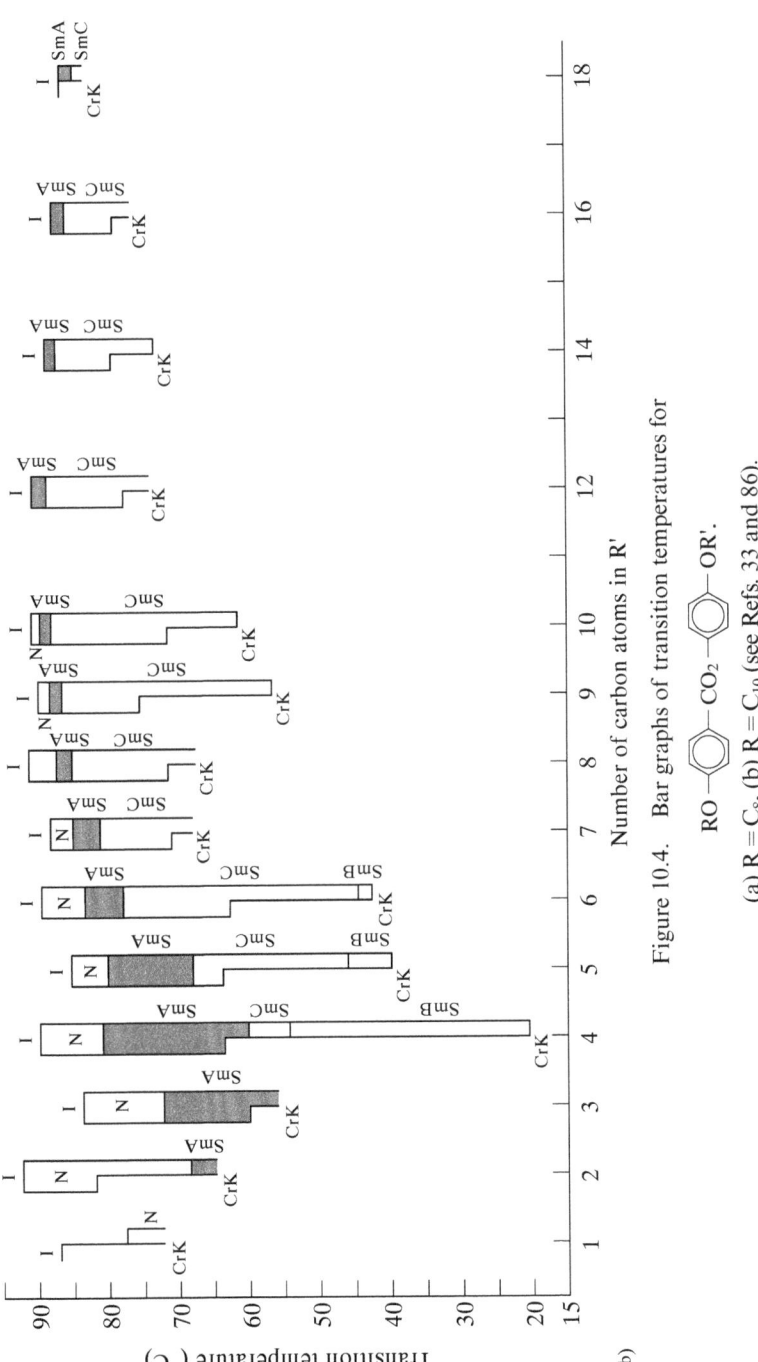

Figure 10.4.   Bar graphs of transition temperatures for

RO—⟨◯⟩—CO₂—⟨◯⟩—OR'.

(a) R = C₈, (b) R = C₁₀ (see Refs. 33 and 86).

included when available as well as both monotropic and enantiotropic phases and all crystal forms. Either a vertical or horizontal format can be used. The heating cycle of crystal to mesophases to isotropic liquid is shown on the left side of the vertical bar, while the cooling cycle of isotropic liquid to mesophases to crystals is indicated on the right portion of the bar.

The gradual evolution of the mesophases with no interruptions is typical of a homologous series of mesogens. In fact, it occurs so often that it is sometimes used to predict mesomorphic properties in a series of compounds. It is the primary reason why the interest in preparing a homologous series has declined dramatically. However, the more recent discovery of some exceptions indicates that this trend is not always reliable. A gradual evolution occurs in nearly all the phenylbenzoate series for which there is enough data to plot an entire series [85]. A typical example is shown in Fig. 10.4(b) for the $X = C_{10}O$, $Y = OR'$ series. However, in two of the series, a seemingly standard gradual evolution of mesophases shows some interruptions in the smectic-A phase. When $R = C_8$ (Fig. 10.4(a)), the homologs $Y = C_3$ and C_5 have smectic-A phases, C_4 and C_{6-10} do not, but the even homologs C_{12-18} do [33]. With $R = C_9$, there are two interruptions: one at $Y = C_8$ and the other at C_{10} with the smectic-A phase having a wider range in the C_9 than in the C_7 homolog. These series also show an unusually long persistence of the nematic phase at long chain length. Transition temperatures of those homologs with crystallization temperatures indicated were rechecked numerous times near the interruptions to be certain that this effect was real. We became convinced that it was. This raises the question as to whether there is some combination of chain lengths that will not follow the trend of a gradual evolution of mesophases throughout the homologous series. An anomaly involving the C_9 homolog in a different series has also been reported but, in this case, it was found in the packing coefficients [88].

The odd-even alternation of the clearing temperatures seen in a homologous series is believed to be due to the rod-like mesogen molecules having their terminal chains in a predominately all-trans (extended) conformation. This alternation was predicted in the theory of liquid crystals when a flexible chain was added to a rigid rod model [89]. X-ray crystallographic studies of crystalline phases of many mesogenic molecules show that the terminal alkyl chains have the all-trans conformation [38, 48, 57, 58, 90–92]. Gauche conformers have also been found in a few mesogens [38]. Seemingly more gauche conformers should occur at the higher temperatures at which mesophases occur. Conceivably, if the temperature is high enough, the energy barrier for rotation around the C–C bond (~3 kcal/mol) could be overcome. Some studies of nematic phases suggest that the chains are not always in the all-trans conformation [93–95]. It is likely that the alkyl chains are in a temperature dependent equilibrium between trans and gauche conformers with deviation from the all-trans conformation increasing with increasing chain length [22b]. This odd-even alternation also occurs in many of the physical properties of mesogens, such as order parameters and entropy values [96–99].

Assuming that the odd-even alternation depends on the terminal chain conformation, then its intensity must be related to the energy differences between the conformations of the odd and even homologs. Generally, this intensity is greatest at a short chain length and diminishes at long chain length until it disappears (Fig. 10.1). This intensity also varies with core structure. For example, it is much greater for the cyclohexane diester series (Fig. 10.5) [100] than for the phenylbenzoates (Fig. 10.1(a)). Modifying the OR chain to OCOR gives a much less obvious odd-even alternation (Fig. 10.1(a)), possibly because this chain favors the formation of more gauche conformers.

Studying the odd-even alternation of clearing temperatures provides another justification for preparing an entire homologous series. There is some indication from the studies of the effect of carbonyl-containing terminal chains on mesomorphic properties that this odd-even alternation can be eliminated by using chains that are more likely to form gauche conformers, supporting the idea that alternation is due to an all-trans chain conformation. More homologous series studies are needed to confirm this idea.

Molecules having more than two rings connected by more than one connecting group usually show a melting curve that has a broader minimum. A typical example is shown in Fig. 10.5. If a minimum is obvious, then it is usually seen at a longer chain length. It still is not possible to predict where the minimum in any specific homologous series will occur. Mesophases tend to continue to occur at longer chain lengths in these series than in two-ring structures.

Systems with more than two rings also tend to show mesophases continuing at longer chain lengths. Both a longer chain and a longer core favor the formation of mesophases and the ranges of those observed. However, the occurrence of mesophases in any compound always depends on a delicate balance between several factors: length vs. width, planar vs. non-planar, linear vs. non-linear, rigid vs. flexible, and polar vs. non-polar. If a linear molecule is either too rigid, planar and polar, or too flexible, non-linear and non-polar, then no mesophases will occur. Thus, too short an alkyl terminal chain (or none at all) often yields molecules that are too rigid

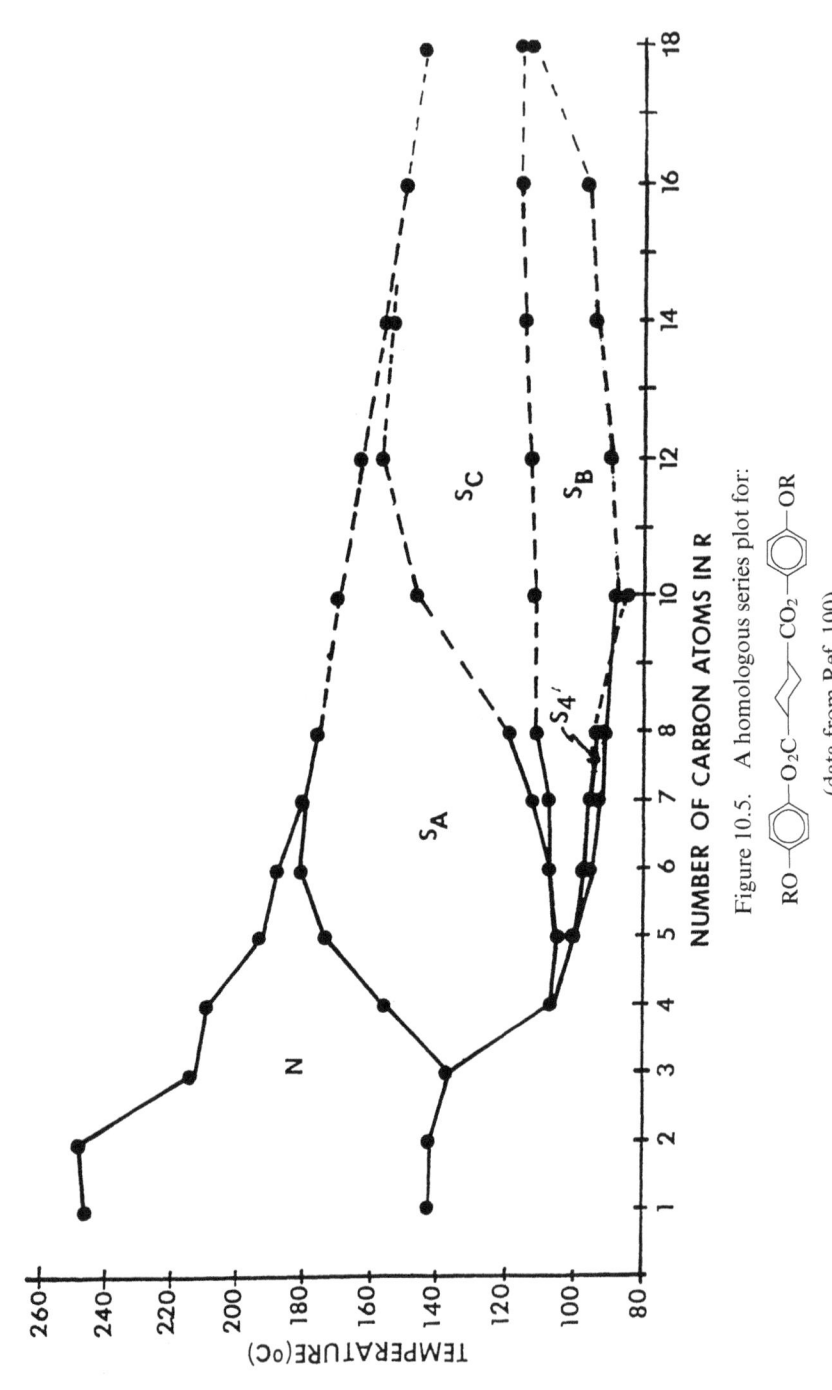

Figure 10.5. A homologous series plot for:

(data from Ref. 100).

and not long enough to form mesophases, while too long an alkyl chain produces too flexible a molecule. Generally, rigidity is provided by the core and flexibility by the terminal chains. A homologous series plot illustrates beautifully the balance in structural features needed in order to observe mesophases. Such a plot of flat-rigid molecules shows a steeply rising curve for melting temperatures as each additional benzene ring is added, whereas a plot of melting temperatures for a homologous series of alkanes shows a gradually rising curve as the number of CH_2 groups increases [101]. A typical homologous series plot for the mesogenic phenylbenzoates (Fig. 10.1(a)) shows a combination of these two curves with melting tempera-tures high at short chain lengths (rigid), decreasing as CH_2 groups are added and flexibility increasing until a minimum is reached, and then rising as the molecule becomes more and more like a flexible molecule until it is no longer a rigid rod and loses mesomorphic properties. Mesophases are the most prevalent in the region where there is a balance between rigidity and flexibility. It is this molecular anisotropy that is now believed to be the most influential factor in determining if mesophases will occur in a com-pound [102].

Obviously, the knowledge of how the molecules can pack would be useful in predicting if a compound would form a mesophase. This, however, is difficult to study in mesophases in the necessary detail. Thus, numerous x-ray determinations of the crystalline phase from which a mesophase is formed have been determined. Usually, the crystal structure that precedes a mesophase shows parallel packing of the molecules and a more layered structure when the crystalline phase precedes a smectic phase. Several studies of the crystalline phases of the homologous series of mesogens indi-cate that the crystal structure changes throughout a homologous series [103, 104]. Although more than one crystalline form often occurs in meso-gens, only the one that melts to the mesophase is useful in suggesting the type of packing that will occur in the mesophase.

Branched alkyl/alkoxy terminal chains

A straight alkyl or alkoxy chain consists of methylene (CH_2) groups aligned all in a row terminating with a methyl group:

$$\left\{ \right. \rangle - (Z)CH_2CH_2CH_2CH_2CH_3$$

Straight Chain (SC)

Z = nothing or oxygen

Chemical structure–property relationships

whereas a branched chain contains at least one methine ($-CH-$) carbon having an additional substituent:

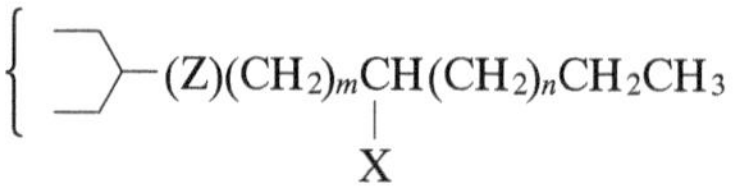

$$\left\{ \rangle\!\!-(Z)(CH_2)_mCH(CH_2)_nCH_2CH_3 \right. \quad \text{Branched Chain (BC)}$$
$$\qquad\qquad\qquad | \atop X$$

X can be a variety of groups; common ones are Me, OEt, CN, Cl and F. This methine carbon is a tetrahedral one consisting of four different substituents. This means that there can now be two different isomers which are mirror images of each other due to two possible arrangements of these four different substituents:

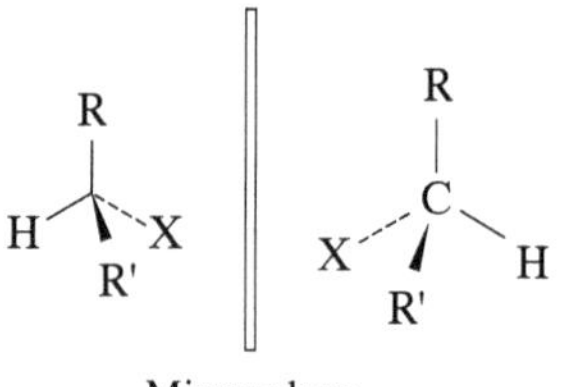

Mirror plane

These two structures are not superimposable and therefore represent different molecules.

If the carbon atom were planar, or contained two identical substituents, no isomers would be possible. These two isomers will show different properties in the presence of polarized light; one will rotate this light to the left (l for levorotatory), and the other to the right (d for dextrorotatory) i.e., they are optical isomers. This is a physical property which can be determined using a polarimeter, but it does not define the absolute configuration (arrangement) of the substituents around the carbon atom. Today, the Cahn–Ingold system using R- and S- to differentiate the isomers is used to define the absolute configuration [104]. Usually, this will be combined with a sign for the rotation: ($-$) for l and ($+$) for d to totally define the isomer that is under consideration such as

(S)-($+$)-2-methylbutyric acid

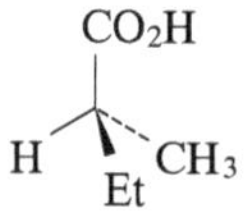

A way of indicating that a carbon atom is chiral without defining the exact isomer is to add a star above the carbon atom:

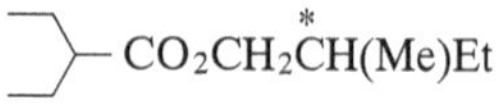

$$\rangle\!\!-CO_2CH_2\overset{*}{CH}(Me)Et$$

This symbol is also used to indicate the chiral mesophases that occur in chiral compounds: N*, SmC*. The N* designation is now preferred over the older Ch for cholesteric.

Interest in branched-chain mesogens increased dramatically with the discovery of ferroelectric properties in the smectic-C phase of the chiral branched cinnamate (DOBAMBC) [105]:

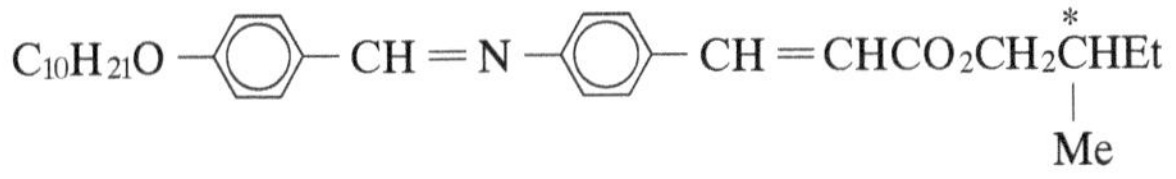

$$C_{10}H_{21}O \text{—} \bigcirc \text{—} CH = N \text{—} \bigcirc \text{—} CH = CHCO_2CH_2\overset{*}{C}HEt$$
$$\underset{Me}{|}$$

(S)-(+)-2-methylbutyl 4-(4-decyloxybenzylideneamino)cinnamate

Numerous papers have been published since then describing the synthesis of a huge variety of branched-chain mesogens. Since a few review articles [106, 107, 108] and a book have been published describing work on ferroelectric (FLC) materials, only a brief discussion will be included here.

Generally, the best ferroelectric properties occur with a branched chain that has a dipole locked in one direction perpendicular to the chain [109]. This can be achieved with a variety of branched chains. Some of those which create the best ferroelectric properties such as a large spontaneous polarization (*Ps*) are as follows:

$$CO \text{ or } CO_2 (CH_2)_n\overset{*}{C}H(X)R \text{ or } -CH_2\overset{*\,*}{\underset{O}{\vee}}R'$$
$$n = 0 \text{ or } 1, X = Me, Cl, F$$

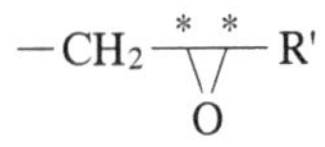

Usually, the presence of nematic and smectic-A phases is useful in aligning the sample in an FLC display. A larger tilt angle occurs with a nematic–chiral smectic-C combination than with a smectic-A–chiral smectic-C combination, but then alignment is much more difficult to obtain. Various chiral phases can be seen in chiral molecules such as chiral nematic and smectics-C, -F and -I. When the chiral nematic phase is present, blue phases (BP) can also be seen. The presence of a chiral smectic-C phase can also be accompanied by ferrielectric and antiferroelectric phases. A good summary of all these chiral phases can be found in Fig. 7 of Ref. 107.

A tetrahedral carbon atom having four different substituents can also produce chiral molecules, when these are located in either the connecting group [110, 111] or in a ring system, such as occurs in cholesterol derivatives [112, 113]:

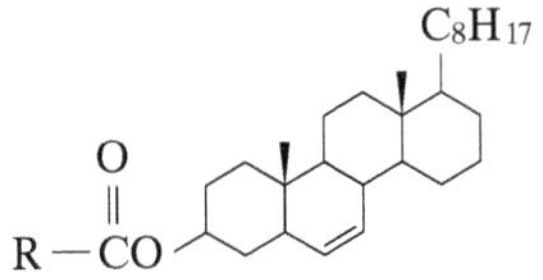

$$C_8H_{17}$$

$$\begin{array}{c} O \\ \| \\ R-CO \end{array}$$

Chiral molecules can also be designed without the presence of a tetrahedral carbon atom [114]. These generally give a much larger chirality since the chirality is built into the entire mesogenic structure.

Some compounds, such as amino acids, fatty acids and alcohols, can be found in nature as a single optical isomer. Usually, only one such isomer occurs. These can be used to synthesize a single isomer of an intermediate needed to prepare a chiral mesogen. In a few instances, such as with 2-octanol, both isomers as well as the racemic material are commercially available. The synthesis of chiral mesogens using these single isomers must be designed to use reactions that either do not occur at the chiral carbon atom or, if they do, still yield a single isomer. Another method is to design a synthesis which will yield predominantly one isomer (stereoselective synthesis). With both methods, it may be necessary to remove small amounts of the minor isomer by separating diastereoisomers or chiral chromatography. Various methods are also available for determining optical purity (enantiomeric excess). If no effort is made to use either optically pure starting materials or to control the synthesis, racemic materials will be formed. These consist of a 1:1 ratio of the two isomers and are called racemates or racemic mixtures. They can be specified by using ($\pm$) before the name, but generally racemic is assumed if no other designation is provided. Currently, it is more practical to synthesize any optically pure mesogen from optically pure starting materials than to try to separate (resolve) a racemic mixture. However, recent efforts in preparing chiral drugs is narrowing this gap which could make resolution more feasible in the future.

Since the synthesis of chiral mesogens is both more time consuming and costly than the preparation of racemic materials, often the racemic material will be synthesized first to determine if it has the desired mesophases before the chiral mesogen is prepared. Some knowledge of the effect of a branched chain on these properties is useful in determining which branched chain to use. It also may be easier to identify smectic phases in the racemic material.

A number of studies of the effect of a branched chain on mesomorphic properties have been reported; all of these study the effect of a methyl branched chain on these properties. Early studies indicate that the smectic-C phase is favored while the smectic-A and nematic phases are not [115–117]. Some exceptions to this trend have been found as can be seen in Table 10.12, but the trend usually predominates. A seemingly very reliable trend occurs in the clearing temperatures in that they always decrease. Melting temperatures usually decrease, but some exceptions are known to occur, as is typical of melting temperatures.

An earlier study also indicated that a branched chain on the acid side of

Table 10.12 *The effect of branched chains on mesomorphic properties*[a] *for* X—⬡—C(=O)—Z—⬡—Y.

	Z	X	Y	S[b]	SmC	SmA	N	Total	Change[c] in mp	Change[c] in clp
1.	O	C_8O	C_5	0	0	0.8	9.6	10.4		
2.	O	C_8O	$3,5^d$	0	5.1	*16.5[e]	0	21.6	−28.3	−17.1
3.	O	$C_{10}O$	OC_5	0	4.3	12.1	5.0	39.0		
4.	O	$C_{10}O$	$O2,5$	0	*0	*13.2	0	23	−1.8	−17.8
5.	O	C_5	OC_7	0	0	0	15.5	15.5		
6.	O	3,5	OC_7	0	*0	0	0.4	0.4	−14.2	−29.3
7.	O	C_5O	C_5	0	0	0	15.3	15.3		
8.	O	3,5O	3,5	0	*0	0	0	0	−23.6	
9.	O	4,5O	3,5	0	*0	0	m[f]	m	−14.5	−52.0
10.	S	C_8O	C5	m	m	4.9	23	27.9		
11.	S	C_8O	3,5	m	24.9	4.2	3.3	7.5	−22.7	−18.2
12.	S	C_8O	OC5	0	m	0	26.4	26.4		
13.	S	C_8O	O3,5	0	5.0	0.2	22.1	27.3	−23.4	−22.5

Notes:
[a] Recorded in °C.
[b] S = smectic phase more highly ordered than SmC; SmC = smectic C, SmA = smectic A, N = nematic and I = isotropic liquid.
[c] Transition temperature for the branched chain compound minus that for its straight chain analog (usually the preceding compound).
[d] The branched chain nomenclature used in Ref. 116 is used with the first number indicating the position of the branch and the second the length of the backbone chain.
[e] *indicates a variation from the earlier reported trends.
[f] m = phase observed which was monotropic and for which there is no mesophase range.

Table 10.13 *The effect of branched chains on the mesomorphic properties*
for X—◯—CO_2—◯—CO_2—◯—Y.

	X	Y	SmC[a]	SmA[a]	N[a]	I[a]	Ref.
1.	C_4	OC_7	67[b]		126	196	118
2.	C_5	OC_7	66[b]		88	150	118
3.	2,4	OC_8	(50.0)		75.6	155.7	117
4.	C_8O	2,4	93.0	100.4	136.4	169.0	117
5.	C_8O	C_5	85	99	149	184	60
6.	C_6	OC_8	89		95	180	60
7.	4,6	OC_8	(79.3)		95.3	168.7	117
8.	C_8O	4,6	78.2	101.5	149.6	162.2	117
9.	C_8O	C_5	85	99	149	184	60
10.	C_4O	OC_6			91	215	29
11.	2,4O	OC_8	82.7		99.3	151.4	117
12.	C_8O	2,4O	95.3	108.5	157.3	174.9	117
13.	C_8O	OC_3	100	106	144	206	60

Notes:
[a] Temperatures are in °C. Designations for phases are SmC = smectic C,
SmA = smectic A, N = nematic and I = isotropic liquid.
[b] The smectic was not identified by data from Ref. 61 for the C_6–OC_8; homolog
suggests this is a SmC phase.

an ester tended to give a smectic-C–nematic combination, whereas one on
the phenolic side seems to prefer a smectic-C-A–nematic combination
[117]. From this, it is easy to conclude that a branched chain on the acid
side causes this change. However, when data for the corresponding straight
chain (chain lengths with 1–2 carbon atoms) analogs are added to this comparison, it becomes obvious that this is not the case (Table 10.13). A comparison of data for the four possible alkyl/alkoxy homologous series of the
phenylbenzoates indicates that the phase sequence observed in each series
depends on three factors: chain length, linkage of the chain directly to the
benzene ring or via an oxygen atom, and the location of an alkoxy chain
when it is present [33]. In Table 10.13, it is obvious that the straight chain
analogs of the branched-chain diesters have the same type of mesophases.
Generally, branched chains are short, and a short chain on the acid side (X)
gives a smectic-C–nematic combination; whereas a short chain on the
phenol side (Y) with a long chain on the acid side (X) gives a smectic-C-
A–nematic combination. Unfortunately, no branched-chain analog long
enough (it would have to be at least C_8) has yet been synthesized.

Table 10.14 *Mesophase ranges (°C) for* $X-\bigcirc-\bigcirc-\overset{O}{\underset{Z}{C}}-\bigcirc-Y$.

	Z	X	Y	S[a]	B	C	A	N	Total
1.	O	C$_8$	C$_4$		10.7	0	59.5	2.7	73.1
2.	O	C$_8$	2,4	16.3	0	15.4	50	*6.1[b]	93.8
3.	O	C$_8$	C$_5$		0.4	0	63.2	4.3	67.9
4.	O	C$_8$	3,5		13.1	7.0	50.5	0	70.6
5.	O	C$_8$	OC$_4$		13.9	0	60.7	12.3	86.9
6.	O	C$_8$	O2,4		22.8	*0	54.8	0	77.6
7.	O	C$_8$	OC$_5$		13.6	0	61.3	6.9	81.8
8.	O	C$_8$	O3,5		17.5	*0	59.1	1.9	78.5
9.	O	C$_8$O	C$_4$		16.7	17.2	64.8	2.3	101
10.	O	C$_8$O	2,4	9.1	0	52.5	37	*5.0	103.6
11.	O	C$_8$O	C$_5$		2.2	14.8	65.7	2.5	85.2
12.	O	C$_8$O	3,5	15.5	0	48.9	33.4	0	97.8
13.	O	C$_8$O	OC$_4$		3	28	57	8	96
14.	O	C$_8$O	O2,4		3.7	42.8	41	0	87.5
15.	O	C$_8$O	OC$_5$		14	34	49	3	100
16.	O	C$_8$O	O3,5		32.3	55.4	33.7	0.7	122.1
17.	S	C$_8$	C$_5$	66.8	0	0	38.6	7.0	112.4
18.	S	C$_8$	3,5	85.2(3)	0	17.7	*42	0	144.9
19.	S	C$_8$	OC$_5$		66.6	26.1	16.2	17.6	126.5
20.	S	C$_8$	O3,5		16.8	31.1	12.7	9.1	69.7
21.	S	C$_8$O	C$_5$	13.5(2)	0	31.7	27.8	5.7	78.7
22.	S	C$_8$O	3,5	90.9(3)	0	45.9	14.9	0	151.7
23.	S	C$_8$O	OC$_5$	58.1(4)	0	54	9.7	11.7	134.2
24.	S	C$_8$O	O3,5	54.5(2)	0	56.6	6.6	7.8	125.5

Notes:
[a] S = smectic with an order greater than that observed in a SmB phase; identified smectic phases are indicated by the appropriate letters B, C, A; N = nematic phase and I = isotropic liquid.
[b] *indicates a deviation from the previously reported trend.

It is important in studying the effect of branched chains on mesomorphic properties to compare the properties of branched-chain mesogens with those of their straight-chain analogs with preferably identical (or nearly so) chain lengths. The addition of such data to that for the branched chains on the phenol side of biphenyl-phenyl esters (Table 10.14), indicates that a branched chain directly attached to the benzene ring tends to favor more highly ordered smectic phases as well as the smectic-C phase, if the branching methyl group is in the 2- or 3- position [119]. Smectic-A and nematic phases are usually not preferred and the smectic-B phase often disappears. A branched chain attached via an oxygen atom, however, tends to favor formation of the smectic-B phase rather than the more highly ordered smectic phases. The smectic-C phase continues to be favored while the smectic-A and nematic phases are not. The R—OR' series shows no smectic phases even when R' is a branched chain in the esters, but not in the thioesters. This again supports the idea that the branched chain does not totally control the type of mesophases observed.

A branching substituent can be located in any of the carbon atoms in a terminal substituent. There are two ways of indicating the position of this group. Both originate by counting the carbon atoms by beginning with the one adjacent to the ring system and using either Arabic numbers or Greek letters:

$$\text{CH}_2\text{CH}_2\text{CH}_2\text{CH}_3$$
$$\begin{matrix} 1 & 2 & 3 & 4 \\ \alpha & \beta & \gamma & \delta \end{matrix}$$

If a functional group containing an atom occurs adjacent to the ring, it is not counted in the Greek system:

$$\text{O}_2\text{CCH}_2\text{CH}_2\text{CH}_2\text{CH}_3$$
$$\begin{matrix} 12 & 3 & 4 & 5 \\ \alpha & \beta & \gamma & \delta \end{matrix}$$

A branch on the carbon atom at the end of the chain cannot be chiral since this carbon atom carries two hydrogen atoms.

A few studies of the effect of location of a branching group in esters on the mesomorphic properties have been done [116–120]. These indicate that a methyl group attached to a carbon atom directly linked to a benzene ring tends to form mesophases only when the other terminal chain is long, and even then it is only a monotropic smectic-A phase. A methyl group on either the 2- or 3-position tends to give the best FLC properties; the 2-position seems to be more favorable for nematic phase formation. Extending the backbone chain does not seem to improve mesomorphic properties [121].

The position of a branching methyl group can affect both the mesophases observed and the transition temperatures. These temperatures are generally higher the further the methyl group is moved along the chain away from the benzene ring. However, even a methyl group on the end carbon atom does not give as high a clearing temperature as the non-branched chain. Two branched terminal chains do not show an additive

Table 10.15 *Transition temperatures (°C) for* C_8O —⟨O⟩—⟨O⟩—Y.

	Y	S	SmC	SmA	I
1.	CO_2Pr	83(B)		64	101
2.	$CO_2CH(Me)Et$		(31)	50	67
3.	CO_2Bu		56	56	86
4.	$CO_2CH(Me)Pr$		(26)	(36)	43
5.	CO_2C_5		62	55	86
6.	$CO_2CH(Me)Bu$			(34)	49
7.	CO_2C_6		72	56	82
8.	$CO_2CH(Me)C_5$			(30)	61
9.	CO_2C_7		(46)	83	87
10.	$CO_2CH(Me)C_6$				61.8
				(37)	75

Source:
Data are from Ref. 32b.

effect either on transition temperatures or the mesophases observed [116]. Longer core molecules, such as the cyclohexane diesters, seem to be better able to handle the presence of two branched terminal alkyl chains showing enantiotropic mesophases, as long as the methyl group is not on the carbon atom adjacent to the benzene ring [116, 122]. The core can also be extended by adding groups that can conjugate with it, such as a cinnamate group: $CH = CHCO_2R$. Two branching groups can then be incorporated into the alkyl chain, R, and mesophases will still occur. If two chiral carbon atoms are incorporated into the terminal chains either in the same or different chains, the chirality of each branch must be such that a racemic material is not formed. There is some indication that a branched chain on the phenolic side of an ester is more favorable for observing mesophases than when it is on the acid side [122].

The effect of branched chains on mesomorphic properties in the disubstituted biphenyls is more difficult to assess due to the absence of data for the simple branched R/OR chains and the corresponding straight chain analogs. More data are available for biphenyls containing an ester branched chain. The comparison with straight chain analogs shown in Table 10.15 shows a continuation of the decrease in both clearing and melting temperatures. However, the trend of enhancement of the smectic-C phase does not seem to occur.

Monovalent substituents

Monovalent substituents, such as halides (F, Cl, Br, I), NO_2, and CN groups can replace alkyl/alkoxy chains and still produce good mesogens. Usually, one of the two terminal substituents must be some sort of an alkyl/alkoxy chain. In the anils and esters, these groups combined with aliphatic chains (usually as alkoxy groups) give good mesomorphic properties. In the disubstituted biphenyls, combinations of alkyl/alkoxy chains with CN and NO_2 groups yield good mesogens but the combinations with halogens do not. Terminal CN and NO_2 compounds usually form bilayer structures [123]. They are also known to form reentrant and incommensurate phases [124]. When both terminal substituents are monovalent groups, no mesophases occur although only a few examples have been reported, and most of these are in the biphenyl series [32b].

Monovalent groups can also be inserted into terminal alkyl chains either as a branch:

$$- CH_2CH - R'$$
$$|$$
$$X$$

or as the terminus of a chain:

$$-(CH_2)_2CN$$

Studies using the latter modification in several types of mesogens (Table 10.16) suggest that saturated spacer groups yield poor mesomorphic properties, usually with a decrease in transition temperatures. Properties are generally better with two than with one spacer group.

Of particular interest is the branched-type substitution using fluorine atoms to replace hydrogen atoms [125]. This substitution tends to enhance smectic phases at the expense of nematic phases. Smectic-A phases tend to be preferred with perfluorinated chains whereas smectic-C phases are enhanced in hybrid chains containing both hydrogen and fluorine atoms. A perfluorinated (all hydrogens replaced with fluorine) chain is more rigid, linear and in the all trans conformation than a hydrocarbon chain, due to the larger size of the fluorine atom. A hybrid chain will then tend to form more gauche conformers which are believed to favor smectic-C phase formation (branched alkyl chains would also favor more gauche isomers). If the fluorinated chains are more rigid than hydrogenated chains, then they should favor the nematic phase. However, the hydrocarbon and fluorinated chains do not like to pack near each other, tending to form more lamellar-like structures. Transition temperatures for mesogens containing

Table 10.16 *Effect of carbon spacer groups on mesomorphic properties (°C) of mesogens containing terminal cyano groups* X—◯—Z—◯—Y.

Z	X	Y	SmA	N	I	Mesophase range
CH = N	$C_{10}O$	CN	65	100	101.5	36.5
	C_9O	CH_2CN			99	
	C_9O	$(CH_2)_2CN$	(71.5)	(72)	84	
	$C_{10}O$	$(CH_2)_2CN$		(80)	84.5	
	C_5O	CH = CHCN		104.7	198.7	94
	$C_{10}O$	CH = CHCN		95	186	91
CO_2	C_6O	CN		70.5	81.0	10.5
	C_6O	CH_2CN			91.5	
	C_6O	$(CH_2)_2CN$		50.5	72	21.5
	C_6O	CH = CHCN		98	138	40
nothing	C_5	CN		24	35.3	11
	C_5	CH_2CN		(8)	84.5	
	C_5	$(CH_2)_2CN$		(16)	66.5	
	C_5	CH = CHCN	80	99	147.1	67.1
	C_5	C ≡ CCN		51	120.2	69.2

Source:
Data for anils and esters are from Ref. 30; biphenyls from Ref. 32b.

fluorinated chains are usually higher than those for the corresponding hydrocarbon chains. The preference for smectic phases makes them of more interest for ferroelectrics than for nematic display materials.

A terminal hydroxy, carboxyl or amide group can produce good mesogens through intermolecular hydrogen bonding with either itself such as in the acids (many examples can be found in Ref. 29, 30)

$$X-\bigcirc-\overset{O\ \cdots\ H-O}{\underset{O-H\ \cdots\ O}{\diagup\diagdown}}-\bigcirc-X$$

or amides [126] or by hydrogen bonding with different molecules, such as an acid (or phenol) and an amine [127].

$$RO-\bigcirc-\overset{O}{\underset{O\ \cdots\ H-N}{\diagup}}-\bigcirc-\bigcirc$$

A few non-linear compounds have been reported to show mesophases possibly due to the formation of hydrogen bonds between like molecules [128].

Divalent functional groups with alkyl chains

Such functional groups as $-CR=CR'-$ (alkenes, olefins), $-C\equiv C-$ (acetylenes), $-O-$ (ethers), $-CO-$ (ketones) and $-CO_2-$ (esters) can be incorporated into alkyl chains in various positions and combinations. Incorporation of an ether group adjacent to the benzene ring gives the very common alkoxy chain ($n=0$).

$$\left\{-\!\!\bigcirc\!\!-(CH_2)_nOR'\right.$$

Numerous studies have been done on the effect on mesomorphic properties of replacing an alkyl with an alkoxy chain. Such a substitution leads to higher transition temperatures and wider mesophase ranges. It can lead to more mesophases, but not necessarily. Two alkoxy chains in a two-ring system separated by a connecting group leads to even higher transition temperatures. In the phenylbenzoates, an alkoxy chain allows mesophases to continue to occur at longer chain lengths in a homologous series [33, 87]. The alkoxy chain also affects the type of mesophases observed. Although smectic-C phases can occur when no alkoxy chain is present, it is much more prevalent when at least one such group is a terminal substituent. In two-ring systems, where the connecting group produces two different substituted rings, the location of a single alkoxy chain also affects the mesomorphic properties.

No study of the effect of incorporating spacer CH_2 groups between the ring and the alkoxy group ($n>0$) has been done. However, the effect of adding an additional ether oxygen atom to an alkoxy chain has been studied in several systems. Enough data for a good comparison is available for a series of acetyl anils (Table 10.17). The melting and clearing temperatures are always lowered and there seems to be a greater preference for smectic than nematic phases. Location of the second ether group also affects these properties, although there is not enough data here to see an obvious trend. Sulfur atoms have also been incorporated into a few terminal chains, but again, not enough data is available to discern any reliable trends.

Double bonds were initially incorporated into terminal alkyl chains to produce monomers for preparing mesogenic polymers [21, 129]. This requires that the double bond be located at the end of the chain. Later studies of the effect of the location of the double bond within the terminal chain indicated that both the position and configuration affect mesomorphic properties. In the series

Table 10.17 *The effect of a second ether group in a terminal alkoxy chain on transition temperatures (°C) for* RO—⬡—CH = N—⬡—COMe.

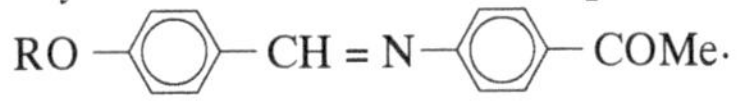

R	S_2	S_1	N	I
C_2H_5			(115)	122
C_3H_7	(67.6)	(90.9)	92	101.7
CH_3OCH_2		(82.4)		94.0
C_4H_9		86.1	102	113.9
$CH_3O(CH_2)_2$	(70.7)	(83.1)	(90.2)	93.7
C_5H_{11}		80.2	104.4	106.3
$C_2H_5O(CH_2)_2$	(55.6)	73.2		76.6
C_8H_{17}	(51.8)	70.0		118.3
$C_6H_{13}O(CH_2)_2$		53.2	7	72.3
$CH_3O(CH_2)_3$		71.6		73.5
C_6H_{13}		70.9	112.1	113.3
$C_2H_5O(CH_2)_3$		58.6		72.8

Source:
Data from Ref. 30 except for $R = C_4$ which is author's data.

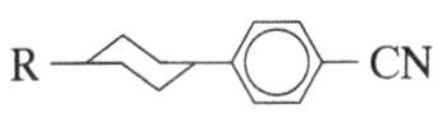

R—◇—⬡—CN

the following trends were observed [130]:

For TTs:

For Δn:

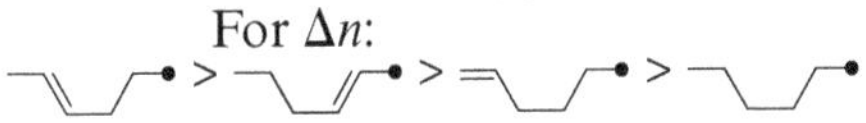

For rotational viscosity:

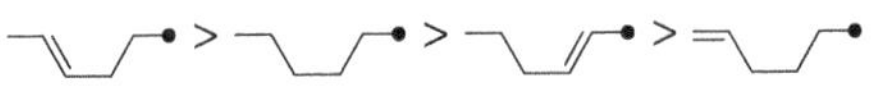

Response times were usually *c.* 30% shorter than in the saturated isomer and the presence of a double bond favors nematic phase formation. Later studies of other systems showed more variations in these trends [131] but often the mesogens with an alkoxy chain having the double bond in the 2–3 position from a benzene ring showed the lowest melting temperatures. Thus, this structure modification appears to be useful for lowering the transition temperatures of a mesogen. Two double bonds seem to lower the temperatures even farther [132]. The presence of double bonds, however,

Table 10.18 *The effect of terminal chain double and triple bonds on mesomorphic properties.*

	X	N (°C)	I (°C)
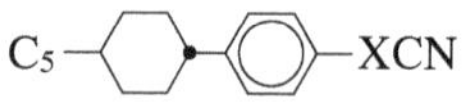 C_5 —⬡—⬡—XCN data from Ref. 136	CH_2CH_2 $CH=CH$ $C\equiv C$	(28.0) 40.6 51	44.2 149 120.2
$X-(CH_2)_3$ —⬡—$(CH_2)_2$—⬡—CN data from Ref. 136	CH_3CH_2 $CH=CH$ $C\equiv C$	(28.0) (17.2)	44.2 25.3 62.5
RX—⬡—⬡—CN data from Ref. 32b	$R=C_3$ CH_2CH_2 $R=C_3$ $CH=CH$ $R=C_4$ $C\equiv C$	30 92.4 64.3	42.8 125.5 39.1

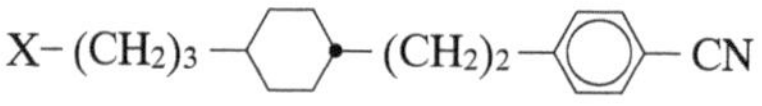

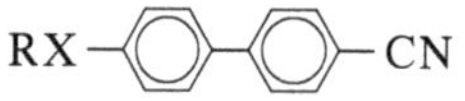

raises both stability concerns due to oxidation, migration of the double bond and the existence of two isomeric cis and trans forms that can inter-convert. Generally, the trans form produces better mesogens because of its linearity, but the cis could be more useful for lowering transition tempera-tures. The use of a double bond as a spacer for a nitrile group generally yields good mesogens (Table 10.16). A terminal double bond with two iden-tical substituents on the end carbon produces swallow-tailed mesogens [133]:

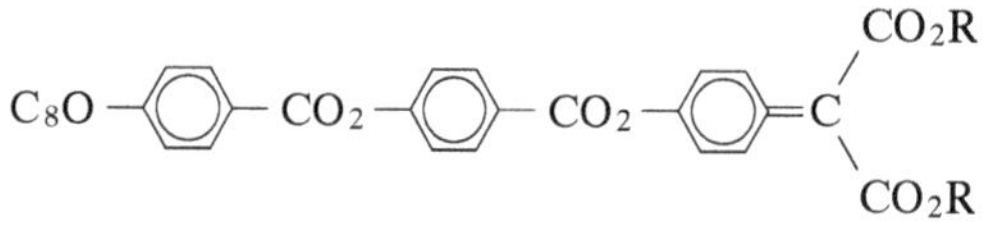

$$C_8O \text{—⬡—} CO_2 \text{—⬡—} CO_2 \text{—⬡=}C\underset{CO_2R}{\overset{CO_2R}{<}}$$

Two such groups give biswallow-tailed mesogens [134]:

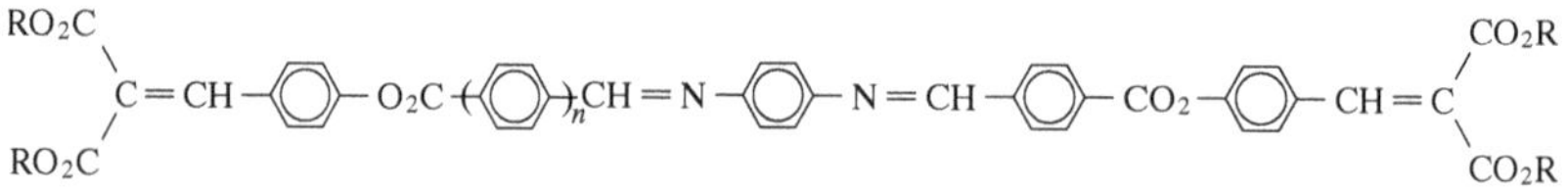

$$\underset{RO_2C}{\overset{RO_2C}{>}}C=CH\text{—⬡—}O_2C\text{—(⬡)}_n CH=N\text{—⬡—}N=CH\text{—⬡—}CO_2\text{—⬡—}CH=C\underset{CO_2R}{\overset{CO_2R}{<}}$$

n = 1 or 2

Such compounds can also be formed using a saturated chain [135].

Triple bonds have been incorporated into terminal chains in a few meso-gens, primarily adjacent to the benzene ring [16c, 136]. No studies of the effect of location within the chain on mesomorphic properties seem to have been reported. From the few comparisons that can be made of the effect of the triple bonds with those of the single and double bonds (Table 10.8), no obvious trend occurs.

The effect of carbonyl containing groups, $C=O$, CO_2 and OCO in terminal chains, on mesomorphic properties has been studied in the phenylbenzoates [82, 137–139]. Generally, the most mesophases occur with a $-O_2C-$ group on the phenolic side with fewer being seen with $-CO_2-$ or $-CO-$. The ketone gives much higher transition temperatures and usually smectic-A and/or -C phases. With both esters, the transition temperatures remain similar to those with just an alkyl or alkoxy chain. The addition of CH_2 spacer groups between the ring and carbonyl containing groups leads to lower transition temperatures. With only one CH_2 group, mesophases usually do not occur, whereas with two they do, but at reduced phase lengths. Some studies have also been done with ester and ketone groups attached directly to the acid benzene ring. Properties were generally found to be similar to those found for the phenolic side substituents, but enough differences occur that these changes could be useful for fine tuning the mesomorphic properties of a compound or a mixture. Studies of phenylthiobenzoates [137] and biphenylphenyl esters [119] show similar results. No additive effect could be found when both terminal chains in the phenylbenzoates contained carbonyl groups [139]. Usually these and biphenyl compounds containing two such substituents [32] showed poor mesomorphic properties. The combination of a substituent yielding poor mesomorphic properties in the monosubstituted parent compound with one that produces good properties will most likely not even give a fair mesogen.

If a divalent functional group is incorporated into a terminal chain between the aromatic ring and the aliphatic chain, this group may conjugate with the ring and actually become a part of the core extending its length. This could explain the poorer mesomorphic properties when these groups are moved farther out along the alkyl chain. Another possible explanation is that these groups generally produce more gauche conformers of the chain when they are not attached directly to the ring.

As is true with connecting groups, more than one functional group can be combined in the terminal chain. These can be the same or different, and also adjacent or not. Some typical examples are:

$$-O(CH_2)_nOCH_2R \quad n>0 \qquad\qquad -C(R)=C(R')CN$$
$$-CO_2CH_2CHR \quad X=\text{halide, CN} \qquad -C\equiv CCN$$
$$\qquad | \qquad\qquad\qquad\qquad\qquad\qquad\qquad$$
$$\qquad X \qquad\qquad\qquad\qquad\qquad -COCH_2O_2CR$$

Numerous other combinations are possible.

Lateral substituents (A, B)
Short substituents

A lateral substituent is attached to the core of a mesogen, usually on a ring, in a position that is not along the molecular axis. Several review articles adequately summarize earlier studies on the effect of small lateral substituents on mesomorphic properties in a variety of mesogens [24a, 25]. Four major factors are involved: the type, size and position of the substituent and the structure of the mesogen. Usually, a lateral substituent will lower both the melting and clearing temperatures by broadening and/or twisting the core of the mesogen. Thus, a larger substituent generally produces a larger decrease. Mesophases can show a decrease in range or even disappear, although increases are also known. A longer mesogenic core which generally produces higher transition temperatures and wider range mesophases can tolerate a lateral substituent better than a short core. This is also true of bent cores such as those containing a naphthalene ring since these provide a hole into which the substituent can lie when packing in a parallel manner. A second substituent on the same side of the ring does not create as large a decrease as the first one since the width of the molecule does not increase. A second substituent on the opposite side broadens the molecule further, giving a larger decrease.

The effect of lateral substituents on the types of mesophases observed is difficult to generalize into a common trend since there are so many variables to consider. For example, nematic and smectic-C phases are preferred over the more commonly seen smectic-A phases in the phenylbenzoates, while the thioesters prefer the nematic phase [140]. These studies also indicate that a lateral substituent decreases the transition temperatures more than a branched chain. A lateral substituent can produce either better nematic phases with suppression of smectic phases, or better smectic-C phases, depending on its location in the mesogen [24a, 141]. Studies of the effect of lateral substituents on the mesomorphic properties for the cyanobiphenyls have shown that one substituent ortho to the terminal cyano group decreases the molecular pairing that occurs in the unsubstituted compounds causing an increase in the $\Delta\epsilon$ values [24a, 142]. Ortho fluorine atoms are particularly useful, since these also tend to favor nematic over smectic phases.

A lateral cyano group produces a $\Delta\epsilon$ value that is negative, but only about half that obtained for a terminal cyano group. Such a large group also gives a large decrease in the transition temperatures along with shorter mesophase ranges. An additional cyano group on the same side of the ring produces a very large negative $\Delta\epsilon$ value, but also creates a viscosity that is much too high and a solubility that is too low to produce useful display materials.

With its small size, tendency to produce wide range nematic phases and low viscosity, the fluorine atom is quite useful in the design of new nematic display materials with lower melting temperatures and good stability [24a, 143]. For these reasons, numerous structural variations incorporating a lateral fluorine atom in a wide variety of mesogens continue to be studied. Much of this work has been made possible by the development of a new method for preparing these compounds [144]. Two fluorine atoms on the same side of the ring in a cyanobiphenyl produce a large (though not as large as a CN group) $-\Delta\varepsilon$, but with little increase in viscosity [23] encouraging modifications using two fluorine atoms. On the other hand, a lateral substituent can also enhance smectic-C phases [141] which has led to numerous modifications of ferroelectric liquid crystals that incorporate lateral fluorine atoms.

A lateral hydroxyl group, when ortho to a connecting group that can hydrogen bond (intramolecular) with it, can also produce good mesogens; better than expected for a bulky lateral substituent. The best known examples are the anils, such as [145]:

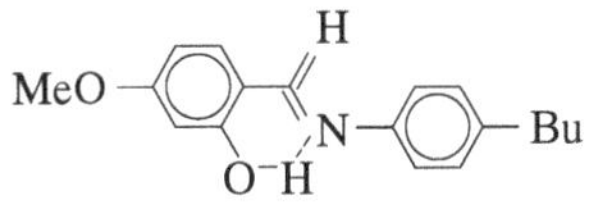

Abbrev.	CrK	N	I	N range (°C)
OHMBBA	<8	44	64.5	20.5
MBBA	11	21	41	20.0

Transition temperatures usually increase and mesophase ranges sometimes do as well. Other modifications include branched chain anils [146] and hydrogen bonded azo compounds. [147] Anils with ortho hydroxyl groups can also form metal complexes such as:

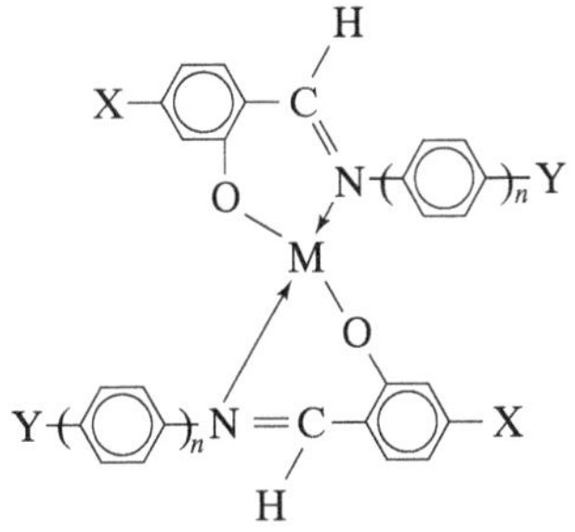

Other structural variations include anils with an additional connecting group, hydroxyl group ortho to a terminal anil, and azo complexes. A recent book discusses these metallomesogens as well as numerous other structure modifications incorporating metals [148].

Long substituents

Early studies suggested that since a lateral ethyl group produced poorer mesogens than a lateral methyl group, longer alkyl chains would yield even poorer mesogens. However, later studies showed that lateral alkyl chains long enough to fold back along the molecular axis could produce good mesogens [149]. Such chains could also contain functional groups. A lateral benzene ring attached directly to the ring even produces monotropic nematic phases [150]. These results opened a new area for structural modifications by attaching a second mesogenic unit to create a new class of mesogens called dimeric mesogens, or twins. The following structure modifications are known:

1. Fused [151]

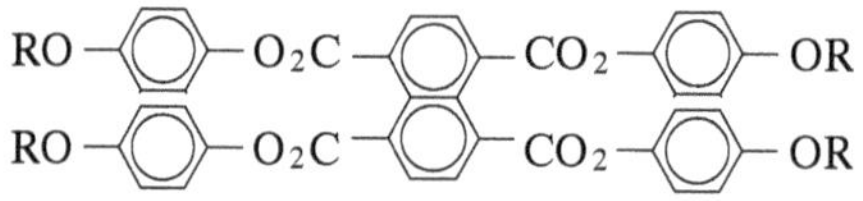

2. Ligated [150, 152, 153] – Many of these can form more than one convertible isomer. Four types are known:

(a)

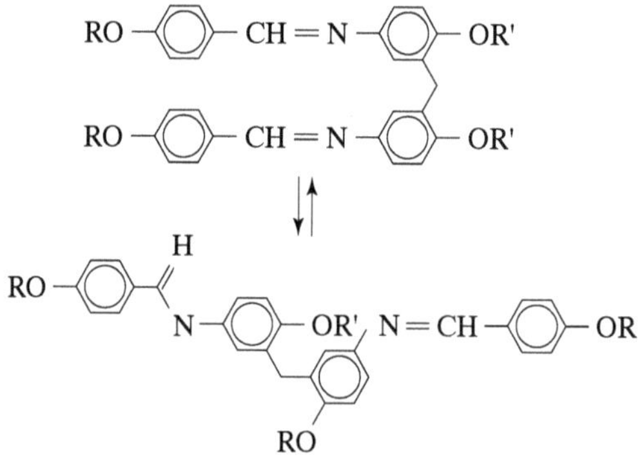

(b)

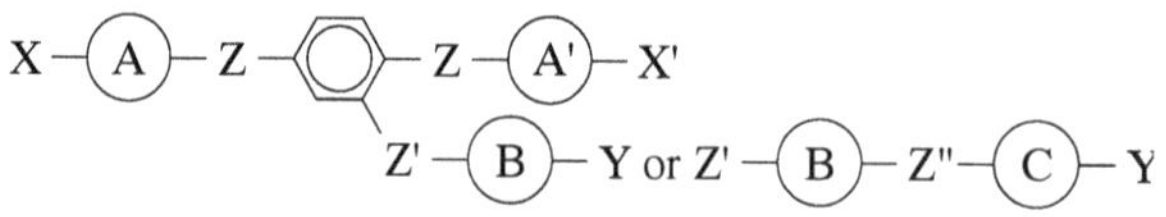

X, X', Y = terminal substituents
Z, Z', Z" = connecting groups but Z' can also be a mesogenic unit
A, A', B, C = ring systems

These fused and ligated twins are board-like rather than rod-like in their shape. Each type of moiety can be the same or different [153, 154]. Example:

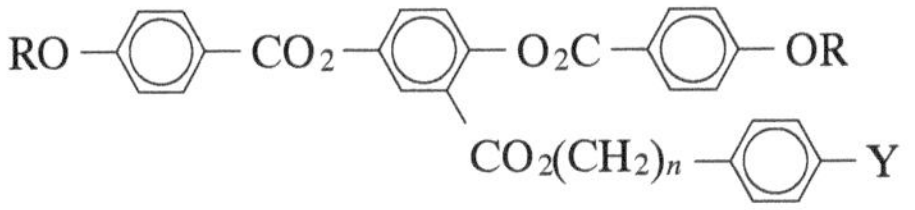

RO—◯—CO_2—◯—O_2C—◯—OR

$CO_2(CH_2)_n$—◯—Y $n = 0,1; Y = OR, CN, NO_2$

(c)

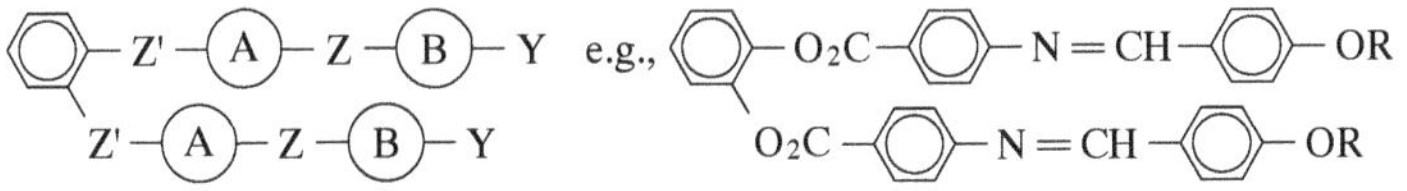

◯—Z′—Ⓐ—Z—Ⓑ—Y e.g., ◯—O_2C—◯—N=CH—◯—OR (Ref. 155)

Z′—Ⓐ—Z—Ⓑ—Y O_2C—◯—N=CH—◯—OR

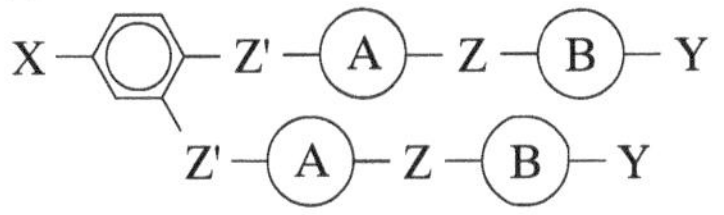

(d)

X—◯—Z′—Ⓐ—Z—Ⓑ—Y (Ref. 156)

Z′—Ⓐ—Z—Ⓑ—Y

e.g., X—◯⟨ $O(CH_2)_5CO_2$—◯—CO_2—◯—◯

$O(CH_2)_5CO_2$—◯—CO_2—◯—◯ ⟩

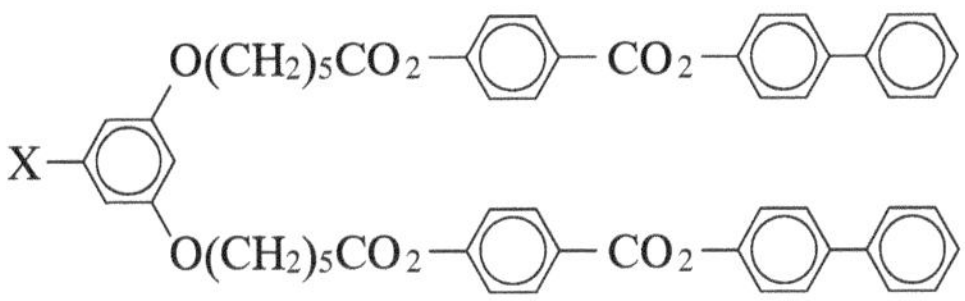

X can be either H or another mesogenic unit (triplet)

3. Tail-to-tail – These twins are more rod-like, but also more flexible than rods:

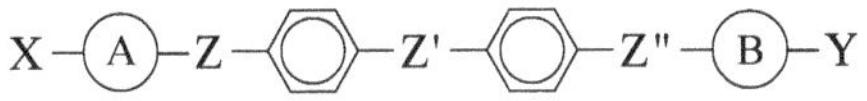

X—Ⓐ—Z—◯—Z′—◯—Z″—Ⓑ—Y

with Z′ usually containing spacer groups. They can be either symmetrical ($Z = Z″$, $A = B$, $X = Y$) [157, 158] e.g.,

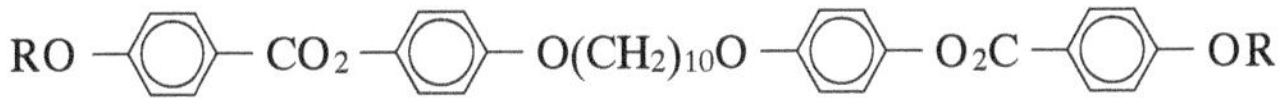

RO—◯—CO_2—◯—$O(CH_2)_{10}O$—◯—O_2C—◯—OR

or unsymmetrical (asymmetrical, non-symmetrical) [159] dimers:

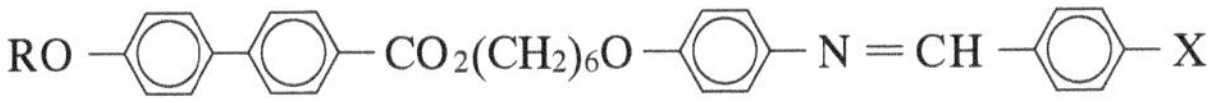

RO—◯—◯—$CO_2(CH_2)_6O$—◯—N=CH—◯—X

Both chiral [160] and discotic [161] twins are also known.

Two lateral mesogenic groups added to a long mesogen para to each other produces a disk-like shape, but these compounds do not show discotic phases [162]. Discotic phases have been reported in a few compounds with only three mesogenic units on a six-membered ring [163]. A huge number of other disk-shaped molecules have been synthesized. Many of these incorporate metal atoms into the disk. Several extensive reviews are available discussing the relationships between structures and the types of discotic phases observed [150, 164–166]. These molecules often pack in columns that then align parallel to each other. Bowls, cones, tubes, stars and chiral discotics are additional modifications that have been reported.

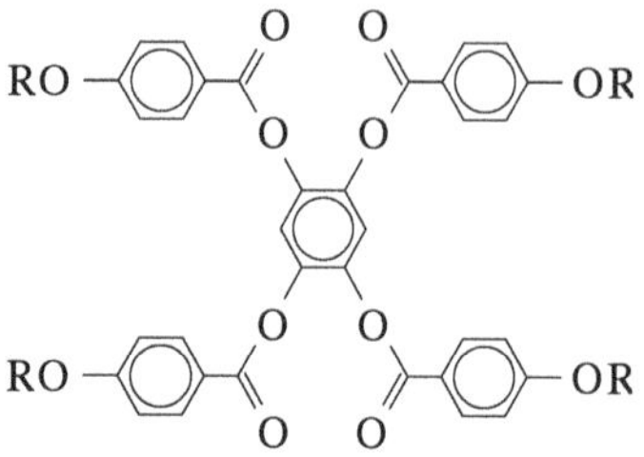

Forked

A lateral substituent can also be placed on an end ring in a rod-like molecule to give forked compounds [167]. The lateral substituent can be either ortho to the terminal chain

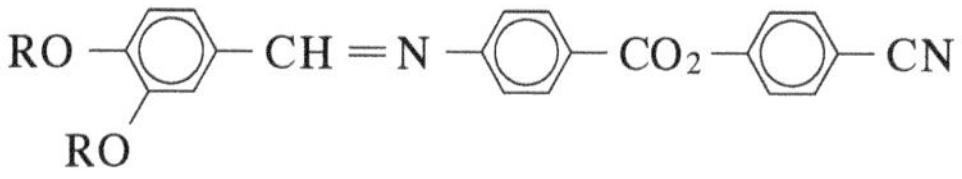

or to the connecting group

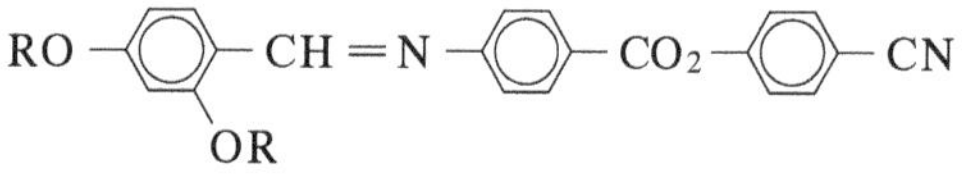

A single lateral substituent on two end rings gives a biforked mesogen [168, 169].

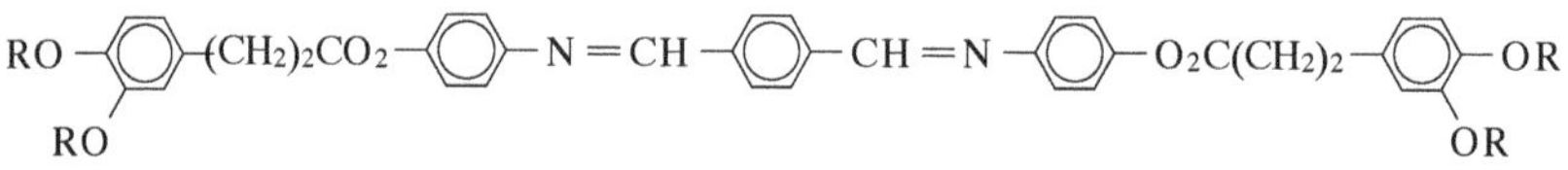

When an additional lateral substituent is added ortho to the terminal chain on both end rings, phasmid mesogens are formed [169, 170].

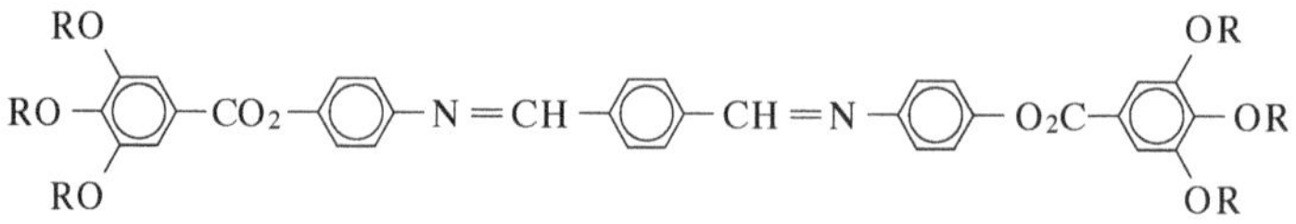

Two lateral chains ortho to the terminal chain on only one ring are also known [171, 172]. One of these was initially thought to form a thermotropic biaxial nematic phase, but this was later shown to have a smectic-C below a nematic phase [172]. Biaxial nematic phases were first observed in lyotropic systems[173] and have been predicted to occur in thermotropic mesogens [174]. At the time of this writing, no reported thermotropic biaxial nematic has withstood rigorous testing for biaxiality [175].

10.4 Additional structure modifications

A very useful modification is the incorporation of deuterium into a variety of mesogens for studying mesophases using NMR and neutron scattering [176]. Other modifications give amphiphilic mesogens which often produce lyotropic phases, sugars which can give discotic phases and those that form glasses. Liquid crystalline polymers has become a very popular and extensive area of study. Space limitations prevent further discussions of these equally interesting structures.

10.5 Molecular theory and phase structure

Many theories exist on the structure of liquid crystal phases. All of these use models which are considerably simplified compared to the actual molecules which form mesophases. Thus, they are not useful for predicting properties quantitatively. However, they can be useful in providing qualitative relationships for designing new mesogens [22b]. Theoretical models have also been used to explain the types of mesophases observed [177]. The importance of shape in the formation of orthogonal phases has been discussed [178] and the arrangement of molecules in mesogens is now being studied by scanning tunneling microscopy [179, 180] and molecular modeling/simulation [180, 181]. Structural features for discotic materials have been compared to those for rod-like ones [182].

10.6 Materials for displays

No single mesogen has all the properties needed to obtain a good display material. This is usually achieved by mixing several components, each of which contributes some of the properties needed for a particular display. These properties vary, depending on the use of the display. The properties considered essential for most nematic displays are as follows:

- wide range nematic phase with a melting temperature well below room temperature and a high clearing temperature
- colorless
- stable to heat, light and electric current
- a large (+) or (−) dielectric anisotropy ($\Delta\epsilon$)
- low viscosity (η)
- good solubility of components
- large ratio k_3/k_2 of elastic constants

For some displays, a large optical birefringence (Δn) is of interest, and for FLC displays, a large tilt angle and spontaneous polarization are needed. Several reviews discuss the properties needed for good display materials in more detail [24a,b, 26, 183–186].

Unlike transition temperatures, there is no comprehensive collection of these physical properties for known mesogens. The effect of molecular structure on dielectric behavior has been reviewed [187]. Numerous small collections of data have been used for making comparisons between structures and properties [13, 16d, 24a,b, 26, 183, 184, 187–190]. These have been used to develop the following discussions.

Viscosity

A comparison of the literature viscosity data is difficult since different types of viscosity (rotational and bulk) along with different measuring units are reported for various mesogens. Generally, replacing a benzene ring with a cyclohexane one produces a lower viscosity, but also decreases the birefringence [191]. A pyrimidine ring, however, can lower the viscosity without decreasing the birefringence [22]. Terminal substituents also influence viscosity with a longer chain giving a higher viscosity. Polar substituents which produce the desirable large dielectric anisotropy ($\Delta\epsilon$) values also tend to increase viscosity as shown by the following order [24a, 192]:

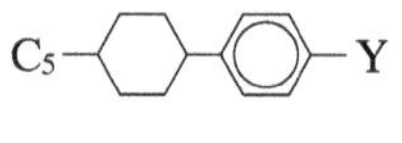

$$C_5 - \bigcirc - \bigcirc - Y \qquad \begin{array}{l} Y = CN \gg NCS > CF_3 > OMe > OCHF_2 \\ \simeq C_3H_7 > OCF_3 > F \end{array}$$

Lateral substituents also tend to increase viscosity although a lateral fluorine atom has only a small effect. These various effects can be seen in the following two orders of viscosity [24b, 193]:

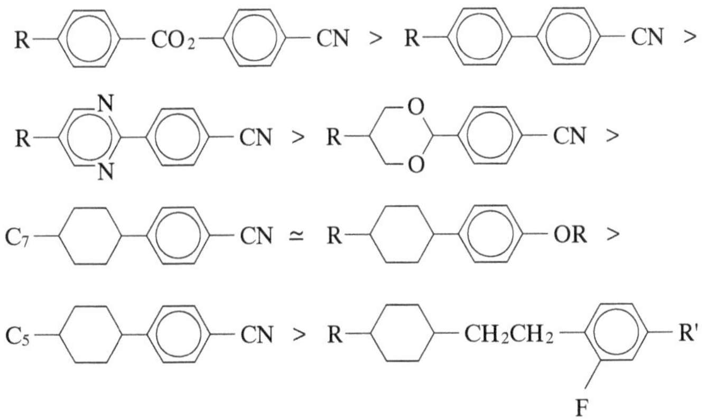

R—⬡—CO₂—⬡—CN > R—⬡—⬡—CN >

R—⬡(N,N)—⬡—CN > R—(dioxane)—⬡—CN >

C₇—⬡(cyclohexyl)—⬡—CN ≈ R—⬡(cyclohexyl)—⬡—OR >

C₅—⬡(cyclohexyl)—⬡—CN > R—⬡(cyclohexyl)—CH₂CH₂—⬡(F)—R′

and

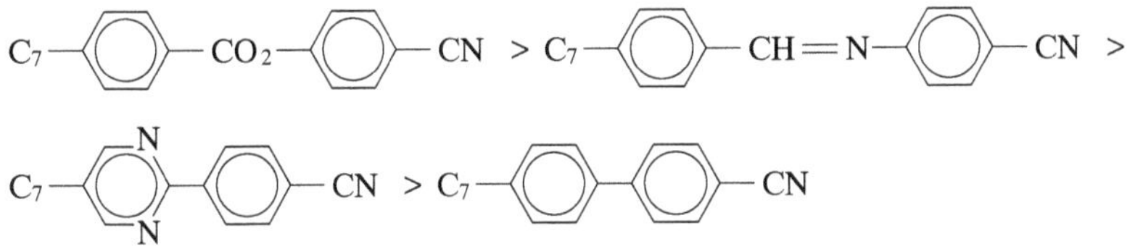

C₇—⬡—CO₂—⬡—CN > C₇—⬡—CH=N—⬡—CN >

C₇—⬡(N,N)—⬡—CN > C₇—⬡—⬡—CN

Dielectric anisotropies (Δε)

Dielectric anisotropies for a variety of mesogens are given in Tables 10.19–10.21. With some mesogens, different values were reported with some differing significantly, such as the $\Delta\epsilon = 32$ initially reported for the dioxane analogue of 5CB. Several sets of data suggest it is 13.3, only a little larger than the value for 5CB. Obviously, a number of structural features affect $\Delta\epsilon$: ring system, polar substituents, connecting groups and lateral substituents. A polar group along the molecular axis gives a larger, positive value, whereas a dipole across the polar axis yields negative values. The role of lateral substituents on such properties has already been discussed in the lateral substituent section. An early tabulation of dielectric properties of some mesogens is available [198].

Elastic constants (k)

The ratio of splay constants, k_3/k_1, needs to be small in order to obtain steep transmission characteristics in twisted nematic liquid crystal displays (TN-LCDs) [189]. Along with this, a large $+\Delta\epsilon$ is needed to obtain low threshold voltages. Some useful comparisons of individual elastic constants in various structures have been made [24b].

Table 10.19 *Some biphenyl type compounds having positive dielectric anisotropy.*

Mesogen	$+\Delta\varepsilon$	Δn	ΔH_f (kJ/mol)	Ref.
C₆O—⟨benzene⟩—⟨pyrimidine(N,N)⟩—C₆	0.8	0.15	19.22	13,32a
R—⟨dioxane(O,O)⟩—⟨benzene⟩—OR'	0.1			194
C₅—⟨cyclohexane⟩—⟨benzene⟩—OMe	2.3	0.07		24b
C₅—⟨cyclohexane⟩—⟨cyclohexane⟩—CN	4.2	0.05	26.8	195
R —	9.7			194
C₅—⟨cyclohexane⟩—⟨cyclohexane⟩—OCF₃	7.1	0.05		24b
C₅—⟨cyclohexane⟩—⟨benzene⟩—CN	9.9 / 12.9	0.12	19.6	24b,32, 195
C₅—⟨benzene⟩—⟨benzene⟩—NCS	10.8	0.18	8.8	24b,32
C₅—⟨benzene⟩—⟨benzene⟩—CF₃	10.9	0.04		24b
C₅—⟨benzene⟩—⟨benzene⟩—CN	11.0 / 16.1	0.18 / 0.20	17.2	24b,195
R —	11.73	0.20		194
C₅—⟨benzene⟩—⟨benzene⟩—⟨benzene⟩—CN	12.5		17.2	195
C₅O—⟨benzene⟩—⟨benzene⟩—CN	9.9 / 13.7		28.9	194,195
C₅—⟨cyclohexane⟩—⟨benzene(F)⟩—CN	17.7 / 19.7	0.09		22b,194
C₅—⟨pyrimidine(N,N)⟩—⟨benzene⟩—CN	19.7 / 20.9 / 34.0	0.22		13,24b,194
C₅—⟨dioxane(O,O)⟩—⟨benzene⟩—CN	32 / 13.3 / 17.4	0.14 / 0.09 / 0.14	21.0	13,24b,32 194

Table 10.20 *Mesogens with a connecting group between two rings having a positive dielectric anisotropy.*

Mesogen	$\Delta\varepsilon$	Δn	Ref.
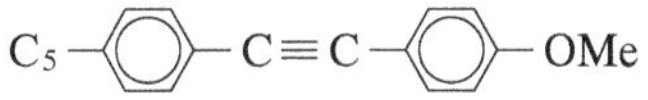 C_5 —◯— $C\equiv C$ —◯— OMe	0.0	0.26	24b
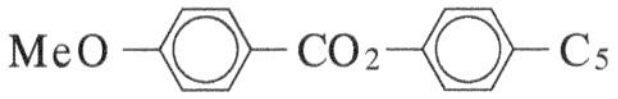 MeO —◯— CO_2 —◯— C_5	0.08		187
 C_4 —◯— CO_2 —◯— C_5	0.14		187
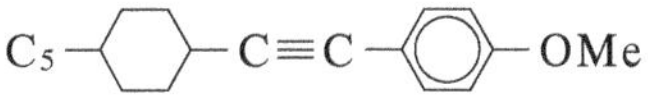 C_5 —⬡— $C\equiv C$ —◯— OMe	0.2	0.13	24b
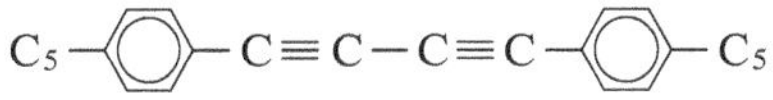 C_5 —◯— $C\equiv C-C\equiv C$ —◯— C_5	0.8	0.36	24b
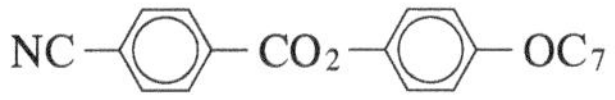 NC —◯— CO_2 —◯— OC_7	4.0		187
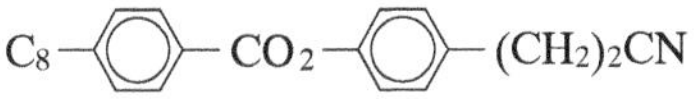 C_8 —◯— CO_2 —◯— $(CH_2)_2CN$	6		187
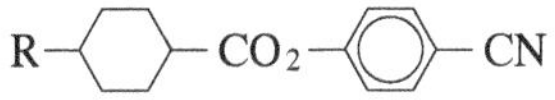 R —⬡— CO_2 —◯— CN	6.4		194
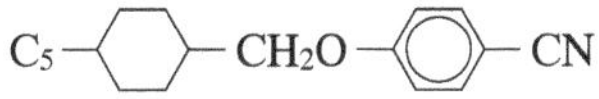 C_5 —⬡— CH_2O —◯— CN	8.2	0.12	24b
 C_5 —⬡— $C\equiv C$ —◯— CN	11.6	0.20	24b
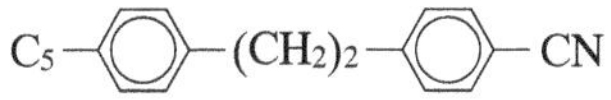 C_5 —◯— $(CH_2)_2$ —◯— CN	12.3	0.12	24b
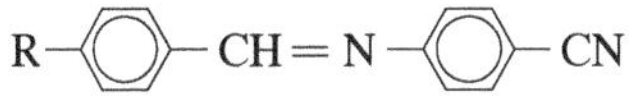 R —◯— $CH=N$ —◯— CN	13.3		196
 C_7O —◯— CO_2 —◯— CN	14.3		187
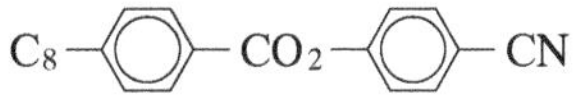 C_8 —◯— CO_2 —◯— CN	17.7 19.7(R)		187 194
 C_5 —◯— COS —◯— CN	33		13
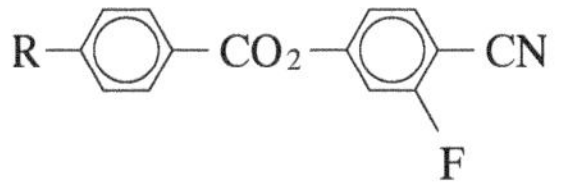 R —◯— CO_2 —◯— CN $\quad\quad\quad$ F	48.9		13

k_1 order:

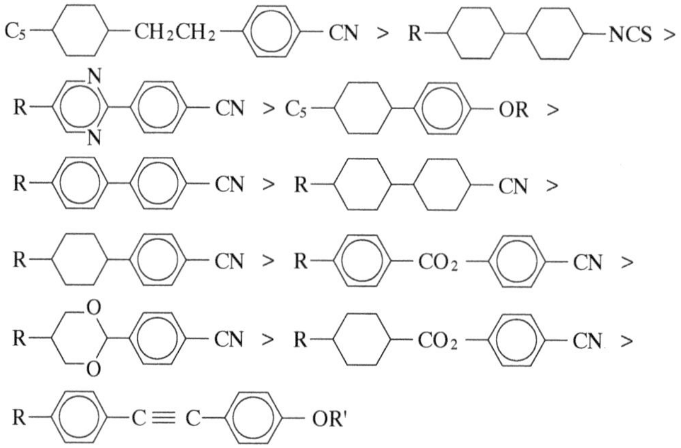

C$_5$—⬡—CH$_2$CH$_2$—⬡—CN > R—⬡—⬡—NCS >

R—[pyrimidine]—⬡—CN > C$_5$—⬡—⬡—OR >

R—⬡—⬡—CN > R—⬡—⬡—CN >

R—⬡—⬡—CN > R—⬡—CO$_2$—⬡—CN >

R—[dioxane]—⬡—CN > R—⬡—CO$_2$—⬡—CN >

R—⬡—C≡C—⬡—OR'

k_2 order:

R—⬡—CH$_2$CH$_2$—⬡—CN > R—[pyrimidine]—⬡—CN >

R—⬡—CO$_2$—⬡—CN ≈ R—⬡—⬡—CN >

R—⬡—⬡—CN ≈ R—⬡—⬡—CN,NCS >

R—⬡—CO$_2$—⬡—CN > R—⬡—⬡—OR' >

R—[dioxane]—⬡—CN > R—⬡—C≡C—⬡—OR'

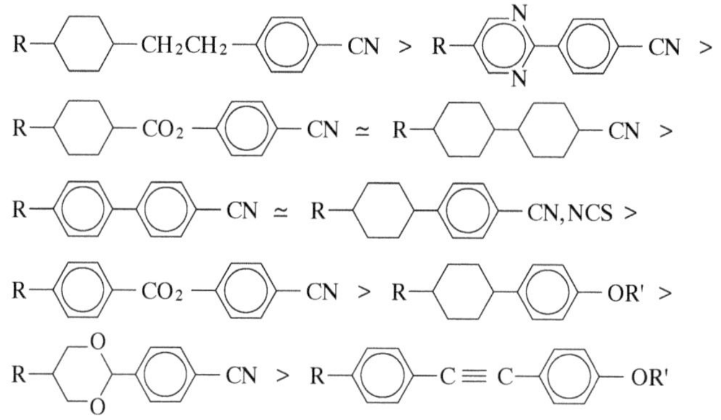

k_3 order:

R—⬡—CH$_2$CH$_2$—⬡—CN > R—⬡—⬡—NCS >

R—⬡—⬡—CN > R—⬡—⬡—CN >

R—⬡—CO$_2$—⬡—CN > R—⬡—CO$_2$—⬡—CN

R—⬡—⬡—CN > R—⬡—C≡C—⬡—OR' >

R—⬡—⬡—OR'

R—[dioxane]—⬡—CN > R—[pyrimidine]—⬡—CN

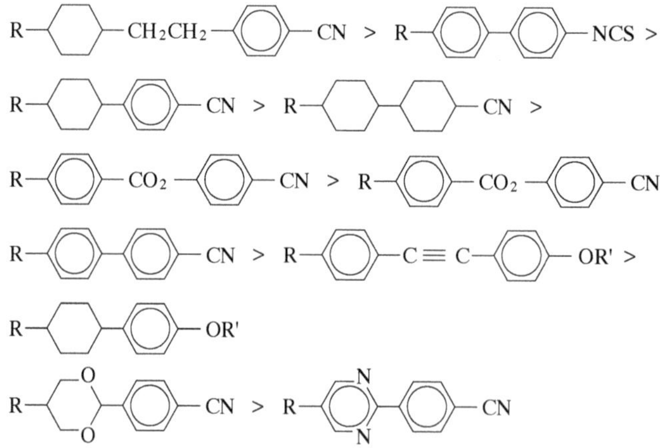

Table 10.21 *Some mesogens having a negative dielectric anisotropy.*

Mesogen	$-\Delta\varepsilon$	Ref.
 C_5O—⬡—CO_2—⬡—OC_8	0.34	13
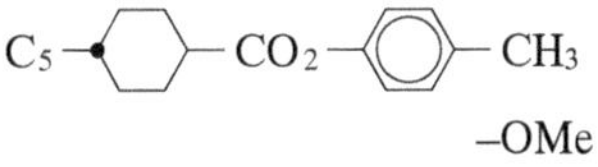 C_5—⬡—CO_2—⬡—CH_3 $-OMe$	0.4 0.7	185 185
 C_3—⬡—CO_2—⬡—C_3	0.9	185
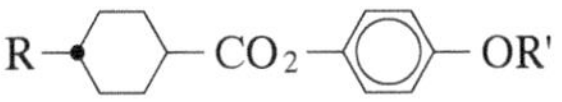 R—⬡—CO_2—⬡—OR'	0.96	194
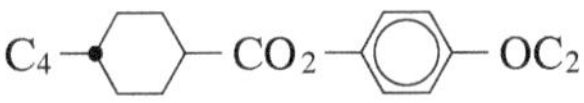 C_4—⬡—CO_2—⬡—OC_2	1.3	13
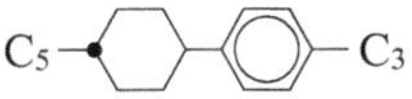 C_5—⬡—⬡—C_3	1.6	24b
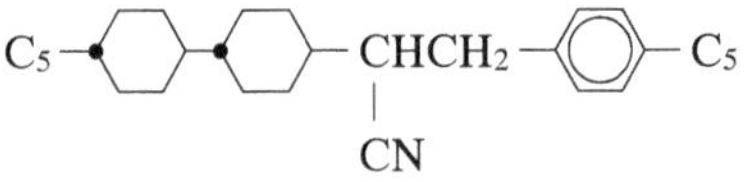 C_5—⬡—⬡—$CHCH_2$—⬡—C_5 $\mid$ CN	3.6	197
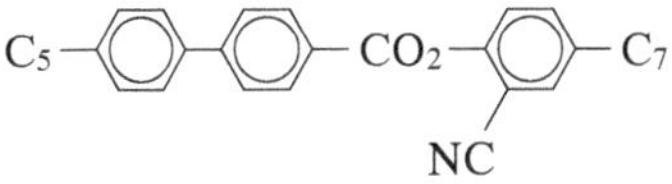 C_5—⬡—⬡—CO_2—⬡—C_7 NC	4	13
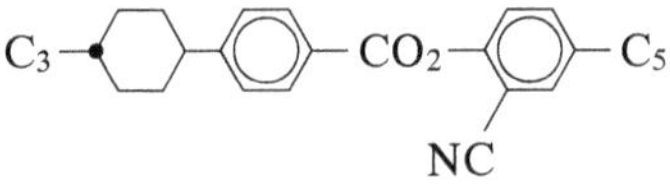 C_3—⬡—⬡—CO_2—⬡—C_5 NC	4	24a
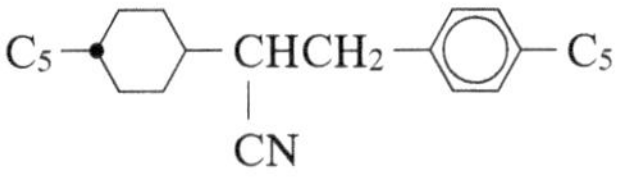 C_5—⬡—$CHCH_2$—⬡—C_5 $\mid$ CN	4.3	197
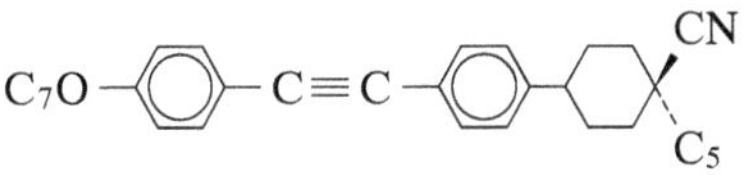 C_7O—⬡—$C\equiv C$—⬡—⬡$\genfrac{}{}{0pt}{}{CN}{C_5}$	4.0	186
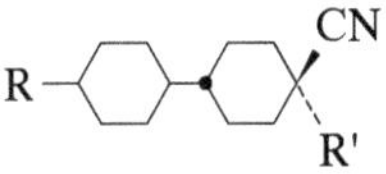 R—⬡—⬡$\genfrac{}{}{0pt}{}{CN}{R'}$	7 to 8	186
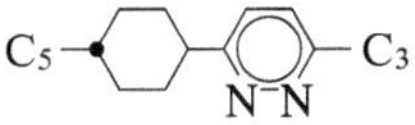 C_5—⬡—⬡($N\text{-}N$)—C_3	9.13	186

Table 10.21 (*cont.*)

Mesogen	$-\Delta\varepsilon$	Ref.
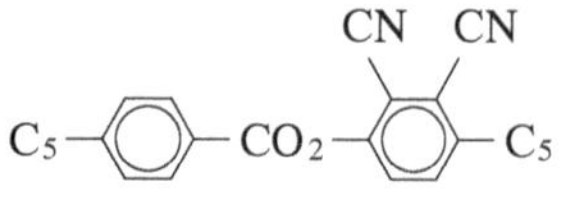 C_5—◯—CO_2—◯(CN CN)—C_5	11.5	186
C_3—◇—◯—CO_2—◯(CN CN)—OC_4	20	24a
C_4—◇—CO_2—◯—O_2C—◯—C_4 (NC CN)	22	13

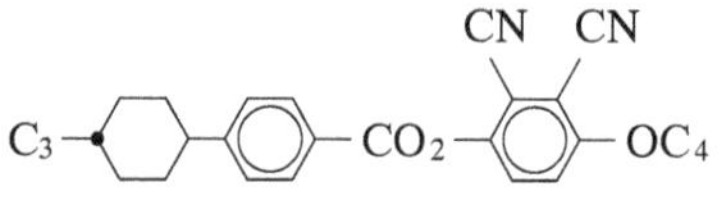

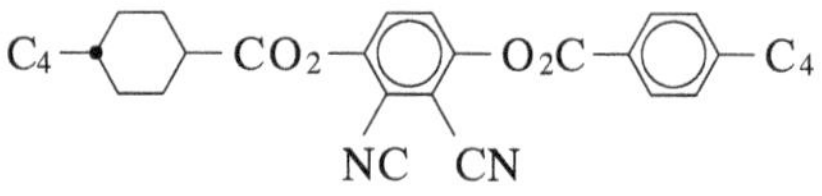

Mesogen	X	Y	$-\Delta\varepsilon$	Ref.
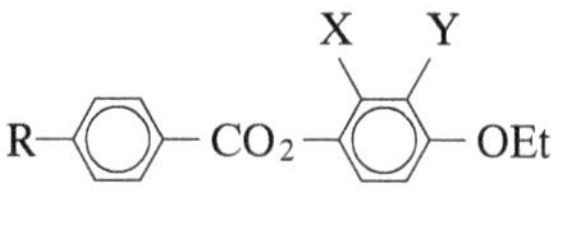 R—◯—CO_2—◯(X Y)—OEt	H	H	1.2	24a
	H	F	1.9	24a
	F	F	4.6	24a

The following order for k_3/k_1 in various ring systems having all other structural features identical applies [26a]:

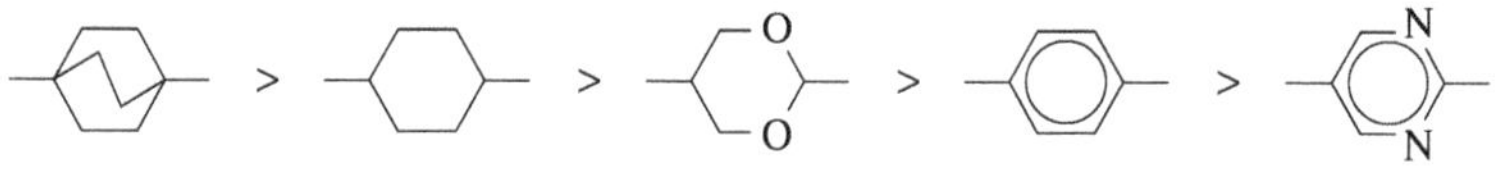

It is obvious from these data and those in Table 10.22 that more than just the ring system influences this ratio. Polar substituents along the molecular axis such as CN, NCS or a double bond increase this elastic constant ratio, as does a lateral F substituent. Studies have also shown asynergetic reduction in this constant when the combination of a terminal double bond, pyrimidine ring, and isocyanide (NCS) are either in the same structure, or in different components of a mixture [189]. The following mixture has a large $\Delta\epsilon$ (21.2) and with a small k_3/k_1 (0.65).

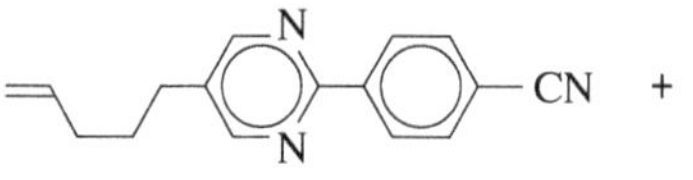

39%   36%

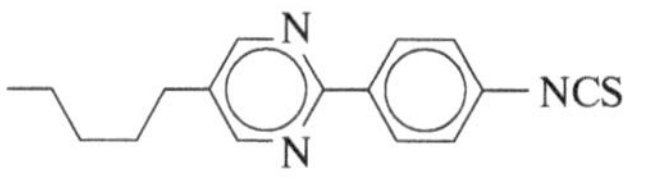

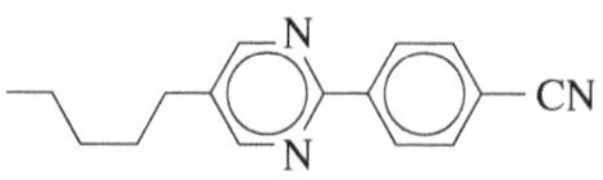

25%

Table 10.22 *Elastic constant ratio k_3/k_1 for a variety of mesogens.*

Mesogen	k_3/k_1
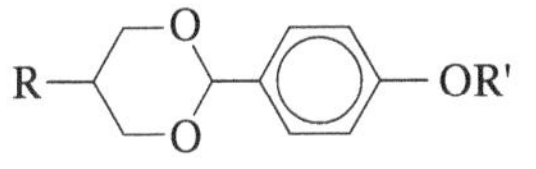 R—(dioxane)—⟨⟩—OR'	0.85
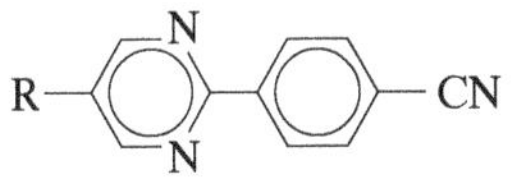 R—(pyrimidine)—⟨⟩—CN	1.16
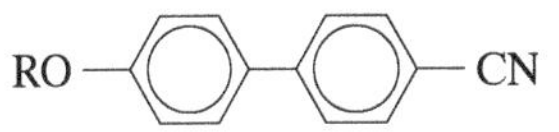 RO—⟨⟩—⟨⟩—CN	1.24
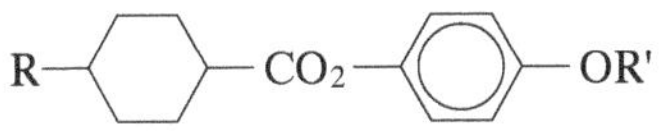 R—⟨cyclohexyl⟩—CO_2—⟨⟩—OR'	1.28
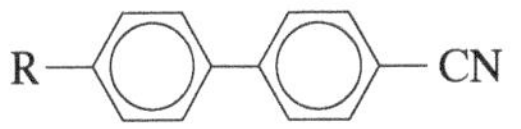 R—⟨⟩—⟨⟩—CN	1.34
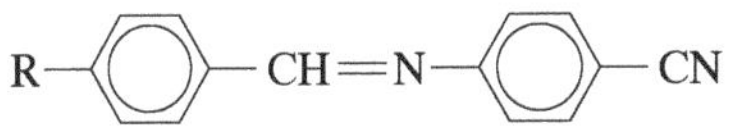 R—⟨⟩—CH=N—⟨⟩—CN	1.39
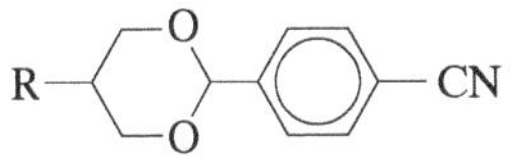 R—(dioxane)—⟨⟩—CN	1.43
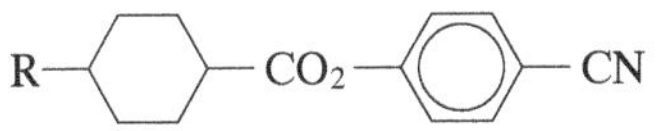 R—⟨cyclohexyl⟩—CO_2—⟨⟩—CN	1.55
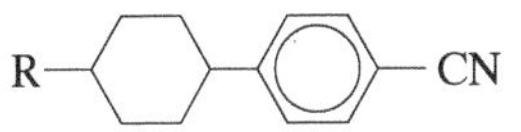 R—⟨cyclohexyl⟩—⟨⟩—CN	1.56
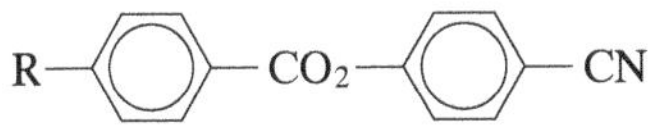 R—⟨⟩—CO_2—⟨⟩—CN	1.67

Source:
Data are from Ref. 194.

Enthalpy values

It has been reported that mesogens with large ΔH_f values show poor solubility in liquid crystalline hosts [24b]. For this reason, some ΔH_f values are given in Table 10.19. This statement is a confusing one, since it does not define the type of mesogenic host. It is true, as indicated, that a larger ΔH_f value indicates stronger forces between the molecules in the solid state. The difference in order between the solid state and a nematic phase is generally large and the ΔH for a crystal-to-nematic transition is, therefore, large. If it is not, then it means that the crystal structure is more fluid-like. If this is true, the interactions would be weaker and the molecules more likely to mix with other molecules. Of course, they are also likely to form solid complexes with these mesogenic hosts, especially if the interactions with the mesogenic hosts are stronger. Considering how easy it is to form solid solutions and complexes in mixtures (see the mixture section), insolubility often may actually be due to either of these phenomena which can be affected by terminal substituents, such as polar mixing with non-polar ones.

The melting enthalpy could also be smaller because the ordered solids melt to a slightly less ordered smectic rather than to a nematic phase. This is more likely to occur than a small ΔH for a CrK–N transition. Several solid modifications can also produce a low ΔH value for the least ordered crystal melting to a mesophase. Of course, all the solid ΔH values should be added together to a given ΔH_f, but all of them might not be seen in a standard DSC scan. Thus, care must be taken in considering only mesogens with small ΔH_f values for mixing with mesogenic hosts.

Optical birefringence (Δn)

With the development of polymer dispersed liquid crystal displays (PDLC), interest in nematic materials having a large optical birefringence has increased considerably. Table 10.19 shows some of the typical low values (generally < 0.20) found in common mesogens. Recently, nematic mesogens with $\Delta n > 0.20$ have been found in compounds containing more than one multiple bond in the connecting link, or at least one triple bond [199] as shown in Table 10.23.

Transition temperatures

Transition temperatures and the right mesophase are the starting point for determining if a material will even have a chance at becoming a display material. Without the right phase and low enough temperatures for this phase, nothing else matters. Most of the structural–property relationship studies deal with how to obtain the desired temperature range and mesophase by structural modifications. Such modification can also affect the other desirable physical display properties, and it is not unusual to find these effects countering each other. For example, replacing a benzene ring with a cyclohexane ring often produces a lower viscosity, but also gives a lower birefringence. Larger molecules can increase the clearing temperature, but will also increase the melting temperature and viscosity of aromatic compounds. Sometimes replacing aromatic rings with saturated ones

Table 10.23 *Optical birefringence (Δn) for*

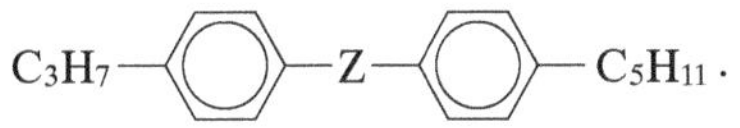

$$C_3H_7 - \bigcirc - Z - \bigcirc - C_5H_{11} \, .$$

Z	Δn
$- C \equiv C - CH_2C \equiv C -$	0.24
$- C \equiv C -$	0.27
$- CH = CH - C \equiv C -$	0.37
$- C \equiv C - C \equiv C -$	0.40
$- C \equiv C - CH = CH - C \equiv C -$	0.42

Source:
Data from Ref. 199.

can compensate for this. Finding compounds with low transition temperatures is more difficult than finding those with temperatures above room temperature. Low melting materials should melt to either a nematic or smectic phase rather than to an isotropic liquid. Increased conjugation along the molecular axis through aromatic rings and/or polar substituents can yield wider range nematic phases with high optical birefringence, but also with high transition temperatures.

Polar groups along the molecular axis increase $\Delta\epsilon$ and make it more positive, but may increase the viscosity, melting temperature, produce no mesophases, and/or cause mixing problems. Lateral polar groups with a dipole perpendicular to the molecular axis produce larger negative $\Delta\epsilon$ values, but also increase viscosity. This effect is smallest for a lateral fluorine substituent which can also increase the melting temperature with little effect on the viscosity. Many properties, such as $\Delta\epsilon$ or Δn, are not strongly affected by alkyl chain length like viscosity. For FLC displays, the addition of F atoms can produce wider range smectic-C and -A phases.

Stability, purity and purification

A mesogen can have the desired display properties but decompose under the conditions in which the display operates. A good example is the decomposition of anils which can hydrolyze to their starting materials – aldehydes and anilines – due to the equilibrium between the anils and these materials [200]:

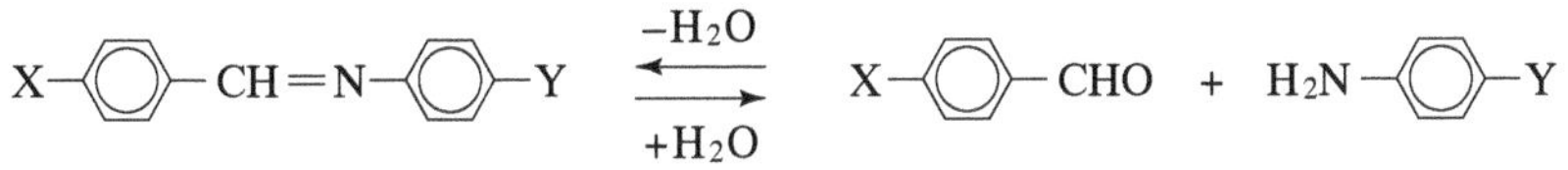

$$X - \bigcirc - CH = N - \bigcirc - Y \underset{+H_2O}{\overset{-H_2O}{\rightleftharpoons}} X - \bigcirc - CHO \; + \; H_2N - \bigcirc - Y$$

Both of these materials can be oxidized by air and an electric current. Although there are methods for preparing cells that minimize these problems, the development of the more stable cyanobiphenyls has virtually eliminated the use of anils, such as MBBA, in displays despite their room temperature nematic phases. Esters can also hydrolyze to their starting phenols and acids, but this generally is more difficult to achieve than anil hydrolysis. The exchange of esters [201] in mixtures can, however, occur and affect the transition temperature.

Another problem arises when more than one geometric isomer can form. Only one of these is linear and therefore mesogenic [13]. Some examples are as follows:

azos

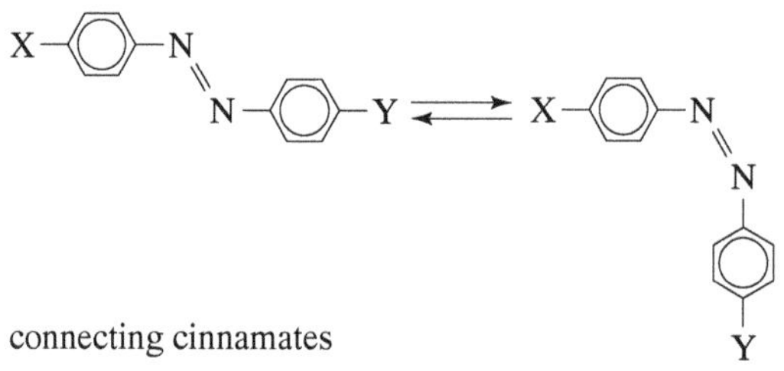

connecting cinnamates

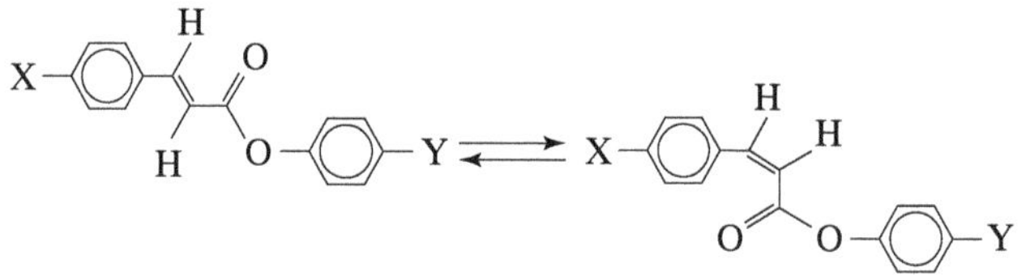

terminal cinnamates

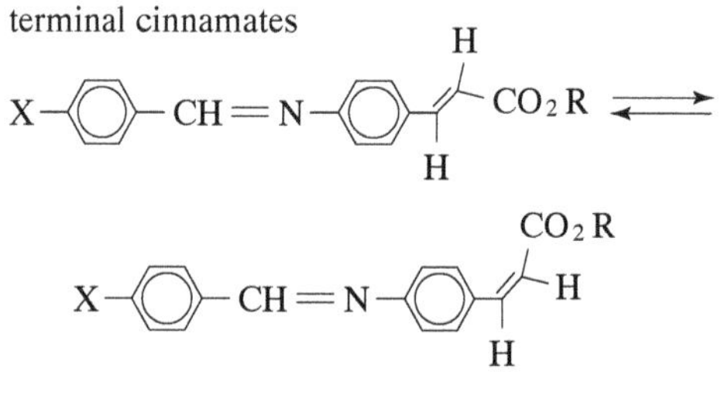

This is a problem when double bonded functional groups are used. However, double bonds in a terminal chain away from the benzene ring and the terminating carbon can form mesogens with lower transition temperatures that are useful in formulating liquid crystal mixtures (see terminal double bond section). A chain terminating in a double bond can polymerize, but has also been reported to yield stable mesogens [189]. Double bonds can, in some instances, migrate.

Mesogens can also be unstable to light or heat. A comparison of several mesogens shows the following near UV photostability [202]:

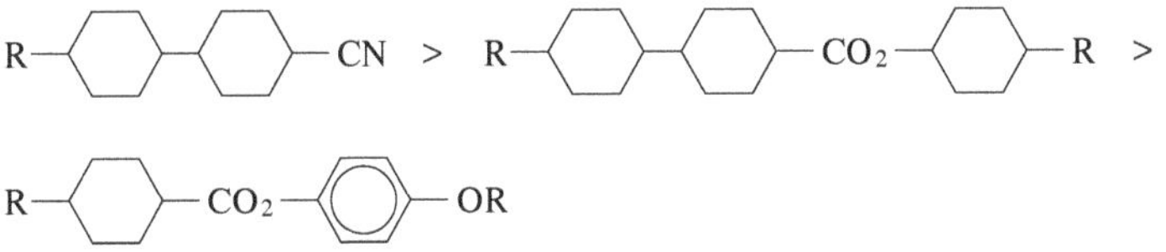

R⬡⬡CN > R⬡⬡CO_2⬡R >

R⬡CO_2◯OR

and

R⬡CO_2◯⬡R > R⬡◯CN >

RO◯◯CN > R◯(N,N)◯CN, R◯◯CN

R⬡(N,N)◯CN > NC◯CO_2◯R >

R◯◯◯CN

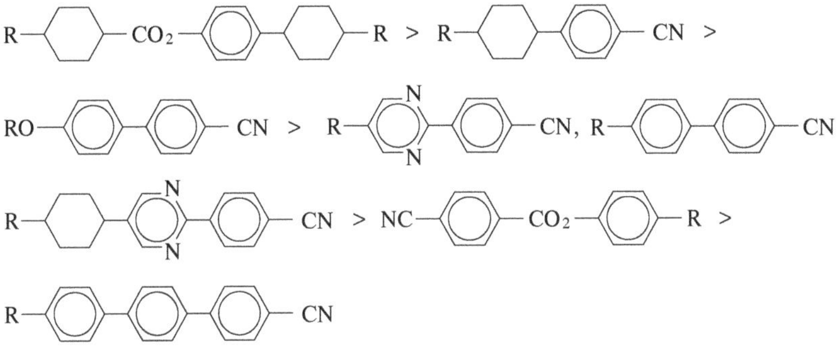

These data indicate that saturated six-membered rings produce more stable mesogens than unsaturated aromatic or heteroaromatic ones, and that the biphenyls are more stable then the phenylbenzoates or the terphenyls. The presence of double bonds can also produce photo-unstable mesogens [203]. Filters can sometimes be used on displays to overcome this problem.

Thermal stability studies on a few compounds suggest that the terminal group has a larger influence on stability than does the connecting group [204]. However, only the ester connecting group was compared with the cyclohexyl-phenyl compounds. A terminal NCS was found to be much less stable than the other terminal substituents studied.

Although the cyanobiphenyls have always been considered to be more stable than mesogens containing a connecting group, early commercial samples of E7 were found to be unstable due to the presence of impurities formed during the synthesis [205]. Organic impurities can be less stable than the mesogen, and lower the transition temperatures. Inorganic impurities can affect conduction of an electric current through the sample. Thus, it is essential that display materials be of a very high purity. Initially, zone refining was used to obtain such high purity [206]. It has the advantage of removing ionic as well as covalent impurities. Today, recrystallization, molecular distillation, flash chromatography and preparatory HPLC are all used. Dust particles can be removed using a filtration of the liquefied mesogen through a millipore filter. Ionic impurities can be removed by treating with a polyamic acid [207]. Methods are available for measuring the ion content [208].

The development of high performance liquid chromatography (HPLC) and high performance gel permeation chromatography (GPC) has made it possible to separate homologs of liquid crystals [209]. All mesogens containing alkyl chains will contain trace amounts of homologs, because the starting hydrocarbon chain materials contain traces of homologs. Since display materials are mixtures anyway, there is no reason to separate homologs for these materials. HPLC can also be used to detect impurities. Other useful methods for doing this are thin layer chromatography (TLC), nuclear magnetic resonance (NMR), mass spectrometry (MS), gas chromatography–mass spectrometry (GC-MS), differential scanning calorimetry (DSC) and hot-stage polarizing microscopy. The more methods that can be used on a sample, the more assurance that the mesogen has a high purity. These methods do not, however, detect ionic impurities which normally exist in most mesogens unless a procedure has been used to remove these. Totally dust free materials can only be prepared in a dust free environment i.e., a clean room. Another use for HPLC is the analysis of liquid crystal mixtures [210].

10.7 Mixtures

There is no known single mesogen that has all of the desired properties needed for displays. Such a material can be achieved only by preparing a mixture of mesogens which collectively have the desired properties. Several review articles on mixtures are available [14b, 18b, 24a, 124].

The primary concern in developing a useful display material is to obtain a mixture that has a very wide range mesophase with a melting temperature far below room temperature, and a clearing temperature much greater than 100 °C. Melting temperatures can be lowered by preparing eutectic mixtures. The composition of these can be calculated using the well known Schröder van Laar equation [211–213]:

$$T_i = \frac{\Delta H_f}{\Delta H_f / T_f - R \ln X_i},$$

where T_i = upper end of the melting range of component i in the mixture, ΔH_f = molar heat of fusion of pure component i, X_i = mole fraction of the component i in the mixture and R = gas constant. When this equation is satisfied for each component, the mixture will be a eutectic one, with the lowest possible melting temperature assuming that an ideal mixture is obtained. The melting temperature (T_f) and ΔH_f values can be obtained from DSC scans. Today, these calculations are done using a computer program making it possible to calculate eutectic compositions for as many components as desired. Usually, however, several simple eutectics are determined and then these are combined to calculate a new eutectic. The clearing temperature of a eutectic mixture can be calculated using the following formula:

$$T_{\text{clp}} = \sum_{i=1}^{n} (X_i T_{ci}).$$

If the observed melting temperature is higher than that calculated (indicating non-ideal behavior), experimental methods, such as partial freezing fractionations of the calculated mixture, are available to obtain a better eutectic composition [214]. Another approach is to use the original van Laar theory to calculate eutectic compositions [215]. Entire phase diagrams have also been calculated for several binary mixtures [216].

Tables of numerous binary mixture data have been published [217]. However, two components will usually not give a wide enough mesophase range to be useful as a display material [24a, 218]. Even a eutectic mixture of several homologs of one type of mesogen is not enough to achieve a useful mesophase range. Formulating a eutectic mixture that consists of both low and high melting mesogens is needed to produce wide enough mesophase ranges to meet the needs of most displays. For example, the best eutectic mixture of 4-alkyl/akoxcyanobiphenyls [218, 219] has a nematic range of 0–60 °C. The addition of 4-alkylcyanoterphenyls increases this range to 5–91 °C.

Still, formulating eutectic mixtures is not quite this simple. Although similar molecules are more likely to form ideal mixtures, they are also more likely to form solid solutions, which then do not yield the lower melting temperatures found in eutectic mixtures [212]. These more often occur with mesogens whose molecular lengths differ by only one CH_2 group [213]. The more different the component molecules are, the less ideal their behavior in a mixture. Mixing two polar or two non-polar mesogens will give a mixture that is closer to ideal than mixing a polar with a non-polar mesogen [220, 221]. Mixtures of the same type are also more likely to show additivity of the physical properties of the components, whereas mixtures of different types do not [220, 221]. Often, induced and/or enhanced mesophases occur in mixtures of polar with non-polar mesogens. These induced phases occur when the interactions between two different molecules are stronger than those between two identical ones. Molecular complexes can form with higher melting temperatures than those of either component and decreased nematic phase range.

Since cyanobiphenyls are polar molecules, mixtures with non-polar mesogens usually produce induced/enhanced mesophases often due to complex formation. Numerous examples have been reported with an enhanced smectic-A phase being observed the most often: with anils [221, 222], azoxy [223], esters [224, 225], tolanes [225], acids [226, 227], and dialkylaminobiphenyls [228]. Examples of the formation of enhanced solids have been reported [229–231], one of them being with MBBA [230]. Other smectic phases can also be induced [232, 233] and two eutectics can occur [229–232].

Mixtures of rods and disks have been studied in an attempt to obtain a thermotropic biaxial nematic phase [234]. However, only two immiscible nematic phases N_R (rods) and N_D (disks) were observed. Mixtures of biswallow-tails with rods or twins can form induced smectic-A and re-entrant nematic phases [134, 235]. Induced smectics have also been observed in mixtures of cetyl alcohol in orthophosphoric acid [236], and of a nematic polymer with a nematic twin [237].

Numerous studies of the effect of structure on mixtures have been done. Many of these are discussed in the review articles already mentioned. Extensive mixture studies have been done with esters/thioesters [150, 238] and a variety of other two benzene ring systems with a central connecting group [239]. Intermolecular interactions in mixtures have been reviewed [240]. Both chain length and molecular length have been implicated as factors affecting the production of induced mesophases. Studies of numerous binary mixtures having a wide variety of component structures, primarily with non-polar tails, suggest that two molecules differing in molecular length by $\sim 8\,\text{Å}$ or more can form induced nematic phases [241]. This is true if either no mesophases or only smectic phases occur in one or both of the components. Considering that molecules having a large difference in molecular lengths will not want to pack in a parallel manner into the layers needed to form smectic phases, it is understandable that they will instead prefer to pack end to end into a nematic phase. Increasing the chain length has been shown to increase the maximum clearing temperature of an induced smectic-A phase in a mixture of polar and non-polar components up to a certain length, after which it decreases [221, 242], possibly because an induced nematic phase becomes more desirable. Chain length has already been mentioned as affecting the ideality of a eutectic mixture. It has also been shown that eutectics are produced without the formation of mixed crystals in the edge regions of the diagram of state when the ratio of molecular lengths (shorter/longer) is equal to or less than ~ 0.80 [212].

Mixtures of a known mesogen with a mesogenic-like compound having no mesophase have been used to determine virtual transition temperatures by extrapolation [243]. A list of useful components for FLC display mixtures has been compiled [244]. The extensive use of mixture studies to identify mesophases in new mesogens is discussed in the characterization section (Chapter 2).

References

1. H. Kelker and B. Scheurle, *Angew. Chem.* **81**, 903 (1969); Internat. Edn p. 884.
2. G. W. Gray, *Mol. Cryst.* **1**, 333 (1966) and in *Liquid Crystals*, G. H. Brown (Ed.), Gordon and Breach, New York (1967), p. 129.
3. G. W. Gray, *Mol. Cryst. Liq. Cryst.* **7**, 127 (1969).
4. G. W. Gray, *Mol. Cryst. Liq. Cryst.* **21**, 161 (1973).
5. G. W. Gray and J. W. Goodby, *Ann. Phys. (Paris)* **3**, 123 (1978).
6. G. W. Gray, *Mol. Cryst. Liq. Cryst.* **63**, 3 (1981).
7. G. W. Gray, *Philos. Trans. R. Soc. London A* **309**, 77 (1983).
8. G. W. Gray, *Proc. R. Soc. London A* **402**, 1 (1985).
9. G. W. Gray, M. Hird, and K. J. Toyne, *Mol. Cryst. Liq. Cryst.* **204**, 91 (1991).
10. D. Demus and H. Zaschke, *Mol. Cryst. Liq. Cryst.* **63**, 129 (1981).
11. D. Demus and H. Sackmann, *Z. Chem.* **26**, 6 (1986)
12. D. Demus, *Mol. Cryst. Liq. Cryst.* **165**, 45 (1988).
13. D. Demus, *Liq. Cryst.* **5**, 75 (1989).
14. H. Kelker and R. Hatz, *Handbook of Liquid Crystals*, Verlag Chemie, Weinhein (1980): a. Chapter 2, p. 34; b. p. 372.
15. R. Eidenschink, *Mol. Cryst. Liq. Cryst.* **123**, 57 (1985).
16. *Advances in Liquid Crystal Research and Applications*, L. Bata (Ed.), Pergamon Press, Oxford (1980): a. V. V. Titov, p. 973; b. H. Zaschke, p. 1059; c. P. Admonenas, V. Butkus, J. Dauguila, J. Dienyte, and D. Girdziuanarte, p. 1029; d. L. Bata and A. Buka, p. 251.
17. L. Chu-Tsin, *Mol. Cryst. Liq. Cryst.* **74**, 25 (1981).
18. *Liquid Crystals and Plastic Crystals*, Vol. I, G. W. Gray and P. A. Winsor (Eds.), Ellis Horwoord Ltd, John Wiley & Sons, Inc., New York (1974: a. G. W. Gray, Chapter 4.1; b. J. S. Dave and R. A. Vora, Chapter 4.2.
19. G. Gray, *Adv. Liq. Cryst.* **2**, 1 (1976).
20. G. W. Gray, in *The Molecular Physics of Liquid Crystals*, G. R. Luckhurst and G. W. Gray (Eds.), Academic Press, London (1979), p. 1.
21. G. W. Gray, in *Polymer Liquid Crystals*, A. Ciferri, W. R. Krigbaum, and R. B. Meyer (Eds.), Academic Press, New York (1982), Chapter 1.
22. *Selected Topics in Liquid Crystals Research*, H.-D. Koswig (Ed.), Akademie-Verlag, Berlin (1990): a. H. J. Deutscher, R. Frach, C. Tschierske, and H. Zaschke, Chapter 1, p. 1; b. D. Demus and A. Hauser, Chapter 2, p. 19.
23. D. Demus, in *Liquid Crystals*, H. Stegenmeyer (Ed.), Springer-Verlag, New York (1994), Chapter 1.
24. *Liquid Crystals Applications and Uses*, Vol. I, B. Bahadur (Ed.), World Scientific, Singapore (1990): a. D. Coates, Chapter 3; b. L. Pohl and U. Finkenzeller, Chapter 4.
25. G. W. Gray, *Molecular Structure and the Properties of Liquid Crystals*, Academic Press, London, 1962.

26. *Thermotropic Liquid Crystals – Critical Reports on Applied Chemistry*, Vol. 22, G. W. Gray (Ed.), John Wiley & Sons, New York (1987): a. I. Sage, Chapter 3; b. D. Coates, Chapter 4.
27. P. Keller and L. Liebert, in *Liquid Crystals – Solid State Physics*, Supplement 14, L. Liebert (Ed.), Academic Press, New York (1978).
28. W. Kast in *Landolt–Börnstein Vol. II*, part 2a, 6th Edn., Springer-Verlag, Berlin-New York (1960, p. 266.
29. D. Demus, H. Demus, and H. Zaschke, 'Flüssige Kristalle in Tabellen,' VEB Deutscher Verlag für Grundstoffindustrie, Leipzig, 1974.
30. D. Demus and H. Zaschke, 'Kristallen in Tabellen II,' VEB Deutscher Verlag für Grundstoffindustrie, Leipzig, 1984.
31. A. Beguin, J. Billard, F. Bonamy, J. M. Bruisine, P. Cuvelier, J. C. Dubois, and P. LeBarny, *Mol. Cryst. Liq. Cryst.* **115**, 1 (1984).
32. V. Vill, a. *Database of Liquid Crystalline Compounds for Personal Computers*, LCI Publishers, Gmb Hi Gr, Hamburg (1995); b. Landolt–Börnstein New Series Group IV, *Macroscopic and Technical Properties of Matter, Vol. 7, Liquid Crystals*, J. Theim (Ed.) Springer-Verlag, New York (1992).
33. T. T. Blair, M. E. Neubert, M. Tsai, and C. Tsai, *J. Phys. Chem. Ref. Data* **20**, 189 (1991).
34. M. E. Neubert, L. T. Carlino, R. D'Sidocky, and D. L. Fishel, in *Liquid Crystals and Ordered Fluids*, Vol. 2, J. F. Johnson and R. S. Porter (Eds.), Plenum Publishing Corp., New York (1974), p. 293.
35. M. E. Neubert, L. T. Carlino, D. L. Fishel and R. M. D'Sidocky, *Mol. Cryst. Liq. Cryst.* **59**, 253 (1980).
36. A. L. Tsykalo, *Thermophysical Properties of Thermotropic Liquid Crystals*, Gordon and Breach Science Publishers, New York (1991), p. 405.
37. W. H. de Jeu and J. van der Veen, *Philips Res. Rep.* **27**, 172 (1972).
38. J. L. Galigne, *Acta Crystallogr.* **26B**, 1977 (1970).
39. J. Shashidhara Prasad and P. K. Rajalakskmi, *J. Phys. (Paris)* **40**, 309 (1979).
40. A. J. Leadbetter and M. A. Mazid, *Mol. Cryst. Liq. Cryst.* **51**, 85 (1978)
41. W. Thyen, F. Heinemann, and P. Zugenmaier, *Liq. Cryst.* **16**, 993 (1994).
42. A. J. Leadbetter and M. A. Mazid, *Mol. Cryst. Liq. Cryst.* **65**, 265 (1981).
43. P. A. C. Gane and A. J. Leadbetter, *Mol. Cryst. Liq. Cryst.* **78**, 183 (1981).
44. U. Baumeister, H. Hartung, and M. Jaskólski, *Mol. Cryst. Liq. Cryst.* **88**, 167 (1982).
45. P. Mandal, S. Paul, H. Schenk, and K. Goubitz, *Mol. Cryst. Liq. Cryst.* **135**, 35 (1986).
46. J. Chrusciel, B. Pniewska, and M. D. Ossowska-Chruściel, presented at the 15th ILCC, Budapest, 1994, abstract no. B-Sb p. 4.
47. I. H. Ibrahim, H. Paulus, M. Mokhles, and W. Haase, *Mol. Cryst. Liq. Cryst. Sci. Technol. A* **258**, 185 (1995).
48. W. Haase, H. Paulus, G. Strobl, and W. Hotz, *Acta Crystallogr.* **C47**, 2005 (1991).
49. L. Waltz, W. Haase, and I. H. Ibrahim, *Mol. Cryst. Liq. Cryst.* **200**, 43 (1991).
50. V. Baumeister, H. Hartung, and M. Jaskólski, *Cryst. Res. and Technol.* **17**, 153 (1983).
51. W. Haase, H. Paulus, and R. Pendialek, *Mol. Cryst. Liq. Cryst.* **100**, 211 (1983).
52. G. V. Vani, *Mol. Cryst. Liq. Cryst.* **99**, 21 (1983).
53. T. Hanemann, W. Haase, I. Svoboda, and H. Fuess, *Liq. Cryst.* **19**, 699 (1955).

54. T. Manisekaran, R. K. Bamezai, N. K. Sharma, and J. Shashidahara Prasad, *Mol. Cryst. Liq. Cryst. Sci. Technol. A* **268**, 45 (1995).
55. M. A. Kravers, V. I. Kilishov, A. I. Polishchuk, and A. S. Tolochko, *Sov. Phys. Crystallogr.* **37**, 375 (1992).
56. P. Mandal and S. Paul, *Mol. Cryst. Liq. Cryst.* **131**, 223 (1985).
57. S. Sinton and A. Pines, *Chem. Phys. Lett.* **76**, 263 (1980); J. W. Emsley, T. J. Horne, H. Zimmermann, G. Celebre, and M. Longeri, *Liq. Cryst.* **7**, 1 (1990); G. Celebre, M. Longeri, E. Sicilia, and J. W. Emsley, ibid. **7**, 731 (1990).
58. T. Kato and T. Uryu, *Mol. Cryst. Liq. Cryst.* **195**, 1 (1991).
59. Y. Sakurai, S. Takenaka, H. Miyoke, H. Morita, and T. Ikemoto, *J. Chem. Soc. Perkin Trans. III*, 1199 (1989).
60. S. Takenaka, Y. Sakurai, H. Takeda, T. Ikemoto, H. Miyaki, S. Kusabayashi, and T. Takagi, *Mol. Cryst. Liq. Cryst.* **178**, 103 (1990); Y. Sakurai, S. Takenaka, H. Suguira, S. Kusabayashi, Y. Nishikata, H. Terauchi, and T. Takogi, ibid. **201**, 95 (1991).
61. N. H. Tinh and C. Destrade, *Nouv. J. de Chimie*, **5**, 337 (1981); N. H. Tinh, *J. Chem. Phys.* **80**, 84 (1983); N. H. Tinh, H. Gasaparoux, J. Malthete, and C. Destrade, *Mol. Cryst. Liq. Cryst.* **114**, 19 (1984); S. Kumar and P. Patel, *Condensed Matter News* **2**, 9 (1993).
62. F. Hardouin, G. Sigaud, N. H. Tinh, and M. F. Achard, *J. Phys. (Paris) Lett.* **42**, 63 (1981); F. Hardouin, N. H. Tinh, M. F. Achard, and A. M. Levelut, ibid. **43**, 331 (1982).
63. M. J. S. Dewar and A. C. Griffin, *J. Am. Chem. Soc.* **97**, 6662 (1975).
64. M. A. Osman, *Z. Naturforsch.* **38a**, 693 (1983).
65. N. Carr, G. W. Gray, and S. M. Kelly, *Mol. Cryst. Liq. Cryst.* **129**, 301 (1985).
66. V. Reiffenrath and F. Schneider, *Z. Naturforsch.* **36a**, 1006 (1981).
67. G. W. Gray and S. M. Kelly, *Mol. Cryst. Liq. Cryst.* **75**, 95 (1981).
68. N. Carr, G. W. Gray, and S. M. Kelly, *Mol. Cryst. Liq. Cryst.* **130**, 265 (1985).
69. N. Carr, G. W. Gray, and D. G. McDonnell, *Mol. Cryst. Cryst.* **97**, 13 (1983).
70. R. Ch. Geivandov, S. O. Lastochkina, I. V. Goncharova, B. M. Bolotin, L. A. Karamysheva, T. A. Gervandova, A. V. Ivashchenko, and V. V. Titov, *Liq. Cryst.* **2**, 235 (1987).
71. M. Petrizilla, R. Buchecker, S. Lee-Schiederer, M. Schadt, and A. Germann, *Mol. Cryst. Liq. Cryst.* **148**, 123 (1987).
72. R. Eidenschink, D. Erdmann, J. Krause, and L. Pohl, *Angew. Chem. Int. Ed. Engl.* **17**, 133 (1978); R. Eidenschink, G. Haas, M. Römer, and B. S. Scheuble, ibid. **23**, 147 (1984); R. Eidenschink, and R. S. Scheuble, *Mol. Cryst. Liq. Cryst. Lett.* **3**, 33 (1986); R. Eidenschink, G. W. Gray, K. J. Toyne, and A. E. F. Wachter, ibid. **5**, 177 (1988).
73. H. J. Deutscher, S. Richter, and H. Zasche, *Mol. Cryst. Liq. Cryst.* **127**, 407 (1985).
74. K. Praefcke and D. Schmidt, *Chem. Ztg.* **105**, 61 (1981).
75. H. Schubert, R. Dehne, and V. Uhlig, *Z. Chem.* **12**, 219 (1972).
76. G. W. Gray and D. Lacey, *Mol. Cryst. Liq. Cryst.* **99**, 123 (1983).
77. J. S. Dave, G. Kurian, and B. C. Joshi, in *Liquid Crystals – Proceedings of an International Conference*, Raman Research Institute, Bangalore, 1979, S. Chandraskhar (Ed.) Heyden, London (1980) p. 549.
78. D. Demus, A. Hauser, A. Isenberg, M. Pohl, Ch. Selbmann, W. Weissflog, and S. Wieczorek, *Cryst. Res. Technol.* **20**, 1413 (1985).
79. C. M. Hudson, M. E. Neubert, A. M. Lackner, J. D. Margerum, and E. Sherman, *Liq. Cryst.* **19**, 871 (1995).

80. M. P. Burrow, G. W. Gray, D. Lacey, and K. Toyne, *Liq. Cryst.* **3**, 1643 (1988).
81. B. K. Sadashiva, *Mol. Cryst. Liq. Cryst.* **55**, 135 (1979).
82. M. E. Neubert, P. J. Wildman, M. J. Zawaski, C. A. Hanlon, T. L. Benyo, and A. de Vries, *Mol. Cryst. Liq. Cryst.* **145**, 111 (1987).
83. W. H. deJeu, J. van der Veen, and W. J. A. Goossens, *Solid State Commun.* **12**, 405 (1973).
84. J. Chruściel, S. Wróbel, H. Kresse, S. Urban, and W. Otowski, *Mol. Cryst. Liq. Cryst.* **192**, 107 (1990).
85. M. E. Neubert and T. T. Blair, unpublished results.
86. B. Heinrich and D. Guillan, *Mol. Cryst. Liq. Cryst. Sci. Technol. A* **268**, 21 (1955).
87. M. E. Neubert, T. T. Blair, and Y. Dixon-Polverine, *Mol. Cryst. Liq. Cryst.* **182b**, 269 (1990).
88. S. Gupta and S. Paul, *Mol. Cryst. Liq. Cryst. Sci. Technol. A* **260**, 483 (1995).
89. S. Marcělja, *J. Chem. Phys.* **60**, 3599 (1974).
90. W. Haase, H. Paulus, and I. H. Ibrahim, *Mol. Cryst. Liq. Cryst.* **107**, 377 (1984).
91. K. Hori, Y. Koma, A. Uchida, and Y. Okashi, *Mol. Cryst. Liq. Cryst.* **225**, 15 (1993).
92. R. Centore, M. R. Ciajolo, A. Roviello, A. Sirigu, and A. Tuzi, *Liq. Cryst.* **9**, 873 (1991).
93. E. N. Keller, E. Nachaliel, D. Davidov, and C. Böffel, *Phys. Rev.* **34**, 4363 (1986).
94. S. Hsi, H. Zimmermann, and Z. Luz, *J. Chem. Phys.* **69**, 4126 (1978).
95. J. R. Lalanne, B. Lemaire, J. Rouch, C. Vaucamps, and A. Proutiere, *J. Chem. Phys.* **73**, 1927 (1980).
96. D. A. Dunmur and W. H. Miller, *J. Phys. (Paris) Colloq. C3* **40**, 141 (1979).
97. A. Pines, D. J. Ruben, and S. Allison, *Phys. Rev. Lett.* **33**, 1002 (1974).
98. J. van der Veen, W. H. de Jeu, M. W. M. Wanninkkof, and C. A. M. Tienhoven, *J. Phys. Chem.* **77**, 2153 (1973).
99. M. Tokahashi, S. Mita, and S. Kondo, *Mol. Cryst.* **147**, 99 (1987).
100. M. E. Neubert, J. P. Ferrato, and R. E. Carpenter, *Mol. Cryst. Liq. Cryst.* **53**, 229 (1979).
101. C. W. Bunn, *J. Polym. Sci.* **16**, 323 (1955).
102. R. F. Bryan, *Z. Strukt. Khim.* **23**, 154 (1982), Engl. transl. p. 128.
103. R. F. Bryan, *J. Chem. Soc. (B)* 1311 (1967); R. F. Bryan and J. J. Jenkins, *J. Chem. Soc.* 1171 (1975); R. F. Bryan and P. Hartley, *Mol. Cryst. Liq. Cryst.* **62**, 259 (1980); R. F. Bryan, P. Hartley, R. W. Miller, and M.-S. Shen, ibid. **62**, 281 (1980); R. F. Bryan, P. Hartley, and R. W. Miller, ibid. **62**, 311 (1980); R. F. Bryan and K. A. Woods, *Trans Am. Crystallogr. Assoc.* **20**, 149 (1984); J. D. Bunning and J. E. Lydon, *J. Chem. Soc. Perkin Trans. II* 1621 (1979).
104. E. L. Eliel, S. H. Wilen, and L. N. Mander, *Stereochemistry of Carbon Compounds*, John Wiley and Sons, Inc., New York (1994), a. pp. 104–106; b. Chapter 7; c. pp. 214–217.
105. R. B. Meyer, L. Liebert, L. Strzelecki, and P. Keller, *J. Phys. (Paris)* **36**, L69 (1975) and P. Keller, *Ann. Phys.* **3**, 139 (1978).
106. J. W. Goodby, *J. Mater. Chem.* **1**, 307 (1991); S. T. Lagerwall, B. Otterholm, and K. Sharp, *Mol. Cryst. Liq. Cryst.* **152**, 503 (1987).
107. J. W. Goodby, A. J. Slaney, C. J. Booth, I. Nishiyama, J. D. Vuijk, P. Styring, and K. J. Toyne, *Mol. Cryst. Liq. Cryst. Sci. Technol. A* **243**, 231 (1994).
108. J. W. Goodby, R. Blinc, N. A. Clark, S. T. Lagerwall, M. A. Osipov, S. A.

Pikin, T. Sakurai, K. Yoshino, and B. Žekő, *Ferroelectric Liquid Crystals: Principles, Properties and Applications in Ferroelectricity and Related Phenomena*, Vol. 7, Gordon and Breach Science Publishers, Philadelphia (1991).

109. D. M. Walba and N. A. Clark, *Ferroelectrics* **84**, 65 (1988); D. M. Walba, H. A. Razavi, A. Horiuchi, K. F. Eidman, B. Otterholm, R. C. Haltiwanger, N. A. Clark, R. Shao, D. S. Parmer, M. D. Wand, and R. T. Volhra, ibid. **113**, 21 (1991).

110. Y. Aoki and H. Noheria, *Chem. Lett.* 113 (1993); E. Górecka, W. Pyżuk, and J. Mieczkowski, *Mol. Cryst. Liq. Cryst. Sci. Technol. A* **249**, 33 (1994); Y. Aoki and H. Nohira, *Liq. Cryst.* **18**, 2197 and **19**, 15 (1995); C. Imrie and C. Loubser, *J. Chem. Soc. Chem. Commun.* 2159 (1994).

111. S. M. Kelly, M. Schadt, and H. Serberle, *Liq. Cryst.* **11**, 761 (1992); M. Marcus, A. Omenat, J. L. Serrano, T. Sierra, and A. Ezcurra, *Adv. Mater.* **4**, 285 (1992); M. Marcus, A. Omenat, and J. L. Serrano, *Liq. Cryst.* **13**, 843 (1993); I. Nishiyama, H. Ishizuka, and A. Yoshizawa, *Ferroelectrics* **147**, 193 (1993); A. Yoshizawa and I. Nishiyama, *J. Mater. Chem.* **4**, 449 (1994); *Mol. Cryst. Liq. Cryst. Sci. Technol. A* **260**, 403 (1995); A. Yoshizawa, Y. Soeda, and I. Nishiyama, *J. Mater. Chem.* **5**, 675 (1995); H. Ishizuka, I. Nishiyama, and A. Yoshizawa, *Liq. Cryst.* **18**, 775 (1995); Y. Zuzuki, T. Isozaki, T. Kusumato, and T. Hiyema *Chem. Lett.* 719 (1995).

112. W. Elser and R. D. Ennulat, *Adv. Liq. Cryst.* **2**, 73 (1976).

113. H. W. Gibson, in *The Fourth State of Matter*, F. D. Saeva (Ed.), Marcel Dekker, New York (1979), Chapter 3.

114. J. C. Bhatt, S. S. Keast, M. E. Neubert, and R. G. Petschek, *Liq. Cryst.* **18**, 367 (1995).

115. M. E. Neubert, K. Leung, S. J. Laskos Jr., M. C. Ezenyilimba, M. R. Jirousek, D. Leonhardt, B. A. Williams, and B. Ziemnicka-Merchant, *Mol. Cryst. Liq. Cryst.* **166**, 181 (1989); M. E. Neubert and I. G. Shenouda, ibid. **205**, 29 (1991).

116. M. E. Neubert, D. Leonhardt, and S. Sabol-Keast, *Mol. Cryst. Liq. Cryst.* **172**, 227 (1989).

117. D. Coates, *Liq. Cryst.* **2**, 63 (1987).

118. H. J. Deutscher, H. Schubert, C. Seidel, D. Demus, and H. Kresse, U.S. Patent 4293434 (1981).

119. M. E. Neubert, S. S. Keast, D. G. Abdallah Jr., and C. E. Law, unpublished results.

120. R. M. D'Sidocky, Ph.D. Dissertation, Kent State University (1978).

121. M. C. Ezenyilimba, M. S. Thesis, Kent State University (1989).

122. M. E. Neubert, S. S. Keast, M. C. Ezenyilimba, C. A. Hanlon, and W. C. Jones, *Mol. Cryst. Liq. Cryst. Sci. Technol. A* **237**, 193 (1993).

123. J. E. Lydon, and C. J. Coakley, *J. Phys. (Paris) Colloq. Cl* **36**, 45 (1975) and A. J. Leadbetter, ibid. 37.

124. D. Guillon, P. E. Cladis, and J. Stamatoff, *Phys. Rev. Lett.* **41**, 1598 (1978); P. E. Cladis, D. Guillon, W. B. Daniels, and A. C. Griffin, *Mol. Cryst. Liq. Cryst.* **56**, 89 (1979); P. E. Cladis, D. Guillon, F. R. Bouchet, and P. L. Finn, *Phys. Rev. A* **23**, 2594 (1981); S. Chandrasekhar, *Mol. Cryst. Liq. Cryst.* **124**, 1 (1985); P. E. Cladis, ibid. **165**, 85 (1988); T. A. Labko and B. I. Ostrovskii, *Mol. Mats.* **1**, 99 (1992).

125. E. P. Januilis Jr., J. G. Novack, G. A. Papapolymerou, M. Tristani-Kendra, and W. A. Huffmann, *Ferroelectrics* **85**, 375 (1988); M. Koden, K. Nakagawa, Y. Ishii, F. Funada, M. Matsuura, and K. Awane, *Mol. Cryst.*

Liq. Cryst. Lett. **6**, 185 (1989); T. Doi, Y. Sakurai, A. Tamatane, S. Takenaka, S. Kusabayashi, Y. Nishihata, and H. Terauchi, *J. Mater. Chem.* **1**, 169 (1991); Y. H. Chiang, A. E. Ames, R. A. Gaudianna, and T. G. Adams, *Mol. Cryst. Liq. Cryst.* **208**, 85 (1991); H. T. Nguyen, G. Sigaud, M. F. Achard, F. Hardouin, R. J. Twieg, and K. Betterton, *Liq. Cryst.* **10**, 389 (1991); R. Twieg, K. Betterton, R. Di Pietro, D. Gravert, C. Nguyen, H. T. Nguyen, A. Babeau, C. Destrade, and G. Sigaud, *Mol. Cryst. Liq. Cryst. Sci. Technol. A* **217**, 201 (1992).

126. M. J. Brienne, J. Gabard, J. M. Lehn, and I. Stibor, *J. Chem. Soc. Commun.* 1868 (1989); J. Malthete, A. M. Levelut, and L. Liébert, *Adv. Mater.* **4**, 37 (1992); V. Beginn and G. Lattermann, *Mol. Cryst. Liq. Cryst. Sci. Technol. A* **241**, 215 (1994).

127. C. M. Paleos and D. Tisiouras, *Angew. Chem. Int. Ed. Engl.* **34**, 1696 (1995).

128. Y. Matsunaga and M. Terada, *Mol. Cryst. Liq. Cryst.* **141**, 321 (1986).

129. C. B. McArdle, *Side-Chain Liquid Crystalline Polymers*, Chapman & Hall, New York (1989); A. Cicerri, *Liquid Crystallinity in Polymers: Principles and Fundamental Properties*, VCH Publishers, Inc., New York (1991), V. P. Shibaev and L. Lam (Eds.), *Liquid Crystalline and Mesomorphic Polymers*, Springer-Verlag, New York (1994).

130. M. Petrzilka, *Mol. Cryst. Liq. Cryst.* **131**, 109 (1985); M. Schadt, M. Petrzilka, R. R. Gerber, and A. Villager, *ibid.* **122**, 241 (1985).

131. M. Petrizilka and A. Germann, *Mol. Cryst. Liq. Cryst.* **131**, 327 (1985); S. M. Kelly, *Liq. Cryst.* **14**, 675 (1993); S. M. Kelly and J. Fünfschilling, *J. Mater. Chem.* **3**, 953 (1993).

132. D. J. Dyer and D. M. Walba, *Chem. Mater.* **6**, 1096 (1994).

133. W. Weissflog, A. Wiegeleben, S. Diele, and D. Demus, *Cryst. Res. Technol.* **19**, 583 (1984); W. Weissflog, G. Pelzl, H. Kresse, and D. Demus, *ibid.* **23**, 1259 (1988); S. Diele, S. Manke, W. Weissflog, and D. Demus, *Liq. Cryst.* **4**, 301 (1989), S. Heinemann, R. Paschke, and H. Kresse, *ibid.* **13**, 373 (1993); H. Kresse, S. Heinmann, R. Paschke, and W. Weissflog, *Ber. Bunsenges. Phys. Chem.* **97**, 1337 (1993); S. Haddawi, S. Diele, H. Kresse, G. Pelzl, W. Weissflog, and A. Wiegeleben, *Liq. Cryst.* **17**, 191 (1994); W. Weissflog, G. Pelzl, I. Lethko, and S. Diele, *Mol. Cryst. Liq. Cryst. Sci. Technol. A* **260**, 157 (1995); W. Weissflog, I. Letko, G. Pelzl, and S. Diele, *Liq. Cryst.* **18**, 867 (1995).

134. S. Diele, K. Ziebarth, G. Pelzl, D. Demus, and W. Weissflog, *Liq. Cryst.* **8**, 211 (1990); G. Pelzl, S. Diele, K. Ziebarth, W. Weissflog, and D. Demus, *ibid.* **8**, 765 (1990).

135. J. Malthete, J. Canceill, J. Gabard, and J. Jacques, *Tetrahedron* **37**, 2823 (1981).

136. M. Petrzilka, *Mol. Cryst. Liq. Cryst.* **111**, 329 (1984).

137. M. E. Neubert, C. Colby, M. C. Ezenyilimba, M. R. Jirousek, D. Leonhardt, and K. Leung, *Mol. Cryst. Liq. Cryst.* **154**, 127 (1988); M. E. Neubert, R. B. Sharma, C. Citano, M. R. Jirousek, and J. L. Paulin, *ibid.* **196**, 145 (1991).

138. M. E. Neubert, F. C. Herlinger, M. R. Jirousek, and A. deVries, *Mol. Cryst. Liq. Cryst.* **139**, 299 (1986); M. E. Neubert, K. Leung, M. R. Jirousek, M. C. Ezenyilimba, S. Sabol-Keast, B. Ziemnicka-Merchant, and R. B. Sharma, *ibid.* **197**, 21 (1991); M. E. Neubert, C. M. Citano, M. C. Ezenyilimba, M. R. Jirousek, S. Sabol-Keast, and R. B. Sharma, *ibid.* **206**, 103 (1991); M. E. Neubert and I. G. Shenouda, *Mol. Cryst. Liq. Cryst. Sci. Technol. A* **210**, 185 (1992).

139. M. E. Neubert, S. S. Keast, M. C. Ezenyilimba, R. B. Greer, W. C. Jones,

D. Leonhardt, and I. Shenouda, *Mol. Cryst. Liq. Cryst. Sci. Technol. A* **237**, 47 (1993).

140. M. E. Neubert, S. S. Keast, Y. Dixon-Polverine, F. Herlinger, M. R. Jirousek, K. Leung, K. Murray, and J. Rambler, *Mol. Cryst. Liq. Cryst. Sci. Technol. A* **250**, 109 (1994).

141. D. Coates, *Liq. Cryst.* **2**, 423 (1987).

142. D. G. McDonnell, E. P. Raynes, and R. A. Smith, *Mol. Cryst. Liq. Cryst.* **123**, 169 (1985); J. E. Fearon, G. W. Gray, I. D. Ifill, and K. J. Toyne, ibid. **124** 89 (1985); G. W. Gray, M. Hird, A. D. Ifill, W. E. Smith, and K. J. Toyne, *Liq. Cryst.* **19**, 77 (1995).

143. Y. Goto, T. Ogawa, S. Sawada, and S. Sugimori, *Mol. Cryst. Liq. Cryst.* **209**, 1 (1991); H. Takatsu, K. Takeuchi, M. Sasaki, H. Ohnishi, and M. Schadt, ibid. **206**, 159 (1991); I. Inoi, in *Organofluorine Chemistry, Principles and Commercial Applications*, R. E. Banks, B. E. Smart, and J. C. Tatlow (Eds.), Plenum Press, New York (1994), Chapter 12.

144. M. Hird, G. W. Gray, and K. J. Toyne, *Mol. Cryst. Liq. Cryst.* **206**, 187 (1991).

145. I. Teucher, C. M. Paleos, and M. M. Labes, *Mol. Cryst. Liq. Cryst.* **11**, 187 (1970); H. Hirata, S. N. Wasman, I. Teucher, and M. M. Labes, ibid. **20**, 343 (1973); P. M. Bolotin, D. S. Pileeva, N. B. Etingen, and Y. S. Narkevich, *J. Org. Chem. USSR* **21**, 326 (1985); *Zh. Organ. Khim.* **21**, 362 (1985).

146. A. Hallsby, A. Nilsson, and B. Otterholm, *Mol. Cryst. Liq. Cryst. Lett.* **82**, 61 (1982); M. Ghedini, D. Pucci, E. Cesarotti, O. Francescangeli, and R. Bartolino, *Liq. Cryst.* **15**, 331 (1993).

147. M. S. Ho, B. M. Fung, and J. P. Bayle, *Mol. Cryst. Liq. Cryst.* **225**, 383 (1993).

148. J. L. Serrano, *Metallomesogens, Synthesis, Properties, and Applications*, VCH, New York (1996).

149. H. J. Deutscher, M. Körber, H. Altmann, and H. Schubert, *J. Prakt. Chem.* **321**, 969 (1979); W. Weissflog and D. Demus, *Cryst. Res. Technol.* **18**, K21 (1983) and **19**, 55 (1984); D. Demus, A. Hauser, C. Selbmann, and W. Weissflog, ibid. **19**, 271 (1984); D. Demus, A. Hauser, A. Isenberg, M. Pohl, C. Selbmann, W. Weissflog, and S. Wieczorek, ibid. **20**, 1413 (1985); D. Demus, S. Diele, A. Hauser, I. Latif, C. Selbmann, and W. Weissflog, *Cryst. Res. Technol.* **20**, 1547 (1985); W. Weissflog and D. Demus, *Mol. Cryst. Liq. Cryst.* **129**, 235 (1985); S. Diele, K. Roth, and D. Demus, *Cryst. Res. Technol.* **21**, 97 (1986); W. Weissflog, G. Pelzl, and D. Demus, ibid. **21**, 117 (1986).

150. *Liquid Crystals and Ordered Fluids*, Vol. 4, A. C. Griffin and J. F. Johnson (Eds.), Plenum Press, New York (1984): a. R. J. Cox, W. Volkensen, and B. L. Dawsen, p. 33; b. A. C. Griffin, G. A. Cambell, and W. E. Hughes, p. 1077; c. J. C. Dubois and J. Billard, p. 1043; d. J. D. Margerum, S. M. Wong, J. E. Jensen, and C. van Ast, p. 111.

151. J. Malthete, J. Billard, and J. Jacques, *Compt. Rend. Acad. Sci. Paris. C* **281**, 333 (1975); J. Malthete, *Compt. Rend. Acad. Sci. Paris. Ser. II*, **296**, 435 (1983).

152. A. C. Griffin, S. F. Thames and M. S. Bonner, *Mol. Cryst. Liq. Cryst.* **34**, 135 (1977); A. C. Griffin, M. L. Steele, J. F. Johnson, and G. J. Bertolini, *Nouv. J. Chim.* **3**, 697 (1979); A. C. Griffin, N. W. Buckley, W. E. Hughes, and D. L. Wertz, *Mol. Cryst. Liq. Cryst.* **64**, 139 (1981).

153. P. Berdagné, J. P. Bayle, M. S. Ho, and B. M. Fung, *Liq. Cryst.* **14**, 667 (1993).

154. S. Diele, W. Weissflog, G. Pelzl, H. Manke, and D. Demus, *Liq. Cryst.* **1**, 101

(1986); H. Kresse and W. Weissflog, *Phys. Stat. Sol. A* **106**, K89 (1988); W. Weissflog and D. Demus, *Liq. Cryst.* **3**, 275 (1988); W. Weissflog, ibid. **5**, 111 (1989); H. Dehne, A. Roger, D. Demus, S. Diele, H. Kresse, G. Pelzl, W. Wedler, and W. Weissflog, ibid. **6**, 47 (1989); S. Takenaka, H. Morito, M. Iwano, S. Kusabayashi, T. Ikemoto, Y. Sakurai, and H. Miyake, *Mol. Cryst. Liq. Cryst.* **166**, 157 (1989); S. Takanaka, Y. Masuda, M. Iwano, H. Morita, S. Kusabayashi, H. Sugiura, and T. Ikemoto, ibid. **168**, 111 (1989); W. Weissflog, D. Demus, and S. Diele, ibid. **191**, 9 (1990); S. Diele, A. Mädicke, K. Knauft, J. Neutzler, W. Weissflog, and D. Demus, *Liq. Cryst.* **10**, 47 (1991); T. Masuda and Y. Matsunaga, *Bull. Chem. Soc. Jpn* **64**, 2192 (1991); M. Loddoch, G. Marowsky, H. Schmid, and G. Heppke, *Appl. Phys. B* **591** (1994); D. Braun, M. Reubold, L. Schneider, M. Wegminn, and J. H. Wendorff, *Liq. Cryst.* **16**, 429 (1994).

155. M. Koboshita, Y. Matsunaga, and H. Matsuzaki, *Mol. Cryst. Liq. Cryst.* **199**, 319 (1991); H. Matzuzaki and Y. Matsunaga, *Liq. Cryst.* **14**, 105 (1993).

156. J. I. Jin, C. S. Kang, and B. Y. Chung, *Bull. Korean Chem. Soc.* **11**, 245 (1990); J. Jin, B. Y. Chung, J. K. Choi, and B. W. Jo, ibid. **12**, 189 (1991); D. Janietz, F. Sundholm, J. Leppänen, H. Karhinen, and M. Bauer, *Liq. Cryst.* **13**, 499 (1993).

157. A. C. Griffin and T. R. Britt, *J. Am. Chem. Soc.* **103**, 4957 (1981); J. Jin, Y. S. Chung, J. S. Kang, and R. W. Lenz, *Mol. Cryst. Liq. Cryst.* **82**, 261 (1982); J. Jin, Y. S. Chung, R. W. Lenz, and C. Ober, *Bull. Korean Chem. Soc.* **4**, 143 (1983); J. Jin, J. S. Kang, B. W. Jo, and R. W. Lenz, ibid. **4**, 176 (1983); J. A. Baglione, A. Roviello, and A. Sirigu, *Mol. Cryst. Liq. Cryst.* **106**, 169 (1984); C. Aguilera, S. Ahmad, J. Bartulin, and H. J. Müller, ibid. **162B**, 277 (1988); L. V. Azároff and A. R. Saini, *Polym. Prepr.* **30**, 515 (1989), C. T. Imrie, *Liq. Cryst.* **6**, 391 (1989); J. Jin, B. Y. Chung, and J. H. Park, *Bull. Korean Chem. Soc.* **12**, 583 (1991); B. W. Jo, J. K. Choi, M. S. Bang, B. Y. Chung, and J. Jin, *Chem. Mater.* **4**, 1403 (1992); R. W. Date, C. T. Imrie, G. R. Luckhurst, and J. M. Seddon, *Liq. Cryst.* **12**, 203 (1992); D. W. Bruce and M. D. Hall, *Mol. Cryst. Liq. Cryst.* **250**, 373 (1994); G. Kossmehl and F. D. Hoppe, *Mol. Cryst. Liq. Cryst. Sci. Technol. A* **257**, 169 (1994).

158. G. S. Attard, C. T. Imrie, and F. E. Karasz, *Chem. Mater.* **4**, 1246 (1992).

159. A. C. Griffin and S. R. Vaidya, *Liq. Cryst.* **3**, 1275 (1988); J. L. Hogan, C. T. Imrie and G. R. Luckhurst, ibid. **3**, 645 (1988); G. S. Attard, S. Garnett, C. G. Hickman, C. T. Imrie, and L. Taylor, ibid. **7**, 495 (1990); J. Jin, H. S. Kim, J. W. Shin, B. Y. Chung, and B. W. Jo, *Bull. Korean Chem. Soc.* **11**, 209 (1990); G. S. Attard, R. W. Date, C. T. Imrie, G. R. Luckhurst, S. J. Roskilly, J. M. Seddan and L. Taylor, *Liq. Cryst.* **16**, 529 (1994); F. Hardouin, M. F. Achard, J. I. Jin, and Y. K. Yun, *J. Phys. II (Paris)* **5**, 927 (1995).

160. K. Sugiyama, K. Kato, and K. Shiraishi, *Chem. Express* **8**, 189 (1993); A. T. M. Marcelis, A. Koudus, and E. J. R. Sudhölter, *Liq. Cryst.* **18**, 851 (1995).

161. C. P. Lillya and Y. L. N. Murthy, *Mol. Cryst. Liq. Cryst.* **2**, 121 (1985); M. Möller, V. Tsukruk, and J. H. Wendorff, *Liq. Cryst.* **12**, 17 (1992).

162. W. D. Amiloprasadh Norbett, J. W. Goodby, M. Hird, K. J. Toyne, J. C. Jones, and J. S. Patel, *Mol. Cryst. Liq. Cryst. Sci. Technol. A* **260**, 339 (1995).

163. S. Takenaka, K. Nishimura, and S. Kusaboyashi, *Mol. Cryst. Liq. Cryst.* **111**, 227 (1984).

164. C. Destrade, N. H. Tinh, H. Gasparoux, J. Malthete, and A. M. Levelut, *Mol. Cryst. Liq. Cryst.* **71**, 111 (1981); G. Destrade, H. Gasparoux, P. Foucher, N. H. Tinh, J. Malthete, and J. Jacques, *J. Chim. Phys.* **80**, 137 (1983); S. Shandrasekhar, *Adv. Liq. Cryst.* **5**, 47 (1982); *Philos. Trans. R. Soc.*

London A **309**, 93 (1983) and *Liq. Cryst.* **14**, 3 (1993); B. Kohne and K. Praefcke, *Chem. Ztg.* **109**, 121 (1985); J. Simon and C. Sirlin, *Pure Appl. Chem.* **61**, 1625 (1989); S. Chandrasekhar and G. S. Ranganath, *Rep. Prog. Phys.* **53**, 57 (1990); S. Bauer, T. Plesnivy, H. Ringsdorf, and P. Schuhmacher, *Makromol. Chem. Macromol, Symp.* **64**, 19 (1992); J. Simon and P. Bassoul, in *Phthalocyanines, Properties and Applications*, C. C. Leznoff and A. B. Lever, Eds., Vol. 2, VCH Publishers, Inc., New York (1993).

165. *Liquid Crystals of One- and Two-Dimensional Order and Their Applications*, Springer Series in Chemical Physics, Vol. 11, W. Helfrich and G. Heppke (Eds.), Springer-Verlag, New York (1980): a. J. Billard p. 38; b. J. W. Goodby, G. W. Gray, A. J. Leadbetter, and M. A. Mazid p. 3.

166. S. Chandrasekhar, in *Liquid Crystals*, 2nd Edn., S. Chandrasekhar (Ed.), Cambridge University Press, Cambridge (1992) Chapter 6.

167. G. Latterman and G. Stauber, *Mol. Cryst. Liq. Cryst.* **191**, 199 (1980); W. Weissflog, S. Diele, and D. Demus, *Mater. Chem. Phys.* **15**, 475 (1986); N. H. Tinh and C. Destrade, *Mol. Cryst. Liq. Cryst.* **6**, 123 (1989); S. Takenaka, H. Morita, M. Iwano, Y. Sakurai, T. Ikemoto, and S. Kusabayashii, ibid. **182B**, 325 (1990); S. Takenaka and H. Yamasu, *Mol. Cryst. Liq. Cryst. Sci. Technol. A* **238**, 157 (1994); S. Takenaka and K. Yamazaki, ibid. **241**, 119 (1994).

168. C. Destrade, H. T. Nguyen, C. Alstermark, G. Lindsten, M. Nilsson, and B. Otterholm, *Mol. Cryst. Liq. Cryst.* **180B**, 265 (1990); Y. Fang, A. M. Levelut, and C. Destrade, *Liq. Cryst.* **7**, 265 (1990); C. Alstermark, M. Eriksson, M. Nillson, C. Destrade, and H. T. Nguyen, ibid. **8**, 75 (1990); H. T. Nguyen, C. Destrade, and J. Malthête, ibid. **8**, 797 (1990); H. Allouchi, J. P. Bideau, and M. Cotrait, *Acta Crystallogr. C* **48**, 1037 (1992).

169. J. Malthête, H. T. Nguyen, and C. Destrade, *Liq. Cryst.* **13**, 171 (1993).

170. J. Malthête, A. M. Levelut, and N. H. Tinh, *J. Phys. (Paris) Lett.* **46L**, 875 (1985); A. M. Levelut, J. Malthête, C. Destrade, and N. H. Tinh, *Liq. Cryst.* **2**, 877 (1987); A. M. Levelut and Y. Fang, *Coll. Phys.* **C7**, Supplement to no. 23, 229 (1990); H. T. Nguyen, C. Destrade, H. Allouchi, J. P. Bideau, M. Cotrait, D. Guillon, P. Weber, and J. Malthête, *Liq. Cryst.* **15**, 435 (1993).

171. M. Ebert, R. Kleppinger, M. Soliman, M. Wolf, J. H. Wendorff, G. Lattermann, and G. Staufer, *Liq. Cryst.* **7**, 553 (1990); W. Paulus, H. Ringsdorf, S. Diele, and G. Pelzl, ibid. **9**, 807 (1991); G. Latterman, G. Staufer, and G. Brezenski, ibid. **10**, 169 (1991).

172. I. G. Shenouda, Y. Shi, and M. E. Neubert, *Mol. Cryst. Sci. Technol. A* **257**, 209 (1994).

173. L. J. Yui and A. Saupe, *Phys. Rev. Lett.* **45**, 1000 (1980).

174. M. J. Freiser, *Phys. Rev. Lett.* **24**, 1041 (1970).

175. S. M. Fan, I. D. Fletcher, B. Gundogān, N. J. Heaton, G. Kothe, G. R. Luckhurst, and K. Praefche, *Chem. Phys. Lett.* **204**, 517 (1993).

176. G. W. Gray and A. Mosley, *Mol. Cryst. Liq. Cryst.* **35**, 71 (1976) and **48**, 233 (1978); N. Boden, R. J. Bushby, and L. D. Clark, *J. Chem. Soc. Perkin. Trans.* **1**, 543 (1983); M. E. Neubert, *Mol. Cryst. Liq. Cryst.* **129**, 327 (1985); H. Zimmermann, *Liq. Cryst.* **4**, 591 (1989); M. D. Ossowska-Chruściel, J. Churściel, A. Suszko-Purzycka, and A. Wiegeleben, ibid. **8**, 183 (1990); K. Foder-Csorba, L. Bata, S. Holly, E. Gács-Baitz and K. Ujszászy, ibid. **14**, 1863 (1993); K. Lipiński, T. Lipińska, J. Chruściel, A. Suszko-Purzycka, and A. Rykowski, *Mol. Cryst. Liq. Cryst. A* **239**, 87 (1994); K. Lipiński, T. Lipińska, A. Suszko-Purzycka, and A. Rykowski, *Polish J. Chem.* **67**, 667 (1993).

177. A. Wiegeleben and D. Demus, *Liq. Cryst.* **11**, 111 (1992).
178. J. W. Goodby, *Mol. Cryst. Liq. Cryst.* **75**, 179 (1981).
179. M. Shigenov, W. Mizutani, M. Suginoya, M. Ohmi, K. Kajimura, and M. Ono, *Jpn. J. Appl. Phys.* **29**, L119 (1990); S. Taki, Y. Tokebayashi, and K. Matsushige, *Mol. Cryst. Liq. Cryst. Sci. Technol. A* **247**, 215 (1994).
180. K. Matsushige, S. Taki, H. Okabe, Y. Takebayashi, K. Hayashi, Y. Yoshida, T. Horiuchi, K. Hora, K. Takehara, K. Isomura, and H. Taniguchi, *Jpn. J. Appl. Phys.* **32**, 1716 (1993).
181. M. R. Wilson and D. A. Dunmur, *Liq. Cryst.* **5**, 987 (1989); A. V. Komolkin, Y. V. Molchanov, and P. P. Yakutseni, ibid. **6**, 39 (1989); A. V. Komolkin and Y. V. Molchanov, *Mol. Cryst. Liq. Cryst.* **192**, 173 (1990); M. R. Wilson and M. P. Allen, *Liq. Cryst.* **12**, 157 (1992); G. Krömer, D. Paschek, and A. Geiger, *Ber. Bunsenges. Phys. Chem.* **97**, 1188 (1993).
182. O. B. Akopova, V. I. Bobrov, and Y. G. Erykalov, *Russ. J. Phys. Chem.* **64**, 784 (1960), *Zh. Fiz. Khim.* 1460; D. J. Cleaver and D. J. Tildesley, *Mol. Phys.* **81**, 781 (1994).
183. G. Eliot, *Chem. Br.* 213 (1973); J. A. Castellano and K. J. Harrison, in *Physics and Chemistry of Liquid Crystal Devices*, G. J. Sprokel (Ed.), Plenum Press, New York (1979), p. 263.
184. M. Schadt, *Chimia* **4**, 347 (1987); *Mol. Cryst. Liq. Cryst.* **165**, 405 (1988); *Liq. Cryst.* **14**, 73 (1993); *Ber. Bunsenges. Phys. Chem.* **97**, 1213 (1993).
185. M. A. Osman, and T. Huynh-ba, *Mol. Cryst. Liq. Cryst.* **82**, 331 (1983); *Z. Naturforsch.* **38b**, 1221 (1983).
186. B. S. Scheuble, *Kontakte* 34 (1989) (Merck).
187. H. Kresse, *Adv. Liq. Cryst.* **6**, 119 (1983).
188. J. Billard, J. C. Dubois, and A. Zann, *J. Phys. (Paris) Colloq.* **36**, C1-355 (1975); V. Reiffenrath, V. Finkenzeller, E. Poetsch, B. Rieger, and D. Coates, *Proc. SPIE Int. Soc. Opt. Eng.* **1257**, 84 (1990); M. Schadt, R. Buchecker, and K. Müller, *Liq. Cryst.* **5**, 293 (1989); M. Schadt, R. Buckecher, and A. Villiger, *Liq. Cryst.* **7**, 519 (1990).
189. M. Schadt, R. Buchecker, A. Villiger, F. Leenhouts, and J. Fromm, *IEEE Trans. Electron Dev.* **33**, 1187 (1986).
190. E. I. Ryumtsev, N. P. Evlampreva, and A. P. Kovshik, *Russ. J. Phys. Chem.* **69**, 848 (1995); *Zh. Fiz. Khim.* 934.
191. R. Eidenschink, J. Krause, L. Pohl, and J. Eichler, in *Liquid Crystals*, S. Chandrasekhar (Ed.), Heyden Publishers, London (1980) p. 515.
192. R. Eidenschink, *Mol. Cryst. Liq. Cryst.* **94**, 119 (1983).
193. A. Boller, M. Cereghetti, M. Schadt, and H. Scherrer, *Mol. Cryst. Liq. Cryst.* **42**, 215 (1977).
194. M. Schadt and P. R. Gerber, *Z. Naturforsch.* **37a**, 165 (1982).
195. Licristal® Liquid Crystals, E. M. Chemicals, pp. 12–13.
196. U. Finkenzeller, T. Geelhaar, G. Weber, and L. Pohl, *Liq. Cryst.* **5**, 313 (1989).
197 M. A. Osman, *Helv. Chim. Acta* **68**, 606 (1985).
198. W. Maier, in *Landolt-Börnstein Zahnwerte und Funktionen aus Physik, Chemie, Astronomie, Geophysik und Technik*, 2nd Edn., Vol. II, Part 6, Springer-Verlag, Berlin (1959), p. 607.
199. Y. Goto, T. Inukai, A. Fuyita, and D. Demus, *Mol. Cryst. Liq. Cryst. Sci. Technol. A* **260**, 23 (1995).
200. A. Denat, B. Gosse, J. P. Gosse, *Chem. Phys. Lett.* **18**, 235 (1973); H. Sorkin and A. Denny, *RCA Review* **34**, 308 (1973); R. Konrad and G. M. Schneider, *Mol. Cryst. Liq. Cryst.* **51**, 57 (1979).

201. J. Koskikallio, in *The Chemistry of Carboxylic Acids and Esters*, S. Patai (Ed.) Interscience Publishers, John Wiley & Sons Ltd, London (1960) Chapter 3.
202. A. M. Lackner, J. D. Margerum, and C. van Ast, *Mol. Cryst. Liq. Cryst.* **141**, 289 (1986).
203. D. Coates and G. W. Gray, *J. Chem. Soc. Chem. Commun.* 514 (1975).
204. J. Szulc and Z. Stolarz, *Mol. Cryst. Liq. Cryst. Sci. Technol. A* **263**, 623 (1995).
205. F. G. Yamagishi, D. S. Smythe, L. J. Miller and J. D. Margerum, in *Liquid Crystals and Ordered Fluids*, Vol. 3, J. F. Johnson and R. S. Porter (Eds.), Plenum Press, New York (1978).
206. J. L. Haberfield, E. C. Hsu, and J. F. Johnson, *Mol. Cryst. Liq. Cryst.* 241 (1973).
207. A. Tashira, Y. Mya, H. Nishizawa, O. Hirai, and O. Watanabe, *Jpn Khai Tokkyo Koho*, JP07292359 (1995), *Chem. Abstr.* **124**, 102686Z (1996); A. Tashira, O. Hirai, Y. Mya, H. Nishizawa, and O. Watanabe, *Jpn Kokai Tokkyo Koho*, JP 07278544 (1995); Y. Mya, A. Tashira, H. Nishizawa, O. Hirai, and O. Watanabe, ibid. JP07270741 (1995).
208. C. Colpaert, B. Maximus, and A. de Meyere, *Liq. Cryst.* **21**, 133 (1996).
209. R. R. Biggers, Ph.D. Thesis, Kent State University (1983); J. K. Swadesk, C. W. Stewart, Jr., and P. C. Uden, *Analyst (London)* **118**, 1123 (1993).
210. R. L. Hubbard, in *Physics and Chemistry of Liquid Crystal Devices*, G. J. Sprokel (Ed.), Plenum Press, New York (1979), p. 331; T. I. Martin and W. E. Haas, *Anal. Chem.* **53**, 593 A (1981).
211. E. C. H. Hsu and J. F. Johnson, *Mol. Cryst. Liq. Cryst.* **27**, 95 (1964); D. E. Hulme, E. P. Rayes, and K. J. Harrison, *J. Chem. Soc. Chem. Commun.* 98 (1974).
212. D. Demus, C. H. Fietkau, R. Schubert, and H. Kehlen, *Mol. Cryst. Liq. Cryst.* **25**, 215 (1974).
213. J. D. Margerum, A. M. Lackner, J. E. Jensen, L. J. Miller, W. H. Smith, Jr., S.-M. Wong, and C. I. van Ast, *Mol. Cryst. Liq. Cryst.* **111**, 103 (1984).
214. J. D. Margerum, C. I. van Ast, G. D. Myer, and W. H. Smith, Jr., *Mol. Cryst. Liq. Cryst.* **198**, 29 (1991).
215. A. Rabinovich and A. Ganelina, *Ferroelectrics* **121**, 335 (1991).
216. J. Sangster, *J. Phase Equilibria* **14**, 340 (1993).
217. E. C.-H. Hsu, J. L. Haberfeld, J. F. Johnson, and E. M. Barrall II, *Mol. Cryst. Liq. Cryst.* **27**, 269 (1974); L. A. Kaszczuk, G. J. Bertolini, J. F. Johnson, and A. C. Griffin, ibid. **88**, 183 (1982).
218. G. W. Gray, K. J. Harrison, and J. A. Nash, *J. Chem. Soc. Chem. Commun.* 431 (1974).
219. D. S. Hulme, E. P. Raynes, and K. J. Harrison, *J. Chem. Soc. Chem. Commun.* 98 (1974); G. W. Gray, *J. Phys. Colloq. (Paris)* **36**, C1-337 (1975).
220. M. A. Osman, Hp. Schad, and H. R. Zeller, *J. Chem. Phys.* **78**, 906 (1983).
221. Hp. Schad and M. A. Osman, *J. Chem. Phys.* **79**, 5710 (1983); N. K. Sharma, B. L. Sharma, and A. Charak, *Ind. J. Chem.* **30A**, 831 (1991).
222. F. Schneider and N. K. Sharma, *Z. Naturforsch.* **36a**, 62 (1981).
223. G. Heppke and E. J. Richter, *Z. Naturforsch.* **33a**, 185 (1978); B. Engelen, G. Heppke, R. Hopf, and F. Schneider, *Ann. Phys. (Paris)* **3**, 403 (1978).
224. B. S. Srinkanta and N. V. Madhusudana, *Mol. Cryst. Liq. Cryst.* **99**, 203 (1983).
225. A. Boij and P. Adomenas, *Mol. Cryst. Liq. Cryst.* **95**, 59 (1983).
226. L. J. Yu and M. M. Labes, *Mol. Cryst. Liq. Cryst.* **54**, 1 (1979).

227. L. A. Batrachenko and L. N. Lisetskii, *Russ. J. Phys. Chem.* **64** 1519 (1990).
228. F. Schneider and N. K. Sharma, *Z. Naturforsch.* **36a**, 1086 (1981).
229. N. K. Sharma, I. Müller, R. Wingen, H. R. Dübal, C. Escher, and D. Ohlendorf, *Mol. Cryst. Liq. Cryst.* **151**, 225 (1987).
230. J. W. Park, C. S. Bak, and M. M. Labes, *J. Am. Chem. Soc.* **97**, 4398 (1975).
231. C. S. Oh, *Mol. Cryst. Liq. Cryst.* **42**, 1011 (1977).
232. J. Szabon, W. Pilz, and H. D. Koswig, *Cryst. Res. Technol.* **18**, 519 (1983).
233. N. K. Sharma and R. K. Bamezai, *Ind. J. Chem. A* **25**, 431 (1986).
234. R. Pratibha and N. M. Madhusudana, *Mol. Cryst. Liq. Cryst. Lett.* **1**, 111 (1985).
235. S. Diele, G. Pelzl, W. Weissflog, and D. Demus, *Liq. Cryst.* **3**, 1047 (1988); G. Pelzl, A. Humke, S. Diele, S. Ziebarth, W. Weissflog, and D. Demus, *Cryst. Res. Technol.* **25**, 587 (1990), G. Pelzl, A. Humke, D. Demus, and W. Weissflog, ibid. **25**, 597 (1990); G. Pelzl, J. Szabon, A. Wiegeleben, S. Diele, W. Weissflog, and D. Demus, ibid. **25**, 223 (1990).
236. M. Marthandappa, M. Nagappa, and K. M. Lakanath Rai, *J. Phys. Chem.* **95**, 6369 (1991).
237. S. Ujiie, H. Uchino, and K. Iumura, *Chem. Lett.* 195 (1995).
238. J. D. Margerum, J. E. Jensen, and A. M. Lackner, *Mol. Cryst. Liq. Cryst.* **68**, 137 (1981); J. D. Margerum, S. M. Wong, A. M. Lackner, and J. E. Jensen, ibid. **68**, 157 (1981); J. D. Margerum, S. M. Wong, A. M. Lackner, and J. E. Jensen, and S. A. Verzwyvelt, *Mol. Cryst. Liq. Cryst.* **84**, 79 (1982).
239. K. Araya and Y. Matsunaga, *Bull. Chem. Soc. Jpn.* **53**, 989, 3079 (1980); M. Fukui and Y. Matsunaga, ibid. **54**, 3137 (1981), K. Araya and Y. Matsunaga, *Mol. Cryst. Liq. Cryst.* **67**, 153 (1981) and *Bull. Chem. Soc. Jpn* **55**, 1710 (1982); E. Chino and Y. Matsunaga, ibid. **56**, 3230 (1983); Y. Matsunaga and I. Suzuki, ibid. **57**, 1411 (1984); N. Homura, Y. Matsunaga, and M. Suzuki, *Mol. Cryst. Liq. Cryst.* **131**, 273 (1985); E. Chino, Y. Matsunaga, and M. Suzuki, *Bull. Chem. Soc. Jpn.* **57**, 2371 (1984).
240. S. Diele, *Ber. Bunsenges. Phys. Chem.* **97**, 1326 (1993).
241. M. E. Neubert, K. Leung, and W. A. Saupe, *Mol. Cryst. Liq. Cryst.* **135**, 283 (1986).
242. J. Szabon, W. Weissflog, G. Pelzl, S. Diele, and D. Demus, *Cryst. Res. Technol.* **21**, 1097 (1986); S. Haddawi, S. Diele, H. Kresse, G. Pelzl, and W. Weissflog, ibid. **29**, 745 (1994).
243. G. W. Gray and J. W. Goodby, *Mol. Cryst. Liq. Cryst.* **37**, 157 (1976).
244. S. Takehara, M. Osawa, K. Nakamura, and T. Kuriyama, *Ferroelectrics* **148**, 185 (1993).

Index

Page numbers in italic, e.g. *255*, signify references to figures. Page numbers in bold, e.g. **397**, denote entries in tables.

For EU product safety concerns, contact us at Calle de José Abascal, 56–1°,
28003 Madrid, Spain or eugpsr@cambridge.org.

www.ingramcontent.com/pod-product-compliance
Ingram Content Group UK Ltd.
Pitfield, Milton Keynes, MK11 3LW, UK
UKHW051007240426
470322UK00018B/551